Game Theory in Action

Concepts and Applications

By Albert Lee, Ph.D, P.E.

Library of Congress Control Number: 2025921473

ISBN: 979-8-89228-424-0 (Paperback)
ISBN: 979-8-89228-425-7 (Hardcover)

Printed in the United States of America

TO THE LOVING MEMORY OF
MY MOTHER CHING - UN LAM, AND
MY FATHER WILLIAM WING - YOU LEE

ACKNOWLEDGMENT

Since my retirement in 2015, after over 25 years as a Software Engineer at Sperry Corp. and as a Systems Engineer and Training Engineer at the Program Management Office (PMO) of the Department of the Navy, I have always aspired to write a book that shares academic insights from my teaching and technical experiences from the engineering field. However, it wasn't until my second retirement in 2021, from my role as an Associate Professor in the Computer Science Department at the New York Institute of Technology, that I found both the topic and title for my book. The inspiration for "Game Theory in Action: Concepts and Applications" emerged from teaching an upper-level undergraduate and first-year graduate course, "Optimization Methods for Data Science (DTSC 615)," during the Fall Semester of 2020, my last semester of teaching.

I am deeply grateful to my students from both the Old Westbury and Metro campuses, particularly David Jacob, Mathew Vasquez, Jun Sang Park, Shue Hua, and Monika Arciszewska. Their support, participation, feedback, and communication made teaching this course an engaging and rewarding experience. I would also like to thank my colleague and Chairperson, Dr. Frank Lee, for the opportunity to teach this course.

While writing this book during my retirement, I often reflect on the enriching experiences of my life as a college and graduate student, professor, and engineer. I extend my heartfelt gratitude to my mathematics and enlightenment teacher, Professor 杜定安教授, at Chu Hai University of Hong Kong, for his profound influence during my freshman year; to my Ph.D. thesis advisor, Professor W.S. Massey at Yale University, for his patience and guidance, which broadened my understanding of mathematics; and to my supervisor, Dr. Fred Kreussling at Sperry Corp., who first introduced me to target tracking algorithms in radar systems and the concepts of Engineering Control Systems. Chapter 9 of this book, "An Introduction to Game Theory and Feedback Control Systems," is dedicated to him.

I also wish to express my sincere appreciation to two retirees, Chairperson Dr. Ted Teng and Dean of Faculty Dr. Mel Drossman, at the Department of Engineering and Computer Science, New York Institute of Technology, for hiring me for my first teaching position as an Associate Professor—a role I never imagined would last over 25 years until my retirement. Lastly, I must thank my late Technical Director and supervisor, Dr. Martin Tai, at the Program Management Office of the Department of the Navy, for hiring me as a Systems Engineer and later as a Training Engineer. These positions allowed me to witness the practical applications of the mathematical and engineering principles I learned in academia, making for a profoundly rewarding experience.

This book was written during the COVID-19 pandemic, and I am immensely grateful to my friends for their caring support, encouraging words, and for lighting up my days during this challenging time. Additionally, this book would not have been possible without the understanding and support of my family. I dedicate this book to my family – my wife Celia, my daughters Ada, Patricia, Christine, and my son Simon – as well as all who have shaped my life and offered their assistance in various ways.

I am deeply grateful to Atticus Publishing for their unwavering support and dedication in bringing this book to life. A special thank you to Ren Shelley, Production Manager, for their careful and creative work on the interior layout. I also sincerely appreciate the guidance and assistance of Sam Robinson, Tina Summers, Vera Estelle, and Rose Cecily over the past six months. Your professionalism, patience, and encouragement have meant a great deal to me.

Table of Contents

PREFACE

"Life is just like a game,
First you must learn the rules of the game,
And then play it better than anyone else."

- Albert Einstein

OBJECTIVE OF THIS BOOK

This book serves as an introduction to game theory, emphasizing the concepts and applications of both non-cooperative and cooperative games. Game Theory explores decision-making scenarios involving two or more participants(players) with conflicting or opposing interests, employing mathematical models to analyze their strategic interactions. Its applications span a broad range of fields, including economics, business strategy, social sciences, political science, computer science, engineering, and artificial intelligence, among others.

Although there are many concepts and definitions in game theory, such as the celebrated Nash Equilibrium and equilibrium strategies, are expressed mathematically, we aim to make our presentations accessible to students with varying levels of mathematical maturity. Our goal is to simplify the mathematics involved, allowing readers to focus on high-level understanding without getting bogged down by intermediate steps. We intentionally incorporate repetition throughout the book, revisiting key definitions and concepts before introducing new ones, thereby reinforcing students' confidence and mastery of the subject.

WHO ARE THE READERS?

This book is tailored for students in mathematics, economics, engineering, and computer science. The mathematical prerequisites are minimal; familiarity with differential and integral calculus, conditional probability, Bayes' theorem, and the Simplex Method in Linear Programming will suffice for understanding most of the material.

In addition to gaining insights into game theory, engineering students will find Chapter 9 particularly rewarding, as it analyzes typical engineering control systems as non-cooperative two-player games. Computer Science students will appreciate Chapter 10, which demonstrates how algorithms can effectively model real-life competitive scenarios like coalition games. Economics and business students will benefit from classical economic models like the Cournot and Betrand models of duopoly, as well as various auction types illustrating Nash Equilibrium applications in real-world scenarios.

Throughout the book, carefully selected examples motivate and facilitate the understanding of new concepts and definitions.

TECHNOLOGICAL INSERTION – USE OF MATLAB IN THE BOOK

Many concepts, definitions, theorems, and algorithms discussed in this books are also presented through MATLAB programs. These scripts serve as teaching tools, enabling students to create additional examples with their own input conditions. By verifying that the expected output aligns with the actual program output, students will deepen their understanding and confidence in the subject matter. Each chapter includes source code for the MATLAB programs, accompanied by a top-level description of their purpose and logical structure, as well as details on input data formats and output listings. Note that while cosmetic aspects and efficiency of the code may not be a priority, correctness and straightforward logic in translating the underlying problems into functional MATLAB scripts are our main objectives.

List of MATLAB programs for this book

- Chapter 2:
  - ComputeSecurityLevelandSaddlePoint.m
    - myMaxLinearProgramExample.m
    - myMinLinearProgramExample.m
  - Chapter2Example1R.m
  - Example2Example1C.m
  - Chapter2Example2C.m
  - Chapter2Example2R.m
  - ComputeSecurityLevelsbySimplexMethod.m
    - mySlacknessTheoremExampleMaxLP.m
    - mySlacknessTheoremExampleMinLP.m
- Chapter 3:
  - NashEquilibriumTest.m
  - ComputeLinearSystemEq.m
  - WorkSheetSecurityLevelsandNashEquilibrium.m
- Chapter 4:
  - ProblemoftheCommons.m
  - ProblemoftheCommonSocialOptimal.m
- Chapter 8:
  - myCrawfordSobel1.m
  - PlotSenderReceiverCombine.m
  - myCheapTalk4.m
- Chapter 9:
  - mycombineMinMaxMatrix.m
  - myminmaxsaddleFunction1.m

myHinfinitiveTesting.m

Chapter 10:

myMinSpanTreeBV.m

MymaximalFlowAlgorithm.m

myAssignmentAlgorithmBV1.m

myFloydWarshallAlgorithm.m

myDijkstraAlgorithmBV.m

Chapter 11:

(1) myShapleyvalueBV.m

(2) myComputeNucleolus112BV.m

(3) myComputeNeuclusBV4444.m

(4) myInsuranceExample.m

(5) myMobilesExample.m

(6) myManufacturingExample.m

Chapter 12:

SecurityLevelsandEquilibriumPairs1.m

CHAPTERS DEPENDENCY AND ORGANIZATION

This mathematics-oriented book covers a board spectrum of important topics in game theory, focusing on concepts and applications across various scientific fields. The book is organized into 12 chapters: Chapters 1 to 9 primarily address non-cooperative games, while Chapters 10-12 focus on cooperative games. Given the extensive information provided, the book can serve as a reference for game theory and can be taught in a two-semester course.

- Chapters 1-4: Foundational concepts in non-cooperative games.
- Chapters 5-8: Static and dynamic games of incomplete information.
- Chapter 9: Engineering feedback control systems.
- Chapter 10-12: Cooperative games.

Chapters 5-8 and Chapters 10-12 are dependent on the concepts introduced in Chapters 1-4, while Chapters 9, 10, 11, and 12 do not reply on Chapters 5-8.

CONTENTS OF THE BOOK

As outlined in the objectives of this book, a sophisticated mathematics background is not required. However, familiarity with basic mathematical tools and concepts—such as simple probability, Bayes' Theorem, differentiation, integration, and linear programming—will significantly enhance the reader's understanding of the topics discussed.

Chapter 1 introduces numerous examples of different types of games and their representations. For instance, Examples 1 and 2 illustrate two-person and three-person games, respectively. Examples 3 and 4 revisit the classical Prisoner's Dilemma in two and three dimensions. Examples 5 and 6 explore non-cooperative and cooperative games, respectively. Example 7 demonstrates how a constant-sum game can model a practical business scenario.

Examples 8 and 10 present both simple and complex games that simulate real-world traffic light systems at street intersections and intricate traffic network systems respectively. Example 9 features the Dating Game scenario from the movie *A Beautiful Mind*.

These real-world examples lead us to a fundamental concept in game theory: the Nash Equilibrium. This concept will be introduced in Chapter 3 and revisited throughout the remainder of the book.

Chapter 2 reviews fundamental concepts from Linear Programming, the Simplex Method, and the MATLAB built-in function linprog(f, A, b). The chapter introduces a very fundamental concept in game theory for the zero-sum games of 2-players. This is the concept of the security level of the row player denoted by sl(1) and the security level of the column player denoted by sl(2). Section 2.4 shows that sl(1) can be expressed as a Minimization Linear Program, and sl(2) as a Maximization Linear Program and both linear programs are solved by the Simplex Method. By utilizing the Duality Theorem from Linear Programming, we prove the Fundamental Theorem of Game Theory which states that we can always find a mixed strategies X* for the row player and a mixed strategy Y* for the column player such that $sl(1) = \underset{1 \le j \le n}{Min}\ X^* M^j$ and $sl(2) = \underset{1 \le i \le m}{Max}\ M_i Y^{*t}$, and sl(1)= sl(2) where M is the payoff matrix. Computation of the security levels is described in Section 2.6. Explanation of the Fundamental Theorem of Game Theory is shown in Section 2.7. Section 2.8 describes The Complementary Slackness Theorem which will be used in the computation of the Nucleolus of Coalition Games discussed in Chapter 11. Ten MATLAB programs are provided to support the understanding of various concepts discussed.

Chapters 3 and 4 delve into the core equilibrium concepts in game theory. These are the basic Nash Equilibrium for static games with complete information, Bayesian Nash Equilibrium for static games of incomplete information, Perfect Bayesian Equilibrium for dynamic games of incomplete information. Section 3.1 gives the mathematical definition of Nash Equilibrium of any strategy pairs (X*, Y*) for a 2-player game with payoff matrix A and B respectively. Per this definition, it will require a huge amount of work in comparison between EV (A, X, Y*) and EV (A, X*, Y*), and between EV (B, X*, Y) and EV (B, X*, Y*) for infinitely many mixed strategies X and Y. But thanks to Lemma 3.1 and Lemma 3.2, we can reduce the amount of work to only when X and Y are the Row strategies for A and the column strategies for B respectively, what a huge difference! Section 3.2 demonstrates the use of the MATLAB program NashEquilibriumTest.m to locate all Nash Equilibriums for pure strategies in several examples including those examples illustrated in Chapter 1 before. Section 3.3 shows by examples, how we could find a mixed strategy pair that is in Nash equilibrium based on a necessary condition of Nash Equilibrium. Section 3.4 is allocated to the proof of the celebrated Nash Equilibrium Existence Theorem which states that in any two-person game, there always exists at least one Nash equilibrium. The proof is based on the Brouwer's Fixed Point Theorem. Section 3.5 shows some useful relationships between the concept of security levels introduced in Chapter 2 and the concept of Nash Equilibrium introduced in this Chapter 3. These relationships are expressed in Theorems 5.1, 5.2, and 5.3. Section 3.6 gives numerical demonstrations of the results claimed in the three theorems using several of the MATLAB programs that were developed before. Section 3.7 demonstrates a practical application of Nash Equilibrium in Economics (The Cournot Duopolistic Competition Model).

In Section 4.1, we show how the conditions for and the method of finding the Nash Equilibrium in games of 2- players described in Chapter 3 can be extended to games of n-players. This is done by solving a set of n Maximization problems and for each player computes the best-response function. Details of this process are shown in equations (E1), (E2), (E3) in Section 4.1.1. We presented three applications of the Nash equilibrium in Social Science, Business, Political Science in Example 1.1 (The Problem of the Commons), Example 1.2 The Hotelling Model), and Example 1.3 (The Voter Median Theorem) respectively. For The Problem of the Commons, we wrote two MATLAB programs to demonstrate numerically that the social optimal solution is better than the individual optimal solution independent of the number of companies around the world. In Section 4.2, we give a formal definition of static Bayesian game and in Definition 2.1 and 2.2, and the formal definition of Bayesian Nash Equilibrium in Definition 2.3. We demonstrate an application of the Bayesian Nash Equilibrium in Economics in Example 2.4. In Section 4.3, we informally discuss Dynamic Games with complete Information and leaving the details to Chapter 6. Use of the Backwards Induction Method to find the subgame perfect Nash Equilibrium was demonstrated in Example 3.5 and 3.6. Example 3.5 also compares the method of Backwards Induction to the direct approach by verifying which existing Nash Equilibrium can be a subgame perfect Nash Equilibrium.

The chapter ends with an informal discussion of Dynamic Games with Incomplete Information in Section 4.4 and leaves the details in Chapter 7 and 8. Definition 4.5 gives an informal definition of perfect Bayesian equilibrium meeting Requirements 1,2,3, and 4. Example 4.7 demonstrates all these four requirements. Example 4.8 and Example 4.9 shows how to find the subgame-perfect equilibrium for a complete and imperfect information game and an incomplete and imperfect information game respectively.

Concepts of the Static Bayesian Game and Bayesian Nash Equilibrium are presented in Chapter 5. Elementary examples and many practical applications of Bayesian Nash equilibrium are presented. Three applications included first price auction, double auction, business trading of goods with different types are presented in Section 5.3 and 5.5. To understand and appreciate the formulation of the Bayesian Nash equilibriums in these applications, familiarity of basic concepts in probability, probability density function, cumulative distribution function, and uniform distribution of a random variable are needed. Section 5.6 presents the celebrated Revelation Principle in game theory and its applications. Understanding this principle requires preliminary concepts like direct mechanism of a Bayesian game, Bayesian Nash Equilibrium for a Mechanism, implementation of a function by a Mechanism, and a function truthfully implementable in Bayesian Nash Equilibrium. These preliminary concepts are described and explained in Sections 5.6.A, 5.6.B. The Revelation Principle can then be summarized in Proposition 5.6.1, Proposition 5.6.2 and Proposition 5.6.3 which states that, under certain conditions, any choice function can be truthfully implementable in Bayesian Nash Equilibrium. Sections 5.6.C shows how this amazing principle applies to a wide spectrum of areas such as the Vickrey-Clarke-Groves (VCG) Mechanism (Section 5.6. C. I), Vickrey Auction (Section 5.5. C. II), Bilateral Trade for a seller and a buyer (Section 5.6.C.III), Public Projects (Section 5.6.C.IV), Computer Networks (Section 5.6.C.V).

Dynamic games with complete information and the subgame perfect Nash Equilibrium concepts are discussed in Chapter 6. As dynamic games are usually represented by a game tree, Section 6.1 gives two simple examples to illustrate three basic terminologies and concepts used for the remaining of this book (1) the information set of each decision note in a game tree, and (2) the action set of each player (3) the definition of a subgame in a game tree. Section 6.2 gives five examples of various complexities in how to represent the pure strategies in these games. Section 6.4 gives a precise definition of the subgame Perfect Nash Equilibrium (Definition 6.4.3) which states that when the strategies in the Nash Equilibrium are restricted to every subgame of the original game, the restricted strategies are also Nash Equilibriums in the subgames. Two examples, Example 6.4.10 and Example 6.4.11, are provided to demonstrate that the stated Nash Equilibrium strategies do satisfy the requirements shown in the definition of a subgame Perfect Nash Equilibrium. Other than to directly verify if a Nash Equilibrium strategies set is a subgame Perfect Nash Equilibrium that meets Definition 6.4.3, Section 6.5 describes and provides two general approaches, the Backward Induction and the Forward Induction, to find the subgame Perfect Nash equilibriums with many illustrated examples. Section 6.6 provides three practical applications of the concept of the subgame

Perfect Nash Equilibrium to Economics via the Cournot and Stackelberg Model (Application 6.6.1), Union and Firm bargaining via the Leontief Model (Application 6.6.2), and Tariff competition between countries (Application 6.6.3). The last two sections of the chapter are devoted to the Multi-stage Games on Sections 6.7 and the Repeated Games on Section 6.8. Representation of strategies in multi-stage games is discussed in Section 6.7.1 followed by the definition of subgame perfect equilibrium of strategies is given in Section 6.7.2. The concept of award and punishment, a common technique used in Multi-stage and Repeated games to find subgame perfect Nash equilibriums, was introduced and illustrated in several examples in Section 6.7.3 and 6.8.1. The concept of the Grim Trigger Strategy for the infinitely repeated games is introduced in Section 6.8.2. Using the Grim Trigger Strategy, the celebrated Folk Theorem (Section 6.9.3) on the infinitely repeated games is established. An application of the Folk Theorem to an economic scenario described by the Cournot and Stackelberg Model is shown in Example 6.9.26.

Chapters 7 and 8 deal with dynamic games of incomplete information and Perfect Bayesian Equilibrium with Chapter 7 focuses on Signaling Games and Chapter 8 on Cheap-Talk Games. Section 7.2 gives the definition of Perfect Bayesian Equilibrium in terms of four requirements, Requirement 1 to Requirement 4, these requirements are illustrated in Examples 7.3.1 to 7.3.6. Section 7.4 introduces Signaling Games, the concept of perfect Bayesian equilibrium for signaling games is shown in Definition 7.4.3, and concept for the Pooling, Separating, and Hybrid strategies in Definition 7.4.4. Section 7.4.5 presents two applications of Signaling Games in two elaborate examples, Example 7.4. 11 (Limit Pricing and Entry Deterrence) and Example 7.4.12 (Application of the Spence Model). Section 7.5 discussed two additional requirements, Requirement 5 and Requirement 6 for sophisticated signaling games. Section 7.6 investigates the existence of possible equilibrium in the Spence Model with respect to the Requirements 5 & 6. Section 8.1 introduces Cheap-Talk Games, and the concept of perfect Bayesian Equilibrium in Section 8.2. Sections 8.4, 8.5 discuss the continuous version of the Crawford Sobel Model, this is a powerful tool used to systematically compute the perfect Bayesian equilibriums in Cheap-Talk games. Section 8.6 presents two MATLAB programs on the continuous and discrete versions of the Crawford Sobel Model with source codes and output listings of illustrative examples.

Chapter 9 shows that the fundamental problem of a control system in engineering can be formulated as a zero-sum two-person game that we discussed earlier. Section 9.1 is a quick overview of a control system and its components; it gives motivation for the need for the H-infinitive (H^∞) norm of a proper transfer function T to be denoted by <<T>>. Section 9.2 shows that, by the Parseval's Identity in Fourier Transform, the H^∞-norm <<T>> of a transformation T and the H^∞-norm of the corresponding matrix representation M of T denoted by ||M|| are equal. Section 9.3 shows how we can derive the explicit transfer function T from the set of disturbances to the set of control outputs that are only depending on the controller K. This H^∞-norm of T, denoted by <<T_K>>, is used to formulate a two-person zero-sum game to solve the original control problem. In Sections 9.4 and 9.5, we review the basic concepts of the L and U values and the cost function of a two-person game. Two simple MATLAB programs were provided in computing these L and U values. Section 9.6 is the core of Chapter 9, it shows that when we consider the two players as the disturbance- player "d" and the controller-player "K", and when the cost function for the game is properly defined as $F\gamma(K,d) = ||T_K(d)||^2 - \gamma^2||d||^2$ for all K and d, then player" K" chooses the control strategy to minimize the U value and player " d " chooses the control strategy to maximize the L value. The original control problem is solved when L =U. Example 9.6.3 illustrates the use of the MATLAB Control Tool Box built-in function hinfsyn to solve a typical control problem by the game theoretical approach discussed. Section 9.7 presents the conventional solution of solving the discrete time control system and shows the implicit interaction between the two players. The result is summarized in Theorem 9.1.

Chapters 10, 11, 12 are devoted to coalition games in this book. Chapter 10 introduces coalition games and representation of these games by means of the characteristic function. Examples 10.1.1, 10.1.2, and 10.1.3 show how the characteristic function can be computed for each coalition by a MaxMin or MinMax optimization processes. Students in Computer Science major would find Section 10.2 familiar and interesting because this section contains five very common yet powerful computer algorithms that are in the core curriculum of any Computer Algorithm Design and Analysis course. We first give a top-level review of these computer algorithms and then show how we could use these algorithms to form meaningful coalitions in many real world situations that can be modelled by coalition games. These algorithms are Minimum Spanning Tree Algorithm, Maximum Flow Algorithm, Assignment Algorithm, Pairwise Shortest Path Algorithm, and Dijkstra's Shortest Path Algorithm. Source codes and output listings of using these algorithms in coalition games are shown at the end of the chapter.

Chapter 11 introduces the concepts of Imputation Set I(v), the Core C(v), the Shapley values S(v), and the Nucleolus N(v) of a coalition game v(G). These concepts are discussed in the respective Sections 11.1, 11.2, 11.3, and 11.4. The content of the Imputation Set, the Shapley values, and the Nucleolus can reveal to us the additional payoff information of each player within every coalition C in a coalition game. This is more informative than what is given in the characteristic function v(G). Two MATLAB programs were written for the computation of Shapley values and the Nucleolus, output listing of respective examples is shown in Example 11.3.6, Example 11.4.7, Example 11.5.8, and Example 11.6.9, Example 11.8.10, Example 11.8.11,and Example 11.8.12.

The final Chapter 12 deals with cooperative games where the players can communicate with each other regarding which strategies they are going to take prior to playing out the game. These games can model many real-life scenarios and with the pre-play agreement, players could usually arrive at a win-win solution. Section 12.1 presents some simple examples and Section 12.2 reviews the applicability of the basic results established in Chapters 2,3 to cooperative games. Section 12.3 presents the MATLAB program SecurityLevelandEquilibriumPairs1.m to support claims from Section 12.2. Section 12.4 introduces the concept of cooperative payoff-set and negotiation-set which are the core concepts of the chapter. Sections 12.5, 12.6, 12.7, and 12.8 presents Nash original solution to the cooperative games by means of the Nash Bargaining Function that was introduced in [18]. Two alternative approaches, since Nash's original approach to cooperative games are due to Raiffa and Braithwaite, are described in Section 12.9.

As this is the first edition, I welcome and value your suggestions for improvement and any corrections you may find. Please send your comments to: gametheoryinaction2025@gmail.com.

The MATLAB source code used for the book's illustrations is published in the Google Drive folder <Game Theory in Action \ my MATLAB programs >. Access it here:
https://drive.google.com/drive/folders/1eVncIMg3sFHzlnUzuhVWNYvJ_XRKuCkL?usp=sharing

A comprehensive overview of the book can be found here at the website:
https://www.ThinkgameTheorywithAlbertLee.com

Chapter 1

What is a game and game theory?

1.0 Introduction

1. In this introductory chapter, we'll explore some fundamental concepts in game theory and uncover just how common and useful they are in everyday life. You might be surprised to see how often strategic thinking, decision-making, and competition play a role in the world around you.

   This chapter focuses on key ideas, and you won't need any advanced mathematics to follow along. However, as we progress through the book, we will introduce essential mathematical tools, including basic probability, calculus for finding maximums and minimums of functions in one or multiple variables, and concepts from linear programming. These mathematical techniques will help us dive deeper into the subject and truly understand its power.

   Like any scientific discipline, mathematics is a crucial tool that allows us to expand and refine our understanding. But don't worry—our approach will always be concept-driven. The key is to grasp the ideas first, without getting bogged down in the calculations. With that mindset, you'll be well-prepared to tackle the more technical aspects of game theory in later chapters.

2. There are countless games in real life and across various fields of study, including economics, politics, business, behavioral sciences, engineering, AI, and many more—far too many to list them all here. Some games are simple and easy to understand, while others are so complex that even describing their rules is a challenge, let alone predicting their outcomes.

   To build a solid foundation, let's start with the following list of familiar games and introduce key terminology that will help us navigate game theory more effectively in the chapters ahead.

   (a) Game of life
   (b) Tik tac toe
   (c) Ma Jong
   (d) Chess
   (e) Black Gammon
   (f) GO

(g) Poker
(h) King's Coalition
(i) Paper Rock and Scissor (reminder: childhood games)
(j) Coordination Game
(k) Traffic lights
(l) Traffic flow in a network
(m) Prisoners' Dilemma
(n) The Dating Game from the movie "A Beautiful Mind"

Examples (a) to (h) above are physical board or card games, while (i) to (n) represent key game theory concepts that we will explore in later chapters. These concepts will help us understand one of the most fundamental ideas in game theory: The Nash Equilibrium.

All the games listed involve either two-players or multiple players. In some, players make their moves simultaneously—these are called static games. In others, players take turns making decisions one after another—these are known as dynamic games).

Games can also be classified as deterministic or stochastic. A game is deterministic when every move follows strict rules with no randomness involved, like Chess and Go. In contrast, aq game is stochastic when randomness plays a role in determining the outcome, such as Backgammon, where dice rolls introduce an element of chance.

These distinctions will help us better analyze different types of games as we move forward in our study of game theory.

3. In game theory, we analyze interactions between two or more players, each with their own interests and a defined set of rules that dictated how the game is played. Every player is given a finite set of actions or strategies, and they have the freedom to choose what they believe to be their best strategy. However, while each player controls their own choices, they cannot control their opponents' decisions. Once all the players have independently selected their strategies, the outcome of the game—or the payoff for each player—is determined. These payoff values then define who benefits the most and shape the overall competitive or cooperative dynamics of the game.

   One of the main goals in game theory is to identify the best strategy for each player. However, as you might expect, what is best for one player may not necessarily be best for another. This brings us to second, deeper objective: finding strategies that not only optimize individual outcomes but also balance the interests of all players in a way that no one has a strong incentive to change their decision.

   In Chapter 2, we will explore how these best strategies are determined using two fundamental concepts in game theory:

   1. Security level for players— a measure of the minimum guaranteed payoff a player can secure, regardless of what others do.

2. Equilibrium strategies, particularly Nash Equilibrium, introduced by John Nash in his 1950 Ph.D. thesis, which describes a stable state where no player can improve their position by unilaterally changing their strategy.

These two topics, along with other key ideas, others will be covered discussed in depth throughout great details in this book, giving you a strong foundation in understanding strategic decision-making.

A few simple examples of games are shown below. These examples illustrate the basic terminologies and notations to be used in the subsequent chapters of the book.

Example 1.

First, we look at a 2-person simple game represented by a 2-dimensional matrix as shown:

$$\begin{array}{cc} & \begin{array}{cccc} A & B & C & D \end{array} \\ M = & \begin{pmatrix} -1 & 5 & -1 & 2 \\ 0 & 3 & 2 & 4 \\ -1 & 2 & 3 & -2 \end{pmatrix} \begin{array}{c} X \\ Y \\ Z \end{array} \end{array}$$

The 2 - person, for easy references, we will call them the Row Player and the Column Player; in the example above, the Row Player has 3 strategies X, Y, Z that he can choose, and the Column Player has 4 strategies A, B, C, D that he can choose. Once both players have made the decision of selecting which of the strategy from his strategy set, then the values shown in the Matrix, called the payoff values for both players are used to determine the payoff value of the two players with respect to the strategy that each player chosen. This is done by reading off the value shown in the intersection of the strategy row of the Row Player selected and the strategy column of the Column Player selected.

For example, if Row Player selects strategy X and Column Player selects strategy B, then the payoff value is 5, denoted by [M, X, B] =5. We will use this notation throughout the rest of the book. Noted that this notation has three parameters, the first parameter is the name of the payoff Matrix, the second and the third is the name of the strategy that the Row Player and the Column Play have chosen, respectively. Similarly, [M,Y,D] = 4, [M,Z,D] = -2, and [M,Y,A] = 0.

In matrix M above, there are positive, negative, and zero payoff values. By convention, we use the positive values to indicate the Row Players' winning payoff value, and the negative values to indicate the Column Players winning payoff value. And 0 payoff value to indicate both Players are break even, neither winning nor losing. But there is no reason that we cannot use a different convention by using the positive sign to indicate the Column Player's winning value and the negative sign to indicate the Row Player winning value. However, we must keep in mind that the correctness of the two important operations Maximin and Minimax on the Matrix M that we will describe in this book depends on which convention we choose.

The payoff matrix of the game above would be as follows if we choose to use positive payoff values for the Column Player.

$$N = \begin{array}{c} \\ \\ \end{array}\begin{matrix} A & B & C & D & \\ \end{matrix}$$

$$N = \begin{pmatrix} 1 & -5 & 1 & -2 \\ 0 & -3 & -2 & -4 \\ 1 & -2 & -3 & 2 \end{pmatrix} \begin{matrix} X \\ Y \\ Z \end{matrix} \quad \text{(columns A, B, C, D)}$$

The two games given by the M and N payoff matrix shown above is an example of a type of game called the zero-sum game of two players. This name zero-sum is because the amount that one player wins shown by a positive value shown in the Payoff Matrix is the same amount that the other player loses shown by the negative value in the Payoff Matrix. In general, there are constant sum games for players. The following example is an example of a 3-person zero sum game.

Example 2.

In this example, each of the three players has two strategies to choose, let us call them GO and NO -GO. The payoff values for each of the three players cannot be represented by a single Matrix as in the example before. We will use a stack of two matrices to store the payoff values of the three players, the matrix in the bottom stack corresponds to the strategy GO (G) and the matrix at the top of the stack corresponds to the strategy NO-GO (N) of the first player. And each matrix in the stack contents the payoff values corresponds to the GO and NO-GO strategies selected for second player and the third player.

$$\begin{matrix} & G & N \\ G & (1,1,-2) & (-4,3,1) \\ N & (3,-4,1) & (-5,-5,10) \end{matrix} \qquad \begin{matrix} & G & N \\ G & (3,-2,-1) & (-6,-6,12) \\ N & (2,2,-4) & (-2,3,-1) \end{matrix}$$

The first matrix corresponds to the G strategy chosen by the first player, and the second matrix corresponds to the N strategy chosen by the first player. The four entries in each matrix are the payoff values for players 1, 2, and 3. respectively. The payoff values for each player are shown in vector form. For example, the vector entry (1, 1, -2) in the above matrices means that Player 1 wins 1, Player 2 wins 1, player 3 lost 2 when Player 1 selects G, Player 2 selects G, and Player 3 selects G. The vector entry (2, 2, -4) means Player 1 wins 2, Player 2 wins 2, Player 3 lost 4 when Player 1 chose N, Player 2 chose N, and Player 3 chose G.

Note that this game is zero-sum, because any player's win is the sum of the other two players lost, and what a player's lost is the sum of the other two players' wins. We can check and verify this by looking at each vector entry from the two matrices.

As we can see in this example, when the number of players gets higher than 3, the data structure (in our present case, it is a stack of two-dimensional arrays) used to store or to represent the players' payoff values gets complicated and it is hard to visualize. Therefore, it is important to come up with a meaningful and convenient data structure and notation to represent the payoff values for the n-players prior to analyzing the n-person game. However, due to the limitation of this book, we will not go there. Interested reader can refer [18].

Example 3.

In this example, I will show you a two person, non-zero-sum game, known as the Prisoner's Dilemma in game theory. This is a representative example of many other games (C[8]).

Prisoner 1 and Prisoner 2 are arrested on suspicion of robbery. The district attorney (DA) is convinced of their guilt but lacks sufficient evidence for conviction at a trial. The DA then promises each prisoner separately that he would apply the following sentence to them in according to the following payoff matrix for each prisoner shown below:

$$\begin{array}{c} \\ C \\ D \end{array} \begin{array}{c} \begin{array}{cc} \text{C} & \qquad\text{D} \end{array} \\ \begin{pmatrix} (-2,-2) & (0,-4) \\ (-4,0) & (-1,-1) \end{pmatrix} \end{array}$$

The matrix above is a short- hand for two payoff matrices, one for Prisoner 1 and one for Prisoner 2. In analyzing the optimal strategy of both players for this type of game, we will treat these two matrices separately (see later section on Chapter 3). The vector entry in the matrix represents Player 1's payoff value in the first component, and Player 2's payoff value in the second component. Here we use the value 0 to mean that is no sentence, -1 means 1 year sentence, -2 means 2 years sentence, and -4 means 4 years sentence. We use negative value in the matrices as these numbers represent undesirable nature of outcomes, for example, -2 is more favorable than -4 because 2 years of sentence is much better than 4 years of sentence. The payoff matrices show that if both prisoners confess (C), there is a two- year sentence for each prisoner, if both defects (D) , both will result a 1- year sentence, and if one confesses and one defects, the one confesses will receive is fee to go and the one defects will receive a 4-years sentence.

Example 4.

In this example, we show an extension of the 2-person Prisoner's Dilemma to the 3-person Prisoner's Dilemma using a stack of two matrices to store the players' payoff values as described in Example 2 above.

$$\begin{array}{ccc} & \text{First Player} & \text{First Player} \\ & \text{C} & \text{D} \\ & \text{Third Player} & \text{Third Player} \\ & \begin{array}{cc} \text{C} & \qquad\text{D} \end{array} & \begin{array}{cc} \text{C} & \qquad\text{D} \end{array} \\ \text{Second Player} \begin{array}{c} C \\ D \end{array} & \begin{pmatrix} (1,1,1) & (0,0,3) \\ (0,3,0) & (-2,2,2) \end{pmatrix} & \begin{array}{c} C \\ D \end{array} \begin{pmatrix} (3,0,0) & (2,-2,\ 2) \\ (2,2,-2) & (-1,-1,-1) \end{pmatrix} \end{array}$$

From the payoff values of the three prisoners, we can see that if all three defects, each loses 1; if all three confess, each gain 1; a single cooperator suffers largest loss -2 and the two defectors gain 2; a single defector gets the largest payoff 3; if one defects and two confess, the defector gains 3, the two confessors each gain 0.

Example 5.

This is an example of a non-cooperative two-person game with the payoff matric shown below.

$$\begin{array}{c} \\ X \\ Y \end{array} \begin{array}{c} \begin{array}{ccc} A & B & C \end{array} \\ \begin{pmatrix} (0,0) & (0,-5) & (10,-15) \\ (-5,-5) & (10,10) & (7,15) \end{pmatrix} \end{array}$$

In non-cooperative two-person games, the two payoff matrices for the two players are independent. For a given pair of strategy selected by both players, the playoff values for the two players could be both positive, could be both negative, or positive for one player and negative for the other player. In this example, if the strategy pair is (Y, C), Player I 's playoff value is 7 and Player 2 's is 15 (both positive); if the strategy pair is (Y, A), both players' payoff is -5 (both negative); and in the strategy pair (X, C), the payoff value for Player 1 is 10 (positive) and that of Player 2 is -15 (negative). Because the two matrices are entirely different, for this type of non-cooperative 2-person game, an optimal strategy pair for one player may not be an optimal strategy pair for the other player. In later sections, we will show that optimal strategies such as security level strategy, equilibrium strategy (both concepts are to be discussed in later sections) that work so well for the zero-sum games does not quite applicable to this type of non-cooperative games.

Example 6.

In this example, I will show you an example of a cooperative two-person game. The payoff matrices for this type of game are like those of the non-cooperative games. What is the difference between the two? In non-cooperative games, each strategy pair selected by the two players is associated with a pair of payoff values (v1, v2) which corresponds to a discreate point in the XY-plane, with v1 the X-axis coordinate and v2 the Y-axis coordinate. As there are only a finite number of strategy pairs, there are only a finite number of such discrete payoff points in the XY-plane. In cooperative games however, because of the cooperative nature of the game, players can discuss in advance as to what strategy they are going to take to benefit both sides before the game is played, all possible outcomes of the game would consist of infinitely many of such (v1, v2) pairs in the XY-plane. This infinite set of (v1, v2) points are bounded by a convex set formed by only a finite number of points each of which corresponds to a discrete strategy pair from the game. This convex set is called the Cooperative Payoff Set (CPS) of the game. Any point inside CPS is the outcome of a suitable manipulation of strategy pairs by both players. The study of cooperative games is to investigate various approaches to locate the coordinate of a point in the CPS that would give the solution to the game. This topic will be discussed in latter chapters.

The payoff matrices of a cooperative game and its CPS is shown below:

Payoff Matrix M:

$$\begin{array}{c} \\ X \\ Y \end{array} \begin{array}{c} \begin{array}{ccc} A & B & C \end{array} \\ \begin{pmatrix} (0,0) & (0,-5) & (10,-15) \\ (-5,-5) & (10,10) & (7,15) \end{pmatrix} \end{array}$$

CPS of M:

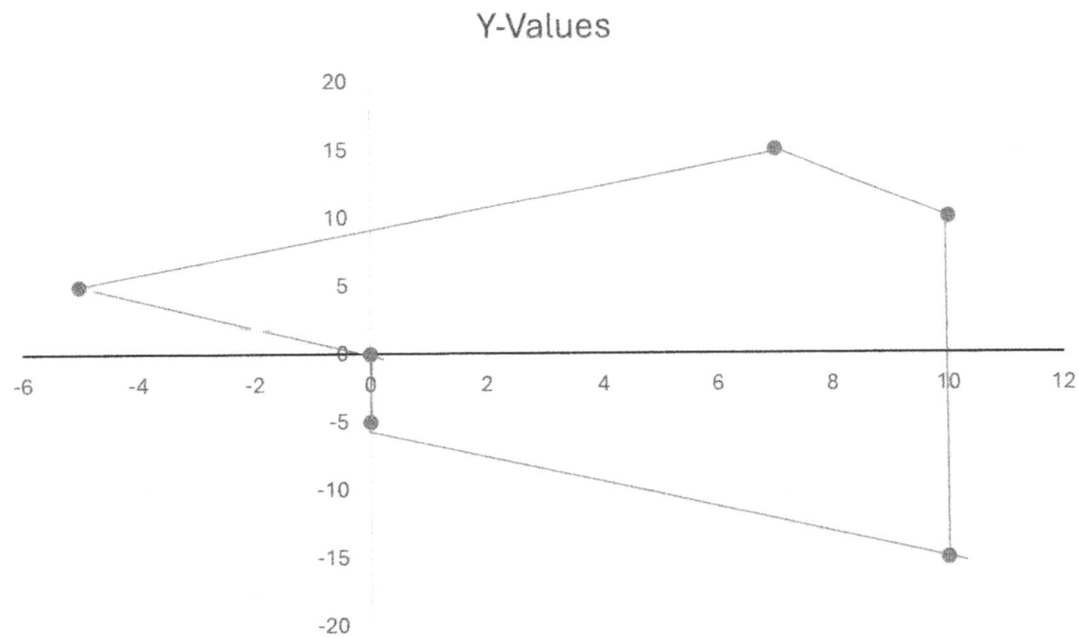

Example 7.

We have seen an example of a 2 -person zero-sum game in Example 3, the last example shown in this section is an example of a constant -sum game. As we recall in a zero-sum game, what one player wins is what the other player loses. However, in the constant-sum games, one player wins more, the other player wins less, but the sum of their wins and lost adds up to a constant depending on the game. Both types of games can be used to model many real-life scenarios and business situations.

Suppose there are two competitive wholesalers X and Y who want to set up their business centers in one of the 7 possible town locations A, B, C …... G. as shown in the graph below. The problem for these wholesalers X and Y is to find out where they should locate their wholesale center that would bring them the most profits.

As shown in the graph below, where the number inside the vertex indicates the population of that town (unit in thousands), and the line joining the two towns indicates that people in one town can drive to another town, where the number above the line indicates the distance (unit in miles) between them. To get the basic concept of game theory easier across to you, I must make some simple assumptions here. Let us assume that people prefer to shop at a nearer wholesale center from their hometown, and if both wholesale centers are in the same town, then half of the population in that town will shop at X – wholesale center and the other half at Y-wholesale center.

Based on these assumptions, we can formulate this business application as a 2-person constant- sum game described by a pair of payoff matrices, and now we are going to show you how to find these payoff matrices.

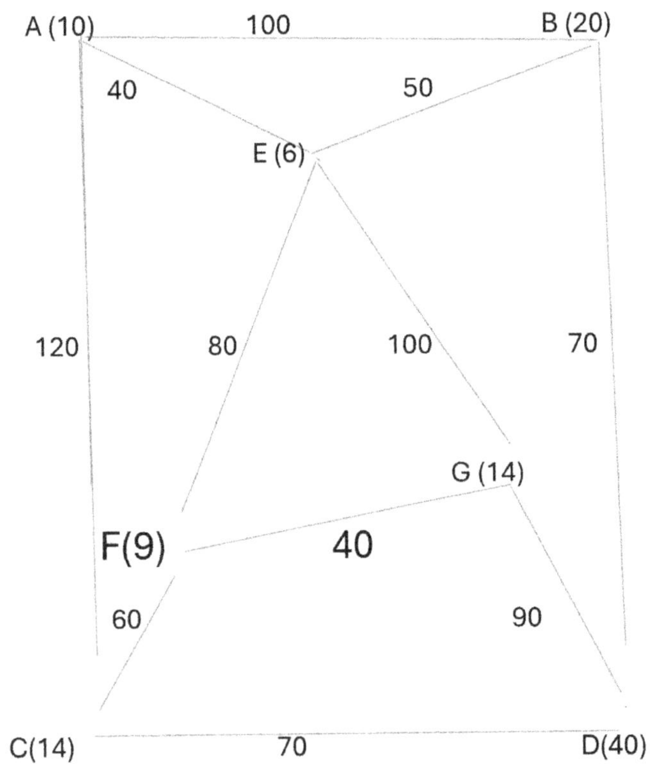

For Wholesaler X and Wholesaler Y, they each have 7 possible choices of a town location to settle their wholesale centers, each choice to them is a strategy to them. There will then be 7*7 =49 that many strategy pairs to consider and each strategy pair will give specific impact or outcome to both wholesalers. How do we measure the impact or outcome in this business problem? In this example, we use the number of customers shopping at the wholesaler's center as an indication of the wholesaler's rewarding or payoff values. Once these payoff values are computed as we will demonstrate it in our next step, the description of the game is complete, and we can then use the principles and concepts in game theory to help the wholesalers to come up with an optimal strategy for their business application.

Payoff Matrices for Wholesaler X and Y:

	A	B	C	D	E	F	G
A	(56.5, 56.5)						
B		(56.5,56.5)					
C			(56.5,56.5)				
D				(56.5,56.5)			
E					(56.5,56.5)		
F						(56.5,56.5)	
G	(97, 16)				(77, 36)		(56.5,56.5)

There are 7 *7 =49 entries in the above payoff matrix pair. I will demonstrate here how the entries at (G, A) and (G, E) which are (97, 16), and (77, 36) are computed. However, as a return demonstration, the reader should compute another entry, say (A, A) or (B, B) or any one on the diagonal of the matrix and be convinced that the result is (56.5, 56.5). Note that the sum of every entry in the above payoff matrix is 113, hence this is a constant sum game.

As an illustration, computation of the payoff value at the (G, A) entry from the above matrix is summarized in the following table:

When Wholesale Center X is located at town G and Wholesale Center Y is located at town A, pertinent data are summarized in the table below:

Towns	Shortest distance from current town to Wholesale Center X	Shortest distance from current town to Wholesale Center Y	Population shopping at Wholesale Center X	Population stopping at Wholesale Center Y
A	140	0	0	10
B	150	100	20	0
C	100	120	14	0
D	90	170	40	0
E	100	40	0	6
F	40	120	9	0
G	0	140	14	0

The entries in each column of the above table are computed as follows:

For each town from town A to town G, we find out how far is that town to town X and town Y, call the respective distances dX and dY, if dX < dY, then the entire population from that town will shop at town X, if dY < dX, then the entire pollution from that town will shop at down Y, when if dX= dY, half the population from that town will shop at both town X and town Y.

Sum over the entries in the third and fourth column gives the pair (97,16), which is the (G, A) entry in the payoff matrix for the game.

As a second illustration, when Wholesale Center X is located at town G and Wholesale Center Y is located at town E, pertinent data are summarized in the table below:

Towns	Shortest distance from current town to Wholesale Center X	Shortest distance from current town to Wholesale Center Y	Population shopping at Wholesale Center X	Population stopping at Wholesale Center Y
A	140	40	0	10
B	150	50	0	20
C	100	140	14	0
D	90	120	40	0
E	100	0	0	6
F	40	80	9	0
G	0	100	14	0

Sum over the entries in the third and fourth column respectively give the pair (77,36) as the entry at the strategy pair (G, E) of the payoff matrix for the game.

Example 8. (Traffic Lights)

This game describes the following street scenario. Two drivers (players) are driving on two respective streets toward an intersection where there is a traffic light controller directing the street traffic to GO or to STOP. Assume the payoff matrix of this game is shown below:

	Driver 2	
Driver 1	GO	STOP
GO	(-100, -100)	(5 , -1)
STOP	(-1, 5)	(-10 , -10)

Interpretation of the payoff values for each strategy pair of the two drivers is as follows. The choice of strategy pair (GO, GO) by both drivers is (-100, -100) because this will cause an accident and will result in a large loss to both drivers, say at a scale of -100. Strategy (STOP, STOP) will cause both drivers some inconvenience because they are not able to go the places where they are supposed to go, hence both drivers receive, say at a scale of (-10, -10). The other strategy pairs (STOP, GO) and (GO, STOP) have similar interpretations. The importance of this example is that the strategy pair (STOP, GO) and (GO, STOP), as we will show in Chapter 3, each turns out to be a Nash Equilibrium for this game.

Example 9. The Dating Game from the movie "A Beautiful Mind" ([10])

This is an example from "A Beautiful Mind", we simplify and summarize the game theory aspect of the movie scenario as follows. Assume there are two men (two players, Nash himself and other Men), and each try to date either a Blonde or a Brunette, but their true preference is to date the Blonde. Suppose the payoff matrix for this game is as follows:

	Other Men	
	Blonde	Brunettes
Nash		
Blonde	(0, 0)	(1, 10)
Brunettes	(10, 1)	(1, 1)

We interpret this game as there are two players, Nash and Other Men, each player has two strategies to choose, the strategy to date the Blonde and the strategy to date the Brunettes. The payoff values of two players Nash and the Other Men are shown above. On the scale of 1 to 10, 10 represents very happy with the dating, 1 represents happy with the dating, and 0 means not at all happy. The strategy pair (Blonde, Blonde) has payoff values (0, 0) because if both Nash and the Other Men approach the Blonde, both men will remain alone as the Blonde would not make the choice of Nash or the Other Men in front of both of them. If both men approach a different Brunette, then each Brunette will enjoy the company of that person, hence the strategy pair (Brunettes, Brunettes) has the payoff value (1,1). The strategy pair (Blonde, Brunettes) and (Brunettes, Blonde) has payoff value (1,10) and (10,1) respectively because the Other Men dates the Blonde with the payoff value of 10 in the first strategy pair, and Nash dates the Blonde with payoff value of 10 in the second strategy pair. Again, the essence of this example is to show that in Chapter 3, there are three Nash Equilibriums in this game corresponding to the strategy pairs (Blonde, Brunettes), (Brunettes, Blonde), and (Brunettes, Brunettes).

Example 10. Traffic flow in a network [Papers [10]]

This example extends the idea of a single traffic light example shown in Example 8 to a traffic management system of many traffic lights located in many road intersections. Each driver on the road is considered as a player of a non-cooperative game, where the traffic flow for each route (from start to destination) is determined by the Nash Equilibrium to ensure that no driver can improve their performance by changing their strategy. Due to the space limitations of this book and the complexity of the traffic management system, however, we will not go on further discussions, interested readers can refer to [10].

1.1 Summary

In this chapter, we explored key concepts, common terminology, and various examples in game theory. Through these examples, we gained clear understanding of zero-sum, constant-sum, and non-zero-sum games, as well as the distinction between cooperative and non-cooperative games.

We also saw how a constant-sum game can be used to model real-world scenarios, such as the business case in Example 7, where game theory helps in solving an underlying business problem. Additionally, we introduced classic games like the Prisoners' dilemma, Traffic Lights, Traffic Flow in a Network, and the Dating Game from the movie "*A Beautiful Mind*." These examples are designed to build your intuition and prepare you for Nash Equilibrium, a fundamental concept that we will explore in depth in Chapter 3.

Chapter 2

2.0 Introduction

In this chapter, we introduce some fundamental yet essential concepts in game theory for a specific type of game known as ***zero-sum games***. This is a two-player game, typically referred to as the row player (Player 1) and the column player (Player 2), where any gain by the row player results in an equivalent loss for the column player.

We explore the concept of pure strategies and mixed strategies, the expected value of a strategy pair (where one strategy is chosen by the row player and the other by the column player), and the security level for both players. We also state and exemplify the proof of the Fundamental Theorem of Game Theory, which guarantees that we can always find a mixed strategy X^* for the row player that ensures their security level sl(1), and similarly, a mixed strategy Y^* for the column player such that guarantees their security level of the column player sl(2).

It turns out that these security strategies X* and Y* are the solutions to a MIN and MAX Linear Programming problem, which can be efficiently solved using the Simplex Method. The Duality Theorem in Linear Programming then ensures that the objective function values of both LP formulations are equal. This common value is known as ***the value of the zero-sum game***.

To enhance readers' understanding of these concepts, we provide multiple examples with detailed numerical solutions, many of which are solved using MATLAB programs to illustrate the application of these theories in practice.

2.1 Pure strategy and mixed strategy in a game

Pure strategies in a game are the strategies that players have the freedom to choose based on their own decisions. These strategies are an integral part of the rules that define a game. In every example presented in the previous chapter, all the strategies used were pure strategies.

On the other hand, there exists another type of strategy known as mixed strategies, where each pure strategy is assigned a probability value. When it is a player's turn to move, they must first determine a probability value and then check which of their pure strategies corresponds to that same probability value to decide which strategy to play.

This leads to two key questions regarding mixed strategies:

1. How does a player determine the matching probability value each time it is their turn to play?
2. Where do these probability values for the pure strategies come from?

To clarify these concepts, we will use the following example.

Example 1:

$$\begin{array}{cc} & \begin{array}{cccc} A & B & C & D \end{array} \\ M = & \begin{pmatrix} -1 & 5 & -1 & 2 \\ 0 & 3 & 2 & 4 \\ -1 & 2 & 3 & -2 \end{pmatrix} \begin{array}{c} X \\ Y \\ Z \end{array} \end{array}$$

In this game the Row Player (Player 1) has three pure strategies X, Y, Z, the Column Player (Player 2) has four pure strategies A, B, C, D. The game is described by the Payoff matrix M consists of all the payoff values (positive or negative) shown above. Each payoff value is associated with a strategy pair. For example, associated with the pairs (X, A), (X, B), (Y, B), (Z, D), the payoff values are [M, X, A] = -1, [M, X, B] = 5, [M, Y, B] =3, [M, Z, D] = -2 respectively. These values are the indicators of how the two players are doing in this game given the strategies they chose.

A mixed strategy, say W of the row player's pure strategies X, Y, Z assigns each of them a probability value, say 0.3 to X, 0.6 to Y, and 0.1 to Z. Similarly, a mixed strategy, say E of the column player's pure strategies A, B, C, D assign each of them a probability value, say 0.2 to A, 0,1 to B, 0.3 to C, and 0.4 to D. When it is time for the row player to play W, how does he decide if he should play A, or B, or C? When it is time for the column player to play E, how does he decide which of the A, B, C, D strategy to play? Based on the meaning of probability, since W is associated with probabilities 0.3, 0.6, 0.1 for the strategies X,Y,Z , the row player can decide which pure strategy to pick by putting 3 red balls, 6 yellow balls, and 1 blue ball in a box with each color corresponds to X,Y,Z strategy, and each time he randomly selects a ball from the box, the color of the ball tells him which of the pure strategy to choose. Similarly, E is associated with probabilities 0.2, 0.1, 0.3, 0.4 for the strategies A, B,C,D, the column player can decide which pure strategy to pick by putting 2 red balls, 1 yellow ball, 3 blue balls, 4 while balls in a box with each color corresponds to the A,B,C,D strategy, and when he randomly selects a ball, the color to ball tells him which pure strategy to choose. It should be noted that the probabilities 0.3, 0.6, 0.1 associated with W does not mean that when we repeatedly playing the game for a fixed number of times, say 100 times, 30 times of which we shall play with strategy X, 60 times with Y, and 10 times with Z. This is because each time we play the game, we are facing with the decision of which strategy to pick for that game at hand right away. We can also view that any pure strategy for a player in a game is a mixed strategy with probability 1 chosen for that pure strategy and probability 0 for all his other pure strategies.

In this example, for the row player, strategy X can be considered as the mixed strategy W with corresponding probabilities (1,0,0). We denote this correspondence by X= W (1,0,0). Strategy Y corresponds to W with probabilities (0,1,0), so Y=W (0,1,0), and similarly, Z= W (0,0,1). For the column player, E is the mixed strategy of A, B, C, and D, so as before we have A= E (1,0,0,0), B=E (0,1,0,0), C=E(0,0,1,0), and D=E (0,0,0,1).

Thus far, we know how to select which pure strategy to select when we are asked to play a mixed strategy, the question remains is how are these probabilities assigned to the pure strategies to come up with a mixed strategy? And why is mixed strategy necessary? In the subsequent chapters, we will answer these questions in detail. For now, it suffices for us to know that for each rational player in a game, his objective is to choose from his strategy set a strategy or strategies that can maximize his gain in terms of his payoff values given in the payoff matrix that defines the game. This gain of a player's strategy can be shown that it is higher with a suitable choice of probabilities in a mixed strategy than with just a pure strategy. Thus, we can think of the probabilities in a mixed strategy are chosen in such a way to maximize the player's gain. How is this done? In the next section, we will introduce the concept of Expected Payoff of a game which will be useful to help us in selecting these probabilities.

2.2 Expected value of a strategy pair between players.

Let us formulate this concept of Expected Value of a strategy pair in a game with payoff values shown below:

$$
\begin{array}{cccccccccc}
 & Y1 & Y2 & & & & Y_{n-1} & Y_{n} & & \\
 & a_{11} & a_{12} & .. & .. & .. & a_{1n-1} & a_{1n} & & X_1 \\
 & a_{21} & a_{22} & .. & .. & .. & a_{2n-1} & a_{2n} & & X_2 \\
 & \vdots & .. & .. & .. & .. & .. & .. & & \\
 & \vdots & .. & .. & .. & .. & .. & .. & & \\
 & a_{m-11} & a_{m-12} & .. & .. & .. & a_{m-1n-1} & a_{m-1n} & & X_{m-1} \\
M = & a_{m1} & a_{m2} & .. & .. & .. & a_{mn-1} & a_{mn} & & X_m
\end{array}
$$

As shown, the row player has the strategy set {X1, X2, …., X m}, the column player has the strategy set {Y1, Y2…, Y n}. The payoff value of M associated with the strategy pair (X i , Y j) is denoted by M(X i, Y j) which is a_{ij} in the diagram shown.

Suppose S is a mixed strategy of X1, X2, ...X n with probabilities p1, p2...p m, we represent S by S= (x1, x2……, x m), and T is a mixed strategy of Y1, Y2…. Y n with probabilities y1, y2… y n, and represent T by T= (y1, y2…, y n).

Having defined what M, S, T meant in a game, we will investigate the following quantity, which is the sum of products of the payoff value of M, the probabilities of S, and the probabilities of T, it is denoted by EV (M, S, T) and it is called the Expected Payoff of the Game M for the strategy pair (S, T), that is:

$$\text{EV (M, S, T)} = \sum_{i=1}^{m}\sum_{j=1}^{n} x_i a_{ij} y_j$$

This is the sum of products of every payoff value aij in the game matrix M with the corresponding probabilities xi y_j that both the row player and the column player will choose this aij. Our objective is to maximize the above EV (M, S, T) by choosing the suitable probabilities in the row player's mixed strategy S and column player's mixed strategy T. This quantity will be used in many of our future analyses, so a clear understanding of it means and to establish some of its properties that we will need are in order. But first we need some simple definitions and notations associated with the following game representation described before.

$$
\begin{array}{cccccccc}
 & Y1 & Y2 & & Yn\text{-}1 & Y\ n & \\
M = & \left(\begin{array}{c} a_{11} \\ a_{21} \\ : \\ : \\ a_{m-11} \\ a_{m1} \end{array}\right. &
\begin{array}{c} a_{12} \\ a_{22} \\ .. \\ .. \\ a_{m-12} \\ a_{m2} \end{array} &
\begin{array}{c} ..\ ..\ .. \\ ..\ ..\ .. \\ ..\ ..\ .. \\ ..\ ..\ .. \\ ..\ ..\ .. \\ ..\ ..\ .. \end{array} &
\begin{array}{c} a_{1n-1} \\ a_{2n-1} \\ .. \\ .. \\ a_{m-1n-1} \\ a_{mn-1} \end{array} &
\left.\begin{array}{c} a_{1n} \\ a_{2n} \\ .. \\ .. \\ a_{m-1n} \\ a_{mn} \end{array}\right) &
\begin{array}{c} X_1 \\ X_2 \\ \\ \\ X_{m-1} \\ X_m \end{array}
\end{array}
$$

(Figure 1.)

This is a m by n matrix of payoff values.

In what follows, we will use the following notations:

Let the ith row vector of M be denoted by $M_{i(r)}$ and the jth column vector by denoted by $M_{j(c)}$.

Let R1, R2, ……R m be the row vectors, each is a 1 by n matrix,

Let C 1, C2…...Cn be the column vectors, each is a m by 1 matrix,

R1 = (1,0,0….0),

R2= (0,1,0…...0),

:

:

Rm = (0,0, 0….1)

And

$$C1= \begin{pmatrix} 1 \\ 0 \\ 0 \\ : \\ : \\ : \\ 0 \end{pmatrix}$$

$$C2= \begin{pmatrix} 0 \\ 1 \\ 0 \\ : \\ : \\ : \\ 0 \end{pmatrix}$$

$$Cn = \begin{pmatrix} 0 \\ 0 \\ 0 \\ : \\ : \\ : \\ 1 \end{pmatrix}$$

Then we can easily verify that the following statements about this EV (M, S, T) are true.

Statement 1. When the strategy S and T are respectively the row vector R i* and

C_{j*} the m*n terms in EV (M, S, T) reduces to a single term $a_{ij.}$ That is:

$$\text{EV (M, Ri*, C j*)} = \sum_{i=1}^{m}\sum_{j=1}^{n} x_i a_{ij} y_j$$

$$= a\ i^*j^*$$

, because all the xi and yj are zero except x i*=1 and y j*=1 from the definition of the vectors R i* and C j*.

Statement 2. EV (M, X, C_j) = X* $M_{j(c)}$ (dot product of row and column vector)

Statement 3. EV (M, R_i, Y) = $M_{i\,(r)}$ * Y

Statement 4. EV (M, X, Y) = $\sum_i EV(M,R_i,Y)x_i$

$$\text{OR} = \sum_j EV(M,X,C_j)y_j$$

We will make use of these properties of the expected payoff value of a strategy pair in the subsequent sections.

2.3 Security level of a strategy and Security level of a Player

Each player has a finite set of pure strategies and an infinite set of mixed strategies. We will begin by defining what is meant by the security level of a player's strategy, followed by the security level of the player.

First, we will define the security level for a pure strategy, and then extend this concept to mixed strategies.

$$\begin{array}{ccccccccc} Y1 & Y2 & \ldots\ldots & & & & Y_n & & \\ \end{array}$$

$$\begin{pmatrix} a_{11} & a_{12} & .. & .. & .. & a_{1n-1} & a_{1n} \\ a_{21} & a_{22} & .. & .. & .. & a_{2n-1} & a_{2n} \\ \vdots & .. & .. & .. & .. & .. & .. \\ \vdots & .. & .. & .. & .. & .. & .. \\ a_{m-11} & a_{m-12} & .. & .. & .. & a_{m-1n-1} & a_{m-1n} \\ a_{m1} & a_{m2} & .. & .. & .. & a_{mn-1} & a_{mn} \end{pmatrix} \begin{matrix} X_1 \\ X_2 \\ \\ \\ X_{m-1} \\ X_m \end{matrix}$$

Suppose row player chooses strategy X_i, then regardless which strategy the column player simultaneously plays, the playoff values for row player will be amongst the possible values a_{i1}, a_{i2}, ……. …… $a_{in.}$ If all the values are positive, as we agree to designate positive as row player's winning (or column player's loosing), and negative as row player's loosing (column player's winning), then the minimum of all these values is the smallest amount that row player is sure or secure to win if he chose strategy X_i. We denote security level of a strategy by

sl (X_i) = Min { a_{i1}, a_{i2}, ……. …… a_{in}} and record that the minimum value took place at a_{ij*}.

If a_{i1}, a_{i2}, ……. …… a_{in} are all negative, then sl (X_i) is the amount that the row player loses at the most. And if a_{i1}, a_{i2}, ……. …… a_{in} is a mixture of both negative and positive, sl (X_i) is also the amount that the row player loses at the most. In summary, sl (X_i) is the minimum amount the row player will win or the maximum amount he loose if he plays strategy X_i. Now if the row player compute his security levels sl(X1), sl(X2),…….sl(Xn) for each strategy X1,X2,…..in his finite pure strategy set, he will have an idea which strategy to choose. If he further computes the maximum of all these sl(X1), sl(X2),…….sl(Xn) and record at which strategy X_{i*} the maximum takes place, denote

sl (row player) = Max { sl(X1), sl(X2)…….sl(Xn)}

, the quantity sl (row player) is called the security level of the row player, it represents the minimum amount he is sure to win or the maximum amount he is sure to lose if he chooses the strategy X_{i*}, regardless of what strategies his opponent the column player is going to choose. This security value for the row player is a_{i*j*}.

For a rational player, he would choose strategies to play so that the outcome of the game would be above his security level.

We will look at our first example below:

Example 2:

$$
\begin{array}{cccccc}
 & Y1 & Y2 & Y3 & Y4 & \\
X1 & 1 & 7 & 1 & 4 & 1 \\
X2 & 2 & 5 & 4 & 6 & 2 \\
X3 & 1 & 4 & 5 & 0 & 1 \\
 & 2 & 7 & 5 & 6 &
\end{array}
$$

sl (X1) = 1, this min occurs at Y1 and Y3

sl (X2) = 2, this min occurs at Y1

sl (X3) = 1, this min occurs at Y1

sl (row player) = Max {sl(X1), sl (X2), sl(X3)}

= Max {1, 2, 1}

= 2, this maximum occurs at X2

This means that if the row player chooses strategy X2, he is sure to win a minimum amount of 2 regardless of what strategies his opponent the column player chooses.

In an analogous manner, we are going to define the security level for the column player when he plays the strategy $Y_{j.}$

sl (Y_j)= Max { a_{1j}, a_{2j}, ... a_{mj}} and this maximum occurs at a $_{i^*j}$

, the security level of the column player is defined as

sl (column player) = Min {sl (Y1), sl(Y2)...... sl(Yn)}, where the minimum take place at strategy Yj* . Because we agree to designate a negative value to mean the column player's winning, so

sl (Yj) is the maximum amount he would lose or the minimum amount he would win if he played Yj, and sl (column player) is the minimum amount he would win or the maximum amount he would lose if he played strategy Yj* and this amount is $a_{i^*j^*}$.

In the above example,

sl (Y1) =max {1,2,1} =2

sl(Y2) = max {7,5,4} = 7

sl(Y3) = max {1,4,5} =5

sl (Y4) = max {4,6,0} =6, and so

sl (column player) = min {sl(Y1), s;(Y2), sl (Y3), sl(Y4)} =2

Note that in this example, the security level of the row player happens to be the same as the security level of the column player, that is!

sl (row player) = sl (column player), this equality however, does not have to hold for any game, the next example shows that it is not. But we will characterize the type of game by its payoff matrix that this equality holds, and whenever this equality holds, this common value is called the value of the game, and if this common value is 0, then we say the game is a fair game. Per this definition, our game is not a fair game since it is 2, and since it is a positive 2, this game is a game in favor of the row player.

Now consider our second example,

Example 3:

$$\begin{array}{cc} & \begin{array}{cccc} Y1 & Y2 & Y3 & Y4 \end{array} \\ \begin{array}{c} X1 \\ X2 \\ X3 \end{array} & \begin{pmatrix} -1 & 1 & 2 & 0 \\ 4 & -2 & -3 & 2 \\ 0 & 3 & 1 & -2 \end{pmatrix} \begin{array}{c} -1 \\ -3 \\ -2 \end{array} \\ & \begin{array}{cccc} 4 & 3 & 2 & 2 \end{array} \end{array}$$

sl (X1) =-1, sl (X2) = -3, sl (X3) = -2,

sl (row player) = max { -1, -3, -2} = -1, this maximum occurs at X1, this means that the row player is sure that he will lose no more than 1 if he plays X1.

sl (Y1) = 4, sl(Y2) = 3, sl(Y3) =2, sl(Y4) = 2,

sl (column player) = min {3,3,2,2} =2, this minimum occurs at Y3 and Y4, this means that if the column player chooses strategy Y3 or Y4, he will not lose more than 2. In order words, column player's security level strategy is to play strategy Y3 or Y4, while row player's security level strategy is to play X1.

In this example, we see that sl (row player ≠ sl (column player).

Definition of a Saddle Point: In any payoff matrix , if sl (row player) = sl (column player) then we say that this common value together with its location in the matrix is called a saddle point for the matrix.

In our second example above, we note that that sl (row player ≠ sl (column player) when only pure strategies are available for both players. But since the set of pure strategies is only a small subset of the mixed strategies, would equality hold when we expand the set of pure strategies to include the mixed strategies? The answer is yes, in fact, this is the well-known theorem called **The Fundamental Theorem of Game Theory** that we will dive into next. But first, we will need to know what it means by the security level of a mixed strategy and the security level of a player when all strategies including the mixed strategies are available for his choice. For the readers' convenience, we have provided a MATLAB program ComputeSecurityLevelsandSaddlePoint.m to compute security levels for pure strategies of any size payoff matrices. Computations of both matrices for the last two examples are shown in the output listing below:

```
>> ComputeSecurityLevelsandSaddlePoint
```

********** Check if given matrix has a saddle point ******************:

```
a =

     1     7     1     4

     2     5     4     6

     1     4     5     0

slx =

     2

sly =

     2


yes =

     1

slevel =

     2

svector =

     0     1     0
```

***:

```
b =

    -1     1     2     0
```

```
    4   -2   -3    2

    0    3    1   -2

slx =

   -1

sly =

    2

yes =

    0

slevel =

   999

svector =

    0    0    0
```

In the above output listing,

slx = the security level for the row player

sly = the security level for the column player

slevel = common value of slx and sly, default to 999 if slx <> sly

yes = the logical variable indicates if the saddle point exists, yes = 1 means the saddle point exist, 0 means it does not exist

svector = a { 0,1} vector with 1 indicates the location of the row number where the saddle point is located

---

Source Codes of the ComputeSecurityLevelsandSaddlePoint.m

```
%fprintf ('********** Input payoff matrices *****************: \n')

a=[ 1 7 1 4; 2  5  4  6; 1  4  5  0];
b=[ -1  1  2  0; 4  -2  -3   2; 0  3  1  -2];

fprintf ('**********  Check if given matrix  has a saddle point *****************: \n')
a
[yes,slevel,svector]= mycomputesaddlepoint(a)

fprintf ('***********************************************************: \n')
b
[yes,slevel,svector]= mycomputesaddlepoint(b)


%****************************************************************
function [ yes, svalue ,svector] = mycomputesaddlepoint(Gmat)
d= size(Gmat);
nrows= d(1);
yes=0;
svector= zeros(1,nrows);
svalue=999;
x=min(Gmat,[], 2);
[slx,index] =max (x);

y= max(Gmat,[],1);
sly=min(y);
slx
sly
if slx==sly
    yes=1;
    svalue = slx;

    svector(index)=1;
end

end
```

2.4 Security level of a mixed strategy and of a player

Since now both row and column players can use mixed strategies in a game defined by the payoff matrix M in Figure 1 before, let S be the set of all mixed strategies defined by the row player's pure strategy set {X1, X2, …., Xm}, and let T be the set of all mixed strategies defined by the column player's pure strategy set {Y1, Y2…. Ym}. Let X* be a mixed strategy from S, and let Y* be a mixed strategy from the set T. We now can define what is meant by the security level sl (X*) of a mixed strategy for the row player and security level sl (Y*) of a mixed strategy for the column player in terms of the expected payoff EV [M, X*, Y*] discussed in section 2 before.

$$sl\,(X^*) = (\text{by definition}) = \min_{Y^* \text{ in } T} EV\,(M, X^*, Y)$$

Note that when X* is any of the pure strategy Ri, and Y* is any of the pure strategies C1, C2.... Cn in T, then by Statement 1 from the last section,

$$sl\ (Xi) = \underset{C_j\ in\ T}{Min}\ EV\,[M, R_i, C_j] = \underset{j}{Min}\ a_{ij}$$

So, this definition coincides with the original definition for security level of a pure strategy on by had before for pure strategies. This means that our definition of security level for a mixed strategy given above is a reasonable one, however, if we were asked to compute the security level sl (X*) of a mixed strategy X* using this definition, we will have a very hard time to compute it simply because the expression on the right hand side calls for computing infinitely many expected value EV (M, X*, Y*), one for each Y* in T and then compute the minimum out of all these infinitely many values. But thanks to the following observation, it simplifies this infinite computation process into a finite process. We summary it into the following statement.

Proposition 1.

$$\underset{Y^*\ in\ T}{Min}\ EV(M, X^*, Y^*) \quad = \quad \underset{C_j\ in\ T}{Min}\ EV(M, X^*, C_j)$$

This proposition states that the result of computing sl (X*) is the same as if we only focus on the column player's pure strategies,

C1, C2.... Cn and there are only finitely many. Before proofing this proposition, we look at an example.

Example 4:

$$\begin{array}{c} \\ X1 \\ X2 \\ X3 \end{array} \begin{array}{c} \begin{array}{cccc} Y1 & Y2 & Y3 & Y4 \end{array} \\ \begin{pmatrix} -1 & 1 & 2 & 0 \\ 4 & -2 & -3 & 2 \\ 0 & 3 & 1 & -2 \end{pmatrix} \end{array} \begin{array}{c} \\ -1 \\ -3 \\ -2 \end{array}$$

Suppose we were asked to compute the security level sl (X*) of a mixed strategy X* = (1/5, 3/5, 1/5), per Proposition 1 above,

Sl (X*) =

min {EV (M, X*, C1), EV (M, X*, C2), EV (M, X*, C3), EV (M, X*, C4)} where the C1, C2, C3, C4 are the column player's pure strategies Y1, Y2, Y3, Y4 corresponds to the mixed strategies,

$$C1 = \begin{pmatrix}1\\0\\0\\0\end{pmatrix} \quad C2 = \begin{pmatrix}0\\1\\0\\0\end{pmatrix} \quad C3 = \begin{pmatrix}0\\0\\1\\0\end{pmatrix} \quad C4 = \begin{pmatrix}0\\0\\0\\1\end{pmatrix}$$

$$EV(M, X^*, C1) = X^* * M * C1 = X^* \begin{pmatrix}-1\\4\\0\end{pmatrix} = 11/5$$

$$EV(M, X^*, C2) = X^* * M * C2 = X^* \begin{pmatrix}1\\-2\\3\end{pmatrix} = -2/5$$

$$EV(M, X^*, C3) = X^* * M * C3 = X^* \begin{pmatrix}2\\-3\\1\end{pmatrix} = -6/5$$

$$EV(M, X^*, C4) = X^* * M * C4 = X^* \begin{pmatrix}0\\2\\-2\end{pmatrix} = 4/5$$

The sl (X*) = min {11/5, -2/5, -6/5, 4/5} = -6/5

Now we are going for the proof of

$$\underset{Y^*\,in\,T}{Min}\, EV(M, X^*, Y^*) \quad = \quad \underset{C_j\,in\,T}{Min}\, EV(M, X^*, C_j)$$

First, the left-hand side is less than or equal to the right-hand side,

that is:

$$\underset{Y^*\,in\,T}{Min}\, EV(M, X^*, Y^*) \quad <= \quad \underset{C_j\,in\,T}{Min}\, EV(M, X^*, C_j)$$

because each EV (M, X*, Y*) and EV (M, X*, C_j) is evaluated on the column player's mixed strategies Y* and pure strategies C_j respectively, and the set of mixed strategies is a superset of the set of pure strategies, since we know the smallest element over a set of numbers is less than or equal to the smallest element over a smaller subset of that original set of numbers, so the above inequality follows.

Next, we need to show that the left-hand side is greater than or equal to the right -hand side,

$$\underset{Y^* \text{ in } T}{Min}\, EV(M,X^*,Y^*) \quad => \quad \underset{C_j \text{ in } T}{Min}\, EV(M,X^*,C_j)$$

By Statement 4 shown in the last section, the left-hand side,

EV (M, X*, Y*) = $\sum_j EV(M,X^*,C_j)y_j$

So $\underset{Y^* \text{ in } T}{Min}\, EV(M,X^*,Y^*) \quad = \quad \underset{Y^* \text{in } T}{Min} \sum_j EV(M,X^*,C_j)y_j$

If the minimum of all the EV (M, X*, C_j) over j take place at C_k, that is,

EV (M, X*, Ck) = $\underset{j}{Min}\, EV(M,X^*,C_j)$

then $\underset{Y^* \text{ in } T}{Min}\, EV(M,X^*,Y^*)$

$$=> \underset{Y^* \text{in } T}{Min} \sum_j EV(M,X^*,C_k)y_j$$

$$=> \underset{Y^* \text{in } T}{Min}\, EV(M,X^*,C_k)\sum_j y_j$$

= EV (M, X*, C_k), as the sum of all the y_i is 1,

= $\underset{j}{Min}\, EV(M,X^*,C_j)$ by the definition of C_k.

Similarly, we can show, if Ri is the mixed strategy corresponds to the pure strategy Xi in the game matrix M as defined before,

Proposition 2.

$$\underset{X^* in\, S}{Max}\, EV(M, X^*, Y^*) \;=\; \underset{i}{Max}\, EV(M, R_i, Y^*)$$

Now that we understand the security level of a row player's mixed strategy, we can define the security level of the row player, denoted as sl(row player). It is simply the maximum of all the row player's security levels across all theirmixed strategies.

sl (row player) = (by definition) = $\underset{all\, X^* in S}{Max}\, sl(X^*)$

, since we know that sl (X*) can be computed by a finite process as shown above, so sl(X*) can be expressed as

sl (row player) = $\underset{all\, X^* in S}{Max}\, \underset{j}{Min}\, EV(M, X^*, C_j)$

As noted above, evaluating the Minimum Process over *j* does not pose any difficulty as we illustrated in Example 1 earlier. However, evaluating the maximization process over an infinite number of strategies X* in *S* is not as straightforward.

To address this challenge, we will present an example below and introduce an optimization technique called Linear Programming. This method, which was independently developed in another field, can be applied here to solve the maximization problem and, consequently, facilitate the computation of the security level sl (row player).

For similar reasoning, we can express sl (column player) as

sl (column player) = $\underset{all\, Y^* in\, T}{Min}\, \underset{i}{Max}\, EV(M, R_i, Y^*)$

Example 5. We will compute sl(row player) and

sl(column player) for the game below.

Y1 Y2 Y3 Y4

$$\begin{pmatrix} 5 & 4 & 5 & 4 \\ 7 & 7 & 4 & 9 \end{pmatrix} \begin{matrix} X1 \\ X2 \end{matrix}$$

sl (row player) = $\underset{X^*\, in\, S}{Max}\, sl(X^*)$

$= \underset{X^*\, in\, S}{Max}\, \underset{j}{Min}\, EV\left(M, X^*, C_j\right)$

where C1, C2, C3, C4 are the column vectors

$$C1= \begin{pmatrix}1\\0\\0\\0\end{pmatrix} \quad C2 = \begin{pmatrix}0\\1\\0\\0\end{pmatrix} \quad C3 = \begin{pmatrix}0\\0\\1\\0\end{pmatrix} \quad C4 = \begin{pmatrix}0\\0\\0\\1\end{pmatrix}$$

Sl (row player)

$$= \underset{X^*\ in\ S}{Max}\ Min\ \{EV (M, X^*, C1), EV (M, X^*, C2),$$

$$EV (M, X^*, C3), EV (M, X^*, C4)\}$$

$$= \underset{X^*\ in\ S}{Max}\ \text{Min } \{(x1, x2)\begin{pmatrix}5\\7\end{pmatrix}, (x1, x2)\begin{pmatrix}4\\7\end{pmatrix}, (X1, X2)\begin{pmatrix}5\\4\end{pmatrix}, (x1, x2)\begin{pmatrix}4\\9\end{pmatrix}\ \}$$

$= \underset{(x1,x2)\ in\ S}{Max}$ Min {5x1+ 7x2, 4x1+7x2, 5x1+4x2, 4x1+9x2}, where x1+x2=1, and both x1, x2 are between 0 and 1 as (x1, x2) is mixed strategy for X*.

After replacing x2 by 1-x1, we get,

sl (row player) = $\underset{x_1\ in[0,1]}{Max}$ Min { -2x1 +7, -3x1+7, x1+4, -5x1 +9)

To evaluate the Max Min operation shown above, we plot the four quantities in the interval between 0 and 1, call them Y1, Y2, Y3, Y4,

Y1= -2x1+7,

Y2=-3x+7,

Y3= x1+4,

Y4=-5x1+9

To find the Min of {Y1, Y2, Y3, Y4}, we can plot them graphically. Select the portion of a line segment in a sub-interval of {0,1] that is below the other three lines; take turn to do this for each line and make sure that each sub-interval come up this way are non-overlapping, and they make up the entire interval [0,1].

Plots of these lines Y1, Y2, Y3, Y4 are shown below.

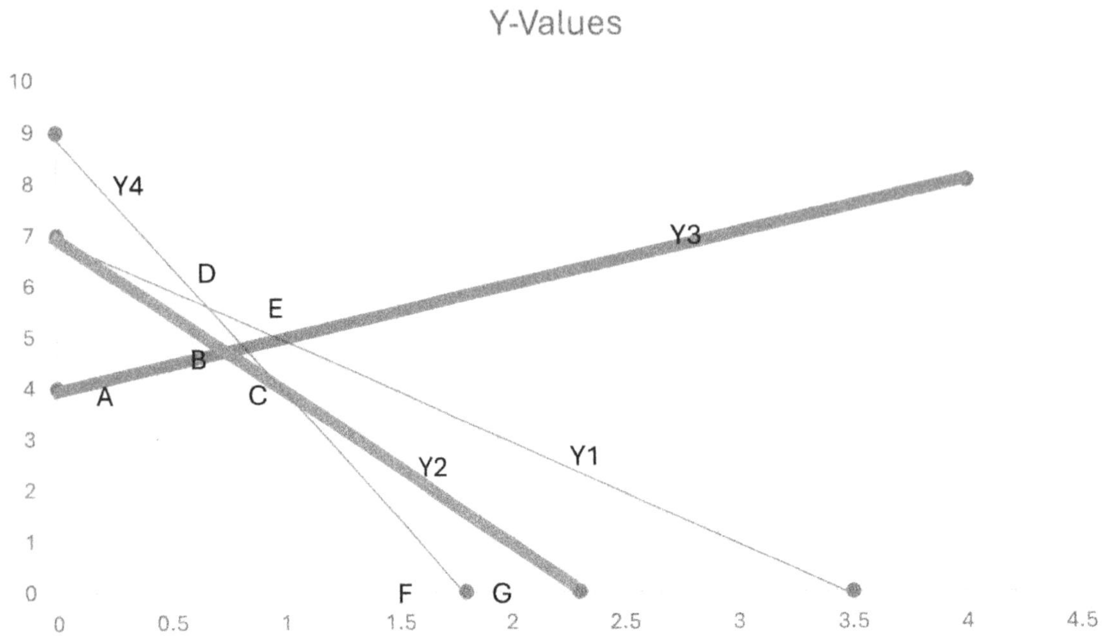

Solving the simultaneous equations between Y2 and Y3, Y2 and Y4, Y1 and Y4, Y1 and Y3 to get the coordinates of the intersection point B (3/4,19/4), C (1,4), D (2/3, 19/3), E (1,5) respectively. So, Min {Y1, Y2, Y3, Y4} consists of segments AB and BC as they are all below the other three segments in the subintervals {0, ¾) and (3/4,1) respectively. Note that we are not interested in point D because its corresponding segment in Y1 does not lie below the other three segments, and we are not interested in E because its x-coordinate is beyond the interval (0,1). Thus Min {Y1, Y2, Y3, Y4} is represented by the segments AB and BC.

So, Max Min (Y1, Y2, Y3, Y4} is Max of (AB, BC}, which is 19/4, which is the y-coordinate of the highest point in segments AB and BC, which is the row player's security level that we are computing for. The x-coordinate of point B (3/4, 19/4), which is 3 /4 in this case, is the probability of a mixed strategy that would give us this maximum-security level. By reviewing the sequential steps shown in this example, we can see that conceptually, we can always use a graphical method to compute the security level of a row player if the number of its pure strategies is 2, and there is no limitation on the number strategies for the column player.

In what follows, we attempt to find the security level of the column player and see if graphical method can be used to solve the problem as well.

By definition,

sl (column player) = $\underset{all\ Y^* in\ T}{Min}\ \underset{i}{Max}\, EV(M, R_i, Y^*)$,

where R1 and R2 are row vectors

, equals to $\underset{allY^*inT}{Min}\ \underset{i}{Max}$ {EV (M, (1,0), Y*), EV (M, (0,1), Y*)}

, equals $\underset{allY*inT}{Min}\ \underset{i}{Max}\ \{(5,4,5,4)\begin{pmatrix} y1 \\ y2 \\ y3 \\ y4 \end{pmatrix}, (7\ 7\ 4\ 9)\begin{pmatrix} y1 \\ y2 \\ y3 \\ y4 \end{pmatrix}\}$

, equals $\underset{allY*inT}{Min}\ \underset{i}{Max}\ \{5y1+ 4y2+ 5y3+ 4y4,\ 7y1+ 7y2 + 4y3+ 9y4\}$

As y1+y2+y3+y4 =1, we can eliminate y4 in terms of y1, y2, y3, the problem then reduces to finding the Max of

{y1+y3+4, -2y1-2y2-5y3+9}, since each of these two quantities represents a plane a 4-dimensional space and it is beyond our graphical visualization, so we can see that when the number of pure strategies for the column player is above 3, we will not be able to utilize the graphical approach to compute sl(column player) even through there is no restriction on the number of pure strategies on the row player. Combining this observation, the restriction on number of strategies for the column player, and that of the number of strategies for the row player noted previously, we see that computation of both the sl(row player) and sl(column player) using the graphical approach is always possible when the number of pure strategies for both players are below 4. So graphical approach for our needs may seem limited by the number of pure strategies in the game.

But is there a general analytical approach that allows us to compute sl (row player) and sl (column player) for any arbitrary size of the game payoff matrix? The answer is yes! This is the well-known Simplex Method; it provides us solution to the computation of both sl (row player) and sl (column player) in the general case. The Simplex Method is a major topic in the field of Linear Programming, which itself is an independent subfield in the huge and fascinating domain of Optimization Techniques. There are so many applications of the Simplex Method in industry and engineering that have proven that the method is powerful, successful, and efficient. In the next section, we will give a quick overview of the Simplex Method and show how it can be used to solve problems in game theory, specifically the computation of sl(row player) and sl(column player).

2.5 A Quick Trip through the Simplex Method

In the Simplex Method, we are concern with maximizing or minimizing a linear function, call it F (x1, x2…Xn) of n-variables, X1, X2., Xn, F is usually called "the objective function", and since F is linear, F is a linear combination of X1, X2…. Xn with constants C1, XC2, … Cn. In symbol,
F (X1, X2…. Xn) = C1 X1 + C2X2+… + CnXn
The variables X1, X2…. Xn satisfies a set of m linear inequalities.

Depending on whether we want to maximize or minimize F, the set of linear constraints takes two different forms (referred to as the standard forms) as shown below: (the a_{ij} and b_j are all constants).

2.5.1 The standard form of the Maximization Linear Programming

Program is shown below:
Maximize F (x1, x2, …., xn) = C1 X1+C2 X2 + ……+CnXn
Subjected to:
(MaxLP)

$$a_{11}x_1 + a_{12}x_2 + \ldots\ldots\ldots\ldots + a_{1n}x_n <= b_1$$
$$a_{21}x_1 + a_{22}x_2 + \ldots\ldots\ldots\ldots + a_{2n}x_n <= b_2$$
$$\vdots$$
$$\vdots$$
$$\vdots$$
$$a_{m-1\,1}x_1 + a_{m-1\,2}x_2 + \ldots\ldots\ldots\ldots + a_{m-1\,n}\,x_n <= b_{m-1}$$
$$a_{m\,1}x_1 + a_{m\,2}x_2 + \ldots\ldots\ldots\ldots + a_{m\,n}x_n <= b_m$$

2.5.2 The Standard Form of the Minimization Linear Programming

Problem is shown below:
Minimize F (x1, x2, …., xn) = C1 X1+C2 X2 + ……+CnXn
Subjected to:
(MinLP)

$$a_{11}x_1 + a_{12}x_2 + \ldots\ldots\ldots\ldots + a_{1n}x_n => b_1$$
$$a_{21}x_1 + a_{22}x_2 + \ldots\ldots\ldots\ldots + a_{2n}x_n => b_2$$
$$\vdots$$
$$\vdots$$
$$\vdots$$
$$a_{m-1\,1}x_1 + a_{m-1\,2}x_2 + \ldots\ldots\ldots\ldots + a_{m-1\,n}\,x_n => b_{m-1}$$
$$a_{m\,1}x_1 + a_{m\,2}x_2 + \ldots\ldots\ldots\ldots + a_{m\,n}x_n => b_m$$

Note that the constraint inequalities in the standard form of a Maximization Linear Program are <= inequalities, whereas those in the standard form of a Minimization Linear Program are => inequalities.

For any optimization problem, once it is converted into either of the standard forms shown above, we can then apply the well-established Simplex Method to determine the specific values X1*, X2*..., Xn* for the variables X1, X2…., Xn such that the value of F (X1*, X2*,….Xn*) attains its maximum or minimum, depending on the form used to solve the underlying optimization problem.

There are many commercial software packages available for solving optimization problems using the Simplex Method. In what follows, and in subsequent chapters, we will focus on using the scientific computing MATLAB package. We will demonstrate the use of the built-in function linprog (f, A, b) in MATLAB to solve optimization problems in game theory, such as computing the security levels of the row and column players mentioned earlier.

Since the Simplex Algorithm is well-established, we will not go into its detailed workings or the theoretical justification behind it. However, we will provide a step-by-step demonstration of how the optimization problem of computing sl(row player) and sl(column player) is translated into a Linear Programming Problem and then solved using the Simplex Algorithm as implemented in the MATLAB function linprog (f, A, b).

2.6 Syntax of the function linprog (f, A, b) in MATLAB

The basic syntax of linprog(f, A, b) is x= linprog(f, A, b) to solve the following minimization problem:

Minimize F (x1, x2, …., xn) = C1 X1+C2 X2 + ……+CnXn
Subjected to:

$$a_{11}x_1 + a_{12}x_2 + \dots\dots\dots\dots + a_{1n}x_n <= b_1$$
$$a_{21}x_1 + a_{22}x_2 + \dots\dots\dots\dots + a_{2n}x_n <= b_2$$
$$\vdots$$
$$\vdots$$
$$\vdots$$
$$a_{m-1\,1}x_1 + a_{m-1\,2}x_2 + \dots\dots\dots\dots + a_{m-1\,n}\,x_n <= b_{m-1}$$
$$a_{m\,1}x_1 + a_{m\,2}x_2 + \dots\dots\dots\dots + a_{m\,n}x_n <= b_m$$

where the parameter f is the row vector of the coefficients C1, C2...Cn of the objective function F; A is the coefficients matrix $\{ a_{ij} \}$; b is the column vector $b_1, b_2, \dots b_m$ of the constraints' upper bounds; the output of the linprog function gives the solution for the unknowns $x_1, x_2, \dots x_n$.

In case there are equalities amongst the set of constraints above, separate those equality constraints into a separate matrix A_{eq} and the corresponding upper bounds by the vector b_{eq}, then the syntax for solving the set of unknown x_i is x = linprog (f,A,b, A_{eq}, b_{eq}). We will give examples for the computations of the Nucleolus using this option in Section 11.4.

Examples 2.1 and 2.2 below show how to use the linprog function to find the max and min of linear programs.

Example 6

This is a typical example of a Maximize Linear Programming problem (Max LP). The objective function f(x1,x2,x3, x4,x5) = 5x1 + 16 x2-4x3 -x4 + 7x5 subjected to the constraint equations

8x1-2x2+ 3x3-2x5 <= 18

2x1 + 4x2 -7x3 +3x4 +x5 <=4

x1 + 3x2 +x3 -x4 +2x5 <=14

x1, x2, x3, x4, x5 => 0

We first show how to use the linprog function to find the optimal value of the given Max LP problem, secondly, we apply the Duality Principle to convert the given Max LP into the corresponding Minimize Linear Programming problem (Min LP) , thirdly, we apply the linprog function to find the optimal value of the Min LP, finally, we demonstrate that the optimal values of both Min LP and Max LP are equal.

Source code of myMaxLinearProgramExample.m is shown below:

```
% This is my Max Linear Program Example in myMaxLinearProgramExample.m


% max f(x1, x2,x3,x4, x5)= 5x1 + 16x2 -4x3 -x4 +7x5

% subject to

%  8x1- 2x2 +3x3 -2x5 <= 18

%   2x1 + 4x2 -7X3 +3x4 +x5 <=4

%   x1 +  3x2+ x3 - x4 + 2x5 <= 14

% x1, x2, x3, x4 >= 0


f= [ 5 16 -4 -1 7]

A= [ 8 -2  3  0  -2

    2  4  -7  3  1 ;

    1  3   1  -1  2

    ]

b= [18;4; 14]

lb=[0;0;0;0;0];


x= linprog( -f, A, b,[],[], lb)

% compute the value of the objective function of the original LP

  f_value= f * x
```

```
% ______________________________________________

% the objective function of the orginal dual Max program is the transpose of b
% which will be a Min program
% the coeffecient matrix of the dual program is the transpose of A
% number of dual variables y is the number of constraint equations of A

g= b'
A_dual = A'
b_dual = f'

lb_dual= [0;0;0];
y= linprog (g,-A',-b_dual, [ ], [ ], lb_dual )
% compute the value of the dual objective function

g_value = g* y
```

Output listing of myMaxLinearProgramExample.m is shown below:

```
>> myMaxLinearProgramExample

f =

   5  16  -4  -1   7


A =

   8  -2   3   0  -2
   2   4  -7   3   1
   1   3   1  -1   2


b =

   18
```

```
     4
    14


Optimal solution found.


x =

         0
    4.0800
    1.7600
         0
         0


f_value =

   58.2400


g =

    18     4    14


A_dual =

     8     2     1
    -2     4     3
     3    -7     1
     0     3    -1
    -2     1     2


b_dual =

     5
    16
    -4
    -1
```

7

Optimal solution found.

y =

0
1.1200
3.8400

g_value =

58.2400

Example 7

This is a typical example of a Minimize Linear Programming problem (Min LP). The objective function f(x1,x2,x3) = 12 x1 + 5x2 +10 x3 subjected to the constraint equations

x1-x2+ 2x3 => 10
-3x1 + x2 + 4x3 => -9
-x1 + 2x2 +3 x3 => 1
2x1- 3x2 => -2
7x1-x2 -5x3 => 34
x1, x2, x3 => 0

We first show how to use the linprog function to find the optimal value of the given Min LP problem, secondly, we apply the Duality Principle to convert the given Min LP into the corresponding Maximize Linear Programming problem (Max LP) , thirdly, we apply the linprog function to find the optimal value of the Max LP, finally, we demonstrate that the optimal values of both Min LP and Max LP are equal. The MATLAB codes are shown below:

The source code of myMinLinearProgramExample.m is shown below:

```
% This is my Min Linear Program example in myMinLinearProgramExample.m
% min f(x1, x2,x3)= 12x1 + 5 x2 + 10x3
% subject to
%  x1-x2 + 2x3 >= 10
% -3x1 + x2 + 4x3 >= -9
%  -x1+2x2 +3x3 >= 1
%   2x1 -3x2  >= -2
%    7x1-x2 -5x3 >= 34
% x1, x2, x3 >= 0

f= [ 12 5 10 ]
A= [ 1  -1  2;
    -3   1   4;
    -1   2   3 ;
     2   -3  0;
     7   -1  -5
    ]
b= [10; -9;  1; -2; 34]
lb=[0;0;0];
% ____________________________________________________________
x= linprog( f, -A, -b, [], [], lb)
% compute the value of the objective function of the original LP
  f_value= f * x
% Next we compute the dual of the given Min Program:

% the objective function of the dual of the original Min program is the transpose of b
% which will be a Max program
% the coeffecient matrix of the dual program is the transpose of A
% number of dual variables y is the number of constraint equations of A

g= b'
A_dual = A'
b_dual = f'
```

```
nr= size ( A,1)

nc= size (A,2)

lb_dual =zeros( 1,nr)'


y= linprog (-g, A_dual, b_dual, [ ], [ ], lb_dual )

% compute the value of the dual objective function


g_value = g* y
```

Output listing of myMinLinearProgramExample.m is shown below:

```
>> myMinLinearProgramExample


f =


  12   5   10


A =


   1  -1   2

  -3   1   4

  -1   2   3

   2  -3   0

   7  -1  -5


b =


  10

  -9

   1

  -2

  34
```

```
Optimal solution found.

x =

    7.0000
         0
    3.0000

f_value =

   114

g =

    10   -9    1   -2   34

A_dual =

     1   -3   -1    2    7
    -1    1    2   -3   -1
     2    4    3    0   -5

b_dual =

    12
     5
    10
```

```
nr =

    5


nc =

    3


lb_dual =

    0
    0
    0
    0
    0


Optimal solution found.


y =

    0
   10
    0
    0
    6
g_value =

   114
```

2.6 Examples of computing security levels using the function linprog (f, A, b)

Example 8

We revisit Example 2 from the last section, with the following payoff matrix:

$$\begin{matrix} Y1 & Y2 & Y3 & Y4 & \\ \end{matrix}$$

$$\begin{pmatrix} 5 & 4 & 5 & 4 \\ 7 & 7 & 4 & 9 \end{pmatrix} \begin{matrix} X1 \\ X2 \end{matrix}$$

In the last section, we have shown how the problem of finding sl (row player) and sl(column player) are reduced to solving the following optimization problems:

$$\underset{x_1\ in[0,1]}{Max}\ \text{Min}\ \{5x1+7x2,\ 4x1+7x2,\ 5x1+4x2,\ 4x1+9x2)$$

and,

$$\underset{allY*inT}{Min}\ \underset{i}{Max}\ \{5y1+ 4y2+ 5y3+ 4y4,\ 7y1+ 7y2 + 4y3+ 9y4\}$$

respectively.

To show how we can translate these problems into a Linear Programming problem, the first one to the standard minimization form, and the second one into the standard maximization form as described earlier in this section.

Start with the first problem, that is to compute Max of a Min of the four quantities in {}. since the minimum of the four quantities in {} is the smallest of the four, let's call it S, the smallest of the four, so we must have the following 4 inequalities:

5x1+7x2 => S

4x1+7x2 => S

5x1+4x2=> S

4x1+9x2 =>S

And one equality, x1+x2 =1, which is the requirement on the mixed strategy of the row player represented by the probability pair (x1, x2).

Next, the problem calls for finding the maximum of all such S each one corresponds to a specific value for x1 and x2. That is to say, the problem is now translated into the following:

Max S (x1, x2)

Subjected to

X1+X2 = 1

5x1+7x2 => S

4x1+7x2 => S

5x1+4x2=> S

4x1+9x2 =>S

Note that our problem has now been translated into a format almost looked like the same as Linear Programming standard maximization format shown before, except that we need to express the objective function S (x1, x2) as a linear function in the variables x1 and x2, and that in all the four constraint inequalities, the right-hand side must be constants. With these two objectives in mind, and with a little bit of ingenuity, we rewrite the above equation and inequalities by dividing both sides by S (x1, x2) to get:

Max S (x1, x2)

Subjected to

X1/S + x2/S = 1/S

5 (x1/S) + 7(x2/S) => 1 (IEQ1)

4(x1/S) + 7(X2/S) => 1

5(x1/S) + 4(x2/S) =>1

4(x1/S) + 9(x2/S) =>1

Now it is clear that if we rename the variables x1/S by x1*, x2/S by x2*, rename the function 1/S by S*, and note that to maximize S is same as to minimize its reciprocal, we have just turned our problem into the following standard Minimization Linear Program format:

Min S* = x1* + x2*

Subjected to

5X1* +7x2* =>1 (IEQ 2)

4 X1* +7X2* =>1

5X1* + 4X2* => 1

4X1* + 9X2* => 1

Now it is just a matter of entering the data represented by the Linear Programming in standard format described above into the standard input format of the linprog function to obtain the values X1* and x2*, and the value for S*. Once these values are obtained, we need to compute the values X1 and X2 for our original problem, which are,

S= 1/S*

X1= S* X1* (EQ 3)

X2= S* x2*

Keep in mind that the values x1 and x2 are the probabilities for the row player's mixed strategy which maximize his security level, and the value S gives the security level of the row player for the original problem.

After we have done the above formulations, we are now in position to show the MATLAB codes Chapter2Example1.m to compute the security levels and strategies of both players followed by the output listings of running the m-codes. The output shows the probabilities are 0.75 and 0.25 for the mixed strategy of the row player with security level 4.75. The answer matches the graphical result shown before. What a neat way for using the linprog and what a difference between the algebraic and geometric approaches!

MATLAB source code Chapter2Example1R.m and its output listing for the computation of the security level and strategy for the row player

Here is the source code:

```
% Computation of security level and strategy for the row player
fprintf(' Computation of security level and strategy for the row player:\n  ')
b=[-1  -1  -1  -1]
a=[ -5  -7; -4  -7; -5   -4;  -4  -9]
f=[1  1  ]
x= linprog(f, a, b)
v=f*x
voriginal= 1/v
xoriginal= (1/v)* x
```

Explanation of the above lines of code is in order. The first three assignment statements translate the set of inequalities in (IEQ2) with greater than or equal to inequalities shown above into inputs for the linprog function. The negatives of each entry in b and in a are because the standard format of linprog requires less than or equal to inequalities. The last three statements in the source code comes from the explanation shown in equations (EQ3).

Here is the output listing:

>> Chapter2Example1R

Computation of security level and strategy for the row player:

b =

-1 -1 -1 -1

a =

-5 -7

-4 -7

-5 -4

-4 -9

```
f =

    1     1

Optimal solution found.

x =

    0.1579

    0.0526

v =

    0.2105

voriginal =

    4.7500

xoriginal =

    0.7500

    0.2500
```

In a very similar way, we now show how the optimization problem of computing the sl(column player)

$\underset{allY*inT}{Min}\ \underset{i}{Max}$ {5y1+ 4y2+ 5y3+ 4y4, 7y1+ 7y2 + 4y3+ 9y4}

is translated into the standard form of the Minimization Linear Programming problem.

First compute the maximum of the two quantities in the {}. Let us call this quantity, the biggest of the two, in symbol M.

So as before, we have:

Y1+Y2 =1

5y1+4y2+5y3+4y4 <= M (IEQ4)

7y1+ 7y2+4y3+ 9Y4 <= M

The problem is now translated into.

Min M

Subjected to

Y1 + y2 +y3 +y4 =1

5y1+4y2+5y3+4y4 <= M (IEQ5)

7y1+ 7y2+4y3+ 9Y4 <= M

As before, divide both sides of the equation and the inequalities by M, we get:

Min M

Subjected to

y1/M + y2/M +y3/M + y4/M =1/M

5 (y1/M) +4(y2/M) +5(y3/M) +4(y4/M) <= 1 (IEQ6)

7(y1/M) + 7(y2/M) +4(y3/M) + 9(Y4/M) <= 1

Now renaming of variables:

1/M= M*

y1/M= y1*

y2/M = y2* (EQ7)

y3/M= y3*

y4/M=y4*,

and note that minimizing M is same as maximizing M*, we have thus translated our original problem into the standard Maximization Linear Program:

Max M* = y1* + y2*+ y3* +y4*

Subjected to

5y1* + 4 y2* + 5y3* + 4 y4* <=1 (IEQ8)

7y1* + 7y2* + 4y3* + 9y4* <=1

Now we can use the Simplex Method to solve for the M*, y1*, y2*, y3*, y4*, then by the renaming of variables shown in (EQ7), we compute the probabilities y1, y2, y3, y4 to obtain the mixed column strategy, and from M*, we compute the security level M of the column player.

Recall that we tried in the last section to use a graphical method for this problem and was not successful. We now show the result of the Simplex Method by the MATLAB codes Chapter2Example1C.m

MATLAB source code Chapter2Example1C.m and its output listing for the computation of the security level and strategy for the column player

```
% Computation of security level and strategy for the column player
fprintf(' Computation of security level and strategy for the column player:\n ')
b=[1 1]
a=[4 5; 7 4]
f=[-1 -1]
y=linprog(f,a ,b)
v=-f*y
voriginal= 1/v
yoriginal= (1/v)*y
```

Explanation of the above lines of code is in order. The first three assignment statements translate the set of inequalities in (IEQ8) into the inputs of the linprog function. The negatives of each entry in f are because the standard format of linprog is for a minimization problem whereas the problem at hand is a maximization problem and maximizing a function is same as minimizing its negative.

In the payoff matrix $\begin{bmatrix} 5 & 4 & 5 & 4 \\ 7 & 7 & 4 & 9 \end{bmatrix}$ shown in (IEQ8), column 2 is dominated by column 4 and column 3 is dominated by column 1, so the column player would never play the strategies corresponds to column 1 (Y1) and column 4 (Y4). Hence the original payoff matrix reduces to the matrix a= $\begin{bmatrix} 4 & 5 \\ 7 & 4 \end{bmatrix}$ which is the input we use for the linprog function.

The last three statements in the source code comes from the explanation shown in equations (EQ7).

Here is the output listing:

```
>> Chapter2Example1C

Computation of security level and strategy for the column player:

b =

   1   1

a =

   4   5
   7   4

f =

  -1  -1

Optimal solution found.
```

```
y =

    0.0526
    0.1579

v =

    0.2105

voriginal =

    4.7500

yoriginal =

    0.2500
    0.7500
```

The output of the linprog function above shows that the values of y1, y2, y3, and y4 are respectively 0, 0.25, 0.75, and 0, and the security level of the column player is also 4.75. The probability (0, 0.25,0.75,0) is the mixed strategy for the column player which gives rise to his security level sl(column player) the value of 4.75., which is same as the security level sl (row player) as we computed before for the row player. In this example, we notice that both the row player and column player have a common value of 4.75 for his or her own security level. This result is not a surprising coincidence, because this, as we shall see, is a consequence of the well-known Fundamental Theorem in Game Theory, a topic which we will discuss in Chapter 4. The probabilities y1 and y4 are 0 as computed here is also consistent with the graphs Y1, Y2, Y3, Y4 shown before when we use the graphical approach to compute sl(row player) in the previous section.

Example 9

Consider the game with the following payoff matrix:

$$\begin{pmatrix} 7 & 1 \\ 2 & 10 \\ 4 & 9 \\ 6 & 1 \end{pmatrix}$$

, where now the row player has four strategies X_1, X_2, X_3, X_4 and the column player has 2 strategies Y_1, Y_2. As X_1 now dominates X_4, we can skip the last row in the above matrix, and the maximization linear program for the column player in (IEQ8) becomes:

Max M* = y1* + y2*

Subjected to

7y1* + y2* <=1 (IEQ9)

2y1* +10y2* <=1

4y1* + 9y2* <=1

The inequalities above correspond to strategy X_1, X_2, and X_3 with payoff matrix:

$$\begin{pmatrix} 7 & 1 \\ 2 & 10 \\ 4 & 9 \end{pmatrix}$$

Let $a_1 = \begin{pmatrix} 7 & 1 \\ 2 & 10 \end{pmatrix}$, $a_2 = \begin{pmatrix} 2 & 10 \\ 4 & 9 \end{pmatrix}$, $a_3 = \begin{pmatrix} 7 & 1 \\ 4 & 9 \end{pmatrix}$ be the payoff matrices correspond to the row player's strategy pair (X_1, X_2), (X_2, X_3),and (X_1, X_3) respectively. The instructions in Chapter2Example2C.m compute the column player's security levels of a1, a2, and a3 which are respectively 4.85, 4, and 5.36 with respective strategies (0.6429, 0.3571), (0,1), and (0.7273, 0.2727). We choose the strategy pair (X_1, X_3) or payoff matrix a3 because it gives us the largest security level 5.36.

Like the derivation for (IEQ2), we obtain the minimization linear program for the row player:

Min S* = x1* + x2* +x3*

Subjected to

7X1* + 2x2* + 4x3* =>1 (IEQ 10)

X1* +10X2* + 9x3*=>1

As we have determined the strategy pair (X_1, X_3) is the optimal strategy pair for the row player above, the minimization linear program simplifies to

Min S* = x1* +x3*

Subjected to

7X1* + 4X3* =>1 (IEQ 11)

X1* + 9X3*=>1

Apply linprog (f, a, b) to (IEQ11) above with a= $\begin{pmatrix} -7 & -4 \\ -1 & -9 \end{pmatrix}$, b=[-1,-1], f=[1 1] gives strategy pair (x1, x3) equals to (0.4545, 0.5454) and security level 5.36 (see the output listing of Chapter2Example2R.m) which agrees with the result of that obtained from the column player as stated before. Note that for the row player's strategies X2 and X4 were never used in the analysis, so the final strategy for the row player is (0.4545, 0, 0.5454, 0) and for the column player is (0.7272, 0.2727) with equal security level 5.36.

MATLAB source code Chapter2Example2C.m

```
% Computation of security level and strategy for the column player
fprintf(' Computation of security level and strategy for the column player:\n ')
b=[-1  -1 ];
a1=[ 7  1; 2 10];
a2=[ 2  10;  4   9];
a3=[7  1; 4  9];

a1= -a1;

a3=-a3;
f=[1  1  ];

a1
x= linprog(f, a1, b);
v=f*x;
voriginal= 1/v
yoriginal= (1/v)* x
fprintf(' ******************************************************:\n ')
a2
[yes, svalue, svector]= mycomputesaddlepoint(a2)

fprintf(' ******************************************************:\n ')
a3
x= linprog(f, a3, b);
v=f*x;
voriginal= 1/v
yoriginal= (1/v)* x
```

```
% ***********************************************************************

function [ yes, svalue ,svector] = mycomputesaddlepoint(Gmat)
d= size(Gmat);
nrows= d(1);
yes=0;
svector= zeros(1,nrows);
svalue=999;
x=min(Gmat,[], 2);
[slx,index] =max (x);

y= max(Gmat,[],1);
sly=min(y);

if slx==sly
    yes=1;
    svalue = slx;

    svector(index)=1;
end

end
```

Here is the output listing:

Computation of security level and strategy for the column player:

```
a1 =

  -7  -1

  -2  -10

Optimal solution found.

voriginal =

  4.8571

yoriginal =

  0.6429

  0.3571
```

***:

a2 =

```
     2    10
     4     9
```

yes =

```
     1
```

svalue =

```
     4
```

svector =

```
     0     1
```

***:

a3 =

```
    -7    -1
    -4    -9
```

Optimal solution found.

voriginal =

5.3636

yoriginal =

0.7273

0.2727

>>

MATLAB source code Chapter2Example2R.m

```
% Computation of security level and strategy for the row player
fprintf(' Computation of security level and strategy for the row player:\n  ')
b=[-1  -1 ];

a3=[7  4; 1  9]
a3=-a3;

f=[1  1  ];

x= linprog(f, a3, b);
v=f*x;
voriginal= 1/v
xoriginal= (1/v)* x
```

Here is the output listing:

>> Chapter2Example2R

Computation of security level and strategy for the row player:

a3 =

```
    7    4

    1    9
```

Optimal solution found.

voriginal =

```
    5.3636
```

xoriginal =

```
    0.4545

    0.5455
```

2.7 The Fundamental Theorem of Game Theory

This celebrated theorem in game theory was originally proved by von Neumann in 1928 using Brouwer fixed-point theorem, a fundamental theorem that Nash used in proofing the existence of Nash Equilibrium that we will demonstrate in Chapter 3. In this section, we will outline another proof of the theorem by G. Dantzig based on the Duality Theorem in Linear Programming. First let us state below what is the Fundamental Theorem of Game Theory.

Theorem 2.7.1 (Fundamental Theorem of Game Theory)

The theorem guarantees that for any game matrix M the existence of two strategies, X^* for the row player and Y^* for the column player, such that

$$sl(row\ player) = \underset{1 \le j \le n}{Min}\ X^* M^j \qquad \text{(EQ12)}$$

and $sl(column\ player) = \underset{1 \le i \le m}{Max}\ M_i Y^{*t}$ (EQ13)

, where M has m-rows and n-columns, and M_i is the ith row of M, M^j is the jth column of M. Moreover, sl(row player) = sl(column player)

To see why this celebrated theorem stated above is a consequence of the Duality Theorem in Linear Programming, let's first state without proof the following .

Theorem 2.7.2 (The Duality Theorem in Linear Programming)

If either of the following Maximization Linear Program or Minimization Linear Program has a finite optimal solution, then so does the other Linear Program, and both optimal values are equal.

Min $f = b\,X^t$ such that $A^t\,Xt \ge c$, $X^t \ge 0$ (IEQ14)

Max $g = c\,Y^t$ such that $A\,Y^t \le b$, $Y^t \ge 0$ (IEQ15)

Min f = Max g (EQ16)

Theorem 2.7.1 follows from Theorem 2.7.2 and can be demonstrated by the following example because the steps in the general proof are parallel to the steps shown in the example.

Example 10

Suppose the game matrix M is shown below with X_1, X_2, X_3 for the row player's strategies and Y_1, Y_2, Y_3 for the column player's strategies.

$$M = \begin{pmatrix} 1 & 8 & 5 \\ 2 & 3 & 7 \\ 4 & 1 & 1 \end{pmatrix}$$

Then like the derivation of the minimization LP in (IEQ2) and maximization LP in (IEQ8) above, we have for the row player the following LP

Min $f = x_1^* + x_2^* + x_3^*$

subjected to

$x_1^* + 2\,x_2^* + 4\,x_3^* \ge 1$

$8x_1^* + 3\,x_2^* + x_3^* \geq 1$ (IEQ17)

$5x_1^* + 7x_2^* + x_3^* \geq 1$

Xoriginal = X^* (1/Min f)

, and the following LP for the column player

Max $g = y_1^* + y_2^* + y_3^*$

subjected to

$y_1^* + 8y_2^* + 5y_3^* \leq 1$

$2y_1^* + 3y_2^* + 7y_3^* \leq 1$ (IEQ18)

$4y_1^* + y_2^* + y_3^* \leq 1$

Yoriginal = Y^* (1/Max g)

Note that both linear programming (IEQ17) and (IEQ18) above are dual to each other with vector b=[1 1 1] and c=[1 1 1] in (IEQ14) and (IEQ15). The minimization LP in (IEQ17) must have a finite optimal solution by the Simplex Method because all entries in M^t are positive, and the objective function f= $x_1^* + x_2^* + x_3^*$ is bound below by 0 and all $x_i^* \geq 0$. Thus, by the Duality Theorem above, the maximization LP (IEQ18) also have a solution and both LP give the same value 2.9615, and the existence of the two strategies X^* and Y^* as described in (EQ12) and (EQ13) in the Fundamental Theorem of Game Theory is given by X^* = (0.23, 0.17, 0.60) and Y^*= (0.65, 0.19, 0.15) . See the output listing of the MATLAB program ComputeSecurityLevelsbySimplexMethod.m for the results stated.

```
>> ComputeSecurityLevelsbySimplexMethod

a =

    1   8   5

    2   3   7

    4   1   1

Result of the Simplex Method:

Simplex method to compute Min LP for the row player  :

Optimal solution found.

xoriginal =
```

0.2308

0.1731

0.5962

Simplex Method to compute Max LP of the dual program for the column player :

Optimal solution found.

yoriginal =

0.6538

0.1923

0.1538

sl1 =

2.9615

sl2 =

2.9615

MATLAB source code for ComputeSecurityLevelsbySimplexMethod.m

```
a= [1  8  5;2  3  7; 4  1  1]
fprintf ('Result of the Simplex Method: \n')

[sl1, sl2 ]= computeSLbySimplexMethod (a)


% ********************************************************
function [sl1, sl2 ]= computeSLbySimplexMethod (a)
fprintf(' Simplex method to compute Min LP for the row player  :\n')

b=a;
fx= [1  1  1];
b= -b';
fx= [ 1  1  1];
bx= [ -1  -1  -1];
[x, vx] = linprog (fx,b,bx);
nu1= 1/(fx *x);
xoriginal= nu1*x
```

```
nu1=1/vx;
sl2=nu1;

fprintf(' Simplex Method to compute Max LP of the dual program for the column player :\n')
fy=[-1 -1 -1]; % because linprog is a MinLP in standard form

by=[ 1  1  1];
y= linprog (fy, a,by);

-fy* y;
nu2= 1/(-fy*y);
yoriginal= nu2*y
sl1=nu2;
end
```

2.8 The Complementary Slackness Theorem

In this section we state without proof the Complementary Slackness Theorem. Like the Fundamental Theorem of Game Theory, this Theorem is also a consequence of the Duality Theorem, and it will be needed in Section 11.4 in the computation of the Nucleolus of coalition games.

Before we can state the theorem, we need to understand what the slack of an optimal solution to a LP is. If the LP is a maximization problem with the standard form described in Section 2.5.1, if $x^* = (x_1^*, \dots\ x_i^*, \dots\ x_n^*)$ is an optimal solution, then the slack of x^* is defined as the (mx1) column vector slack(x^*) = $(b_1, \dots\ b_i, \dots\ b_m)$' - A (x^*)' , where A is the (mxn) matrix . In other words, the jth component of slack (x*) measures the difference between b_j and the jth – component of the column vector A (x^*)'. If the LP is a minimization problem with the standard form described in Section 2.5.2, if $x^* = (x_1^*, \dots\ x_i^*, \dots\ x_n^*)$ is an optimal solution, then the slack of x^* is defined as the (mx1) column vector slack(x^*) = A (x^*)' - $(b_1, \dots\ b_i, \dots\ b_m)$' , where A is the (mxn) matrix . In other words, the jth component of slack (x*) measures the difference between the jth – component of the column vector A (x^*)' and b_j . With the understanding of the slack vector of an optimal solution in an LP, we can state the following :

Theorem 2.8.1 The Complementary Slackness Theorem

If we have a Max LP in standard form, with the (mxn) coefficient matrix A, let $x^* = (x_1^*, \dots\ x_i^*, \dots\ x_n^*)$ be an optimal solution of the Max LP; let Min LP be the corresponding dual program with optimal solution $y^* = (y_1^*, \dots\ y_j^*, \dots\ y_m^*)$ and coefficient matrix A^t of size (nxm) . As the number of constraint inequalities of the Max LP is m, the number of variables in the dual Min LP in also m. The Complementary Slackness Theorem states that, for any positive value component y_i^* in the ith position of the vector $y^* = (y_1^*, \dots\ y_j^*, \dots\ y_m^*)$, the corresponding j^{th} component of the slack vector slack(X^*) must be 0. Similarly, the optimal solution x^* of the Max LP and the number of inequality constraints of the min LP are both vectors of the same length n. The Complementary Slackness Theorem then states that , any positive value x_k^* in the kth position of the optimal solution $x^* = (x_1^*, \dots\ x_i^*, \dots\ x_n^*)$ of the Max LP, the corresponding k^{th} component of the slackness vector slack (y^*) must be 0.

We are going to illustrate these results by the following two examples Example 11 and Example 12 below.

Example 11

```
% This is my slackness Theorem example in mySlacknessTheoremMaxLP.m
% max f(x1, x2,x3,x4)= -5x1 + 18 x2 + 6x3 -x4
% subject to
%  2x1-x3 + 3x4 <= 20
% x2-2x3-x4 <=30
% -3x1 + 6x2+ 3x3 + 4x4 <= 24
% x1, x2, x3, x4 >= 0

f= [ -5 18 6 -1]
A= [ 2  0  -1  3;
    0  1  -2   -1;
    -3  6  3   4;
    ]
b= [20; 30; 24]
lb=[0;0;0;0];
x= linprog( -f, A, b,[],[], lb)
% compute the value of the objective function of the original LP
  f_value= f * x

% compute the slackness of the LP at the optimal solution
  slack =  b- A* x
% ______________________________________________

% the objective function of the original dual Max program is the transpose of b
% which will be a Min program
% the coefficient matrix of the dual program is the transpose of A
% number of dual variables y is same as the number of constraint equations of A

g= b'
A_dual = A'
b_dual = f'
```

```
ny= size ( A);
lb_dual= [0;0;0];
y= linprog (g,-A',-b_dual, [ ], [ ], lb_dual )
% compute the value of the dual objective function

g_value = g* y

% Now compute the slackness of the dual program at the dual optimal solution
slack_dual =  A_dual* y -b_dual
```

Output listing of Example 11

```
>> mySlacknessTheoremMaxLP

f =

  -5   18    6   -1


A =

   2    0   -1    3
   0    1   -2   -1
  -3    6    3    4


b =

  20
  30
  24


Optimal solution found.


x =
```

```
    10
     9
     0
     0


f_value =

   112


slack =

     0
    21
     0


g =

    20    30    24


A_dual =

     2     0    -3
     0     1     6
    -1    -2     3
     3    -1     4


b_dual =

    -5
    18
     6
    -1
```

Optimal solution found.

```
y =

    2
    0
    3


g_value =

   112


slack_dual =

    0
    0
    1
    19
```

Example 12

```
 % This is the slackness Theorem Example for Min LP in the file
% mySlacknessTheoremExampleMinLP.m
% min f(x1, x2,x3)= 12x1 + 5 x2 + 10x3
% subject to
%  x1-x2 + 2x3 >= 10
% -3x1 + x2 + 4x3 >= -9
%  -x1+2x2 +3x3 >= 1
%   2x1 -3x2  >= -2
%    7x1-x2 -5x3 >= 34
% x1, x2, x3 >= 0

f= [ 12 5 10 ]
A= [ 1  -1  2;
    -3   1   4;
    -1  2  3 ;
```

```
    2   -3   0;
    7   -1   -5
    ]
b= [10; -9;  1; -2; 34]
lb=[0;0;0];
x= linprog( f, -A, -b,[],[], lb)
% compute the value of the objective function of the original LP
  f_value= f * x- b

% compute the slackness of the LP at the optimal solution
  slack =  A* x - b
% __________________________________________________

% the objective function of the original dual Max program is the transpose of b
% which will be a Min program
% the coefficient matrix of the dual program is the transpose of A
% number of dual variables y is the number of constraint equations of A

g= b'
A_dual = A'
b_dual = f'
ny= size ( A);
lb_dual= [0;0;0;0;0];
y= linprog (-g, A_dual, b_dual, [ ], [ ], lb_dual )
% compute the value of the dual objective function

g_value = g* y

% Now compute the slackness of the dual program at the dual optimal solution
slack_dual = b_dual - A_dual* y
```

>> mySlacknessTheoremExampleMinLP

```
f =

    12     5    10


A =

     1    -1     2
    -3     1     4
    -1     2     3
     2    -3     0
     7    -1    -5


b =

    10
    -9
     1
    -2
    34


Optimal solution found.


x =

    7.0000
         0
    3.0000


f_value =

   104
   123
   113
   116
    80
```

```
slack =

    3.0000
   -0.0000
    1.0000
   16.0000
         0


g =

    10    -9     1    -2    34


A_dual =

     1    -3    -1     2     7
    -1     1     2    -3    -1
     2     4     3     0    -5


b_dual =

    12
     5
    10


Optimal solution found.


y =

     0
    10
     0
     0
     6
```

```
g_value =

  114


slack_dual =

   0
   1
   0
```

2.9 Summary

In this chapter, we introduce, among many other concepts, a fundamental and essential idea in game theory: the security level of an individual player in a game.

We provide a step-by-step explanation of how to define and understand the security level, demonstrating that it corresponds to the optimal value of the objective function in an appropriately formulated Linear Programming problem.

To reinforce these concepts, we present several examples of computing security levels by solving the corresponding Linear Programming problems using the Simplex Method, as implemented in MATLAB codes.

Section 2.7 presents the classical Fundamental Theorem of Game Theory. This theorem allows us to conclude that the security levels for both players are equal.

Section 2.8 describes the Complementary Slackness Theorem which will be used to compute the Nucleolus of Coalition Games described in Chapter 11.

Chapter 3

Concepts of Nash Equilibrium in a game

3.0 Introduction

In this chapter, we introduce the fundamental concept of Nash Equilibrium and define its meaning in terms of two strategies: X^* for the row player, and one strategy Y^* for the column player. Nash Equilibrium is a crucial concept in game theory, with far-reaching implications and applications across various types of games. Throughout this book, we will explore this important notion and its extensions in depth.

Recalling from the previous chapter, we studied two distinct strategies for players—security level strategies. In later sections, we will present some fascinating results that establish a connection between security level strategies and the Nash Equilibrium. To aid in understanding these results, we provide MATLAB codes to generate numerical examples that illustrate these concepts.

It is important to remember that when we refer pure or mixed strategies in this chapter:

- Pure strategies are represented as row or column vectors with 0s and 1s as entries.
- Mixed strategies are represented as probability vectors, with probabilities p_i entries.

The climax of this chapter is the proof of the celebrated Theorem on the Existence of at least one Nash Equilibrium in any game.

3.1 Equilibrium of a strategy pair in 2-person game

In the previous chapter, we introduced the key concept of the security level for both the row and column players. In this chapter, we introduce another fundamental concept in game theory: the equilibrium of a row player's strategy X* and a column player's strategy Y*. This concept applies not only to two - person games but also extends to n-person games, as we will see.

Before defining this concept formally, let us denote the relationship between X* and Y* using the symbol NE (X*, Y*), which stands for Nash Equilibrium. This notation helps distinguish it from other equilibrium concepts that will appear in later chapters, such as:

- SPNE (X*, Y*) – Subgame Perfect Nash Equilibrium
- BNE(X*, Y*) – Bayesian Nash Equilibrium
- PBNE(X*, Y*) – Perfect Bayesian Nash Equilibrium

The notation NE (X*, Y*) is used throughout this book to represent Nash Equilibrium, as this equilibrium concept was first introduced by John Nash in his Ph. D. thesis in 1950. The other equilibrium concepts—Subgame Perfect Nash Equilibrium, Bayesian Nash Equilibrium, and Perfect Bayesian Nash Equilibrium—were later developed by researchers such as John Harsanyi, Reinhard Selten, and others.

Here is the formal definition for NE (X*, Y*):

Y1 Y2 ...... Yn

$$A = \begin{pmatrix} a_{11} & a_{12} & \cdots & a_{1n-1} & a_{1n} \\ a_{21} & a_{22} & \cdots & a_{2n-1} & a_{2n} \\ \vdots & \vdots & & \vdots & \vdots \\ \vdots & \vdots & & \vdots & \vdots \\ a_{m-11} & a_{m-12} & \cdots & a_{m-1n-1} & a_{m-1n} \\ a_{m1} & a_{m2} & \cdots & a_{mn-1} & a_{mn} \end{pmatrix} \begin{matrix} X_1 \\ X_2 \\ \\ \\ X_{m-1} \\ X_m \end{matrix}$$

$$B = \begin{pmatrix} b_{11} & b_{12} & \cdots & b_{1n-1} & b_{1n} \\ b_{21} & b_{22} & \cdots & b_{2n-1} & b_{2n} \\ \vdots & \vdots & & \vdots & \vdots \\ \vdots & \vdots & & \vdots & \vdots \\ b_{m-11} & b_{m-12} & \cdots & b_{m-1n-1} & b_{m-1n} \\ b_{m1} & b_{m2} & \cdots & b_{mn-1} & b_{mn} \end{pmatrix}$$

Definition 1: As shown above, let A be the payoff matrix for the row player, B be the payoff matrix for the column player, the strategy pair (X*, Y*) is Nash Equilibrium, in symbol NE (X*, Y*), if for all the row player strategy X, and all the column player strategy Y, the following two conditions are satisfied,

(a) EV (A, X, Y*) <= EV (A, X*, Y*)
(b) EV (B, X*, Y) <= EV (B, X*, Y*)

Remember, X* is a row vector, Y* is a column vector above.

Condition (a) says that if the column player selected the strategy Y*, then there is no reason for the row player to switch from X* to any other strategies, because he already got the largest expected value EX (A, X*, Y*), that is, EV (A, X*, Y*) is larger than any EV (A, X, Y*). Similarly, condition (b) says that if the row player already selected X*, then there is no reason that the column player to switch from Y* to other strategy Y because he already got the largest expected value EV (B, X*, Y*). Combining (a) and (b), we can say that if NE (X*, Y*), both row and column players are not willing to switch to other strategies from what they already have, that is, the strategies X* and Y*, thus in some sense, the pair (X*, Y*) is a stable pair, a win-win strategy pair for both players.

Note (1) that, in this definition of NE (X*, Y*), both A and B are any payoff matrices for the row and column player respectively for any 2-person game, but in the special case when it is a zero-sum 2-person game,

B= -A, as what the row player won is the same as the column player lost. So, the condition (a), (b) for NE (X*, Y*) become,

(a) EV (A, X, Y*) <= EV (A, X*, Y*)

(b) EV (A, X*, Y) => EV (A, X*, Y*), note the change of the direction of the inequality

Note (2) that, if we kept in mind both row player and column player each has different pay-off matrix A and B, from here on in this book, we will compactly write the payoff values for each player in a single payoff matrix, where each entry is a pair (x, y), where x is the payoff value for the row player and y is the payoff value for the column player.

Note (3) that in our definition of NE (X*, Y*) above, there is no restriction if the pair should be pure strategies or mixed strategies, however, the requirements (a) and (b) must be satisfied by infinitely many strategies X and Y including all the other strategies, so seemingly there is an infinite amount of checking. The following two lemmas tell us why this infinite amount of work can be reduced to a finite one.

Lemma 1. If the pair (X*, Y*) are pure strategies from the game, to see if NE (X*, Y*), the strategies X and Y can be limited to only pure strategies in the conditions (a) and (b). In order words, we can relax (a) and (b) to (a)' and (b)' below as,

(a)' EV (A, R, Y*) <= EV (A, X*, Y*), R is any other pure strategy represented by the row vector R= (0, 0..., 1....,0,0)

(b)' EV (B, X*, C) <= EV (B, X*, Y*), C is any other pure strategies represented by the column vector C= (0, 0..., 1....,0,0) ^{T}

In other words, we need only check n*m that many inequalities, not infinitely many as required by the original definitions (a) and (b)

Proof. We will demonstrate that (a)' implies (a) as follows. (b)' implies (b) is similar. Since X is any mixed strategy, X= (x1, x2..., xm), where xi is the probability of playing the pure strategy Ri, so, X= x1R1 +x2R2 +.... +xmRm,

EV (A, X, Y*) = EV (A, x1R1+ x2R2+...+xmRm, Y*)

$$= \sum_i x_i EV(A, R_i, Y^*)$$

$$\leq \sum_i x_i EV(A, X^*, Y^*) \quad \text{by (a)'}$$

$$\leq EV(A, X^*, Y^*) \sum_i x_i$$

$$\leq EV(A, X^*, Y^*) \quad \text{which is (a)}$$

Since the X* and Y* are pure strategies, which corresponds to row vectors Ri* and column vector Cj*, EV(A, X*, Y*) is just the entry a(i*,j*) in the A matrix, so, condition (a) simply says that a(I*,j*) is the largest value in the i*-th row of the A matrix, and (b) says that b(i*,j*) is the largest value in the j*-th column of the B matrix.

Lemma 2. If the pair (X*, Y*) are mixed strategies from the game, to check if NE (X*, Y*), the strategies X and Y can be limited to only pure strategies Ri and Cj in the conditions (a)" and (b)". In order words, we can relax from checking the original requirement (a) and (b) to checking (a)" and (b)" below instead.

(a)" EV (A, Ri, Y*) <= EV (A, X*, Y*), Ri is any other pure strategy represented by the row vector Ri= (0, 0..., 1....,0,0)

(b)" EV (B, X*, Cj) <= EV (B, X*, Y*), C is any other pure strategies represented by the column vector Cj= (0, 0..., 1....,0,0) ^{T}

In other words, as there are only m inequalities in (a), and n inequalities in (b), we need only check n+m that many inequalities, not infinitely many as required by the original definitions (a) and (b).

Proof. As in the last lemma, we will demonstrate (a)' implies (a) as follows. Proof of (b)" implies (b) is similar.

EV (A, X, Y*) = EV (A, x1R1+ x2R2+...+xmRm, Y*)

$$= \sum_i x_i EV(A, R_{i,} Y^*)$$

$$\leq \sum_i x_i EV(A, X^*, Y^*) \quad \text{by (a)''}$$

$$\leq EV(A, X^*, Y^*) \sum_i x_i$$

$$\leq EV(A, X^*, Y^*) \quad \text{which is (a)}$$

Example 1. Consider the Prisoner's Dilemma example first described in Chapter 1, Example 3.

$$G = \begin{array}{c c} & \begin{array}{cc} C & D \end{array} \\ \begin{array}{c} C \\ D \end{array} & \begin{pmatrix} (-2,-2) & (0,-4) \\ (-4,0) & (-1,-1) \end{pmatrix} \end{array}$$

From note (3) above, it is easily checked that the strategy pair (D, D) is in NE. The A-matrix in this case is:

	C	D
C	-2	0
D	-4	-1

The entry corresponds to strategy pair (D, D) is the entry -1, and -1 is the largest value in row D (-1 > -4). Similarly, the B-matrix in this case is:

	C	D
C	-2	-4
D	0	-1

The entry corresponds to strategy pair (D, D) is the entry -1, and -1 is the largest value in column D. (-1 > -4)

Using the observation in Note (3), we can check, entry by entry, of the 4 entries in the game matrix G above, and found (D, D) is the only NE pair. This simple method as described in Note (3) can be used to check if any strategy pair is a NE pair for any size game say, m by n, as the one shown in the beginning of this section before. Now a natural question to ask is then, is there any finite game that there is no NE pair? The answer is yes, as the following example shows>

Example 2. Consider the following 2-person zero-sum game with game matrix shown as

		Y	N
G =	Y	(5, -5)	(-5, 5)
	N	(-5,5)	(5, -5)

We can interpret this game as a 2-person matching game. For the row player, if both players say YES or both say NO, the row player wins \$ 5, if both players say different, then the row player loses \$ 5. For the column player, if both players say YES or both say NO, the column player loses \$ 5, and if both players say different, the column player wins \$5. We can check each entry from the above game matrix, one at a time, to see if the two conditions (a), (b) for NE can be met.

1. At entry (Y, Y): In the matrix for the row player, 5 => -5 (a) satisfied, but in the matrix for the column player, -5 => 5 is not true, (b) is not satisfied
2. At entry (Y, N): In the matrix for the row player, -5 => 5 is not true, so (a) is not satisfied.
3. At entry (N, Y): In the matrix for the row player, -5 => 5 is not true, so (a) is not satisfied.
4. At entry (N, N): In the matrix for the row player, 5 => -5 so (a) is satisfied, but in the matrix for the column player, -5=> 5 is not true, so (b) is not satisfied.

So, we see that this game does not have a strategy pair (X*, Y*) in NE when X* and Y* are pure strategies. The question is, does this game have a mixed strategy pair in NE? The answer is yes, in fact, this is guaranteed by the Nash Existence Theorem which states that every finite game has at least one NE. In what follows, we will prove this important theorem in game theory and show that such an equilibrium pair always exists, and after we know it exists, our next goal is to find it.

3.2 A MATLAB program NashEquilibriumTest.m to locate all possible Nash Equilibrium(s) for pure strategies

The MATLAB program NashEquilibriumTest.m below takes two matrices a and b as inputs, a matrix contains the payoffs of the row players (player 1) and b contains the payoffs of the column players(player 2). The row vectors X_i are the row players' i^{th} pure strategy and the column vectors Y_j are the column players' j^{th} pure strategy. The program outputs the location of the Nash Equilibriums if there is any. The expected value for the row and column players is EV(X_i,a, Y_j) and EV(X_i,b, Y_j) respectively. We give three examples below to demonstrate this important concept. But first, let us look at the logic we use in the function `myNashEquilibriumTest(matrixRow,` matrixCol, irow,jcol) to determine if the location (irow, jcol) is a Nash Equilibrium per the definition of Nash Equilibrium we described in section 3.1 above. For (irow, jcol) to be a Nash Equilibrium, if we focus only on column jcol of the matrixRow, the payoff at matrixRow (irow, jcol) must be greater than or equal to any of the other payoffs on the other rows along the $jcol^{th}$ column. And, if we focus only on row irow of the column player's payoff matrix matrixCol, the payoff at matrixCol(irow, jcol) must be greater than or equal to any of the other column player's payoff matrix matrixCol along the $irow^{th}$ row.

Example 3

Suppose the game matrix for the row and column players is shown below

$$M = \begin{pmatrix} (7,7) & (4,2) & (1,8) \\ (2,4) & (5,5) & (2,3) \\ (8,1) & (3,2) & (0,0) \end{pmatrix}$$

The row and column players' payoff matrices are respectively:

$$a = \begin{pmatrix} 7 & 4 & 1 \\ 2 & 5 & 2 \\ 8 & 3 & 0 \end{pmatrix} \quad \text{and} \quad b = \begin{pmatrix} 7 & 2 & 8 \\ 4 & 5 & 3 \\ 1 & 2 & 0 \end{pmatrix}$$

The Nash Equilibrium in this simple example is (5,5) located at the (2,2), the second pure strategy for the row player and the second pure strategy for the column player. (5,5) is a NE because 5 is greater than 4 and 3 along the second column in matrix a, and 5 is greater than 4 and 3 along the second row in b. The output of NashEquilibriumTest.m matches our intuition.

Output listing of NashEquilibriumTest

```
>> NashEquilibriumTest

a =

   7   4   1

   2   5   2

   8   3   0
```

```
b =

     7     2     8
     4     5     3
     1     2     0

s =

     3     3

nrow =

     3

mcol =

     3
```

********** Check if playoff matrix has a pure Nash Equilibrium strategy pair ******************:

the pair 2, 2 is in Nash Equilibrium

payoff of the pair is 5, 5

Source Code for NashEquilibriumTest.m

```
a=[7  4  1;2  5  2;8  3  0]
b=[7  2  8; 4   5  3;1  2  0]

s=size(a)
nrow=s(1)
mcol=s(2)

fprintf ( '**********  Check if playoff matrix has a pure Nash Equilibrium  strategy  pair
*****************: \n')
flag=0;
for ix= 1: nrow
    for jx=1: mcol

        yes =myNashEquilibriumTest(a,b,ix,jx);
        if (yes==1)
            fprintf(' the pair %4d, %4d is in Nash Equilibrium \n', ix, jx)
            fprintf (' payoff of the pair is  %4d, %4d  \n', a(ix,jx), b(ix,jx))
            flag=1;
        end
    end
end

if (yes==0 & flag ==0)
    fprintf ('there is no Nash Equilibrium found for the payoff matrices \n')
end

% *****************************************************************
function yes = myNashEquilibriumTest(matrixRow, matrixCol, irow,jcol)
d= size(matrixRow);
d=size(matrixCol);
nrow= d(1);
mcol= d(2);

i=1;
j=1;

yes =1;
while (i<= irow-1 )

    if matrixRow(i,jcol) > matrixRow(irow,jcol)
        yes= 0; break;
    else
    i = i+1;
    end

end

i=irow+1;
while (i<= nrow)

    if matrixRow(i,jcol) > matrixRow(irow,jcol)
        yes= 0; break;
    else
    i = i+1;
    end
end

while (j<= jcol-1 )

    if matrixCol(irow,j) > matrixCol(irow,jcol)
```

```
            yes= 0; break;
        else

                j = j+1;

        end
    end

    j=jcol+1;
    while (j<= mcol)

        if matrixCol(irow,j) > matrixCol(irow,jcol)
            yes= 0; break;
        else
        j = j+1;
        end

    end




    %if yes==0
     %   fprintf(' there is no pure Nash Equilibrium pair found : \n')
    %else
       % fprintf(' the given pair is in Nash Equilibrium  %4d, %4d \n ', irow, jcol);
       % fprintf(' the payoffs are  %4d, %4d \n', matrixRow( irow, jcol), matrixCol(irow,
    jcol));
    %end

    end
```

Example 4

The input game matrix in this example is adapted from Tadelis [22]. The six Nash equilibria obtained through our MATLAB program NashEquilibriumTest.m correspond to those presented in ([22], page 164).

$$M = \begin{pmatrix} (0,0) & (0,0) & (0,0) & (0,0) \\ (0,0) & (0,0) & (0,0) & (0,0) \\ (10,-10) & (10,-10) & (-5,-5) & (-100,-100) \\ (10,-10) & (10,-10) & (-100,\ -100) & (-100,-100) \end{pmatrix}$$

The output listing of NashEquilibriumTest.m is shown below:

>> NashEquilibriumTest

a =

```
     0     0     0     0

     0     0     0     0

    10    10    -5  -100

    10    10  -100  -100


b =


     0     0     0     0

     0     0     0     0

   -10   -10    -5  -100

   -10   -10  -100  -100


s =


     4     4


nrow =

     4

mcol =

     4
```

**************** Check if playoff matrix has a pure Nash Equilibrium strategy pair *****************:

the pair 1, 3 is in Nash Equilibrium

payoff of the pair is 0, 0

the pair 1, 4 is in Nash Equilibrium

payoff of the pair is 0, 0

the pair 2, 3 is in Nash Equilibrium

payoff of the pair is 0, 0

the pair 2, 4 is in Nash Equilibrium

payoff of the pair is 0, 0

the pair 4, 1 is in Nash Equilibrium

payoff of the pair is 10, -10

the pair 4, 2 is in Nash Equilibrium

payoff of the pair is 10, -10

Example 5

Suppose the familiar Rock - Paper - Scissors game matrix is show below:

$$M = \begin{pmatrix} (0,0) & (-2,-2) & (1,\ -1) \\ (2,-2) & (0,\ 0) & (-1,\ 1) \\ (-1,\ 1) & (1,\ -1) & (0,\ 0) \end{pmatrix}$$

The strategies are named R, P, S for the row player corresponds to the rows 1,2, 3 strategies, and R,P,S for the column player corresponds to the column player strategies.

The output listing of NashEquilibriumTest.m is shown below:

```
>> NashEquilibriumTest

>> NashEquilibriumTest


a =


     0    -2     1

     2     0    -1

    -1     1     0



b =


     0    -2    -1

    -2     0     1

     1    -1     0



s =
```

3 3

nrow =

3

mcol =

3

****************** Check if playoff matrix has a pure Nash Equilibrium strategy pair ******************:

there is no pure Nash Equilibrium found for the payoff matrices

As shown in the last example, a game does not always have a pure Nash Equilibrium, but by the Nash Equilibrium Theorem shown in the next sections below, any game must have at least one mixed Nash Equilibrium as demonstrated in Example 6 and 7 below.

Example 6

3.3 Nash Equilibrium for Mixed Strategies

Consider the following game $M = \begin{pmatrix} (2,4) & (4,1) \\ (3,2) & (1,3) \end{pmatrix}$

Let X_1, X_2 be the pure strategies for the row player with payoff matrix $a = \begin{pmatrix} 2 & 4 \\ 3 & 1 \end{pmatrix}$, and let Y_1, Y_2 be the pure strategies for the column player with payoff matrix $b = \begin{pmatrix} 4 & 1 \\ 2 & 3 \end{pmatrix}$. There are different ways to find the Nash Equilibrium using mixed strategies. For now, we are going to describe one approach as follows. Suppose $X^* = (p, q)$ and $Y^* = (r, s)$ are two mixed strategies in Nash Equilibrium, where $p + q = 1$ and $r + s = 1$ with $p, q, r, s > 0$.

When X^* is held fixed, Y^* does not want to deviate to any other strategies and this happens if all the expected values of each column strategies are equal, in which case, there is no improvement in deviation, that is to say

$$EV(b, X^*, Y_1) = EV(b, X^*, Y_2) \qquad \text{(EQ1)}$$

Similarly, when Y^* is held fixed, X^* does not want to deviate to any other strategies and this happens if all the expected values of each row strategies are equal, that is to say

$$EV(a, X_1, Y^*) = EV(a, X_2, Y^*) \qquad \text{(EQ2)}$$

(EQ1) gives:

$4p + 2q = 1p + 3q$

As $p + q = 1$, so we can solve $p=1/4$ and $q=3/4$

Similarly, (EQ2) gives:

$2r + 4s = 3r + 1s$, and as $r + s = 1$, we find $r= 3/4$, $s= 1/4$

Thus, we have found the two equilibrium mixed strategies in Nash Equilibrium are $X^*=(1/4,\ 3/4)$ and $Y^*=(3/4,\ 1/4)$

Example 7

Consider the following game $M = \begin{pmatrix} (3,3) & (1,5) & (4,2) \\ (5,1) & (3,3) & (2,4) \\ (2,4) & (4,2) & (3,3) \end{pmatrix}$ As before,

$a = \begin{pmatrix} 3 & 1 & 4 \\ 5 & 3 & 2 \\ 2 & 4 & 3 \end{pmatrix}$ and $b = \begin{pmatrix} 3 & 5 & 2 \\ 1 & 3 & 4 \\ 4 & 2 & 3 \end{pmatrix}$

Let X_1, X_2, X_3 be the pure strategies for the row player with payoff matrix a, let Y_1, Y_2, Y_3 be the pure strategies for the column player with payoff matrix b.

Let $X^* = (p, q, r)$ and $Y^* = (t, u, v)$ a mixed strategy pair that is in Nash Equilibrium.

Follow the same explanation as in Example 4, then we have the following equalities of expected values

$EV(b, X^*, Y_1) = EV(b, X^*, Y_2)$ (EQ3A)

$EV(b, X^*, Y_2) = EV(b, X^*, Y_3)$ (EQ3B)

$EV(a, X_1, Y^*) = EV(a, X_2, Y^*)$ (EQ4A)

$EV(a, X_2, Y^*) = EV(a, X3, Y^*)$ (EQ4B)

(EQ3A) gives:

$3p + 1q + 4r = 5p + 3q + 2r$ (EQ5A)

(EQ3B) gives:

$5p + 3q + 2r = 2p + 4q + 3r$ (EQ5B)

(EQ4A) gives:

$3t + 1u + 4v = 5t + 3u + 2v$ (EQ6A)

(EQ4B) gives:

$5t + 3u + 2v = 2t + 4u + 3v$ (EQ6B)

(EQ5A), (EQ5B), $p + q + r = 1$ gives the system of 3 equations in 3 unknowns:

$2p + 2q - 2r = 0$

$3p - q - r = 0$ (EQ7)

$p + q + r = 1$

Solving the above system (EQ7)by the built-in function linsolve in MATLAB, we find the mixed strategy

$X^* = (p, q, r) = (1/4, ¼, ½)$ (see output listing of ComputeLinearSystemEq.m below)

Similarly, (EQ6A) and (EQ6B) together with $t + u + v = 1$ gives rise to the linear system,

$2t + 2u - 2v = 0$

3t - u - v = 0 (EQ8)

t + u + v = 1

Again, linsolve gives us the solution $Y^* = (t, u, v) = (1/4, ¼, ½)$. Now the pair (X^*, Y^*) is a mixed Nash Equilibrium pair and we can check that

$EV(a, X^*, Y^*) = (p,q,r)\begin{pmatrix} 3 & 1 & 4 \\ 5 & 3 & 2 \\ 2 & 4 & 3 \end{pmatrix}\begin{pmatrix} t \\ u \\ v \end{pmatrix} = (p,q,r)\begin{pmatrix} 3 \\ 3 \\ 3 \end{pmatrix} = 3$, and similarly check that $EV(b, X^*, Y^*) = 3$.

Output listing of ComputeLinearSystemEq.m below:

```
>> ComputeLinearSystemEq

m =

   2   2  -2

   3  -1  -1

   1   1   1


c =


   0

   0

   1


x =

   0.2500
```

0.2500

0.5000

Source Code for ComputeLinearSystemEq.m

```
m=[2  2  -2;  3  -1  -1; 1  1  1] % coefficients matrix for the linear system
c=[0 0 1]' % equations'right hand side constants
x=linsolve(m,c)
```

3.4 Existence of the Nash Equilibrium strategy pair NE (X^*, Y^*) in a 2-person game

In this section, we will prove the existence of a strategy pair (X*, Y*) that is in Nash Equilibrium for a 2-person game. The proof of this theorem for n-person games will be discussed later in this book. Proof of the Nash Equilibrium is based on an incredibly famous theorem due to Brouwer in Algebraic Topology (Brouwer's Fixed Point Theorem). We will not prove this theorem, but we will state what it is and explain how it is applied to prove the existence of the strategy pair (X*, Y*) in Nash Equilibrium.

Brouwer's Fixed Point Theorem Any continuous mapping F from a n-dimensional cube I^n itself has at least one fixed point. (Recall that I^n is the unit interval from 0 to 1, I^2 is the unit square, I^3 is the unit cube, I^n is the Cartesian Product of I^1 n times.)

F: I^n —> I^n

That is, if F is a continuous function, then there exists at least one point, say x in I^n such that F(x) = x. Note that this theorem only guarantees us that there is at least one such fixed point, but it does not show us how to find it.

For our application, we can think of the set S of all mixed strategies of a player that has n pure strategies as I^n, the set T of all mixed strategies of another player that has m pure strategies as I^m.

The original Brouwer's theorem can be extended to say,

F: I^n x I^m —> I^n x I^m has at least one fixed point. Consider S and T as subsets or homeomorphic images in I^n and I^m respectively, then the version of Brouwer's fixed point theorem that we need for our application is:

If F: SXT ➔ SXT is continuous , the F has at least one fixed point.

Before we go into the proof of the existence of the Nash Equilibrium, we outline the steps of the proof and show how the Brouwer's Theorem is applied.

We construct a continuous function F: SXT ➔ SXT, this function carries the information of the 2-person game matrices A and B.

By Brouwer's Theorem, F has a fixed point.

We use proof by contradiction to show that this fixed point must be the Nash Equilibrium strategy pair.

Theorem: Nash Equilibrium Theorem
In any two-person game with game matrices A and B respectively, there always exists at least one strategy pair (X*,Y*) in equilibrium.

$$
\begin{array}{ccccccccc}
 & & Y1 & Y2 & & \ldots\ldots & & Y_n & \\
 & & & & & & & & \\
\end{array}
$$

$$
A = \begin{pmatrix}
a_{11} & a_{12} & \cdots & \cdots & \cdots & a_{1n-1} & a_{1n} \\
a_{21} & a_{22} & \cdots & \cdots & \cdots & a_{2n-1} & a_{2n} \\
\vdots & \cdots & \cdots & \cdots & \cdots & \cdots & \cdots \\
\vdots & \cdots & \cdots & \cdots & \cdots & \cdots & \cdots \\
a_{m-11} & a_{m-12} & \cdots & \cdots & \cdots & a_{m-1n-1} & a_{m-1n} \\
a_{m1} & a_{m2} & \cdots & \cdots & \cdots & a_{mn-1} & a_{mn}
\end{pmatrix}
\begin{matrix} X_1 \\ X_2 \\ \\ \\ X_{m-1} \\ X_m \end{matrix}
$$

$$
B = \begin{pmatrix}
b_{11} & b_{12} & \cdots & \cdots & \cdots & b_{1n-1} & b_{1n} \\
b_{21} & b_{22} & \cdots & \cdots & \cdots & b_{2n-1} & b_{2n} \\
\vdots & \cdots & \cdots & \cdots & \cdots & \cdots & \cdots \\
\vdots & \cdots & \cdots & \cdots & \cdots & \cdots & \cdots \\
b_{m-11} & b_{m-12} & \cdots & \cdots & \cdots & b_{m-1n-1} & b_{m-1n} \\
b_{m1} & b_{m2} & \cdots & \cdots & \cdots & b_{mn-1} & b_{mn}
\end{pmatrix}
$$

Step 1 details.

To describe how the function F is constructed, we first translate the conditions (a)" and (b)" in Lemma 2 for NE (X*, Y*) into two sets of numbers {c1, c2, c3…. Cm} for the m -inequalities in (a) and the set {d1, d2….,dn } for the n-inequalities in (b).

Each C i is defined as ci = Max {0, EV (A, Ri, Y) - EV(A, X,Y)} and similarly each d j is defined as d j= Max {0, EV(B, X, C j)- EV (B, X,Y)}. Then by the definition, the requirement for (X, Y) be in NE that satisfies (a)" and (b)" is equivalent to saying all the ci and d j are zeros. If EV (A, Ri,Y) > EV(A, X,Y) , then c i > 0, and if EV(A, R j, Y) <= EV(A, X, Y) , then c j=0.

Now we define F from S X T to SXT by assigning any point in SXT to another point in SXT. A point in SXT is a pair of strategy (x1,x2…xm) and (y2,y2….yn), the function F takes this pair of strategies into a new pair of strategies denoted by (x1',x2'….xm') and (y1', y2'…yn'),

F((x1,x2…xm), (y1,y2…yn)) = ((x1',x2'….xm'), (y1',y2'…yn')) where we impose this new pair to be

$$x_i' = \frac{x_i + c_i}{1 + \sum c_k} \quad \text{and} \quad y_j' = \frac{y_j + d_i}{1 + \sum d_k}$$

We must verify that this new point in SXT is a strategy pair, and function F is continuous.

(x1', x2'…, xm') , (y1', y2'…yn') represent another mixed strategy pair for the row and column players if

the sum x1'+ x2'+…., + xm'=1 and y1' + y2' +…. +yn'=1. But this is easily checked that,

$$\sum x_i = \sum \frac{x_i + c_i}{1 + \sum c_k} = \frac{1}{1 + \sum c_k}\left(\sum x_i + \sum c_k\right)$$

$$= \frac{1}{1 + \sum c_k}\left(1 + \sum c_k\right)$$

$$= 1$$

Next, we want to check if the function F defined this way is continuous.

Without delving into the rigorous definition of a continuous function as defined in topology, it suffices to understand that a function is continuous if its trajectory or disjointed sections. In other words, a continuous function changes smoothly without sudden jumps.

Algebraically, any functional composition of continuous functions results in a continuous function. The function *F* defined above is obtained through the composition of a linear function, identity function, and constant function, all of which are inherently continuous.

Step 2 details:

Now that F: SX T to SX T is defined and continuous, so by Brouwer's fixed point theorem, F has a fixed point, let's call this fixed point,

((x1*, x2*......., Xm*) (y1*, y2*....yn*))

Since it is a fixed point, this means that, the initial coordinates and the new coordinates of the fixed point are the same, that is,

$$x_i* = \frac{x_i * + c_i}{1+\sum c_k} \quad \text{and} \quad y_j* = \frac{y_j * + d_i}{1+\sum d_k}$$

A little algebra then shows the following relationship between the xi* and the (C1, c2...cm}, the yj* and the {d1, d2...dn} by

$$c_i = x_i \sum c_k \quad \text{and} \quad d_j = y_j \sum c_k$$

which we will used in Step E details. Note that the xi and yj are positive by choice, the C_k could be 0 or positive. Our claim is that this fixed point ((x1*, x2*, ..., xm*), (y1*, y2*....,Yn*)) whose existence in SXT guaranteed by the Brouwer's fixed point theorem with the strategy pair (X1*, x2*,... xm*) and (y1*, y2*, ... yn*) is a Nash Equilibrium. We will prove this claim by contradiction. So, we assume on the contrary, this pair (x1*, x2*... xm*) and (y1*, y2*, ... yn*) is not NE and we will show how this assumption can lead to a contradiction.

Step 3 details: We will show how a contradiction rises.

Let X* to denote the mixed strategy (x1*, x2*...., xm*), and Y* to denote the strategy (y1*, y2*, ..., Yn*) and that the strategy pair (X*, Y*) is not in NE. This means that not all the conditions (a)" and (b)" stated in (Step 1 details) are satisfied, say, when i = k1, the condition is not satisfied in (a)". This means that,

EV (A, $R_{k1,}$ Y*) > EV (A, X*, Y*), and by the definitions of the constants {C_I} described earlier, $C_{k1} > 0$.

But $C_{k1} = x_{k1} \sum c_k$ from Step 2, and $x_{k1} > 0$ by the choice of mixed strategy,

So $\sum c_k > 0$. Since EV (A, X*, Y*) can be written as a weighted average of the EV (A, R1, Y*), EV (A, R2, Y*), ..., EV (A, Rm,Y*) with weights x1*, x2*....,x m*, that is:

EV (A, X*, Y*) = x1*EV (A, R1, Y*) + x2*EV (A, R2, Y*) +......+ xm*EV (A, Rm,Y*), the condition EV (A, R k1, Y*) > EV(A, X*,Y*) implies there is some i = k2 such that EV (A, $R_{k2'}$ Y*) < EV (A, X*, Y*) and $x*_{k2} > 0$. This is because if a component in a weighted average is bigger than the entire weighted average of all the components, then there must be another component that is smaller than the entire weighted average. But then, $C_{k2} = 0$ by the definitions of the $\{C_l\}$ and because C_{k2} is related to {X*I} by $C_{k2} = x^*_{k2} \sum c_k$, so we have, 0 is the product of two non-zero quantities x^*_{k2} and $\sum c_k$ ... which is impossible. Note that $\sum c_k > 0$ results from our assumption that the pair (X*, Y*) is not in NE when (X*, Y*) is a fixed point of F from S X T to itself. The above contradiction shows that when (X*, Y*) is a fixed point of F, then it must be NE (X*, Y*), or they are in NE.

3.5 Relationships between security levels and equilibrium strategy pair

In section 2.1, we discussed the meaning of the security level of a player, and the security level of a strategy associated with that player. We show how the security level of the row player sl(row player) and the security level of the column player sl(column player) are computed for a 2-person zero-sum game. As now we know the concept of Nash Equilibrium strategy pair, so in this section, we wish to investigate, in a general 2-person game with payoff matrix A and B for the row and column player respectively, if there is any relationship between the expected value of any Nash Equilibrium strategy pair (X*, Y*), that is EV(A,X*,Y*) and EV (B, X*,Y*) and the two security levels, that is sl (row player), and sl (column player).

Theorem 1. If (X*, Y*) is any strategy pair that is in NE. then the following is true

(a) Sl (row player) <= EV(A, X*, Y*)

(b) Sl (column player) <= EV (B, X*, Y*)

Proof of (a),

As before, let S be the set of all strategies including pure or mixed strategies of the row player, and T be the set of all strategies for the column player.

Then sl (row player) = by definition = $\underset{X\ in\ S}{Max}\ \underset{Y\ in\ T}{Min}\ EV(A,X,Y)$

Because (X*, Y*) is in NE, so EV (A, X, Y*) <= EV (A, X*, Y*) and

EV (B, X*, Y) <= EV (B, X*, Y*),

So, sl (row player) = $\underset{X\ in\ S}{Max}\ \underset{Y\ in\ T}{Min}\ EV(A,X,Y)$ by definition

$<= \underset{X\ in\ S}{Max}$ EV (A, X, Y*) because Y was selected to minimize

EV (A, X, Y)

= EV (A, X*, Y*) by the definition of EV (A, x*, Y*)

Proof of (b),

Sl (column player) = $\underset{Y\ in\ T}{Max}\ \underset{X\ in\ S}{Min}\ EV(B,X,Y)$ by definition,

<= $\underset{Y\ in\ T}{Max}$ EV (B, X*, Y) because X was selected to minimize EV (B, X, Y)

= $\underset{Y\ in\ T}{Max}$ EV (B, X*, Y*) because (X*, Y*) is in NE and that is given

= EV (B, X*, Y*)

Thus, we see that in a general two-person game, any Nash Equilibrium pair (X*, Y*) has their individual expected value above their respective security level, that is, EV (A, X*, Y*) => sl(row player), and EV (B, X*, Y*) => sl(column player). But sl(row player), is, the maximum of all the security levels of infinitely many mixed strategy X for the row player, that is,

sl (row player) = $\underset{X\ in\ S}{Max}$ sl (X), where sl (X) = $\underset{Y\ in\ T}{Min}\ EV(A,X,Y)$, and thanks to Lemma 1 in section 2,1 taking minimum over infinitely many strategies Y in 'I' can be reduced to taking minimum over only finitely many pure strategies because,

$\underset{Y\ in\ T}{Min}\ EV(A,X,Y)$ was shown to be,

$\underset{j=1\ to\ n}{Min}\ EV(A,X,C_j)$

So sl (row player) = $\underset{X\ in\ S}{Max}\ \underset{j=1\ to\ n}{Min}\ EV(A,X,C_j)$. (****)

Similarly, sl (column player)

sl (column player) = $\underset{Y\ in\ T}{Min}$ sl (Y), where sl (Y) is by definition, equals to, $\underset{X\ in\ S}{Max}\ EV(A,X,Y)$,and in turn this was shown to be $\underset{i=1\ to\ m}{Max}\ EV(A,R_i,Y)$, so

sl (column player) = $\underset{Y\ in\ T}{Min}\ \underset{i=1\ to\ m}{Max}\ EV(A,R_i,Y)$ (****)

These two expressions (***) tell us that the problem of computing sl(row player) and sl(column player) can be translated into solving two optimization problems in Linear Programming. The Fundamental Theorem in Game Theory simply states that we can always find a strategy X* that has the largest sl(X*) amongst all the sl(X) and find a strategy Y* that has the smallest sl(Y*) amongst all the sl(Y), and that sl(X*) = sl(Y*). Henceforth, we call these two strategies X* and Y* by more meaningful names X_{sl} and Y_{sl}, X_{sl} is the row player security level strategy and Y_{sl}

is the column player security level strategy. To reiterate, $sl(X_{sl}) = sl(Y_{sl})$, and $sl(X_{sl})$ is the largest amongst all the sl (X) and sl (Y_{sl}) the smallest amongst all the sl(Y). We will give a formal description of the Fundamental Theorem of Game Theory in the next section, but for now, it suffices for us to know the existence of these two strategy level strategies X_{sl} and Y_{sl} are guaranteed by the Fundamental Theorem of Game Theory, and for now, we will investigate if these two strategies X_{sl} and Y_{sl} are actually in Nash Equilibrium, that is NE (X_{sl}, Y_{sl}) in the next theorem. The answer is yes as we will show in the next Theorem.

Theorem 2. `The security level strategy X_{sl} for the row player and the security level strategy Y_{sl} for the column player is in Nash Equilibrium.

Proof. To show NE (X_{sl}, Y_{sl}), we must show, by definition, that the following two conditions are true.

(a) $EV(A, X, Y_{sl}) <= EV(A, X_{sl}, Y_{sl})$ for all X in S
(b) $EV(A, X_{sl}, Y_{sl}) <= EV(A, X_{sl}, Y)$ for all Y in T

But these two inequalities follow directly from the following inequalities,

$sl(X_{sl})$ = by definition,

$$= \underset{Y\ in\ T}{Min}\ EV(A, X_{sl}, Y)$$

$$<= EV(A, X_{sl}, Y_{sl})$$

$$<= \underset{X\ in\ S}{Max}\ EV(A, X, Y_{sl})$$

$$= sl(Ysl)$$

But by The Fundamental Theorem of Game Theory, $sl(X_{sl}) = sl(Y_{sl})$, so all the inequalities in between become equality and so we have,

$$\underset{Y\ in\ T}{Min}\ EV(A, X_{sl}, Y) = EV(A, X_{sl}, Y_{sl})$$

and $$EV(A, X_{sl}, Y_{sl}) = \underset{X\ in\ S}{Max}\ EV(A, X, Y_{sl})$$

The first equality implies, $EV(A, X_{sl}, Y) => EV(A, X_{sl}, Y_{sl})$ for all Y in T, which is (b). This is because Ysl corresponds to taking the smallest Y in T.

The second equality implies $EV(A, X_{sl}, Y_{sl}) => EV(A, X, Y_{sl})$ for all X in S, which is (a). This is because X_{sl} corresponds to taking the largest X in S.

Now we know that, in a 2-person zero-sum game, the security level strategy pair (X_{sl}, Y_{sl}) is in Nash Equilibrium. The next theorem states that the converse is also true in a 2-person game, that is to say, if (X*, Y*) are any strategy pair that is in Nash Equilibrium, then this strategy pair (X*, Y*) will also behave like the security level strategy pair (X_{sl}, Y_{sl}) whose existence was guaranteed by the Fundamental Theorem of Game Theory. That is to say,

Sl (X*) = $\underset{Y\ in\ T}{Min}\ EV(A, X^*, Y)$, the same definition as sl(X sl) and

Sl (Y*) = $\underset{X\ in\ S}{Max}\ EV(A, X\ , Y^*\ )$, the same definition as sl (Ysl)

Sl(X*) = sl(Y*).

Theorem 3. If (X*, Y*) are strategy pairs in Nash Equilibrium, then

sl (X*) = sl(X_{sl}), sl (Y*) = sl(Y_{sl}), and sl(X*) = sl(Y*) = EV(A, X*,Y*).

Proof. Since we already know that sl(row player) = sl(column player) by the Fundamental Theorem of Game Theory, we start with

$$\text{Sl (row player)} = \underset{X\ in\ S}{Max}\ \underset{Y\ in\ T}{Min}\ EV(A, X, Y)$$

$$= \underset{Y\ in\ T}{Min}\ EV(A, X_{sl}, Y)$$

$$=> \underset{Y\ in\ T}{Min}\ EV(A, X^*, Y)$$

$$=> EV(A, X^*, Y^*)\ ,$$

On the other hand,

$$\text{sl (column player)} = \underset{Y\ in\ T}{Min}\ \underset{X\ in\ S}{Max}\ EV(A, X_i, Y)$$

$$= \underset{X\ in\ S}{Max}\ EV(A, X, Y_{sl})$$

$$<= \underset{X\ in\ S}{Max}\ EV(A, X, Y^*)$$

$$<= EV(A, X^*, Y^*)\ ,$$

As sl(row player) = sl(column player), so all the terms in between are equal, that is to say,

sl(row player) = $\underset{Y\ in\ T}{Min}\ EV(A, X^*, Y)$, and sl (column player) = $\underset{X \text{ in } S}{Max}\, EV(A, X, Y^*)$, and

sl(row player) = sl(column player) = EV(X*,Y*).

3.6 Example illustrations of the theorems from 3.5

Example 8: Suppose the payoff matrices for the two players are

$$a = \begin{pmatrix} 1 & 3 & 5 \\ 5 & 2 & 2 \\ 4 & 7 & 1 \end{pmatrix} \quad b = \begin{pmatrix} 1 & 4 & 6 \\ 8 & 5 & 5 \\ 5 & 8 & 2 \end{pmatrix}$$

Using the MATLAB program ComputeSecurityLevelsbySimplexMethod33.m we find the security level strategy for matrix a (player 1) to be aoriginal= [0.4286 0.4286 0.1429]

and the security level strategy for its column player is

acolumn= [0.3816 0.1667 0.4524] and the security level for player 1 is sl1=3.1429 ; the security level strategy for matrix b (player 2) is

boriginal =[0 0.5 0.5], the security level strategy for the dual player is

bcolumn = [0.75 0 0.25], and the security level for player 2 is sl2= 5.

Follow the method described in Examples 4 and 5 (Section 3.2), using the MATLAB program ComputeLinearSystemEq.m, we can always find a Nash Equilibrium pair (x_1, x_2) for the game with payoff matrix a and b, where

x_1= (0.375, 0.5, 0.125) and x_2= (0.381, 0.1667, 0.4524) is a strategy pair for a and b respectively. The MATLAB program WorkSheetSecurityLevelsandNashEquilibrium.m verifies the theorems stated in Section 3.4 hold true in our numerical example.

```
>> WorkSheetSecurityLevelsandNashEquilibrium

Security level strategies for Matrix A and B:

a =

   1   3   5
```

```
   5  2  2

   4  7  1

b =

   1  4  6

   8  5  5

   5  8  2
```

**** results of the simplex method for the security levels sl1 and sl2:

this is the result from ComputeSecurityLevelsbySimplexMethod33.m

```
aoriginal =

   0.4286   0.4286   0.1429

sl1 =

   3.1429

acolumn =

   0.3816   0.1667   0.4524
```

Information below is for the transpose of matrix b:

```
boriginal =

        0   0.5000   0.5000

sl2 =

     5

bcolumn =
```

0.7500 0 0.2500

Nash Equilibrium strategies X1 and X2 :

this is the result from ComputeLinearSystemEq.m ********* Solution for system m1 ***************:

m1 =

3 -3 3

2 0 -6

1 1 1

x1 =

0.3750 0.5000 0.1250

********* Solution for system m2 ***************:

m2 =

4 -1 -3

1 -5 1

1 1 1

x2 =

0.3810

0.1667

0.4524

Verifications of relationships between security levels and Nash Equilibrium Pairs :

verify that sl1 <= EV (a, x1,x2) and sl2 <= EV(b, x1,x2):

sl1 =

3.1429

EV1 =

3.1429

sl2 =

5

EV2 =

5.0000

verify the computation of sl1 and sl2 directly from its definition :

verify the fundamental theorem of game theory for player 1 Matrix A :

nu1a =

3.1431

nu2a =

3.1462

verify the fundamental theorem of game theory for player 2 matrix b:

nu1b =

5

nu2b =

5

Source Code for WorkSheetSecurityLevelsandNashEquilibrium.m

```
fprintf('Security level strategies for matrix a and b:\n')
a=[1 3 5;5  2  2; 4  7  1]
b=[1  4  6;8 5  5;5  8  2]

fprintf('****  results of the simplex method for the security levels sl1 and sl2 :\n')
fprintf('this is the result from ComputeSecurityLevelsbySimplexMethod33.m')

aoriginal= [0.4286  0.4286   0.1429]
sl1= 3.1429
acolumn=[ 0.3816  0.1667  0.4524]

fprintf('Information below is for the transpose of matrix b: \n')
boriginal= [0  0.5   0.5]
sl2 =5
bcolumn = [0.75  0   0.25]


fprintf('Nash Equilibrium strategies X1 and X2 :\n')
fprintf('this is the result from ComputeLinearSystemEq.m')


fprintf(' *********   Solution for system m1    ***************:\n')
m1= [3 -3  3;2 0  -6; 1 1 1]
x1= linsolve( m1,c);
x1=x1'

fprintf(' *********   Solution for system m2    ***************:\n')

m2= [4  -1  -3; 1  -5  1; 1  1  1]

x2= linsolve(m2,c)


fprintf(' Verifications of relationships between security levels and Nash Equilibrim Pairs
:\n')
fprintf(' verify that  sl1 <=  EV (a, x1,x2) and sl2 <= EV(b, x1,x2): \n')
sl1
EV1 = x1*a*x2
sl2
EV2= x1*b*x2

fprintf(' verify the computation of sl1 and sl2 directly from its definition :\n')
fprintf('verify fundamenatal theorem of game theory for player 1 matrix a :\n')
c1a = [1  5  4]';
c2a = [3  2  7]';
c3a = [5  2  1]';
temp = [aoriginal*c1a; aoriginal*c2a; aoriginal*c3a ];
nu1a = min ( temp )
row1a = [1  3  5];
row2a =[5  2  2];
row3a = [4  7  1];
temp=[row1a* acolumn'  row2a* acolumn'    row3a* acolumn'];
nu2a= max ( temp)

% ***********************************************************************
fprintf('verify fundamenatal theorem of game theory for player 2 matrix b:\n')
c1b = [1 4  6]';
c2b = [8  5  5]';
c3b = [5  8  2]';
temp = [boriginal*c1b; boriginal*c2b; boriginal*c3b ];
nu1b = min ( temp )
```

```
row1bt = [ 1  8  5];
row2bt = [ 4  5  8];
row3bt = [6  5  2];
temp = [row1bt* bcolumn'  row2bt* bcolumn'    row3bt* bcolumn'];
nu2b = max ( temp)
```

3.7 An application of the Nash Equilibrium concept in Economic

In this final section of the chapter, we introduce a practical application of the Nash Equilibrium concept in a simplified economic environment, where two competing companies determine the quantities of goods they produce.

Before discussing Nash Equilibrium, we first establish the basic assumptions of this simplified economic scenario, as captured by the *Cournot Duopolistic Competition Model.*

In this model, suppose there are two companies, A and B, each producing q_1 and q_2 units of a certain good, respectively. Let the per unit production costs for these companies be c_1 and c_2 respectively. The Cournot Model states that the market selling price " p", regardless of which company sells the good, is given by the equation

$$p = Q_0 - (q_1 + q_2) \tag{E15}$$

, where Q_0 is a constant parameter.

We now formulate the above situation as a two-person game with each of their payoff values described below. Let P_1 and P_2 denote the respective profit of company A and B, then :

$$P_1(q_1, q_2) = p\, q_1 - c\, q_1 \tag{E16}$$

$$= [Q_0 - (q_1 + q_2)]\, q_1 - c_1\, q_1$$

$$= [Q_0 - c_1]\, q_1 - q_1\, q_2 - q_1^2 \tag{E17}$$

Similarly,

$$P_2(q_1, q_2) = p\, q_2 - c_2\, q_2$$

$$= [Q_0 - c_2]\, q_2 - q_1\, q_2 - q_2^2 \tag{E18}$$

Thus, the profits P_1 and P_2 for the respective companies A and B are functions in both the production quantities q_1 and q_2. Now apply the Nash Equilibrium concept to the above situation and think that the strategy of each player is to choose the right production quantity. Following the definition of Nash Equilibrium pair, we want to find for Company A production quantity q_1 that would maximize its profit P_1 when Company B is holding a fixed production quantity $q_2^*$and to find for Company B production quantity q_2 that would maximize her profit P_2 when Company A is holding a fixed production quantity q_1^*. Thus, the Nash Equilibrium pair (q_1, q_2) pair is equivalent to the following:

$$\text{Max } P_1(q_1, q_2^*) = [Q_0 - c_1]\, q_1 - q_1\, q_2^* - q_1^2 \tag{E19}$$

$$\text{Max } P_2(q_1^*, q_2) = [Q_0 - c_2]\, q_2 - q_1^*\, q_2 - q_2^2 \tag{E20}$$

Now differential calculus tells us that differentiate P_1 with respect to q_1 when q_2^* is considered constant, and setting the derivative to zero, we have

$$[Q_0 - c_1] - q_2^* - 2\, q_1 = 0$$

, which gives

$$q_1 = \tfrac{1}{2}\,(Q_0 - c_1 - q_2^*) \tag{E21}$$

Similarly, differentiate P_2 with respect to q_2 when q_1^* is considered constant, and setting the derivative to zero, we have

$$[Q_0 - c_2] - q_1^* - 2\, q_2 = 0$$

, which gives

$$q_2 = \tfrac{1}{2}\,(Q_0 - c_2 - q_1^*) \tag{E22}$$

The equations (E21) and (E22) above tell us that if Company A knows in advance how many units of good q_2^* Company B produced, then the optimal unit Company A would produce is given by (E21). And if Company B knows in advance how many units of good q_1^* Company A produced, then the optimal unit Company B would produce is given by (E22). But since neither company A nor B knows what the other produces, so a win-win solution for both companies A and B is to solve the simultaneous equations (E21) and (E22) which gives us the following:

$$q_1^{ne} = \frac{1}{3}[Q_0 - 2c_1 + c_2] \tag{E23}$$

$$q_2^{ne} = \frac{1}{3}[Q_0 - 2c_2 + c_1] \tag{E24}$$

Now, the above q_1^{ne}, q_2^{ne} are Nash Equilibrium pairs they must satisfy the Nash Equilibrium condition,

$$P_1 (q_1^{ne}, q_2^{ne}) \geq P_1 (q_1, q_2^{ne}) \text{ for all } q_1 \text{ for a fixed } q_2^{ne}$$

and $$P_2 (q_1^{ne}, q_2^{ne}) \geq P_2 (q_1^{ne}, q_2) \text{ for all } q_2 \text{ for a fixed } q_1^{ne} \qquad \text{(E25)}$$

Example 9: As a numerical example, suppose the parameters in the Cournot Model are given as $p = 1200$ and $c_1 = 15$, $c_2 = 30$, then (E23) and (E24) give $q_1^{ne} = 400$ and $q_2^{ne} = 385$.

Although q_1 and q_2 are continuous variables and take on infinitely many values; to help us visualize the concept of Nash Equilibrium, we can think of each discrete value q_1 or q_2 takes represents a strategy that the one player (Company A) chooses and a strategy that the other player (Company B) takes.

A finite portion of the game's infinite payoff matrix shown below gives us an idea of what a strategy pair of Nash Equilibrium (q_1^{ne}, q_2^{ne}) should satisfy.

E(25) is translated into saying that, when company B is playing q_2^{ne}, company A has the largest payoff P_1 amongst all other rows when company B is choosing column q= q_2^{ne}; And when company A is playing q_1^{ne}, company B has the largest payoff P_2 amongst all other columns when company A is choosing column q= q_1^{ne}.

Table 3.6.1

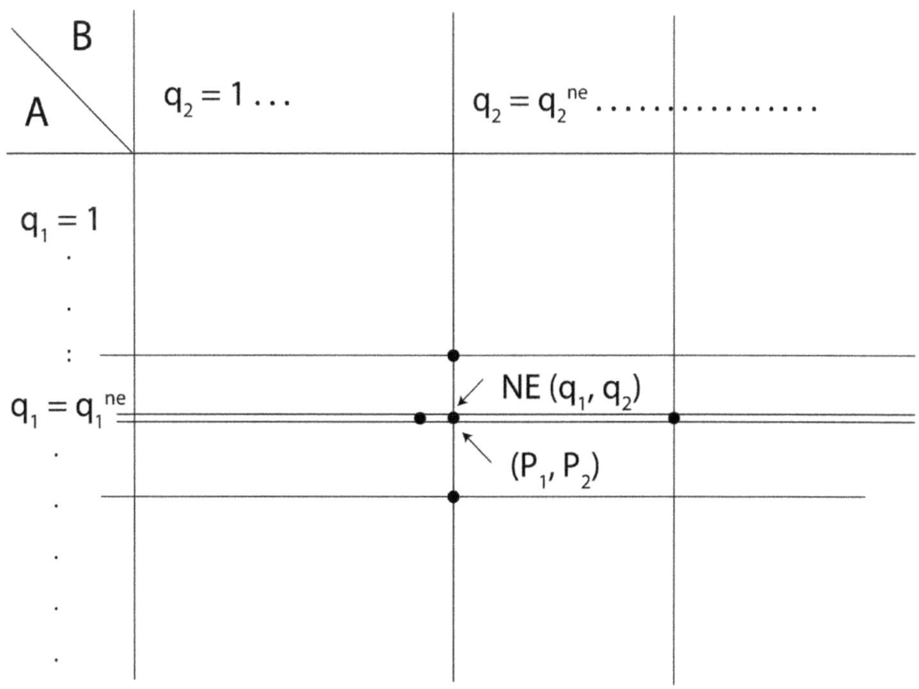

3.8 Summary

This chapter introduces the fundamental concept and definition of Nash Equilibrium between two strategies—one for the row player (Player 1, Player A) and one for the column player (Player 2, Player B) in a two-person game.

Identifying a pair of strategies in Nash Equilibrium provides valuable insights into the game itself. As such, the search for Nash Equilibrium is a major topic that we will explore across various classifications of games throughout the remaining chapters of this book.

In this chapter, we present a MATLAB program to search for all Nash Equilibria of pure strategy pairs and outline a procedure for finding the Nash Equilibrium of mixed strategy pairs. We also provide a step-by-step proof of the celebrated Nash Equilibrium Existence Theorem, which guarantees that every game has at least one Nash Equilibrium.

The chapter concludes with a simple yet insightful application of the Nash Equilibrium concept in Economics.

Chapter 4

Classifications of games and Applications of Nash Equilibrium

4.0 Introduction

In this chapter, we examine other common types of games and their classifications. In the previous two chapters, we explored games in which the payoff values for each player are known when all players simultaneously choose their strategies. The payoff matrices or functions specify these values. From this point forward, we refer to such games as *Static Games with Complete Information*.

The term 'static' indicates that each player makes a decision without knowing the choices of other players. This contrasts with dynamic games, where players make decisions sequentially—each later player knows the choices of those who played before them but not vice versa.

The term 'Information' refers to players' knowledge of payoff values. In a game with 'Complete' Information, every player knows with certainty the payoff values of all participants, including their own.. In contrast, a game with 'Incomplete' Information, means that players do not know the exact payoff values of others but instead rely on probability distribution to estimate them.

By understanding these distinctions, we build a stronger foundation for analyzing different types of strategic interactions.

The other types of games we will explore in the subsequent chapters are

Static Games with Incomplete Information
Dynamic Games with Complete Information
Dynamic Games with Incomplete Information

4.1 Static Games with Complete Information

As previously stated, in static games with complete information, each player chooses their strategy simultaneously, and the corresponding payoff values are common knowledge among all players. These payoff values are typically represented by a payoff matrix or, for continuous values, by known payoff functions.

In this section, I will present several examples illustrating the applications of the Nash Equilibrium concept. First, it is important to note that the Nash Equilibrium Existence Theorem, which was stated and proved in Section 3.3, is not limited to two-player games. In fact, it holds for any number of players (n-players). Given its broad applicability, we will restate the theorem below.

***4.1.1 Definition of Nash Equilibrium**: Suppose the n-player game $G = \{S_1, \ldots, S_n ; u_1, \ldots, u_n\}$, where S_i is the strategy set for player i, and where u_i is the payoff function for player i. We say the set of strategies $(s_1^* \ldots s_n^*)$, where each of the s_i^* is taken from S_i, are Nash Equilibrium, if the following n- inequalities hold for i = 1 to i = n.*

$$u_i\,(s_1^* \ldots s_{i-1}^*, s_i^*, s_{i+1}^*, \ldots s_n^*) \geq u_i\,(s_1^* \ldots s_{i-1}^*, s_i, s_{i+1}^*, \ldots s_n^*) \tag{E1}$$

for all possible s_i in S_i.

The expressions (E1) above are equivalent to saying that for each player, say player i, if the rest of players 1,2, ,i-1, i+1 ...n is holding their strategies unchanged, then the strategy s_i^* is the best strategy for player i. So, to find the Nash Equilibrium $(s_1^* \ldots s_n^*)$, we need find the solution for s_i^* by solving the maximization problem:

$$\underset{s_i \text{ in } S_i}{Max}\; u_i\,(s_1^* \ldots s_{i-1}^*, s_i, s_{i+1}^*, \ldots s_n^*) \tag{E2}$$

The solution of s_i from (E2) above will be expressed as a function in the remaining variables:

$s_1^* \ldots s_{i-1}^*, s_{i+1}^*, \ldots s_n^*$, that is,

$$s_i^* = bR_i\,(s_1^* \ldots s_{i-1}^*, s_{i+1}^*, \ldots s_n^*) \tag{E3}$$

for a suitable function bR( ), which we may name as the best-response function because it is indeed the player i 's best strategy when the other players are holding their strategies constant.

Now (E3) is a system of n best-response equations in n unknowns, solving the system will give us the Nash Equilibrium $s_1^* \ldots s_{i-1}^*, s_i^*, s_{i+1}^*, \ldots s_n^*$ for the problem.

4.1.2 Existence Theorem of Nash Equilibrium:

There always exists at least one Nash Equilibrium solution to the system E3

Next, we will see a few important applications of this Nash Equilibrium concept.

Example 1: The Problem of the Commons (Hardin 1968)

The problem of the Commons refers to a phenomenon in which individuals within a group overused shared resources due to self-interest, ultimately leading to catastrophic consequences for the group as a whole. These commons may include natural resources such as the atmosphere, oceans, land, parks, birds, rivers, fish stocks, and wildlife. The group affected by this overuse could be humanity itself.

This phenomenon has occurred in various real-world scenarios. In the following discussion, we take a Game Theory Approach and apply the concept of Nash Equilibrium to analyze a specific example.

Suppose there is n companies C_1, C_2 … C_n around the world and suppose the clean air resource in the whole world is given by a constant amount Q_0. Assume each company C_1, C_2 … C_n is using the clean air amount q_1, q_2, …q_n for production respectively. Assume v(q) is a profit function that gives the amount of profit to the company C_i when it consumes the amount of clean air $q = q_i$ for production.

The total profit P_i for each C_i can be expressed as follows: for i=1 to n

$$P_i = v(q_i) + v\,(Q_0 - \sum_{i=1:n} q_i) \qquad \text{(E4)}$$

The second term above represents the profit for C_i by consuming the remaining amount of the clear air.

To find the Nash Equilibrium pair (q_1^*… q_{i-1}^*, q_i^*, q_{i+1}^*… q_n^*), follow the steps outlined in (E2) and (E3) above, we compute derivative of P_i with respect to q_i, setting it to zero, and solve for q_i , we then have the following n equations in n unknowns (q_1… q_{i-1}, q_i, q_{i+1}… q_n),

$$v'(q_i) = v'(Q_0 - \sum_{i=1:n} q_i) \qquad \text{(E5)}$$

As a numerical example illustration, we assume the benefit function is a natural logarithmic function ln(x), Q_0 =1, and n takes the values 2,3,4,5 for comparisons, we find:

$$q_i = Q_0 - \sum_{i=1:n} q_i \qquad \text{(E6)}$$

Thus, for n=2, (E6) gives a linear system of equations in q_1 and q_2

2q1 + 1q2 =1

1 q1 + 2 q2 =1

We just solve the above linear equations by hand to obtain the Nash Equilibrium solution (q_1, q_2).

Similarly for n=3,4,5, we employ the MATLAB function linsolve to find the Nash Equilibrium solution for each case n=3,4 and 5. (See the output listing of ProblemoftheCommons.m)

In summary, we find:

for n=2: $q_1 = q_2 = 0.333 = 1/3$,

for n=3: $q_1 = q_2 = q_3 = 0.25 = 1/4$

for n=4: $q_1 = q_2 = q_3 = q_4 = 0.2 = 1/5$

for n=5: $q_1 = q_2 = q_3 = q_4 = q_5 = 0.1667 = 1/6$

We conclude that when n=k, $q_n = 1/(k+1)$, and as the number of companies gets large, q_n approaches 0

Output listing of ProblemoftheCommons.m

```
>> ProblemoftheCommons

q2 =

    2   1

    1   2

c2 =

    1

    1

*********  Solution for system x2  ***************:

x2 =
```

```
  0.3333

  0.3333

********* Solution for system x3 ***************:

q3 =

   2   1   1

   1   2   1

   1   1   2

c3 =

   1

   1

   1

x3 =

  0.2500

  0.2500

  0.2500

********* Solution for system x4 ***************:

q4 =

   2   1   1   1

   1   2   1   1

   1   1   2   1

   1   1   1   2
```

```
c4 =

     1

     1

     1

     1

x4 =

    0.2000

    0.2000

    0.2000

    0.2000

 *********   Solution for system x5    ***************:

q5 =

     2     1     1     1     1

     1     2     1     1     1

     1     1     2     1     1

     1     1     1     2     1

     1     1     1     1     2

c5 =

     1

     1
```

1

1

1

x5 =

0.1667

0.1667

0.1667

0.1667

0.1667

The MATLAB source code for the above output listing is ProblemoftheCommons.m shown below:

```
q2=[2 1;  1   2 ]      % coefficients matrix for the linear system
c2=[1 1]'               % equations'right hand side constants
fprintf( ' *********   Solution for system x2     ***************:\n')
x2=linsolve(q2,c2)

fprintf( ' *********   Solution for system x3     ***************:\n')
q3=[2 1 1 ;1  2  1;1  1  2]
c3=[1 1 1]'
x3=linsolve(q3,c3)
fprintf( ' *********   Solution for system x4     ***************:\n')

q4=[ 2 1 1 1; 1 2 1 1; 1 1 2 1; 1 1 1 2]
c4=[1 1 1 1]'
x4= linsolve(q4,c4)


fprintf( ' *********   Solution for system x5     ***************:\n')

q5=[ 2 1 1 1 1; 1 2 1 1 1; 1 1 2 1 1; 1 1 1 2 1; 1 1 1 1 2]
c5=[1 1 1 1 1]'
x5=linsolve(q5,c5)
```

Now that we have found our solution for the Nash Equilibrium based on the individual interest incentives of each company (P_i in the expression (E4)) alone, now we must compare our solution to the social (Pareto) optimal solution based on the totality of profits of each company, that is to say, we must find $(q_1^* \ldots q_{i-1}^*, q_i^*, q_{i+1}^* \ldots q_n^*)$ that optimize

$$\sum_{i=1:n} P_i = \sum_{i=1:n} v(qi) + v(Q0 - \sum_{i=1:n} q_i)$$

, which is $\sum_{i=1:n} v(qi) + n * v(Q0 - \sum_{i=1:n} q_i)$ (E7)

As before, when the modelling function v(q) is a logarithmic function, taking derivatives with respect to q_i, setting it to 0 , we get :

$$n * q_i = K - \sum_{i=1:n} q_i \qquad \text{(E8)}$$

Let K =1 and solve the system (E8) of n-linear equations as before, we have:

In summary, as K=1, we find

(See output listing of ProblemofCommonsSocialOptimal.m below)

for n=2: $q_1 = q_2 = 0.25 = 1/4$,

for n=3: $q_1 = q_2 = q_3 = 0.1666 = 1/6$

for n=4: $q_1 = q_2 = q_3 = q_4 = 0.125 = 1/8$

for n=5: $q_1 = q_2 = q_3 = q_4 = q_5 = 0.1 = 1/10$

We conclude that when n=k, q_n = 1/2k , and as the number of companies gets large, q_n approaches 0.

Thus, by comparing the values of qi for each i = 1 to n in the case of Nash Equilibrium and in the case of Social Optimal, we find the Social Optimal solution consumes less usage of the clean air commons Q_0.

Output listing of ProblemoftheCommonsSocialOptimal.m

```
>> ProblemoftheCommonsSocialOptimal

q2 =
```

```
     3     1

     1     3

c2 =

     1

     1

 *********  Solution for system x2   ***************:

x2 =

    0.2500

    0.2500

 *********  Solution for system x3   ***************:

q3 =

     4     1     1

     1     4     1

     1     1     4

c3 =

     1

     1

     1

x3 =

    0.1667
```

```
   0.1667

   0.1667

 *********   Solution for system x4    ***************:

q4 =

     5     1     1     1

     1     5     1     1

     1     1     5     1

     1     1     1     5

c4 =

     1

     1

     1

     1

x4 =

    0.1250

    0.1250

    0.1250

    0.1250

 *********   Solution for system x5    ***************:

q5 =
```

6 1 1 1 1

1 6 1 1 1

1 1 6 1 1

1 1 1 6 1

1 1 1 1 6

c5 =

1

1

1

1

1

x5 =

0.1000

0.1000

0.1000

0.1000

0.1000

Source Codes of ProblemoftheCommonsSocialOptimal.

```
q2=[3 1;  1   3 ]       % coefficients matrix for the linear system
c2=[1 1]'                % equations'right hand side constants
fprintf( ' *********   Solution for system x2    ***************:\n')
x2=linsolve(q2,c2)

fprintf( ' *********   Solution for system x3    ***************:\n')
q3=[4 1 1 ;1  4  1;1  1  4]
```

```
c3=[1 1 1]'
x3=linsolve(q3,c3)
fprintf( ' *********   Solution for system x4    ***************:\n')

q4=[ 5 1 1 1; 1 5 1 1; 1 1 5 1; 1 1 1 5]
c4=[1 1 1 1]'
x4=linsolve(q4,c4)


fprintf( ' *********   Solution for system x5    ***************:\n')

q5=[ 6 1 1 1 1; 1 6 1 1 1; 1 1 6 1 1; 1 1 1 6 1; 1 1 1 1 6]
c5=[1 1 1 1 1]'
x5=linsolve(q5,c5)
```

The concept of Nash Equilibrium can explain many things the way they happen in everyday life. For example, multiple gas stations are located on the same street intersection, fast food chains are in nearby vicinity, and so are different business stores located in the same shopping center. The question to ask is why these businesses are not spreading out but rather flock together to some extent. The following example from the Hotelling provides a mathematical explanation.

Example 2: The Hotelling Model (1973)

To simplify the description of this model, let's pretend that the street is the interval [0,1], and the customers or consumers are distributed uniformly along [0,1]. Assume that there are two shops (sellers) A and B located at A=a and B= b. Hers are the input parameters to the model.

c = transportation cost by unit of distance to either seller

p_A, p_B = the price of a unit good at A and B respectively sells

Let q_A, q_B be the number of units buying the goods from A and B, respectively. Since the customers are uniformly along [0,1], they are represented by segments in the unit interval buying the good from A and B, respectively.

Let P_A, P_B be the profit of seller A and B respectively.

Thus, we see that $P_A = q_A\, p_A$ and $P_B = q_B p_B$ and note that p_A , p_B and q_A, q_B are all functions in a and b, that is,

$p_A = p_A\,(a,b)$, $q_A = q_A\,(a,b)$, $p_B = p_B\,(a,b)$, $q_B = q_B\,(a,b)$

The purpose of the Hotelling Model is to find out the locations $a = a^*$ and $b = b^*$ of the shops A and B such that $P_A = q_A\,(a^*, b^*)\, p_A\,(a^*, b^*)$ and

$P_B = q_B(a^*, b^*)p_B(a^*, b^*)$ is maximum, such that the pair (a^*, b^*) is a Nash Equilibrium. It turns out from [1] a^* and b^* is found to be:

$a^* \leq 1/4$ and $b^* \geq 3/4$

This result shows that $a^* = b^* = 1/2$ as stated in Hotelling's original paper [3] is correct only in the special situation when both p_A and p_B are 0.

Instead of repeating the analysis in [1] and repeating the tedious algebra manipulations here, we outline the procedure and emphasize the application of the Nash Equilibrium concept below.

First, we need to find out the location of the indifferent consumer. There are various possible locations of x, such as x in different intervals [0, a], (a, b), [b,1], for example, when x is in (a,b), then

$$p_A + c(x-a) = p_B + c(b-x) \text{ and so } x = \frac{p_B - p_A}{2c} + \frac{a+b}{2} \tag{E1}$$

and the various possible ranges of $| p_B - p_A |$ such as $| p_B - p_A | < c(b-a)$,we find the following:

$$P_A(a,b) = q_A\, p_A = \frac{p_A p_B - p_A^2 + cp_A(a+b)}{2c} \tag{E2}$$

$$P_B(a,b) = q_B\, p_B = p_B - \frac{p_A p_B - p_B^2 + cp_B(a+b)}{2c} \tag{E3}$$

Two Nash Equilibrium processes in taking partial derivatives are involved here. First, we take the partial derivatives (E2) with respect to p_A and (E3) with respect to p_B, setting both derivatives to 0, and solve for p_A and p_B from both equations in terms of variables a, b. We get,

$$p_A = p_A(a, b) \text{ and } p_B = p_B(a, b) \tag{E4}$$

Next, we substitute (E4) back to (E2) and (E3), and we obtain the expression of P_A and P_B in functions of both a and b alone. Now, by taking partial derivative of $P_A(a,b)$ with respect to a, and taking the partial derivative of $P_B(a,b)$ with respect to b, setting both derivatives to 0 to solve for results of a* and b* stated before. Note that (a*, b*) is now a Nash Equilibrium pair when the payoffs for A and B are $P_A(a,b)$ and $P_B(a,b)$, respectively.

Example 3: The Median Voter Theorem

The Median Voter Theorem is a key result in voting theory. Without going into the details of voting theory and the many applications of politicians using this theorem as a tool to analyze their policies, we state and prove the theorem and show that it is a result of the concept of Nash Equilibrium.

Analogous to the [0,1] interval and shops A and B in the Hotelling Model, here in the Median Voter Theorem, we assume that we have a single-dimensional policy space P, and we have two candidates, C1 and C2. We assume that each policy p in P can be represented as points in a line where the ordering relation <,=,> is defined. And assume that each voter has a so-called *single-peaked preference* over the policy space P. By a single-peaked preference, we mean that the voter has utility function u such that u( ) has a maximum (peak point) at some point p on the line and slopes away from the maximum(peak) on either side. For the policies in P that C1 and C2 commits, each voter can report the peak p_i from u( ). Each of these p_i is a point in P, Now sort these values $\{p_i\}$ in ascending order and let their median be denoted by m.

$$m = \text{median } (p_1, \ldots p_n) \qquad \text{(E5)}$$

Under the above circumstances, we state and prove the following well-known theorem:

Median Voter Theorem:

Assume winning is by majority rule, the voters have single-peaked preference, and the candidates C1 and C2 only care about winning and simultaneously announce to implement the policy p1 and p2 respectively as their strategy pair. The Median Voter Theorem states that: if (p1, p2) is in Nash Equilibrium, then it is necessary that p1 = p2 = m.

Proof: We prove by contradiction.

Suppose the above condition p1 = p2 = m is false, then without loss of generality, we can assume p1 > p2 (p1 has more votes than p2) and that p1 < m. We are going to prove that (p1, p2) cannot be in Nash Equilibrium.

Now p2 will deviate because p1> p2, p2= p1 + d for some small d, such that p2<m. Because of the single-peaked preference assumption, all voters with their peak points in the interval [p2, ∞] prefer p2 to p1. Since p2 < m, p2 is more than half of the voters. So p2 would win, and thus p2 would prefer to deviate. So, it is not an equilibrium for p1 to win with p1< m. Thus, the only strategy pair (p1,p2) that is in Nash Equilibrium is when p1=p2= m.

4.2 Static Games with Incomplete Information

Static games with incomplete information are also called Bayesian games. In a game of incomplete information, at least one player is uncertain about another player's payoff function. This contrasts with a game of complete information in which the players' payoff functions are common knowledge. In static Bayesian Games, each player is associated with different types, and each player knows his or her payoff function but may be uncertain about the other players' payoff functions. In Chapter 5, we will study games of incomplete information in great detail. In this chapter, however, we shall introduce the reader to some basic concepts and definitions on the representation of this type of game. We shall show the concept of Bayesian Nash Equilibrium below.

Normal-Form Representation of Static games with incomplete information (Static Bayesian game)

Denote a n-player game of incomplete information by $G = [S_{1,} S_{2...} S_{n;}\ u_{1,} u_{2, ...} u_{n}]$

, where S_i is player i's strategy space and $u_i(s_1, \ldots, s_n; t_i)$ is player i's payoff when the players choose the strategies $(s_1, \ldots, s_n)$, where t_i is one of player i's type that belongs to a set T_i of all the player i's possible types. If $t_a, t_b \in T_i$, then the payoffs $u_i(s_1, \ldots, s_n; t_a)$ and $u_i(s_1, \ldots, s_n; t_b)$ may be different.

As player i in a Static Bayesian game is associated with the type space T_i, to say that player i is uncertain about the other players' payoff functions is equivalent to saying that player i is uncertain about the types of the other players. Likewise, saying that player i knows his or her own payoff function is equivalent to saying the player i knows his or her own type.

Let $t = (t_1, t_2 \ldots t_n)$ be the type vector for each player. Let t_{-i} be the type vector $(t_1, t_2, \ldots, t_{i-1}, t_{i+1} \ldots t_n)$, and let T_{-i} be the set of all possible values of t_{-i}. Assume that the prior probability distribution p(t) is given and when $p(t_i)$ is known, then the degree of beliefs, $p_i(t_{-i} | t_i)$ can be computed using the Bayes' Theorem below,

$$p_i(t_{-i} \mid t_i) = \frac{p(t_{-i}, t_i)}{p(t_i)} = \frac{p(t_{-i}, t_i)}{\sum\limits_{t_{-i} \in T_{-i}} p(t_{-i}, t_i)} \qquad \text{(E6)}$$

(E6) above is player i's belief $p_i(t_{-i} \mid t_i)$ that describes i's uncertainty about the n-1 other players' possible types, t_{-i} given player i's own type, t_i

Now we have all the ingredients for a formal definition of a static Bayesian game.

Definition. (Representation of static Bayesian game) A static Bayesian game is represented by $G = \{A_1, A_2, \ldots A_n; T_1, \ldots T_n; p_1, p_2, \ldots, p_n; u_1, u_2 \ldots u_n\}$, where $A_1, A_2, \ldots, A_n$ are the action spaces, $T_1, \ldots T_n$ are the type spaces, $p_1, p_2, \ldots, p_n$ are the beliefs, $u_1, u_2 \ldots$un are the payoff functions.

Definition. (Pure Strategy in static Bayesian game)

A strategy for player i is a function $s_i : T_i \xrightarrow{s_i} A_i$ that assigns each type t_i in T_i, an action $a_i = s_i(t_i)$ A specifies an action a_i from the action set A_i.

Now we can define the Nash Equilibrium in the static Bayesian game

Definition. (pure-Strategy Nash Equilibrium in static Bayesian game)

Let $G = \{A_1, A_2, \ldots A_n; T_1, \ldots T_n; p_1, p_2 \ldots, p_n; u_1, u_2 \ldots u_n\}$ be a static Bayesian Game, the strategies $s^* = (s_1^*, s_2^* \ldots s_n^*)$ is a pure-strategy Bayesian Nash Equilibrium if for each player i and for each i's types t_i in T_i, each of the $s_i^*(t_i)$ is a solution to the following maximization problem (E7) when the rest of the strategies

$s_1^*(t_1),\ldots,s_{i-1}^*(t_{i-1}),s_{i+1}^*(t_{i+1}),\ldots,s_n^*(t_n)$ are holding fixed. Since a strategy for player i is a function $s_i : T_i \xrightarrow{s_i} A_i$, this means that no player wants to change his or her strategy even if the change involves only one action by one type.

$$\max_{a_i \in A_i} \sum_{t_{-i} \in T_{-i}} u_i(s_1^*(t_1),\ldots,s_{i-1}^*(t_{i-1}),a_i,s_{i+1}^*(t_{i+1}),\ldots,s_n^*(t_n);t_{-i})p_i(t_{-i} \mid t_i) \tag{E7}$$

For each player i, the factor $p_i(t_{-i} \mid t_i)$ can be thought of as the weighting factors of the payoff function $u_i\ (s_1, \ldots, s_n; t)$ for player i whose type is $t = t_{-i}$.

Proof of the existence of the above Bayesian Nash Equilibrium $s^* = (s_1^*, s_2^* \ldots s_n^*)$ is very similar to that of the static game with complete information shown in Chapter 3 before. We will see many examples and applications on the existence of Bayesian Nash Equilibrium in Chapter 5, but now, we need to focus on the understanding of the optimization process for each of the n players shown in (E7).

Example 4: An application of the Nash Equilibrium Concept in Economic

To facilitate the understanding of the concept of " type of players" in static Bayesian game, consider the same example of the Cournot Duopolistic Competition Model described in section 3.6, this time, however, this time, we let the values of the cost parameter c_1 be a single type of company A (player 1), and let c_{2L}, c_{2M}, c_{2H} be the types for company B (player 2), then the notations shown in (E6) and (E7) will be as follows:

$T_1 = \{c_1\}$ and $T_2 = \{c_{2L}, c_{2M}, c_{2H}\}$ as type spaces.

There will still be one payoff function for player 1, but three payoff functions for player 2, one for each - type.

$$P_1(q_1, q_2 ; c_1) = [Q_0 - c_1]\, q_1 - q_1q_2 - q_1^2 \tag{E8}$$

$$P_2(q_1, q_2 ; c_{2L}) = [Q_0 - c_{2L}]\, q_2 - q_1q_2 - q_2^2$$

$$P_1(q_1, q_2 ; c_{2M}) = [Q_0 - c_{2M}]\, q_2 - q_1q_2 - q_2^2$$

$$P_1(q_1, q_2 ; c_{2H}) = [Q_0 - c_1]\, q_2 - q_1q_2 - q_2^2$$

In determining the Bayesian Nash Equilibrium (q_1^*, q_2^*) defined in (E7), we will employ the payoff functions with the respective types shown above (E8) accordingly. (See details from examples in Chapter 5)

4.3 Dynamic Games with complete Information

Dynamic games with complete information are games in which players' payoff functions are common knowledge, but players choose their strategies sequentially rather than simultaneously. Much of the foundational work on dynamic games is attributed to Reinhardt Selton, a Nobel Prize-winning game theorist.

In any dynamic game, having complete and perfect information means that, at each move, the player making a decision knows the full history of the game up to that point. In contrast complete but imperfect information means that, at some stage in the game, a player does not have access the full history.

In Chapter 6, we will study Dynamic games with complete information in detail. In this chapter, we focus on the fundamental concepts and key issues of this type of game.

One central issue in dynamic games is the credibility of a player's claimed actions during the sequential progression of the game. Since the game unfolds in steps, a natural way to represent its structure is through a game tree—a tree-like diagram that captures the sequential nature of the player's choices. The Nash Equilibrium concept, which we explored in static games with complete information, extends to this game tree structure with appropriate modifications (see details in Chapter 6).

In general, a powerful method for determining Nash Equilibrium strategies in a game tree is the so-called *Backwards Induction process*. In the example below, we illustrate Backwards induction to determine the Nash Equilibrium in a game tree. We also provide a formal definition of Nash Equilibrium in this context and examine the credibility of certain threat strategies that a player might claim..

Example 5: An example used to illustrate subgame perfect Nash Equilibrium, backwards induction, and credibility of threat strategies.

Given the following 2- players game matrix,

Figure 4.1

1 \ 2	X	Y
A	5, 3	0, 0
B	4, 4	4, 4

Figure 4.2

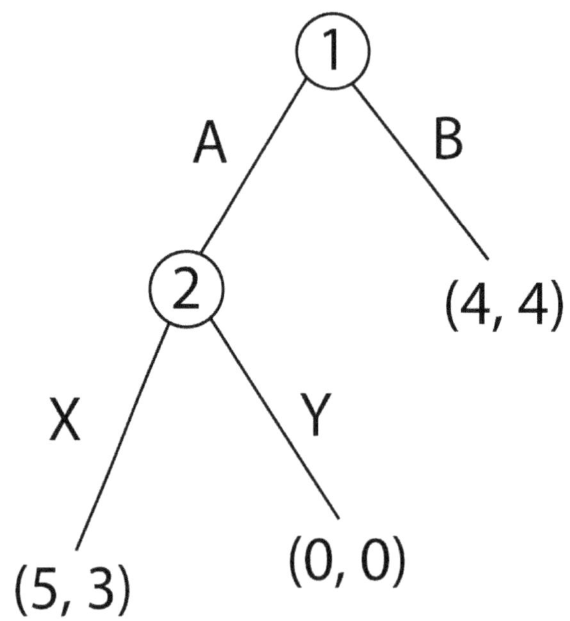

Figure 4.1 above is the usual matrix representation and Figure 4.2 is the dynamic tree representation of the same game. Let's first review all the Nash Equilibriums we have for this game so far. From the definition of static Nash

Equilibrium, we can directly verify that (A,X), (B,Y) are pure Nash Equilibriums from the given game matrix (see the discussion of pure Nash Equilibrium in Section 3.2, and using the method outlined in Examples 6 and 7 from Section 3.2, we can find the Nash Equilibrium with mixed strategies as follows:

Let p and (1-p) be the probability of playing actions A and B respectively. Let q and (1-q) be the probability of playing actions X and Y respectively.

Setting the expected- values of X and Y are equal when player 1 is playing A and B with respective probabilities p and (1-p), we have:

$5p + 4(1-p) = 0p + 4(1-p)$, which gives us $p=0$, or $1-p=1$, which means that player 1 will always play strategy B. Similarly, setting the expected -values of A and B are equal when player 2 is playing X and Y with respective probabilities q and (1-q), we have:

$$5q + 0(1-q) = 4q + 4(1-q).$$

Since in player 2's payoff matrix $\begin{pmatrix} 3 & 0 \\ 4 & 4 \end{pmatrix}$, the second row dominates the first row, the expected value for B is larger than the expected value for A, and so

$5q + 0(1-q) \leq 4q + 4(1-q)$ instead, so $q \leq 4/5$, and so the pair (B, X) where player 1 always selects B, and player 2 plays X with any probability that less than or equal to 4/5 gives an infinitely many mixed Nash Equilibrium as stated.

Amongst the three types of Nash Equilibriums as found, Selton introduced a new kind of Nash Equilibrium called Subgame Perfect Nash Equilibrium (SPE) that satisfies one additional condition: The usual Nash Equilibrium must be also a *Nash Equilibrium in every subtree of the game tree*. In Figure 2, the game tree has its root at node 1, and there is a subtree with its root at node 2. Let us take a closer look at each of the three Nash Equilibriums and check which of these three meet the stated condition.

(a) The pair (A,X) is a SPE because if player 1 chooses A, then player 2 would choose X rather than Y because 3 is better than 0. Now suppose player 2 chooses X, then player 1 would not want to deviate from A to B, because if player 1 chooses A, and player 2 stays with X, player 1 would get 5, whereas if player 1 deviates to B, he would only get 4. Thus (A,X) is a SPE.

(b) The pair (B,Y) is not a SPE because although (B,Y) is a Nash Equilibrium for the entire tree rooted at node 1, it is not a Nash Equilibrium in the subtree rooted at node 2. Suppose player 2 chooses Y, then player 1 would never choose A but to choose B because that would reduce his (her) payoff from 4 to 0. Now suppose player 2 chooses Y, then player 1 would choose B rather than A, because player 1 choosing A would give him (her) 0 and choosing B would give him (her) 4, so player 1 would choose B. Now look at the subtree rooted at node 2, player 2 playing Y gives he(she) 0, but playing X would give he(she) 5, so player 2 playing Y is not the best strategy for player 2, and thus Y cannot be a Nash Equilibrium in the subtree rooted at node 2 (player 2). Thus, (B,Y) is not a SPE.

(c) The mixed strategy Nash Equilibrium pair (B,X) , where player 1 plays B all the time, and player 2 plays X with probability less than or equal to 4/5, cannot be a SPE. This is because if player 1 changes B to A, player 2's would not hold on to its mixed strategy of playing X with probability less than or equal to 4/5, player 2's best response is to play the pure strategy X alone, which gives player 1 the payoff of 5 and player 2 the payoff of 3. Hence player 1 is willing to deviate from B to A as its payoff value increases from 4 to 5. So, the mixed strategy for player 2 is not the best response in the subtree rooted at node 2.

Now that we have verified that the pair (A, X) is a subtree perfect Nash Equilibrium (SPE) from the list of Nash Equilibriums (a), (b), (c) above. Next, we describe the process of Backwards Induction to find this SPE, (A, X) directly from the game tree shown in Figure 2. Starting from the subtree rooted at node 2, select the best strategy for player 2, which is strategy X, because choosing X gives the payoff of 3 and choosing Y gives the payoff of 0. Now we move backwards up from the subtree rooted at node 2 to the tree rooted at node 1. At node 1, player 1 will choose A rather than B. Because choosing B, player 1 will get the payoff value of 4. If player 1 chooses A, since player 2 chooses X, the payoff values for both players are (5,3) , and 5 is better than 4, and so player 2 would choose to play A. This entire backwards process gives the strategy pair (A, X) the SPE in this game. We will come back to more details on the Backwards Induction process in Chapter 6.

Finally, let us look at the creditability of threat strategies in this game. The strategy Y can be a threat strategy of player 2 to player 1 because when player 2 plays Y, it will cause player 1 a significant loss of 0 including player 2 himself (herself). However, if there is a better choice of strategy for player 2 when it is his (hers) turn to move, he (she) will not execute the threat strategy Y. In the game tree Figure 2 above, even through player 1 claims that he(she) can take the strategy Y against player 1, player 1 knows that player 2 's claim is not creditable because when it is player 2's turn to move, player 2 has a better strategy X to choose rather than execute the threat strategy Y. Thus, when player 1 is choosing strategy A , he(she) knows that player 2's claim of executing Y is not creditable. In the next example, we will show how to make threat strategies creditable.

Example 6 An example used to illustrate subgame perfect Nash Equilibrium (SPE), backwards induction, and credibility of threat strategies.

This example is the classic example of Buring Bridges due to Thomas Schelling [ ]described as follows. The game deals with two countries, 1 and 2 that are at war. A small island is located between the two countries and connected to each country by a bridge, as shown in Figure 3 above. For either country to initiate an invasion over the other, it must first cross its bridge to occupy and control the island. Afterward, the soldiers must decide whether to burn the bridge behind them. The other country decides whether to attack or concede to the country whose soldiers are at the island. If the first country has no bridge to use as an escape route, it must fight a battle, but if the bridge is not burned, the first country can decide whether to fight or retreat.

Figure 4.3 above shows a game tree that represents the following scenario: Country 1 has initiated an invasion and moved the soldiers across the bridge to the island. Country 2 then can either choose attack or concede with country

1. If country 1's bridge is burned, then this action of player 1 to burn the bridge makes the threat of player 1 to control the island and fight with player 2 creditable. And player 2 must respond to player 1 with either attack or concede and end the game. If, however, country 1's bridge is not burned, country 2 can respond with either attach or concede. If player 2 choose to concede, the game ends. If player 2 chooses to attack, country 1 can choose to fight or to retreat. We are going to find a Subgame Perfect Nash Equilibrium (SPE) using the process of backwards induction described below.

In Figure 4.3, we start with the subtree rooted at node 1, this subtree is shown inside Box A with dotted lines. As it is 1's turn to choose the actions, 1 would choose retreat as it gives value 2 rather than 1 by choosing fight. Moving up the subtree at Box A to a larger subtree rooted at 2 inside Box B with dotted lines. As 2 is making the decision, if 2 choose to concede, 2's payoff is1, if 2 choose attack, since 1 already chose retreat, so 2's payoff is 3. As 3 is better than 2, so 2 choose to attack in Box B. Next consider the subtree rooted at 2 shown in Box C, as 2 is making the decision, 2 would choose to concede as it would give a value of 2 as opposed to getting a value of 1 when 2 chooses attack. Finally, we look at the entire tree rooted at node 1. Since 1 is making the decision, 1 choose to burn gives the value of 3, and 1 choose not-burn gives the value of 2. Thus, 1 chooses to burn. Thus, the strategies chosen for the SPE are for player 1: burn , retreat and for player 2: concede, attack. In notation, we denote the SPE by < (burn, retreat), (concede, attack)>. See the strategies with the checked marks shown in Figure 4.3

Figure 4.3

Schelling game of burning bridges

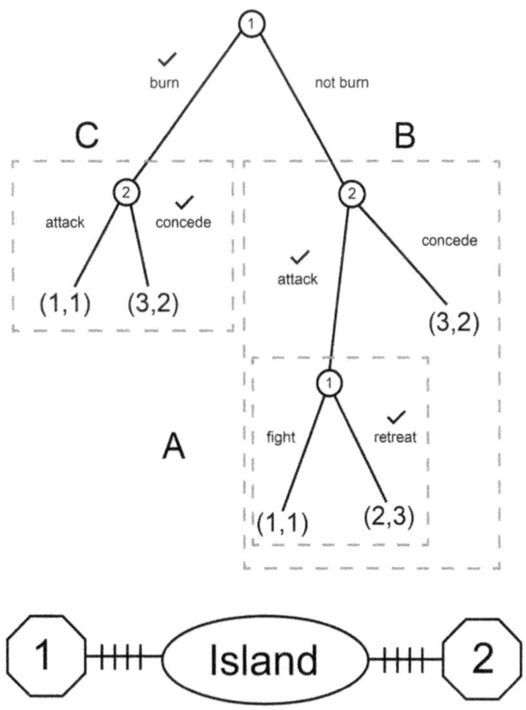

4.4 Dynamic Games with Incomplete Information

Dynamic Games with Incomplete Information is the class of games where the players are interacting sequentially (dynamic games) , and incomplete information in this context could mean there is at least one player when it is his (her) turn to move, he(she) does not know the complete history of the move, that is to say, he (she) does not know which action his (her) previous player has taken; or, incomplete information could also mean that there is at least one player in the game that has more than one type and the different types of that player is described by a probability distribution.

For this class of games as stated, we introduce yet another new equilibrium concept called the *perfect Bayesian equilibrium*, and we will discuss this new equilibrium concept, mostly due to Harsanyi (1967), in detail in Chapter 7. In this section, however, we will give an informal definition of perfect Bayesian equilibrium and give examples to show its application.

Before doing so, let us recall that there are four equilibrium concepts. Nash Equilibrium in static games of complete information is the most fundamental concept we already discussed in detail in Chapters 3, Bayesian Nash Equilibrium in static games of incomplete information to be discussed in detail in Chapter 5, subgame - perfect Nash Equilibrium in dynamic games of complete information in Chapter 6, perfect Bayesian equilibrium in dynamic games of incomplete information in Chapter 7. This chapter, Chapter 4 serves as an overview of the four equilibrium concepts each address to the fundamental problem in the four types of games, it shows the fundamental Nash equilibrium concept compared to that of the other three equilibrium concepts, and it shows the conceptual difference between the Bayesian Nash Equilibrium concept in static and dynamic games. We also recall that from previous writings, a fundamental task in game theory is to analyze threat strategies from the players and players to make their incredible strategies credible. The original Nash equilibrium concept does not rule out incredible tasks, but the strengthened subgame perfect equilibrium does. As we shall see in Chapter 7, the new concept of perfect Bayesian equilibrium concept informally introduced in this chapter successfully excludes or eliminates the implausible equilibrium with incredible tasks which we are not able to identify using the subgame perfect or Bayesian Nash equilibrium concept alone. We will give two examples below to demonstrate our claims. But before we proceed, we need to have an informal notion and definition of what perfect Bayesian equilibrium is.

Definition 4.4.1 An informal definition of perfect Bayesian equilibrium

In a game tree, a perfect Bayesian equilibrium consists of strategies and beliefs (probabilities assigned to each tree node) satisfying the following 4 requirements: Refer to the following Figure 4.4 for the explanation of these four requirements.

Example 7 An example used to illustrate the 4 requirements for subgame perfect equilibrium.

Figure 4.4

Perfect Bayesian Equilibrium

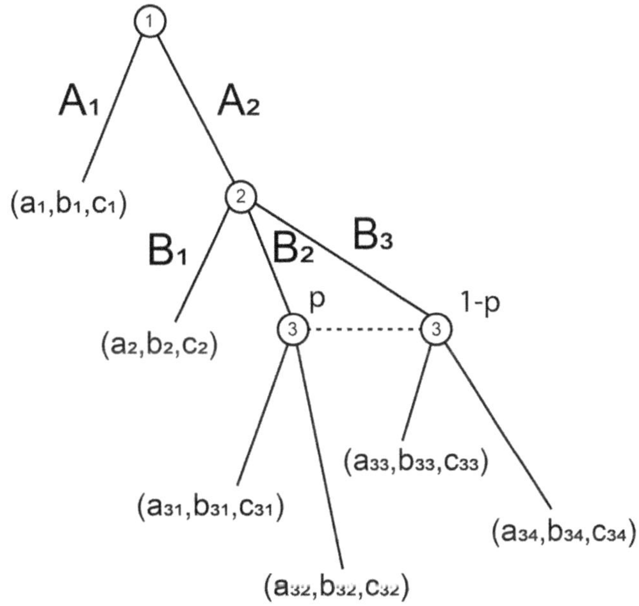

The figure above shows a three-player game. Let's assume that we start with the game that involves only two players 2 and 3 with payoff values (b_j, c_k), with strategies B_1, B_2, B_3 for player 2 and strategies C_1, C_2 for player 3. Then the third player (player 1 with strategies A_1, A_2 and payoff a_i) joins the game and makes it a three-player game.

Requirement 1:

Each player in the game tree with the move must have a belief (probability) about which of his (her) previous node is. The belief is 1 if the player is certain above his previous node, and a probability of p if not certain.

In the three-stage game above, node 1 and node 2 each have a belief of 1 and node 3 has a belief of p as shown, because player 3 is uncertain if he (she) is coming from player 2 by taking strategy B_2, or by taking strategy B_3.

Requirement 2:

Given each player's belief, the player with the move must choose his (her) subsequent strategy rationally.

In the game shown above, player 3 has the move, he (she) must choose either C_1 or C_2 rationally. Rationally means, he (she) must choose either C_1 or C_2 with the larger expected payoff. In the figure above, the expected payoff of C_1 denoted by $EV(C_1) = p\,c_{31} + (1\text{-}p)\,c_{33}$ and $EV(C_2) = p\,c_{32} + (1\text{-}p)\,c_{34}$

Requirement 3:

The belief of any node on the equilibrium path is determined by the Bayes' rule and the players' equilibrium strategies.

In the game shown above, suppose the equilibrium path is $A_2\,B_2C_2$ in the tree, then p must be 1 because B_2 is chosen. Since C_2 was chosen by player 3, then $EV(C_2) \geq EV(C_1)$, which means $c_{32} \geq c_{31}$ in the figure because p is 1 in this case.

If player 2 were to choose a mixed strategy, playing B_1 with probability r, playing B_2 with probability s, and playing B_3 with probability t, by Bayes Theorem, we should belief that the probability of node 3 coming from node 2 by taking the action B_2 is:

$$p = P(B_2 \mid 3)$$

$$= \frac{P(3 \mid B_2)P(B_2)}{P(3 \mid B_1)P(B_1) + P(3 \mid B_2)P(B_2) + P(3 \mid B_3)P(B_3)}$$

$$= \frac{1s}{0r + 1s + 1t}$$

$$= \frac{s}{s+t}$$

Requirement 4:

If a node is off the equilibrium path, the belief of that node must be determined by the Bayes' rule and by the players' equilibrium strategies where possible.

But sometimes there are scenarios where the belief of a node off the equilibrium path cannot be determined pending on the equilibrium strategies. Again, referring to Figure 1, we use the three-player game above to illustrate this scenario as an example.

Suppose for proper choices of the payoff values (a_i, b_j, c_k), the strategies $A_1B_1C_1$ is a Nash equilibrium, then node 3 is off the equilibrium path, and Requirement 4 above may not determine node 3's belief from node 2's strategy B_1, that is to say, p is undetermined. But if node 2's strategy is to play a mixed strategy such that B_1 with probability r, B_2 with probability s , B_3 with probability t, then Requirement 4 would force the belief of node 3 by $p = \frac{s}{s+t}$ as explained previously.

Now that we know the 4 requirements for a perfect Bayesian equilibrium, we may informally think of perfect Bayesian equilibrium as a set of strategy one for each player, that are Nash Equilibrium, subgame -perfect Nash equilibrium in the game tree, together with a set of beliefs satisfying four requirements as described before. The following two examples will exemplify more on this new concept.

Example 8 A dynamic game of complete but imperfect information

Figure 4.5

A Complete But Imperfect Game

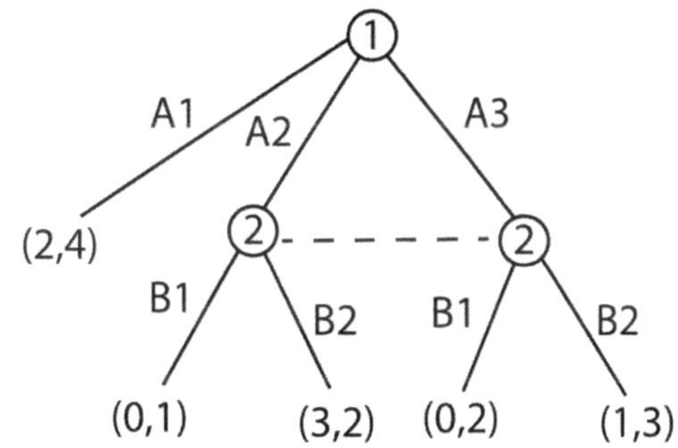

(Extensive Form Representation)

①		B1 ②	B2
	A1	(2,4)	(2,4)
	A2	(0,1)	(3,2)
	A3	(0,2)	(1,3)

(Matrix or Normal Form Representaion)

Figure 4.5 above shows an example of a two persons game with complete but imperfect information, complete because the playoffs are deterministic, and imperfect because player 2 does not know for sure if he (she) is coming from his (her) previous node 1 through strategy A_2 or A_3. From the matrix representation of the game, we see that there are two Nash equilibriums, A_1B_1 and A_2B_2. From the tree or extensive form representation, we note that player 2 has a non-credible threat strategy B_1 to player 1. B_1 is non-credible because player 2 has a better strategy B_2 to choose when it is his (her) turn to move. Note that the original Nash equilibrium A_1B_1 includes this threat strategy B_1 which is an undesirable feature in an equilibrium strategy. Note also that there is only one subgame in the entire game in Figure 1, namely the whole game itself, so the Nash equilibriums, A_1B_1 and A_2B_2 are also subgame -perfect equilibriums in our case. Now we want to distinguish which of the subgame-equilibriums, A_1B_1 or A_2B_2 can be further qualified to be a perfect7Bayesian equilibrium. Let p be the belief of 2 coming from 1 through A_2, and 1-p be the belief of 2 coming from 1 through A_3. By Requirement 2, we need to compare the expected values of B_1 and that of B_2 for 2 to choose the right action.

$$EV(B1) = 1p + 2\,(1\text{-}p) = 2\text{-}p$$

$$EV(B2) = 2p + 3\,(1\text{-}p) = 3\text{-}p$$

As $EV(B_2)$ is always larger than $EV(B_1)$, thus player 2 will always choose B_2 for any value of belief p we made, thus the subgame-perfect equilibrium A_1B_1 cannot be a perfect Bayesian equilibrium as it did not satisfy Requirement 2. As for A_2B_2, we can choose p =1 as node 2 is on the equilibrium path A_2B_2. As there is no other node that is outside the equilibrium path A_2B_2, Requirement 4 is automatically satisfied. Thus, the game has only one perfect Bayesian equilibrium A_2B_2 and it excludes player 2's non-credible threat B_1.

Example 9 A dynamic game of incomplete and imperfect information

Figure 4.6

An incomplete and imperfect information game

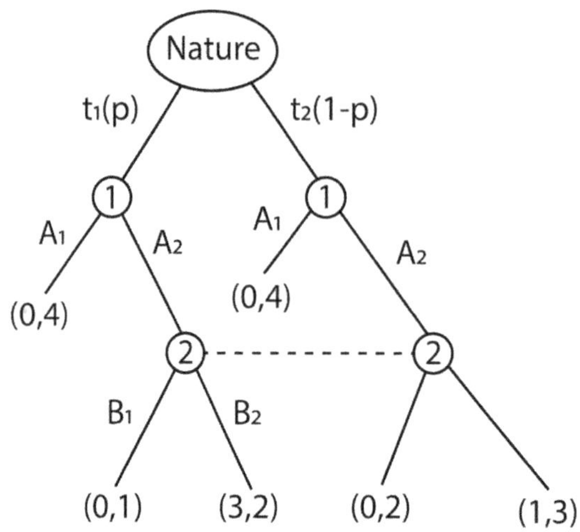

	B_1	B_2
A_1 A_1	(0,4)	(0,4)
A_1 A_2	(0,2+2p)	(1-p,3+p)
A_2 A_1	(0,4-3p)	(3p,4-2p)
A_2 A_2	(0,2-p)	(1+2p,3-p)

matrix representation of the Baysian game above

This game has incomplete information because player 1 has more than one type, type t_1 and type t_2, and the player Nature picks up type t_1 with probability p, and type t_2 with probability 1-p; player 2 has only one type, however, player 2 is not sure which type of node 1 he (she) came from; hence this game is also a game of imperfect information. The payoff matrix of this game considered as a Bayesian game is shown above. Recalling that the entries in the payoff matrix of a Bayesian game are the expected values of each strategy pair. We now demonstrate the computation of a couple of entries, say entry $(A_1 A_1, B_1)$ and entry $(A_2 A_2, B_2)$ in the table above.

$$EV (A_1 A_1, B_1) = p\, V(A_1, B_1) + (1\text{-}p)\, V(A_1, B_1)$$

$$= p\,(0,4) + (1\text{-}p)\,(0,4)$$

$$= (0,4)$$

$$EV (A_2 A_2, B_2) = p\, V(A_2, B_2) + (1\text{-}p)\, V(A_2, B_2)$$

$$= p\,(3,2) + (1\text{-}p)\,(1,3)$$

$$= (1+2p,\ 3\text{-}p)$$

The rest of the entries are computed similarly. From the payoff matrix above, we can easily check that the entries $(A_1 A_1, B_1)$ and $(A_2 A_2, B_2)$ are Nash equilibriums. $(A_2 A_2, B_2)$ is a Nash equilibrium because its payoff is (1+2p, 3-p), and $1+2p \geq 3p$, $1+2p \geq 1\text{-}p$, and $1+2p \geq 0$, and $3\text{-}p \geq 2\text{-}p$ for any value of p. Like Example 8, there is only one sub-tree in the game tree, namely the entire game tree itself, so both $(A_1 A_1, B_1)$ and $(A_2 A_2, B_2)$ are subgame-perfect equilibria, with $(A_1 A_1, B_1)$ included the incredible threat strategy B_1. If we dictate the belief p at node 2 be 1 per Requirement 1 and Requirement 3, then the equilibrium $(A_1 A_1, B_1)$ cannot satisfy Requirement 2 because at node 2, the expected value of B_2, which is 2, is greater than the expected value of B_1, which is 1. Thus, in this example, only the Nash equilibrium is a subgame-perfect equilibrium.

4.5 Summary

This chapter (Chapter 4) gives an informal introduction to all four equilibrium concepts discussed in this book. These four important equilibrium concepts are static games with complete information, static games with incomplete information, dynamic games with complete information, and dynamic games with incomplete information. As applications of Nash equilibriums in static games on problems we encountered in everyday life, we showed three scenarios on the Problem of the Commons, the Hotelling Model, and the Median Voter Theorem. Details of the other three equilibrium concepts will be discussed in subsequent chapters (Chapters 5, 6, 7). Examples 7,8, and 9 serve as an informal introduction to these concepts.

Chapter 5

Concept of Bayesian Nash Equilibrium

5.0 Introduction

This chapter provides a detailed exposition and a precise mathematical definition of static Bayesian games and the Bayesian Nash Equilibrium (BNE) associated with these games. A fundamental result established in Bayesian game theory is the Revelation Principle, which is formalized in Propositions 5.6.1, 5.6.2, and 5.6.3.

In Section 5.6, we explore vaious applications of this celebrated principle in the decision-making processes of social choice functions across multiple practical scenarios. These include public projects, voting systems, marketing operations, and applications in computer networks, auctions, trading, and more.

By the end of this chapter, readers will gain a clear understanding of how the Revelation Principle shapes strategic decision-making in both economic and computational settings.

5.1 What does it mean by incomplete information in a game?

In a game, if each player knows not only their own payoff value, but also the payoff values of all other players, then the game is said to have complete information. In other words, all players' payoff values are common knowledge among them.

In contrast, a game has incomplete information if at least one player is uncertain about another player's payoff value or if a player knows other players' payoff values only in terms of probability distributions.

5.2 What is a static Bayesian game with n players?

First let us recall from previous chapters, a static (simultaneous) game with n players consists of the following information:

a. A1, A2..., An are the action spaces of each of the n players
b. V1(a1, a2, an), V2(a1, a2, an), ..., V n (a1, a2, ..., an) are the playoff values of the n players 1, 2, ..., n, where the a1, a2...., a n are the actions in the corresponding action spaces A1, A2...A n.
This simply says that each player, say player i's payoff value Vi depends simultaneously not only on the action player i takes but also on the actions the other players each simultaneously takes.

Now a static Bayesian game has two additional levels of complexity than a static game, the first additional level of complexity is that each player i is now associated with a type-space Ti, by that we mean each player i can be of diverse types in the game say type a, b, c, d, ... etc. In symbol, we denote a typical type in Ti by t_i, so this variable ti can take on specific types a, b, c, d...for example. The second additional level of complexity is that we need to know how often the players of these types t1, t2, ..., t n is engaged in the game. And for this purpose, we need to know the probability distribution of these types specified by the probability functions in c* below.

Because each player can come from a different type from its type- space, and the playoff values previously shown in b. above must now replace by:

b.* V1(a1, a2, an; t1), V2(a1, a2, an; t2), ..., V n (a1, a2, ..., an; tn) are the playoff values of the n players 1, 2, ..., n , where the a1, a2...., a n are the actions in the corresponding action spaces A1, A2...A n, and corresponds to the types t1, t2, ..., t n respectively.

c.* For each player 1, 2, ..., n, we need to specify the following n conditional probability functions:

p1 (t2, t3, ..., t n | t1), p2 (t1, t3, ..., t n | t 2), ..., pn (t1, t2...t (n-1) | tn)

Refer back to the definition of games of incomplete information stated earlier, we can say that a Bayesian game is of incomplete information because the playoff function Vi is uncertain as it depends on the type ti which is in turn described by the probability functions above.

A static Bayesian game is an n-person game that satisfies the requirements a, b*, and c* described above. In what follows, we will give examples to demonstrate the above requirements and to show the existence of Bayesian Nash Equilibrium in these games.

5.3 Examples of static Bayesian game

Example 5.3.1

Suppose we have two players, player 1 has two types, say type t_{1A} and type t_{1B}, player 2 has only a single type say t_2. Suppose the playoff values of the two players are show in the two tables below:

For type t_{1A}:

	L	R
T	7,9	3,1
B	9,7	5,1

For type t_{1B}:

	L	R
T	13,5	1,9
B	11,3	-1,9

Using the notations in section 2, we have the action space for player 1, A1= {T, B}, with action T and B, and for player 2, the action space A2= {L, R} with action L and R. Suppose the probability of choosing t_{1A} is p and the probability of choosing t_{1B} is (1-p), then the probability functions of the two players are:

$p1\ (t2 \mid t_{1A}) = 1$, $p1\ (t2 \mid t_{1B}) = 1$, $p2\ (t_{1A} \mid t2) = p$, $p2\ (t_{1B} \mid t2) = 1 - p$

$V1\ (T, L; t_{1A}) = 7$, $V2\ (T, L; t_{1A}) = 9$, $V1\ (B, R; t_{1A}) = 5$, $V2\ (B, R; t_{1A}) = 1$,

$V1\ (T, L; t_{1B}) = 13$, $V2\ (T, L; t_{1B}) = 5$, $V1\ (B, R; t_{1B}) = -1$, $V2\ (B, R; t_{1B}) = 9$, etc.

The above two players, their action spaces A1, A2, their player off functions V1, V2, and their probability functions p1 and p2 all together defines what we call a Bayesian game.

Example 5.3.2

Instead of Player1 having two types as in the last example, we assume Player2 has two types and Player 1 has a single type in this example. Suppose the two types of Player2 are t t_{2A} and type t_{2B}, and player 1 has only the single type t_1. Suppose the playoff values of the two players are show in the two tables below:

For type t_{2A}:

	L	R
T	(0,0)	(0,8)
B	(16,0)	(4,4)

For type t_{2B}:

	L	R
T	(0,0)	(0,0)
B	(16,0)	(4,-24)

Assume the probability that player 1 engages with player 2 of t_{2A} is p and engages with type t_{2B} is (1-p). Per the notations in section 2, we have the action space for player 1, A1= {T, B}, with action T and B, and for player 2, the action space A2= {L, R} with action L and R. The probability functions are

$p1\ (t_{1A|}t_{2)} = p$, $p1(t_{1B}|t_2) = 1-p$, $V1(T, L; t_{2A}) = 1$, $V2(T, L; t_{2A}) = 5$, $V1(B, R; t_{2A}) = 3$, $V2(B, R; t_{2A}) = 3$,

$V1(T, L; t_{2B}) = 1$, $V2(T, L; t_{2B}) = 5$, $V1(B, R; t_{2B}) = 3$, $V2(B, R; t_{2B}) = 3$, etc.

The two players, their action spaces A1, A2, their player off functions V1, V2 pending on their types, and their probability functions p1 and p2, all together define a Bayesian game.

Example 5.3.3

This example differs from Example 5.3.2 in that, instead of a single type, player 2 also has two types, say t_{2A} and t_{2B}. So, when the two players engage in the game, it would involve 4 different playoff values shown below:

For type t_{1A} and t_{2A}:

	L	R
T	(0,0)	(0,0)
B	(16,0)	(4,-24)

For type t_{1B} and t:$_{2A}$:

	L	R
T	(0,0)	(0,16)
B	(16,0)	(-24, 4)

For type t_{1B} and t_{2B}:

	L	R
T	(0,0)	(0,16)
B	(16,0)	(-24,-24)

The playoff value of the two players dependents not only on the two players' actions but also depends on their types. For example,

V1 (B, L; t_{1A}, t_{2A}) = 16, V2 (B, L; t_{1A}, t_{2A}) = 0,

V1 (B, L; t_{1B}, t_{2A}) = 16, V2 (B, L; t_{1B}, t_{2A}) = 0,

V1 (T, R; t_{1B}, t_{2A}) = 0, V2 (T, R; t_{1B}, t_{2A}) = 16,

V1 (B, R; t_{1B}, t_{2B}) = - 24, V2 (B, R; t_{1B}, t_{2B}) = - 24,

We also need to specify how often these different types of players will engage in the game by specifying their probability functions below:

p (t_{1A}, t_{2A}) = 1/4,

p (t_{1A}, t_{2B}) = 1/4

$p(t_{1B}, t_{2A}) = 1/4$

$p(t_{1B}, t_{2B}) = 1/4$

The above two players, their action spaces A1={T, B}, A2 = {L,R}, their playoff functions V1, V2 depending on both the players' actions and types, and their probability functions on the distribution of these types, all together defines what we call a Bayesian game.

Example 5.3.4 Player with infinite types

The number of types of a player need not be a finite number as is the case in the last two examples. Player 1 in Example 5.3.1 has 2 types and Player 2 has only 1 type, and both players have 2 types in Example 5.3. 2. In general, the number of types of a player can be infinite; each type is a value represented by a number within the finite interval [a, b]. In this example, the two players each have their types is a random number uniformly distributed inside a finite interval [0, c], where c is an arbitrary chosen small constant, and the players' playoff values are depending on their types again each of which is a random number in [0, c]. The action space for Player 1 is A1 = {T, B}, and for Player 2 is A2 = {L, R}. The type spaces are T1 = [0, c] for player 1, T2 = [0, c] for Player 2. Let t_1 be a random variable in T1 and t_2 be a random variable in T2. The payoff values for Player 1 and Player 2 are shown in the table below:

	L	R
T	(3 + t1, 2)	(1,1)
B	(0,0)	(2, 3 + t2)

Let the probability functions for t1 and t2 be p1(t2 | t1) = 1/c

and p2(t1 | t2) = 1/c, respectively. The action spaces, the payoff functions, the type spaces, and the probability distribution for the types described before constitutes a Bayesian game.

Example 5.3.5 (First prize auction)

This example shows that the first prize, sealed - bid auction between two bidders is a Bayesian game between the two bidders. Recall that in this kind of auction, each bidder has his own private valuation v of the good and bids at his own price b. The two bidders simultaneously submit their bids. The higher the bidder wins the good and pays the price he bid, the playoff for the winner is the difference between the price he bid (b) and his private valuation of the good (v), namely (v-b). The lower bidder loses, gets, and pays nothing. If both players bid the same price, the winner is determined by flipping a coin, and each player gets the payoff of (v-b)/2.

To formulate this auction problem as a static Bayesian game, we must identify what is the Action Spaces, Type Spaces, the probability functions, and the payoff functions for each bidder.

An action for Player 1 is to submit a non-negative bid b1 based on his valuation v1 where v1 is a random value chosen between [0,1]. So, the Action Space A1= [0, R) and the Type Space T1= [0,1]. Similarly, an action for Player 2 is to submit a non-negative bid b2 based on her valuation v2, a random value chosen between [0,1] so, the Action Space A2= [0, R) and the Type Space T2= [0,1]. Because the valuations V1, v2 from each bidder are independent, to simply matters, we assume in this example that the probability function for Bidder1, p1 (v2 |v1) is a uniformly distributed function on [0,1], and the probability function for Bidder2, p2(v1|v2) is also a uniformly distributed function on [0,1]. Based on the rules of auction described earlier we can now express the playoff function V1 for Bidder 1 and V2 the playoff function V2 for Bidder 2 as follows:

$$V1(b1,b2;v1,v2) = \begin{Bmatrix} v1-b1 & if\ b1>b2 \\ (v1-b1)/2 & if\ b1=b2 \\ 0 & if\ b1<b2 \end{Bmatrix} \tag{E1}$$

$$V2(b1,b2;v1,v2) = \begin{Bmatrix} v2-b2 & if\ b2>b1 \\ (v2-b2)/2 & if\ b2=b1 \\ 0 & if\ b2<b1 \end{Bmatrix} \tag{E2}$$

The information given above {A1, A2, T1, T2, p1(v2|v1), p2(V1|v2), V1, V2} defines a Bayesian game.

Example 5.3.6 (Double Auction)

There are two players, a seller and a buyer in the double auction trading game. The seller wants to sell a good and the buyer wants to buy it at a price that both players agree.

To formulate this trading process as a Bayesian game, let p_s be the price that the seller names and asks, let p_b be the price that the buyer simultaneously names and offers. If $p_b >= p_s$ then both players agree at the trading price $p_t = ½ (p_b + p_s)$; if $p_b < p_s$ then no trade occurs. The seller's price p_s that the seller offers depends on his valuation v_s of the good, and similarly, the buyer's offering price p_b depends also on her valuation v_b. of the good, and these valuations v_b and v_s are private- information to each player respectively, and they are independent and uniformly distributed on the interval [0,1]. If trading occurs, then the payoff value V_b for the buyer is $V_b = v_b - p_t$ and the payoff value V_s for the buyer is $V_s = p_t - v_s$. If there is no trade, then each player's payoff value is zero, respectively. In summary, this is a Bayesian game, the action space for the seller A_s is the set of prices that the seller asks, and the action space for the buyer A_b is the set of prices that the buyer offers, the type spaces T_s and T_b for both players are respectively the interval [0,1], and the probability function for each player is the the uniformly distribution function on [0,1] respectively.

5.4 Formal Definition of the Bayesian Nash Equilibrium (BNE)

Now we know that a Bayesian game G consists of the following information

G= {A1, A2, ……., An; T1, T2, …., Tn; p1, p2, …., pn; V1, V2, …., Vn}

To define the important concept of BNE in the Bayesian game, we need to know what is meant by a strategy for a player.

Definition. A strategy for player i denoted by si is a function from the type- space T_i to the action - space A_i, s_i: $T_i \rightarrow A_i$, so $s_i(t_i)$ is an action in A_i. Because a strategy is a function, it means that, by the definition of what a function is, we must assign every type t_i in the type - space T_i an action a_i in A_i, different types can be assigned to the same action, but any type t_i cannot be assigned to two different actions in A_i.

Now, suppose for each player in this game the strategies are s_1 s_2, ………, s_n, we say that these strategies are in BNE if the following conditions are satisfied for each player. In what follows, we use the shorthand notations t for the type- vector

$(t_1, t_2 \ldots, t_i \ldots, t_n)$, and for the type space T_i of player i, use the notation t_{-I} to include those types in T_i that is different from t_i.

For player 1:

$$\max_{a_1 \in A_1} \sum_{t_{-1} \in T_{-1}} V_1(a_1, s_2(t_2), \ldots\ldots\ldots\ldots, s_n(t_n); t) p_1(t_{-1} | t_1)$$

The above expression says that, when all the strategies for the rest of the players 2 to player n remain fixed, action a_1 is selected amongst all the actions in the action space A_1 so that player 1's expected payoff value V1 is the largest. The summation in the above expression is the expected value of player 1's payoff value V1, and the probability $p_1(t_{-1} | t_1)$ tells us how often the types in t_{-1} could affect player 1's payoff value $V_1(a_1, s_2, \ldots\ldots, s_n; t)$

For player 2:

$$\max_{a_2 \in A_2} \sum_{t_{-2} \in T_{-2}} V_2(s_1(t_1), a_2, \ldots\ldots\ldots\ldots, s_n(t_n); t) p_2(t_{-2} | t_2),$$

Pick a_2 in A_2 so that the expected value of V_2 is the largest when the strategies of players 1, 3, …, n remain fixed

For player i:

$$\max_{a_i \in A_i} \sum_{t_{-i} \in T_{-i}} V_i(s_1(t_1), \ldots, s_{i-1}(t_{i-1}), a_i, s_{i+1}(t_{i+1}), \ldots, s_n(t_n); t) p_i(t_{-i} | t_i)$$

Pick a_i in A_i so that the expected value of V_i is the largest when the strategies of the players 1, 2, …, i-1, i+1, …, n remain fixed.

For player n:

$$\max_{a_n \in A_n} \sum_{t_{-n} \in T_{-n}} V_n(s_1(t_1), s_2(t_2), \ldots\ldots\ldots\ldots, s_{n-1}(t_{n-1}), a_n; t) p_n(t_{-n} | t_n)$$

Pick a_n in A_n so that the expected value of V_n is the largest when the strategies of player 1, 2, 3, ..., (n-1) remain fixed.

A set of strategies s_1 s_2, ………, s_n satisfies the n conditions above is said to be in BNE. In BNE no player wants to change his or her strategy because they already selected the best strategy that gives them the maximum expected payoff.

If we define

$$E_{t_{-i}}[V_i(a_i, s_{-i}(t_{-i}); t_i) | t_i] = \sum_{t_{-i} \in T_{-i}} V_i(s_1(t_1), \ldots, s_{i-1}(t_{i-1}), a_i, s_{i+1}(t_{i+1}), \ldots, s_n(t_n); t) p_i(t_{-i} | t_i) \quad \text{(D1)}$$

$$E_{t_{-i}}[V_i(s_i(t_i), s_{-i}(t_{-i}); t_i) | t_i] = \sum_{t_{-i} \in T_{-i}} V_i(s_1(t_1), .., s_{i-1}(t_{i-1}), s_i(t_i), s_{i+1}(t_{i+1}), .., s_n(t_n); t) p_i(t_{-i} | t_i) \quad \text{(D2)}$$

Then the conditions for a set of strategies s1 s2, ………, sn in BNE can be restated as follows:

$$E_{t_{-i}}[V_i(s_i(t_i), s_{-i}(t_{-i}); t_i) | t_i] >= E_{t_{-i}}[V_i(a_i, s_{-i}(t_{-i}); t_i) | t_i] \; for\ all\ a_i\ in\ A_i \quad \text{(D3)}$$

In Section 5.5 that follows, we will show how to find the BNE for each of the examples of static Bayesian games described in Section 5.3

But does such a set of strategies s_1 s_2, ………, s_n always exist? The answer is yes! We state without proof the following theorem:

Theorem 5.4: In any static Bayesian game, there exists at least one BNE in pure strategies or mixed strategies.

The proof is like that of the Existence of Nash Equilibrium in Section 3.3 in Chapter 3.

5.5 Examples and applications of BNE

Example 5.5.7

To show how to find the BNE step by step, first recall that there is only 1 type t_2 for player 2, there are two types t_{1A} and t_{1B} for Player 1, and the payoff values for both players are:

For type t_{1A}:

	L	R
T	7,9	3,1
B	9,7	5,1

For type t1B:

	L	R
T	13,5	1,9
B	11,3	-1,9

The action sets are $A_1 = \{T, B\}, A_2 = \{L, R\}$

For Player 1, there would be 4 strategies because as mentioned before, a strategy is a function s: T1➔ A1 and there are 4 functions from a set of 2 objects to another set of 2 objects. Call these 4 strategies $s_1^1, s_1^2, s_1^3, s_1^4$ defined as follows:

$s_1^1 (t_{1A}) = T, s_1^1 (t_{1B}) = T,$

$s_1^2 (t_{1A}) = T, s_1^2 (t_{1B}) = B,$

$s_1^3 (t_{1A}) = B, s_1^3 (t_{1B}) = T,$

$s_1^4 (t_{1A}) = B, s_1^4 (t_{1B}) = B,$

For Player 2, the strategies are functions s: $T_2 \rightarrow A_2$, so there are only 2 strategies s_2^1, s_2^2 defined as follows:

$s_2^1 (t_2) = L$, $s_2^2 (t_2) = R$.

Now we can compute the expected values of every possible pair of strategies between Player 1 and Player 2 as shown in the following table:

Player1\Player2	s_2^1	s_2^2
s_1^1	13-6p, 5+4p*	2p+1, 9-8p
s_1^2	11-4p, 3+6p	4p-1, 9-8p
s_1^3	13-4p, 5+2p*	4p+1, 9-8p*
s_1^4	11-2p, 4p+3	-1+6p, 9-8p

We demonstrate how the expected values EV (s_1^1, s_2^1), EV (s_1^1, s_2^1) and EV (s_1^3, s_2^2) are computed in the above table for the * entries, the rest of the entries are computed similarly and is left as an exercise.

EV $(s_1^1, s_2^1) = (7,9)(p) + (13,5)(1-p)$

$= (13-6p, 5+4p)$

EV $(s_1^1, s_2^2) = (3,1)(p) + (1,9)(1-p)$

$= (2p+1, 9-8p)$

EV $(s_1^3, s_2^1) = (9,7)(p) + (13,5)(1-p)$

$= (13-4p, 5+2p)$

EV $(s_1^3, s_2^2) = (5,1)(p) + (1,9)(1-p)$

$= (4p+1, 9-8p)$

Now, to find the Nash Equilibrium of the above payoff table for the two players, we proceed as follows.

For each Player 2's strategy, find Player 1's largest expected value amongst all his strategies, this is same as checking for each column, find the row with the largest first component of each entry in the table. There are two pairs (13-4p,

5+2p) and (4p+1, 9-8p) that satisfies this requirement. The first pair with Player 1's largest expected value 13-4p corresponds to the strategy pair (s_1^3, s_2^1) in column 1 and the second pair with Player 1's largest expected value 4p+1 corresponds to the strategy pair (s_1^3, s_2^2) in column 2. We look at these two pairs separately.

First, the pair (13-4p, 5+2p) can be NE only if Player 2' s expected value, which is 5+2p, is also the largest amongst her other strategies, that is, 5+2p >= 9-8p, which is same as saying p>= 2/5. Thus, the entry (13-4p, 5 + 2p) corresponds to the strategy pair (s_1^3, s_2^1)

Is a BNE if p >=/5. Similarly, the pair (4p+1, 9-8p) can be a NE if 9-8p is the largest amongst her strategies, that is, 9-8p>= 5+2p, which means p<= 2/5. Thus, the strategy pair (s_1^3, s_2^2) is a NE if p<=2/5. When p=2/5, both strategy pairs are NEs with expected values (57/5,29/5) and (13/5,29/5) respectively.

Example 5.5.8

Recall that the payoff values of both players are shown below:

For type t_{2A}:

	L	R
T	1,5	1,5
B	-1,-1	3,3

For type t_{2B}:

	L	R
T	1,5	1,5
B	-1,5	3,3

Player 1 has only 1 type, and player 2 has 2 types, as in Example 1, we have two strategies for Player 1:

$s_1^1(t_1) = T$, $s_1^2(t_1) = B$,

and four strategies for Player 2:

$s_2^1(t_{2A}) = L$, $s_2^1(t_{2B}) = L$,

$s_2^2(t_{2A}) = L$, $s_2^2(t_{2B}) = R$,

$s_2^3 (t_{2A}) = R$, $s_2^3 (t_{2B}) = L$,

$s_2^4 (t_{2A}) = R$, $s_2^4 (t_{2B}) = R$,

Now we can compute the expected values of every possible pair of strategies between Player 1 and Player 2 as shown in the following table:

		s_2^1	s_2^2	s_2^3	s_2^4
	Player 1\Player 2	LL	LR	RL	RR
s_1^1	T	(1,5)	(1,5)*	(1,5)	(1,5)
s_1^2	B	(-1,5-6p)	(3-4p, 3-4p)*	(-1+4p,5-2p)*	(3,3)

We show here how the * entries in the table above are computed, the rest of the table entries are done similarly.

$EV (s_1^1, s_2^2) = (1,5) (p) + (1,5) (1-p)$

$= (p,5p) + (1-p, 5-5p)$

$= (1, 5)$

$EV (s_1^2, s_2^2) = (-1, -1) (p) + (3,3) (1-p)$

$= (3-4p, 3-4p)$

$EV (s_1^2, s_2^3) = (3,3) (p) + (-1,5) (1-p)$

$= (-1 +4p, 5-2p)$

If we pick p= 1/2, the table above reduces to:

Player 1\Player 2	LL	LR	RL	RR
T	(1,5)	(1,5)	(1,5)	(1,5)
B	(-1,2)	(1,1)	(1,4)*	(3,3)

It is easily seen that the entry marked* is a NE

because Player 1 with the expected value 1 is already the best, and Player 2 with the expected value 4 is also the best amongst all her strategies corresponding to the strategy pair $(s_1^{\,2}, s_2^{\,3})$.

If we pick p=1/2, the table above reduces to:

Player 1\Player 2	LL	LR	RL	RR
T	(1,5)	(1,5)	(1,5)*	(1,5)
B	(-1,7/2)	(2,2)	(0,9/2)	(3,3)

Then it is easily seen that the entry (1,5) marked* is a BNE because Player 1 does not want to change his current strategy with expected value 1 to another strategy with expected value 0, and Player 2 does not need to change her strategy with expected value 5 to her other strategies having the same expected values 5.

Example 5.5.9

As there are two different types for each of the two players, so there are 4 possible type to type combinations when these two players are engaging in the games, the playoff values for both players are given for each of the four scenarios.

I. For type t_{1A} and t_{2A}, that is, when Player 1 is of type $t_{1A,}$ and Player 2 is of type t_{2A}:

	L	R
T	(0,0)	(0,16)
B	(16,0)	(4,4)

When player 1 is of type t_{1A} and Player 2 is of type t_{2A}, V(T, R) =(0,16), V(B,R) = (4,4),…etc.

II. For type t_{1A} and t_{2B}:

L R

T	(0,0)	(0,16)
B	(16,0)	(4, -24)

III. For type t_{1B} and t_{2A}:

	L	R
T	(0,0)	(0,16)
B	(16,0)	(-24, 4)

(When player 1 is of type t_{1B} and player 2 is of type t_{2A})

IV. For type t_{1B} and t_{2B}:

	L	R
T	(0,0)	(0,16)
B	(16,0)	(-24, -24)

When player 1 is of type t_{1B} and player is of type t_{2B}, V (T, L)=(0,0) and

V (B, R) = (-24, -24), etc…

We also need to know the probability distribution of these types; in order words, we need to know how often these two players of different types will be engaged in the game. For simplicity, let us assume in this example that the probability distribution of t_{1A}, t_{1B}, t_{2A}, and t_{2B} are shown in the table below:

	t_{2A}	t_{2B}
t_{1A}	1/4	1/4
t_{1B}	1/4	1/4

that is $p(t_{1A}, t_{2A}) = p(t_{1A}, t_{2B}) = p(t_{1B}, t_{2A}) = p(t_{1B}, t_{2B}) = 1/4$.

As shown in the last two examples, each player has 4 strategies, so we will compute the 16 expected values for each of the strategy pair (s_1^i, s_2^j) for i = 1 to 4, and j = 1 to 4 as shown in the table below:

	s_2^1	s_2^2	s_2^3	s_2^4
S_1^1	(0,0)	(0,8)	(0,8)	$(0,16)^*$
S_1^2	(8,0)	(-2,-2)	(-2,5)	(-12,3)
S_1^3	(8,0)	(5,-2)	$(5,5)^*$	(2,3)
S_1^4	(16,0)	(3,-12)	(3,2)	$(-10, -10)^*$

where S_1^1: T_1 -> A1 = {T, B} is the strategy defined by $S_1^1(t_{1A})$ = T and $S_1^1(t_{1B})$ = T. From now on, we compact the strategy notation S_1^1 to mean $S_1^1(T, T; t_{1A,} t_{1B})$, where the first action T from player 1 is of type t_{1A}, and the second action T is of type t_{1B}.

Similarly, S_2^1: T_2-> A2= {L, R} is the strategy defined by $s_2^1(t_{2A})$ = L and by $s_2^1(t_{2B})$ = L, the notation $S_2^1(L, L; t_{2A,} t_{2B})$ means that the first action L from player 2 is of type t_{2A} and the second action L is of type t_{2B}. We summarize the exact meaning of these strategies below. For player 1,

$s_1^1 = s_1^1$ (T, T; $t_{1A,}$ t_{1B}).

$s_1^2 = s_1^2$ (T, B; $t_{1A,}$ t_{1B}).

$s_1^3 = s_1^3$ (B, T; $t_{1A,}$ t_{1B}).

$s_1^4 = s_1^4$ (B, B; $t_{1A,}$ t_{1B}).

and for Player 2,

$s_2^1 = s_2^1$ (L, L; $t_{2A,}$ t_{2B}).

$s_2^2 = s_2^2$ (L, R; $t_{2A,}$ t_{2B}).

$s_2^3 = s_2^3$ (R, L; $t_{2A,}$ t_{2B}).

$s_2^4 = s_2^4$ (R, R; $t_{2A,}$ t_{2B}).

Now we show how the expected values in the above table entries marked * are computed, computation of the expected values for the rest of the table entries are left as an exercise.

$EV (s_1^{\,1}, s_2^{\,4})$

$= EV (s_1^{\,1} (T, T; t_{1A,} t_{1B}), s_2^{\,4} (R, R; t_{2A,} t_{2B}))$

$= EV (T, R; t_{1A}, t_{2A}) + EV (T, R; t_{1A}, t_{2B}) + EV (T, R; t_{1B}, t_{2A}) + EV (T, R; t_{1B}, t_{2B})$

$= p (t_{1A}, t_{2A}) V (T, R; t_{1A}, t_{2A}) + p (t_{1A}, t_{2B}) V (T, R; t_{1A}, t_{2B}) + p (t_{1B}, t_{2A}) V (T, R; t_{1B}, t_{2A}) + p (t_{1B}, t_{2B}) V (T, R; t_{1B}, t_{2B})$

$= ¼ (0,16) + 1/4 (0,16) + 1/4 (0,16) + 1/4 (0,16)$

$= (0,16)$

$EV (s_1^{\,3}, s_2^{\,3})$

$= EV (s_1^{\,3} (B, T; t_{1A,} t_{1B}), s_2^{\,3} (R, L; t_{2A,} t_{2B}))$

$= EV (B, R; t_{1A}, t_{2A}) + EV (B, L; t_{1A}, t_{2B}) + EV (T, R; t_{1B}, t_{2A}) + EV (T, L; t_{1B}, t_{2B})$

$= p (t_{1A}, t_{2A}) V (B, R; t_{1A}, t_{2A}) + p (t_{1A}, t_{2B}) V (B, L; t_{1A}, t_{2B}) + p (t_{1B}, t_{2A}) V (T, R; t_{1B}, t_{2A}) + p (t_{1B}, t_{2B}) V (T, L; t_{1B}, t_{2B})$

$= ¼ (4,4) + 1/4 (16,0) + 1/4 (0,16) + 1/4 (0,0)$

$= (5,5)$

$EV (s_1^{\,4}, s_2^{\,4})$

$= EV (s_1^{\,4} (B, B; t_{1A,} t_{1B}), s_2^{\,4} (R, R; t_{2A,} t_{2B}))$

$= EV (B, R; t_{1A}, t_{2A}) + EV (B, R; t_{1A}, t_{2B}) + EV (B, R; t_{1B}, t_{2A}) + EV (B, R; t_{1B}, t_{2B})$

$= p (t_{1A}, t_{2A}) V (B, R; t_{1A}, t_{2A}) +p (t_{1A}, t_{2B}) V (B, R; t_{1A}, t_{2B}) + p (t_{1B}, t_{2A}) V(B,R; t_{1B}, t_{2A}) + p (t_{1B}, t_{2B}) V (B, R; t_{1B}, t_{2B})$

$= ¼ (4,4) + 1/4 (4, -24) + 1/4(-24,4) + 1/4(-24, -24)$

$= (-10, -10)$

After we have computed the expected values of all the strategy pairs for the two players, it is easily found and verified from the table that the pair (s_1^3, s_2^3) with payoff value (5,5) is a NE and, hence, a BNE for the original problem.

Before we move on to the next three examples, we need to know some basic facts about the expected value of random variables and some properties of a uniform probability distribution on an interval [a, b]. We summarize this information in the following proposition. You can look them up in any elementary statistics text.

Proposition 5.5.1 (Random variable and probability density function)

1. If X, Y are random variables on an interval [a, b], then the expected value E(X) of X is a linear operator, that is, E (aX +bY] =a E(X) +b E(Y), where a, b are any constants
2. If X is a uniform random variable in the interval [a, b] with the probability density function p(x)= 1/R(b-a), then the mean of X, or, the expected value of X, E(X) = (a+b)/2
3. If p(x) is a probability distribution function of a random variable X between the interval (a, b), then the probability that X lies in between a and c, where c is in (a, b), is $P(a < X < c) = \int_{t=a}^{t=c} p(t)dt$

Example 5.5.10 (BNE for players with infinite types)

Recall that in Example 5.3.4, each player has infinite number of types each of which is represented by a random number uniformly distributed inside the interval [0, c], and each player's playoff values depend on his or her type which is a random number in [0, c], where c is an arbitrary chosen small number. The action space for Player 1, A1 = {T, B}, and for Player 2, A2 = {L, R}. The type - space for Player 1 is T1 = [0, c], and for player 2 be T2 = [0, c]. Note that if we let c approach to 0, then each player will have only one single type instead of infinitely many types. This means that we will end up with a regular game of two players, the kind of games that we discussed before in earlier chapters. Let t_1 be a random variable in T1 and t_2 be a random variable in T2. The probability density function for t1 and t2 are both p(t1) = 1/c and p(t2) = 1/c by the Proposition 5.5.1 stated above. Let the payoff values for Player 1 and Player 2 be given in the table below:

	L	R
T	(3+ t1, 2)	(1,1)
B	(0,0)	(2, 3 +t2)

We now show how to construct a strategy pair (s_1, s_2). This is a BNE for the game defined above. Before doing so, let's look at the original single-type game with the playoff matrix shown below:

	L	R
T	(3, 2)	(1,1)
B	(0,0)	(2, 3)

It is easily seen that this game has a pure NE at (T, L) and at (B, R), and it is also having a mixed NE at (T, R) by playing T with probability 3/4 and playing R with probability 3/4. This can be verified as follows.

If T is playing with probability ¾, then the expected value of L, E(L) = 2*3/4 + 0*1/4 = 3/2, and the expected value of R, E(R) = 1*3/4 + 3*1/4 = 3/2, so that E(L) = E(R). If R is playing with probability 3/4, then the expected value of T, E(T) = 3*1/4+ 1*3/4 = 3/2, and the expected value of B, E(B) = 0*1/4 + 2*3/4 = 3/2, so that E(T) = E(B).

Recall a strategy is a function from the type - space to the action - space. We define Player 1's strategy s_1: T1 ➔ A1 as follows. Pick an appropriate threshold t_1^* in T1 = (0, c) so that for any type t_1 of Player 1, if $t_1 >= t_1^*$ then set $s_1(t_1)$ = T, and if $t_1 < t_1^*$ then set $s_1(t_1)$ = B. In other words, we break T1 = [0, c] into two partitions using a suitable chosen threshold t_1^* such that for any type t_1 of Player 1 if it is to right of t_1^* Player 1 will choose action T and chooses action B if it's type t_1 is to the left of t_1^*. We will see how to pick such a threshold t_1^* coming up next. But for now, note that the probability of t_1 be to the right of t_1^* is

$\int_{t=t_1^*}^{t=c} \frac{1}{c}dt = (c - t_1^*)/c$ and the probability of t_1 to the left of t_1^* is

$\int_{t=0}^{t=t_1^*} \frac{1}{c}dt = t_1^*/c$ by Proposition 1(3).

Similarly, we define Player 2's strategy s_2: T_2 ➔ A_2 as follows. Pick an appropriate threshold t_2^* in T_2= (0, c) so that for any type t_2 of Player 2, if $t_2 >= t_2^*$ then set $s_2(t_2)$ = R, and if $t_2 < t_2^*$ then set $s_2(t_2)$ = L. In other words, we break T_2= [0, c] into two partitions using a suitable chosen threshold t_2^* such that for any type t_2 of Player 2 if it is to right of t_2^*, Player 2 chooses action R and chooses action L if it's type t_2 is to the left of t_2^*. As before, we will show how to determine t_2^* coming up next. Again, by Proposition 1 (3), the probability of t_2 be to the right of t_2^* is $(c - t_2^*)/c$ and the probability of t_2 to the left of t_2^* is t_2^*/c.

Having formulated the probability of playing the strategies T and B for Player 1, and the probability of playing the strategies L and R for Player 2 in terms of the two thresholds t_2^* and t_2^*, we can compute the expected values of each strategy for Player 1,

$E_1(T) = (3+ t_1)^* t_2^*/c + 1^* (1 - t_2^*/c)$

E_1 (B) = $0^* t_2^*/c + 2^* (1 - t_2^*/c)$

and for Player 2,

$E_2(L) = 2^* (1 - t_1^*/c) + 0^* t_1^*/c$

$E_2(R) = 1^* (1 - t_1^*/c) + (3+ t_2)^* t_1^*/c$;

Example 5.5.11 (The game of simple auction)

To find a Bayesian Nash Equilibrium in this game, first recall that a BNE is a pair of strategies s_1: T_1 ➔ A_1 and s_1: T_2 ➔ A_2 such that $s_1(v_1)$ is what player 1 bids, rename it $b_1(v_1)$, when player 1's valuation of the good is v_1, and similarly $s_2(v_2)$, rename it as $b_2(v_2)$. These pair ($b_1(v_1)$, $b_2(v_2)$) must satisfies the following conditions:

1. When $b_2(v_2)$ holds unchanged, $b_1(v_1)$ is the best response to player 2's strategy $b_2(v_2)$.
2. When $b_1(v_1)$ holds unchanged, $b_2(v_2)$ is the best response to player 1's strategy $b_1(v_2)$.

By best response we mean, each player gets the largest expected value when the other player's strategy holds unchanged, this means $b_1(v_1)$ and $b_2(v_2)$ must maximize player 1's and player 2's expected values simultaneously. Note that the expected value for each player follows from the payoff values given for this game before.

$$\underset{b_1(v_1)}{Max}(v_1 - b_1)P[b_1 > b_2(v_2)] + \frac{1}{2}(v_1 - b_1)P[b_1 = b_2(v_2)] \quad \text{(D1)}$$

$$\underset{b_2(v_2)}{Max}(v_2 - b_2)P[b_2 > b_1(v_1)] + \frac{1}{2}(v_2 - b_2)P[b_2 = b_1(v_1)] \quad \text{(D2)}$$

These two expressions are symmetrical; in the first one, we hold $b_2(v_2)$ fixed, while in the second one, we hold $b_1(v_1)$ fixed. In what follows we will show how to maximize $b_1(v_1)$ and the maximization of $b_2(v_2)$ is done similarly.

To further simplify our discussion for this auction game, we assume that the players only take linear strategy. That is both $b_1(v_1)$ and $b_2(v_2)$ are of the form

$b_1(v_1) = c_1 v_1 + d_1$, $b_2(v_2) = c_2 v_2 + d_2$, where the c's and d's are constants. We further assume that c_1 and c_2 are >= 0. This is because for each player, his or her bidding on a higher valuation should be at least as much as the lower valuation's bid. These constants c_1, c_2, d_1, d_2 are to be determined. We should, however, keep in mind that there is no reason that we only limit ourselves to consider linear function forms and not others such as non-linear functional forms for the biding functions $b_1(v_1)$ and $b_2(v_2)$.

Now to maximize (v1- b1(v1)) P [b1(v1) > b2(v2)] + ½ ((v1- b1(v1)) P [b1(v1) = b2(v2)] with respect to b1(v1), let

$Y = (v_1 - b_1(v_1))\, P[\, b_1(v_1) > b_2(v_2)] + ½\, ((v_1 - b_1(v_1))\, P[\, b_1(v_1) = b_2(v_2)]$,

apply the differentiation technique in Calculus, we will compute the derivative of Y with respect to $b_1(v_1)$, then set it zero to solve for b1. But before doing so, we must first evaluate the two probability terms $P[b_1(v_1) > b_2(v_2)]$ and $P[b_1(v_1) = b_2(v_2)]$ in terms of $b_1(v_1)$ as follows.

Since $b_2(v_2) = c_2 v_2 + d_2$ is assumed fixed, we will express c_1 and d_1 in terms of the c_2 and $d_{2,}$ and

$$P[b_1(v_1) > b_2(v_2)] = P[b_1(v_1) >= c_2 v_2 + d_2] \qquad \text{(D3)}$$

$$= P[b_1(v_1) - d_2 >= c_2 v_2]$$

$$= P[v_2 <= (b_1(v_1) - d_2)/c_2)$$

$$= (b_1(v_1) - d_2)/c_2) \text{ by Proposition 3 (c)}$$

And $P[b_1(v_1) = b_2(v_2)] = 0$ as $b_2(v_2)$ is also randomly distributed in (0,1), the type spaces of both players.

So, Y is simplified as $Y = (v_1 - b_1(v_1))\,(b_1(v_1) - d_2)/c_2)$, and

$dY/db_1 = 1/c_2\{v_1 + d_2 - 2b_1(v_1)\}$

Setting dY/db_1 to 0 gives $b_1(v_1) = ½\,(v_1 + d_2)$. Note that $b_1(v_1)$ also depends player's parameter $d_{2,}$ which is yet to determine.

Now, we must put some range restrictions on the above expression $b_1(v_1)$. Since

$b_2(v_2) = c_2 v_2 + d_2$ player 2's bid must be in between d_2 and $c_2 + d_2$, we must have

$d_2 <= b_1 <= d_2 + c_2$ because it would not help for player I to bid outside this range.

So, player 1's best strategy $b_1(v_1)$ is shown below, assuming that player 2's strategy is given as $b_2(v_2) = c_2 v_2 + d_2$.

$$b_1(v_1) = \begin{cases} (v_1 + d_2)/2 & if\ v_1 >= d_2 \\ d_2 & if\ \ v_1 < d_2 \end{cases} \qquad \text{(D4)}$$

The value of the constant d2 is any constant in player 2's biding function b2(v2), so it could be anywhere in the interval (0,1), or above and including 1 in the interval [1, Θ), or below 1 in the interval (-Θ, 0], we will show that d2 cannot be in the first two intervals and can only be in the interval (-Θ, 0] for the following reasons.

d_2 cannot be in the interval (0,1) because then from the definition of $b_1(v_1)$ defined in (D4) above, any distinct values of v_1 that are less than d_2, $b_1(v_1)$ would have the same constant value d_2, which means $b_1(v_1)$ is not a linear function, which contradicts the original assumption that $b_1(v_1)$ and $b_2(v_2)$ are linear function.

d_2 cannot be in [1, Θ], because $b_2(v_2) = c_2 v_2 + d_2 >= c_2 v_2 + 1 >= v_2$, as c_2 is >= 0. This contradicts that $b_2(v_2)$ is an optimal bid for v_2.

therefore d_2 must be >=0.

In summary, when player 2's strategy is $b_2(v_2) = c_2 v_2 + d_2$,

Then the best response that player 1 can do is to choose the strategy

$b_1(v_1) = c_1 v_1 + d_1$,

$$b_1(v_1) = \left\{ \begin{array}{ll} (v_1 + d_2)/2 & if\ v_1 >= d_2 \\ d_2 & if\ \ v_1\ < d_2 \end{array} \right\} \tag{D5}$$

And now we know $d_2 <= 0$, and $c_1 = ½$, $d_1 = 1/2\ d_2$. As the above analysis is entirely symmetric, we can reverse the role of player 1 and player 2, when player 1's strategy is $b_1(v_1) = c_1 v_1 + d_1$, the best response that player 2 can do is to choose the strategy $b_2(v_2) = c_2 v_2 + d_2$,

$$b_2(v_2) = \left\{ \begin{array}{ll} (v_2 + d_1)/2 & if\ v_2 >= d_1 \\ d_1 & if\ \ v_2\ < d_1 \end{array} \right\} \tag{D6}$$

And the preceding analysis shows $d_1 <= 0$, and $d_1 = ½\ d_2$, $c_2 = ½$. Combining the two results gives $d_1 = d_2 = 0$, $c_2 = c_2 = ½$, and thus the two Bayesian Nash Equilibrium pairs

Is simplified to $b_1(v_1) = 1/2\ v_1$, and $b_2(v_2) = 1/2\ v_2$.

Example 5.5.12 (The game of Double Auction)

To find a Bayesian Nash Equilibrium of this game, first recall that trading can occur only if the buyer's offering price is higher than the seller's asking price, that is if $p_b >= p_s$, then the trading price is $p_t = ½ (p_b + p_s)$ and no trade otherwise. Recall also that a strategy is a function from the type - space to the action space. So, if

s_1 is a strategy for the buyer, s_1: (0,1) ➔ R, it takes buyer's valuation v_b into the buyer's offering price $p_b(v_b)$, and if s_2 is a strategy for the seller, s_2: (0,1) ➔ R, it takes the seller's valuation v_s into the seller's asking price $p_s(v_s)$, we denote $s_1(v_b) = p_b(v_b)$ and $s_2(v_s) = p_s(v_s)$ and we denote strategy pair (s_1,s_2) as (p_b, p_s) from here on.

The strategy pair (p_b, p_s) is a BNE if the following two conditions hold simultaneously,

(1). When the seller's strategy is presented the strategy $p_s(v_s)$, then the best response of the buyer is to offer a strategy $p_b(v_b)$ that would maximize the buyer's expected value for every v_b in (0,1), the expected value of the buyer is given by,

$E_b(p_b, v_b) = (v_b - p_t(v_b))\, P[p_b >= p_s(v_s)]$, where the trading price $p_t(v_b)$ is given by $p_t(v_b) = 1/2\, \{p_b + E_s[p_s(v_s) \mid p_b \geq p_s(v_s)]\}$, where the conditional probability

$E_s[p_s(v_s) \mid p_b \geq p_s(v_s)]$ is the expected price the seller will demand conditional on the demand is less than the buyer's offer p_b.

(2). When the buyer's strategy is presented the strategy $p_b(v_b)$, then the best response of the seller is to offer a strategy $p_s(v_s)$ that would maximize the seller's expected value for every v_s in (0,1), the expected value of the seller is given by,

$E_s(p_s, v_s) = (p_t(v_s) - v_s)\, P[p_b(v_b) \geq p_s]$, where the trading price $p_t(v_s)$ is given by $p_t(v_s) = 1/2\, \{p_s + E_b[p_b(v_b) \mid p_b(v_b) \geq p_s]\}$, where the conditional probability

$E_b[p_b(v_b) \mid p_b(v_b) \geq p_s]$ is the expected price the buyer will offer conditional on the offer is greater than the seller's demand p_s.

Thus, the problem of finding the two BNE pair (p_b, p_s) is translated into performing $\max_{p_b} E_b(p_b, v_b)$ and $\max_{p_s} E_s(p_s, v_s)$.

So, evaluation of the four probability expressions from E_b and E_s expression is in order. To do so, we need to know functional form of the $p_s(v_s)$ and $p_b(v_b)$, and to simplify the analysis, let's assume that they are both linear as was the case in the last example, that is, $p_b(v_b) = c_b v_b + d_b$ and $p_s(v_s) = c_s v_s + d_s$, where c_b, d_b, c_s, d_s are all constants. Now we are going the following probabilities in terms of these constants.

(i) $P[p_b \geq p_s(v_s)]$, where p_b is treated as a constant for the time being,

as $p_s(v_s) = c_s v_s + d_s$, the probability above is same as,

$P[p_b \geq c_s v_s + d_s)$, which is equal to, $P[(p_b - d_s)/c_s \geq v_s]$, as v_s is uniformly distributed on the interval (0,1), $P[(p_b - d_s)/c_s \geq v_s] = (p_b - d_s)/c_s$ by Proposition 5.5.1 before, so $P[p_b \geq p_s(v_s)] = (p_b - d_s)/c_s$.

(ii) P [pb(vb) ≥ ps], where ps is treated as constant. As before,

$$P[p_b(v_b) \geq p_s] = P[c_b v_b + d_b \geq p_s] \tag{D7}$$

$$= P[v_b \geq (p_s - d_b)/c_b]$$

$$= 1 - P[v_b \geq (p_s - d_b)/c_b]$$

$$= 1 - (p_s - d_b)/c_b \quad \text{, by Proposition 3 (a)}$$

$$= (c_b + d_b - p_s/c_b$$

(iii) Es [ps (vs) | pb ≥ ps (vs)]

As before, $p_s(v_s) = c_s v_s + d_s$, the uniform random variable v_s ranges between 0 and 1 with probability density function 1, but now because of the restriction that $p_b \geq p_s(v_s)$, or, $p_b \geq c_s v_s + d_s$, or, $(p_b - d_s)/c_s \geq v_s$, so v_s ranges from 0 to $(p_b - d_s)/c_s$ instead, so it's probability density function is now, by Proposition 5.5.1 over the length of the interval, which is $1/[(p_b - d_s)/c_s]$, so

$E_s[p_s(v_s) \mid p_b \geq p_s(v_s)]$

$$= \int_{v_s=0}^{v_s=(p_b-d_s)/c_s} \frac{1}{(p_b - d_s)/c_s}(d_s + c_s v_s)dv_s$$

$$= d_s + \frac{1}{(p_b - d_s)/c_s} c_s \int_{v_s=0}^{v_s=(p_b-d_s)/c_s} v_s dv_s$$, and the last integral is evaluated to be

$1/2[(p_b - d_s)/c_s]^2$, simplifying the algebra gives

$E_s[p_s(v_s) \mid p_b \geq p_s(v_s)] = (p_b + d_s)/2.$

(vi) $E_b[p_b(v_b) \mid p_b(v_b) \geq p_s]$

As $p_b(v_b) \geq p_s$, so $c_b v_b + d_b \geq p_s$, so the range of v_b is between $(p_s - d_b)/c_b$ and 1, so the range of the interval $(p_s - d_b)/c_b$ is $(c_b + d_b - p_s)/c_b$, so the probability density function is $p(v_b) = 1/[(c_b + d_b - p_s)/c_b]$, so

$$E_b [\, p_b(v_b) \mid p_b(v_b) \geq p_s \,] = \int_{v_b=(p_s-d_b)/c_b}^{v_b=1} \frac{1}{(c_b+d_b-p_s)/c_b}(d_b+c_b v_b)dv_b$$

$$= d_b + c_b \frac{1}{(c_b+d_b-p_s)/c_b} \int_{v_b=(p_s-d_b)/c_b}^{v_b=1} v_b dv_b$$

$$= (c_b+d_b+p_s)/2$$

after some algebra simplifications.

Putting these probabilities back to (1) and (2) to find the strategy pair (p_b, p_s)

by performing the two maximizations $\max_{p_b} E_b(p_b, v_b)$ and $\max_{p_s} E_s(p_s, v_s)$

To maximize $E_b(p_b, v_b)$, using the results of (i) and (iii),

$$\text{Eb (pb, vb)} = \max_{p_b}\{v_b - \frac{1}{2}(p_b + \frac{d_s+p_b}{2})\}\frac{p_{b-d_s}}{c_s}$$

Let $Y = v_b - 1/2\,(p_b + (d_s + p_b)/2)(p_b - d_s)/c_s$, differentiate Y with respect to p_b, $dY / dp_b = 2\,d_s + 4\,v_b - 6\,p_b$, setting dY / dpb to 0 gives

$p_b = 1/3\ d_s + 2/3\ v_b$ (A)

Similarly, to maximize $E_s(p_s, v_s)$, using the results of (ii) and (vi), we get

$$E_s(p_s, v_s) = \max_{p_s}\{\frac{1}{2}(p_s + \frac{c_b+d_b+p_s}{2}) - v_s\}\frac{c_b+d_b-p_s}{c_b}$$

Let $Y = \{\frac{1}{2}(p_s + \frac{c_b+d_b+p_s}{2}) - v_s\}\frac{c_b+d_b-p_s}{c_b}$,

Setting $dY/d\,p_s = 0$ gives $6p_s - 4v_s - 2\,d_b - 2c_b = 0$, so

$p_s = 2/3v_s + 1/3(c_b + d_b)$ (B)

Combining (A) and (B), we find

$c_b = 2/3$,

$d_b = 1/3\ d_s$,

$c_s = 2/3$,

$d_s = 1/3(c_b + d_b)$,

Solving these simultaneous equations gives $d_b = 1/12$, $d_s = ¼$, and hence

$p_b = 2/3\ v_b + 1/12$,

$p_s = 2/3\ v_s + 1/4$.

Recall that trading occurs only if $p_b \geq p_s$, this means $v_b \geq v_s + ¼$ is the condition for trading for the buyer and the seller in terms of their private valuations of the good. The trade region is shown in the region in between the two lines below:

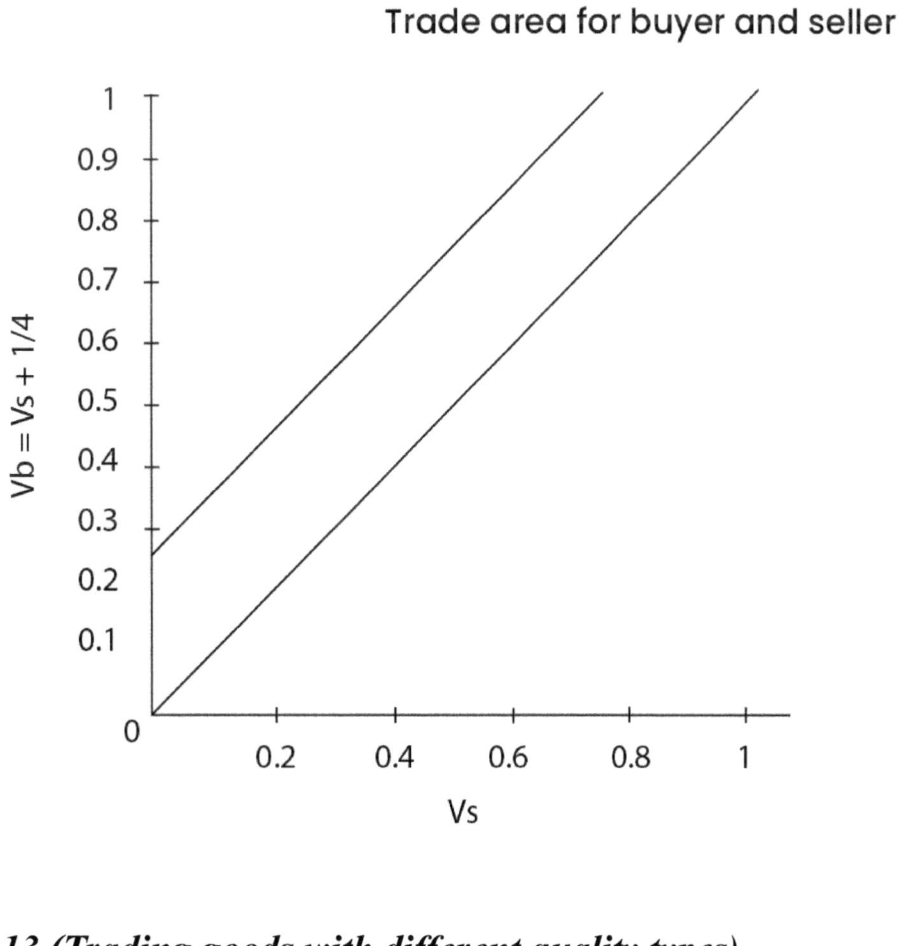

Example 5.5.13 (Trading goods with different quality types)

In this example, we consider a type of game where the types of one player effects the payoff values of the other player.

Consider the following two-player game scenario. There is a seller we will call player 1, who would like to sell his goods, say, a bag of rice, that has equal probabilities 1/3 of being a low-quality type, a mediocre-quality type, and a high-quality type. As Player 1 (the seller) knows about the quality types of his good, we can associate the types of the good with his types. Player2 (the buyer) wants to buy the good from Player1 by offering a reasonable price, however, she only knows Player 1's quality type is 1/3 evenly distributed amongst the three types, however, she does not know which exact type she is dealing with. To formulate this scenario as a Bayesian game, we are given the following information. The type-space for Player1, $T_1 = \{t_1\}$, t_1 = L, M, H is the low, mediocre, and high-quality type respectively, and the type-space for Player2 = $\{t_2\}$ is only a single type. The monetary value of the good given by each player is shown below:

$$v_1(t_1) = \begin{Bmatrix} 15 \;\; if \;\; t_1 = L \\ 33 \;\; if \; t_1 = M \\ 51 \; if \; t_1 = H \end{Bmatrix}$$

$$v_2(t_1) = \begin{Bmatrix} 20 \;\; if \;\; t_1 = L \\ 38 \;\; if \; t_1 = M \\ 65 \;\; if \; t_1 = H \end{Bmatrix}$$

The problem here is, the buyer (Player2) needs to offer a reasonable price, say p, and then up to the seller (Player1) to decide if the trade will go through. Let T be the action for Player 1 decides to trade and let N be the action that he decides not to trade, let P be the action that Player 2 decides at which price p she is going to offer, then the payoff values for both players are given as follows:

When t_1 = L:

	P
T	(p, 20-p)
N	(15 ,0)

When t_1 = M:

	P
T	(p,38-p)
N	(33,0)

When t_1 =H:

	P
T	(p,65-p)
N	(51,0)

As was mentioned before, the evenly distributed types of Player1 are given as P(L)= P(M) = P(H)= 1/3. Now that all the information given for the problem is formulated, we are going to find the best strategy pair, the Bayesian Nash Equilibrium pair, for both players. First recall that a strategy is a function from the type -space to the action - space, let s_1^1, s_1^2, s_1^3 be the strategy for Player1 corresponds to his L, M, H types respectively, as Player 1 must maintain his minimal monetary value of the good for 15, 33, 51 for the three types L, M, H respectively, so we can define

s_1^1: {L} X [0, ∞) ➔ {T, N} by

$$s_1^1(L,p)=\begin{pmatrix} T \;\; if \; p \geq 15 \\ N \;\; otherwise \end{pmatrix}$$

S_1^2: {M} X [0, ∞) ➔ {T, N} by

$$s_1^2(M,p)=\begin{pmatrix} T \;\; if \; p \geq 33 \\ N \;\; otherwise \end{pmatrix}$$

S_1^3: {H} X [0, ∞) ➔ {T, N} by

$$s_1^3(H,p)=\begin{pmatrix} T \;\; if \; p \geq 51 \\ N \;\; otherwise \end{pmatrix}$$

Similarly, Player 2 cannot offer the price p that is bigger than 30,72, 96 when Player 1's type is L, M, H respectively because Player 2 also wants to maintain her minimal monetary value on the goods she trades. So, the strategy for Player 2 can be defined as

S_2^1: $\{t_2\}$ ➔ P, $S_2^1(t_2) = P$, where P is the action of offering the price p to Player 1 satisfying the 30, 72, 96 restrictions just described.

To find the Bayesian Nash Equilibrium pair, we need to compute the expected value of each strategy pairs between the two players and pick up the pair that is optimal to both players.

The possible strategy pairs are summarized in the table below:

Payoff value of strategy pairs	Player 2 strategy S_2^1	Condition for p under which the strategies apply
s_1^1	(p, 20-p)*	p in (15, 20)
S_1^2	(p, 38-p)	$33 \le p \le 38$
S_1^3	(p, 65-p)	$51 \le p \le 65$

First note that the average monetary-equivalent value of the good for Player 1 is 1/3(15+33+51)=33, and that of Player 2 is 1/3(20+38+65)= 41, so Player 2 is reasonable to offer Player 1 at the lowest price of p=33, but then Player 1 knows that he would accept the offer only if his type is either L or M, this in turn means that Player 2 she is buying the good that is of L or M quality, which means her expected value from the trading is:

P(L|L or M)* 20 + P(M| L or M)* 38.

To work out this expected value, we need to first compute the two conditional probabilities P (L|L or M) and P(M| L or M) by the Bayes Theorem as shown below:

P (L| L or M) = P [L and (L or M)] / P(L or M)

= P (L) /[P(L) + P(M) – P(L and M)]

= 1/2 because P (Land M) = 0 as L and M are mutually exclusive events and that P(L)=P(M)= P(H) by assumption. Thus Player 2's expected value is 1/2(20+38) which is 29. As 29 is less than 33, so Player 2 would not offer the price of 33 for the trading.

Furthermore, Player 2 would not offer any price in the range (33, 51) because her expected value is still 29, and any price offer in this interval is larger than 29 which implies she is losing than gaining. Thus Player 2 can only offer the price of 51 to get the good for all three types L, M, and H. But then the expected value for Player 2 is only 41 as noted before and 41 is less than the price of 51 that she offers to Player 1.

The above analysis shows that Player 2 would not offer any price equal or above Player 1's average monetary value of the good which is 33 in this case. Thus, the expected value of Player 2 for various p values she offers to Player 1 can be summarized as follows:

Expected value of strategy pairs	Expected value for Player 2 S_2^1	Condition for p under which the strategies apply
s_1^1	$(20\text{-}p)^*$	$p \in (15, 20)$
S_1^2	29 - 33 = -6	$33 \leq p \leq 38$
S_1^3	41 - 51 = -10	$51 \leq p \leq 65$

Thus, the Bayesian Nash Equilibrium strategy pair is (s_1^1, s_2^1) shown in the above table with the * entry. The pair of strategies are mutual best responses to each other. The expected value for Player 2 is positive rather than negative as is in the other two entries. The payoff value for the equilibrium pair is (p, 20-p), where p is in the interval range (15, 20).

A Final Note on this Example 5.5.13

This example illustrates the concept of adverse selection in economics, an important contribution due to George Akerlof in 1970 for which he earned the Nobel Prize. In economics, markets allocate goods to the people who value them the most. If a good is misallocated to people, who have it value less than people who do not, then the price of the good will be increase to a level at which the current owners will prefer to sell it rather than hold on to it, and the people who value it more will be willing to pay the price. Example 7 shows the determination of such price by using the concept of Bayesian Nash Equilibrium. In our example, Player 2 obviously values the good more than Player 1. The concept of adverse selection means that, for mutual best responses between the two players, the trade goes through between the buyer and the seller when the quality of the good is the lowest (type L in our example for the seller). When the buyer is willing to pay a price equal to the seller's average value (type M in our example) of the good, then the type of the seller who is willing to sell at this price is below average (type L in our example), this is because the best types (type H in our example) select not to sell for an average price, hence the sellers adversely select the lower-than-average price.

Section 5.6 The Revelation Principle

After we have seen many of the typical Bayesian games and their associated Bayesian Nash equilibriums from previous sections, we are now in a position to introduce the most important accomplishment and a powerful tool in the theory of Bayesian games, the Revelation Principle. This principle is due to Myerson 1979([16]), it is an important tool in designing Bayesian games to solve many problems happening in economic, social, and political situations. To state this principle in the language of the Bayesian game and demonstrate how this principle works, we choose the following order of presentation. We define and introduce along the way some necessary definitions and concepts such as Direct Mechanism, social choice function, and implementation of social function by a Mechanism. All these definitions and concepts would lead us to this principle.

5.6.A What is a Direct Mechanism associated with a Bayesian game and a Bayesian Nash equilibrium in the game?

Recall that a static Bayesian game G can be specified by

$G = \{A_1, A_2, \ldots\ldots , A_n; T_1, T_2, \ldots\ldots , T_n; p_1, p_2, \ldots\ldots , p_n; v_1, v_2, \ldots\ldots , v_n\}$ and a strategy

for player i is a function $s_i(t)$ for each type t in T_i, and each $v_i = v_i (a_1, a_2, \ldots, a_n)$. Suppose the given Bayesian Nash Equilibrium is given as $s^* = (s_1^*, \ldots s_i^*, s_{i+1}^*,\ldots s_n^*)$. We define a new game denoted by G^{DM} called it the Direct Mechanism associated with G and its Nash Equilibrium s^* as follows:

Definition 5.6.0 Direct Mechanism of a game

The type spaces and beliefs in G^{DM} are the same as that in G, but the action spaces in G^{DM} are the claims about player i 's possible types, T_i. The term Direct Mechanism refers to a mechanism (game) whose only action is for the player to submit a claim about his or her type, as opposed to Indirect Mechanism where the action spaces are different than the type spaces. The payoff value of G^{DM} is defined in terms of the given equilibrium s^* in G as follows:

Let $\Theta = (\Theta_1, \Theta_2, \ldots. \Theta_n)$ be the players' reported types in the new game G^{DM}, then player i 's payoff value is defined as,

$$v_i^{DM}(\Theta, t) = v_i (s^* (\Theta), t) \text{ where } t = (t_1, t_2, \ldots. ,t_n) \qquad \text{(D1)}$$

This definition simply says that the payoff function of G^{DM} is obtained by substituting the players' type reported into their equilibrium strategies s* from G, and then substituting the resulting actions in G, into the payoff functions in G.

Now let Θ_i^j be the strategies for player i in G^{DM}, then $\Theta_i^j : T_i \rightarrow T_i$, j = 1,2,3… m, where m is the possible number of functions from T_i to T_i, or the possibly number of strategies for player i, the strategy Θ_1^1 that assigns any type t in T_i to itself is called the truth-telling strategy for player i. Any other strategies that differ from this identity mapping are considered as possibly dishonest claims about their types for player i.

From the above definition of G^{DM}, we can see that no matter what the original game G is, the associated game G^{DM} is always a directed mechanism, and we shall see from the examples below that no matter what the original Nash equilibrium strategy s^* in G is, the equilibrium strategy for G^{DM} is always the truth-telling strategy Θ^1, the payoff values in the new equilibrium in G^{DM} are identical to those payoff values in the old equilibrium in G.

Example 5.6.14: Recall the game G in Example 5.5.7 discussed in Section 5.5 before.

The BNE was found as for the game G in $s^* = (s_1^3, s_2^1)$, $T_1 = \{t_{1A}, t_{1B}\}$, and $T_2 = \{t_2\}$. Let $\Theta_1^1, \Theta_1^2, \Theta_1^3, \Theta_1^4$ be the strategies for player 1 of G^{DM}, and let Θ_2^1 be the strategy for player 2 of G^{DM} defined below:

$\Theta_1^1(t_{1A}) = t_{1A}$, $\Theta_1^1(t_{1B}) = t_{1B}$,

$\Theta_1^2(t_{1A}) = t_{1A}$, $\Theta_1^2(t_{1B}) = t_{1A}$,

$\Theta_1^3(t_{1A}) = t_{1B}$, $\Theta_1^3(t_{1B}) = t_{1B}$,

$\Theta_1^4(t_{1A}) = t_{1B}$, $\Theta_1^4(t_{1B}) = t_{1A}$,

$\Theta_2^1(t_2) = t_2$.

We can compute the payoff values of the two players in G^{DM} by the expression (D1) defined above. When player 1's type is t_{1A} and player 2's type is t_2, we find:

Player 1\Player 2	Θ_2^1
Θ_1^1	(9,7)
Θ_1^2	(9,7)
Θ_1^3	(7,9)
Θ_1^4	(7,9)

Computation of the entry for $(\Theta_1^1, \Theta_2^1) = (9,7)$ and $(\Theta_1^4, \Theta_2^1) = (7,9)$ is demonstrated here, the rest of the table entries are done similarly.

$$v^{DM}(\Theta_1^1, \Theta_2^1; t_{1A}, t_2) = v\,(s^*\,(\Theta_1^1, \Theta_2^1); t_{1A}, t_2)$$

$$= v\,[\,(s_1^3, s_2^1)\,(\Theta_1^1(t_{1A}), \Theta_2^1(t_2)\,); t_{1A}, t_2]$$

$$= v\,[\,(s_1^3, s_2^1)\,(t_{1A}, t_2)\,); t_{1A}, t_2]$$

$$= v\,[\,s_1^3(t_{1A}), s_2^1(t_2); t_{1A}, t_2]$$

$= V(B, L; t_{1A}, t_2)$

$= (9,7)$

Now the entry for (Θ_1^4, Θ_2^1) :

$v^{DM}(\Theta_1^4, \Theta_2^1; t_{1A}, t_2) = v(s^* (\Theta_1^4, \Theta_2^1); t_{1A}, t_2)$

$= v[(s_1^3, s_2^1) (\Theta_1^4(t_{1A}), \Theta_2^1(t_2)); t_{1A}, t_2]$

$= v[(s_1^3, s_2^1) (t_{1B}, t_2)); t_{1A}, t_2]$

$= v[s_1^3(t_{1B}), s_2^1(t_2); t_{1A}, t_2]$

$= v(T, L; t_{1A}, t_2)$

$= (7,9)$

When player 1's type is t_{1B} and player 2's type is t_2, playoff values for both players in G^{DM} are similarly computed as before,

Player 1\Player 2	Θ_2^1
Θ_1^1	(13,5)
Θ_1^2	(11,3)
Θ_1^3	(13,5)
Θ_1^4	(11,3)

As demonstration, computation of vDM(Θ12, Θ21; t 1B, t 2) and vDM(Θ13, Θ21; t 1B, t 2) are shown below, the rest of the entries are left as an exercise

$v^{DM}(\Theta_1^2, \Theta_2^1; t_{1B}, t_2) = v (s^* (\Theta_1^2, \Theta_2^1); t_{1B}, t_2)$

$= v[(s_1^3, s_2^1) (\Theta_1^2(t_{1B}), \Theta_2^1(t_2)); t_{1B}, t_2]$

$= v[(s_1^3, s_2^1) (t_{1A}, t_2)); t_{1B}, t_2]$

$= v[s_1^3(t_{1A}), s_2^1(t_2); t_{1B}, t_2]$

$= v(B, L; t_{1B}, t_2)$

$= (11, 3)$

$v^{DM}(\Theta_1^{\,3}, \Theta_2^{\,1}; t_{1B}, t_2) = v\,(s^*\,(\Theta_1^{\,3}, \Theta_2^{\,1}); t_{1B}, t_2)$

$= v[(s_1^{\,3}, s_2^{\,1})\,(\Theta_1^{\,3}(t_{1B}), \Theta_2^{\,1}(t_2)\,); t_{1B}, t_2]$

$= v[(s_1^{\,3}, s_2^{\,1})\,(t_{1B}, t_2)\,); t_{1B}, t_2]$

$= v[s_1^{\,3}(t_{1B}), s_2^{\,1}(\,t_2); t_{1B}, t_2]$

$= v(T, L; t_{1B}, t_2)$

$= (13, 5)$

When the probability $p(t_{1A}, t_2) = 2/5$ and $p(t_{1B}, t_2) = 3/5$, we can combine the two tables before to compute the expected values of the strategy pairs $(\Theta_1^{\,i}, \Theta_2^{\,j})$ and from which we can find the Bayesian Nash Equilibrium in G^{DM}.

Player 1\Player 2	$t_{1A}, t_2\Theta_2^{\,1}$	$t_{1B}, t_2\Theta_2^{\,1}$	Expected Values
$\Theta_1^{\,1}$	(9,7)	(13,5)	(57/5, 29/5)*
$\Theta_1^{\,2}$	(9,7)	(11,3)	(51/5, 23/5)
$\Theta_1^{\,3}$	(7,9)	(13,5)	(53/5,33/5)
$\Theta_1^{\,4}$	(7,9)	(11,3)	(47/5,27/5)

The * entry in the table above corresponds to the Nash Equilibrium of the strategy pair $(\Theta_1^{\,1}, \Theta_2^{\,1})$ in G^{DM}, it has the largest expected value for both players, and it is also the truth-telling strategy pair for both players. Note also that the expected value of the BNE in G^{DM} is the same as the expected value of the BNE s* in G, both have the same value (57/5, 29/5). As an example, the * entry in the table above is computed as $(9^*\ 2/5 + 13^*\ 3/5,\ 7^*\ 2/5 + 5^*3/5) = (57/5, 29/5)$. The rest of the expected values are computed similarly.

We know from Example 1 in Section 3 that G has two BNEs $s^* = (s_1^{\,3}, s_2^{\,1})$ and

$s^* = (s_1^{\,3}, s_2^{\,2})$ If we were to switch to a new BNE $s^* = (s_1^{\,3}, s_2^{\,2})$, then as an exercise, we can construct the above table below for G^{DM} with this new s^*.

Player 1\Player 2	$t_{1A}, t_2\Theta_2^{\,1}$	$t_{1B}, t_2\Theta_2^{\,1}$	Expected Values

Θ_1^1	(5,1)	(1,9)	(13/5, 29/5)*
Θ_1^2	(5,1)	(-1,9)	(7/5, 29/5)
Θ_1^3	(3,1)	(1,9)	(9/5, 29/5)
Θ_1^4	(3,1)	(-1,9)	(3/5,29/5)

We also note that in the table above, the * entry corresponds to the Nash Equilibrium of the strategy pair (Θ_1^1, Θ_2^1) in G^{DM}, it has the largest expected value for both players , and it is also the truth-telling strategy pair for both players. Note also that the expected value of the BNE in G^{DM} is the same as the expected value of the BNE $s^* = (s_1^3, s_2^2)$ in G, both have the same value (13/5, 29/5).

Example 5.6.15: Recall the game G in Example 5.5.8 discussed in Section 5.5 before.

The BNE was found as for the game G in $s^* = (s_1^2, s_2^3)$, $T_1 = \{t_1\}$, $T_2 = \{t_{2A}, t_{2B}\}$. Let Θ_1^1 be the strategy for player 1 of G^{DM}, and let $\Theta_2^1\, \Theta_2^2, \Theta_2^3, \Theta_2^4$ be the strategies for player 2 of G^{DM} defined below:

$\Theta_1^1(t_1) = t_1$,

$\Theta_2^1(t_{2A}) = t_{2A}$, $\Theta_2^1(t_{2B}) = t_{2B}$,

$\Theta_2^2(t_{2A}) = t_{2A}$, $\Theta_2^2(t_{2B}) = t_{2A}$,

$\Theta_2^3(t_{2A}) = t_{2B}$, $\Theta_2^3(t_{2B}) = t_{2B}$,

$\Theta_2^4(t_{2A}) = t_{2B}$, $\Theta_2^4(t_{2B}) = t_{2A}$.

As in the last example, we compute the payoff values of the two players in G^{DM} by the expression when player 1's type is t_1 and player 2's type is t_{2A}, we find:

Player 1\Player 2	Θ_2^1	Θ_2^2	Θ_2^3	Θ_2^4
Θ_1^1	(3,3)	(3,3)	(-1,-1)	(-1,-1)

As a demo, we show the computation of entries (Θ_1^1, Θ_2^1) and (Θ_1^1, Θ_2^3) here, the rest of the table entries are done as an exercise.

$v^{DM}(\Theta_1^1, \Theta_2^1; t_1, t_{2A}) = v(s^*(\Theta_1^1, \Theta_2^1); t_1, t_{2A})$

$\quad = v[(s_1^2, s_2^3)(\Theta_1^1(t_1), \Theta_2^1(t_{2A})); t_1, t_{2A}]$

$= v[(s_1^{\,2}, s_2^{\,3}) (t_1, t_{2A})); t_1, t_{2A}]$

$= v[s_1^{\,2}(t_1), s_2^{\,3}(t_{2A}); t_1, t_{2A}]$

$= v(B, R; t_1, t_{2A}) = (3,3)$

$v^{DM}(\Theta_1^{\,1}, \Theta_2^{\,3}; t_1, t_{2A}) = v(s^* (\Theta_1^{\,1}, \Theta_2^{\,3}); t_1, t_{2A})$

$= v[(s_1^{\,2}, s_2^{\,3}) (\Theta_1^{\,1}(t_1), \Theta_2^{\,3}(t_{2A})); t_1, t_{2A}]$

$= v[(s_1^{\,2}, s_2^{\,3}) (t_1, t_{2B})); t_1, t_{2A}]$

$= v[s_1^{\,2}(t_1), s_2^{\,3}(t_{2B}); t_1, t_{2A}]$

$= v(B, L; t_1, t_{2A}) = (-1, -1)$

Similarly, when player 1's type is t_1 and player 2's type is t_{2B}, we find:

Player 1\Player 2	$\Theta_2^{\,1}$	$\Theta_2^{\,2}$	$\Theta_2^{\,3}$	$\Theta_2^{\,4}$
$\Theta_1^{\,1}$	(3,3)	(3,3)	(-1,-1)	(-1,-1)

Let the probability of $p(t_1, t_{2A}) = p(t_1, t_{2B}) = ½$, we can compute the expected values of the strategy pairs $(\Theta_1^{\,1}, \Theta_2^{\,i})$, $i = 1,2,3,4$ as follows:

Player 1\Player 2	$\Theta_2^{\,1}$	$\Theta_2^{\,2}$	$\Theta_2^{\,3}$	$\Theta_2^{\,4}$
For type (t_1, t_{2A}) $\Theta_1^{\,1}$	(3,3)	(3,3)	(-1,-1)	(-1,-1)
For type (t_1,t_{2B}) $\Theta_1^{\,1}$	(-1,5)	(3,3)	(-1,5)	(3,3)
Expected values	(1,4)*	(3,3)	(-1,2)	(1,1)

Note that the BNE for G^{DM} is the truth-telling strategy pair $(\Theta_1^{\,1}, \Theta_2^{\,i})$ with payoff value (1,4), and this is the same as the payoff value at the BNE strategy pair $s^* = (s_1^{\,2}, s_2^{\,3})$ of the original game G.

Example 5.6.16: Recall Example 5.5.9 discussed in Section 5.5 before.

Let G be the game in Example 3, the BNE was found as $s^* = (s_1^{\,3}, s_2^{\,3})$, $T_1 = \{t_{1A,} t_{1B}\}$, $T_2 = \{t_{2A}, t_{2B}\}$. Let $\Theta_1^{\,1}$, $\Theta_1^{\,2}$, $\Theta_1^{\,3}$, $\Theta_1^{\,4}$ be the strategies for player 1 in G^{DM}, and let $\Theta_2^{\,1}$ $\Theta_2^{\,2}$, $\Theta_2^{\,3}$, $\Theta_2^{\,4}$ be the strategies for player 2 of G^{DM} defined below:

$\Theta_1^1(t_{1A}) = t_{1A}, \Theta_1^1(t_{1B}) = t_{1B},$

$\Theta_1^2(t_{1A}) = t_{1A}, \Theta_1^2(t_{1B}) = t_{1A},$

$\Theta_1^3(t_{1A}) = t_{1B}, \Theta_1^3(t_{1B}) = t_{1B},$

$\Theta_1^4(t_{1A}) = t_{1B}, \Theta_2^4(t_{1B}) = t_{1A}.$

$\Theta_2^1(t_{2A}) = t_{2A}, \Theta_2^1(t_{2B}) = t_{2B},$

$\Theta_2^2(t_{2A}) = t_{2A}, \Theta_2^2(t_{2B}) = t_{2A},$

$\Theta_2^3(t_{2A}) = t_{2B}, \Theta_2^3(t_{2B}) = t_{2B},$

$\Theta_2^4(t_{2A}) = t_{2B}, \Theta_2^4(t_{2B}) = t_{2A}.$

As in the last example, we compute the payoff values of the two players in G^{DM} by the expression when player 1's type is t_{1A} and player 2's type is t_{2A}, we find:

Player 1\Player 2 (t_{1A}, t_{2A})	Θ_2^1	Θ_2^2	Θ_2^3	Θ_2^4
Θ_1^1	(4,4)	(4,4)	(16,0)	(16,0)
Θ_1^2	(4,4)			
Θ_1^3	(0,16)			
Θ_1^4	(0,16)			(16,0)

As a demo, we show the computation of entries (Θ_1^1, Θ_2^1) and (Θ_1^4, Θ_2^4) here, the rest of the table entries are done as an exercise.

$$v^{DM}(\Theta_1^1, \Theta_2^1; t_{1A}, t_{2A}) = v\,(s^*(\Theta_1^1, \Theta_2^1); t_{1A}, t_{2A})$$

$$= v[(s_1^3, s_2^3)(\Theta_1^1(t_{1A}), \Theta_2^1(t_{2A})); t_{1A}, t_{2A}]$$

$$= v[(s_1^3, s_2^3)(t_{1A}, t_{2A})); t_{1A}, t_{2A}]$$

$$= v[s_1^3(t_{1A}), s_2^3(t_{2A}); t_{1A}, t_{2A}]$$

$= v(B, R; t_{1A}, t_{2A}) = (4,4)$

$v^{DM}(\Theta_1^4, \Theta_2^4; t_{1A}, t_{2A}) = v\,(s^*(\Theta_1^4, \Theta_2^4); t_{1A}, t_{2A})$

$= v[(s_1^2, s_2^3)\,(\Theta_1^4(t_{1A}), \Theta_2^4(t_{2A})); t_{1A}, t_{2A}]$

$= v[\,(s_1^2, s_2^3)\,(t_{1B}, t_{2B})); t_{1A}, t_{2A}]$

$= v[\,s_1^2(t_{1B}), s_2^3(t_{2B}); t_{1A}, t_{2A}]$

$= v(B, L; t_{1A}, t_{2A}) = (16, 0)$

For player 1's type is t_{1A}, and player 2's type is $t_{2B,}$ we find:

Player 1\Player 2 $(t_{1A,}\ t_{2B})$	Θ_2^1	Θ_2^2	Θ_2^3	Θ_2^4
Θ_1^1	(16,0)	(4,-24)	(16,0)	(4,-24)
Θ_1^2	(16,0)			
Θ_1^3	(0,0)			
Θ_1^4	(0,0)			

Similarly, when player 1's type is t_{1B} and player 2's type is t_{2A}, we find:

Player 1\Player 2 $(t_{1B,}\ t_{2A})$	Θ_2^1	Θ_2^2	Θ_2^3	Θ_2^4
Θ_1^1	(0,16)	(0,16)	(0,0)	(0,0)
Θ_1^2	(-24,4)			
Θ_1^3	(0,16)			
Θ_1^4	(-24,4)			

, when player 1's type is t_{1B} and player 2's type is t_{2B}

Player 1\Player 2 (t_{1B}, t_{2B})	$\Theta_2^{\,1}$	$\Theta_2^{\,2}$	$\Theta_2^{\,3}$	$\Theta_2^{\,4}$
$\Theta_1^{\,1}$	(0,0)			
$\Theta_1^{\,2}$	(16,0)			
$\Theta_1^{\,3}$	(0,0)			
$\Theta_1^{\,4}$	(16,0)			

The blank entries are computed similarly. The main point we want to stress here is that the payoff value of the old game, which is (5,5) at the equilibrium pair $(s_1^{\,3}, s_2^{\,3})$ has the same payoff value of the new game at the truth telling strategy pair $(\Theta_1^{\,1}, \Theta_2^{\,1})$ as the following computation shows:

$EV(\Theta_1^{\,1}, \Theta_2^{\,1}) = ¼\,(4,4) + ¼\,(16,0) + ¼\,(0,16) + ¼\,(0,0) = (5,5)$.

5.6.B What is the overall problem with Mechanism design?

The Mechanism G^{DM} is a special kind of game associated with the original game G, as we recall in G^{DM}, the action sets and the type sets are the same, and the payoffs of the players are expressed in terms of the payoffs of the players from the old game G and a Nash Equilibrium that existed in G. Now we extend this definition of a mechanism without these restrictions as follows. But first, we review our notations:

Let $A_1, A_2, \ldots ,A_n$ be the action sets of players 1 ,2, …, n as before,

Let Y be a set called the set of outcomes and there is given a function

$g: A_1 \text{ X } A_2 \text{ X } \ldots, A_n \rightarrow Y$.

Let $T_1, T_2, \ldots ,T_n$ are players respective type sets, and the strategy are functions $s_i^{\,j}: T_i \rightarrow A_i$ is player i' s j^{th} strategy.

For each outcome y in Y, each player i, and each player type t_i has a payoff value $v_i\,(y, t_i)$.

X is called the set of alternatives. The meaning of the sets X and Y will become clear after we see the applications below.

Definition 5.6.2 What is a mechanism ?

A mechanism M is a set of action sets $A_1, A_2, \ldots, A_n$ together with a function g (which we call the outcome function for the Mechanism M) that assigns each set of actions $(a_1, a_2, \ldots, a_n)$ to an object y in Y, that is, $y = g(a_1, a_2, \ldots, a_n)$, and each object y in Y is given the payoff values $v_i(y, t_i)$.

In symbol, a mechanism is represented as $M = (A_1 \times A_2 \times \ldots \times A_n, g)$ with payoffs given by $v_i(y, t_i)$. Refer to Figure 5.1 below:

Figure 5.1

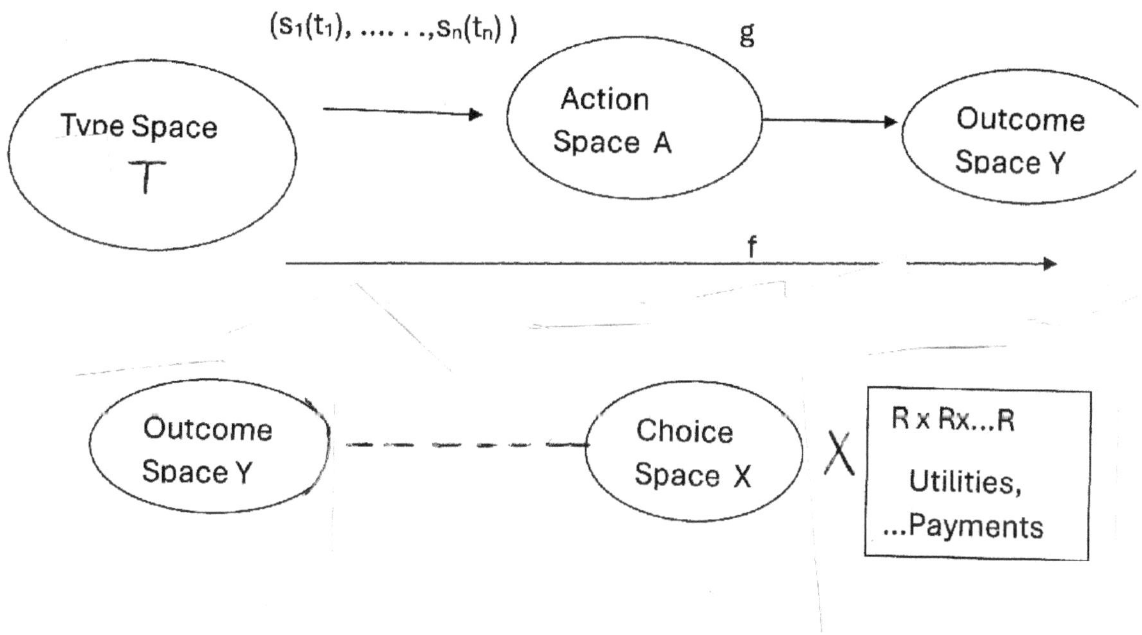

Figure 5.1 above can be interpreted as follows. Each abstract object in the Outcome Space Y is the result of some possible actions taken by the N players and it can be represented as an alternative chosen from the Choice Space X together with N meaningful tuples of numbers called the utilities or payments for the players.

Definition 5.6.3. Bayesian Nash Equilibrium for $M = (A_1 \times A_2 \times \ldots \times A_n, g)$

A strategy profile $(s_1^*, s_2^* \ldots, s_n^*)$ is a Bayesian Nash Equilibrium if for all t_i in T_i, the following condition holds,

$$E_{t_{-i}}[v_i(g(s_i^*(t_i), s_{-i}^*)), t_i) \mid t_i] \geq E_{t_{-i}}[v_i(g(a_i, s_{-i}^*(t_{-i})), t_i) \mid t_i] \quad where\ a_i \in A_i$$

Example 5.6.17: Consider the following game with $A_1 = \{T, B\}$ and $A_2 = \{L, R\}$,

$T_1 = \{t_{1A}, t_{1B}\}$, $T_2 = \{t_2\}$, and $p(t_{1A} \mid t_2) = 1/2$, $p(t_{1B} \mid t_2) = 1/2$,

$Y = \{y_1, y_2, y_3, y_4\}$ and the map $g: A_1 \times A_2 \rightarrow Y$ defined as $g(T,L) = y_1$, $g(T,R) = y_2$, $g(B,L) = y_3$, $g(B,R) = y_4$, and the payoff values

(v1,v2) for both players are defined by the following table:

(v1, v2)	(t_{1A}, t_2)	(t_{1B}, t_2)
y_1	(7,9)	(13,5)
y_2	(3,1)	(1,9)
y_3	(9,7)	(11,3)
y_4	(5,1)	(-1,9)

Let s_1^1, s_1^2, s_1^3, s_1^4 be the strategies for player 1 and s_2^1 , s_2^2 be the strategies for player 2 as defined as in Example 5.5.7, the expected values of the strategy pairs (s_1^i, s_2^j) are computed and summarized below:

E (v1, v2)	s_2^1	s_2^2
s_1^1	(10,7)	(2,5)
s_1^2	(9,6)*	(11,5)
s_1^3	(11,6)	(3,5)
s_1^4	(10,5)	(2,5)*

We show the computation in the * entries as examples.

$E[v(g(s_1^2 , s_2^1); t_{1,} t_2]$

$$= ½\, v(g(s_1^2 (t_{1A}), s_2^1 (, t_2)); t_1^A, t_2) + ½v(g(s_1^2 (t_1^B), s_2^1(, t_2)); t_1^B, t_2)$$

$$= ½\, v(g(T,L); t_{1A,} t_2) + ½\, v(g(B,L); t_{1B,} t_2)$$

$$= ½\, v(y_1;t_{1A,} t_2) + ½\, v(y_3;t_{1B,} t_2)$$

$$= ½\, (7,9) + ½(11,3)$$

$$= (9,6)$$

$E[v(g(s_1^4, s_2^2); t_{1,} t_2]$

$= ½ v(g(s_1^4 (t_{1A}), s_2^2(t_2)); t_{1A,} t_2) + ½ v(g(s_1^4 (t_{1B}), s_2^2(t_2)); t_{1B,} t_2)$

$= ½ v(g(B,R); t_{1A,} t_2) + ½ v(g(B,R); t_{1B,} t_2)$

$= ½ v(y_4 ;t_{1A,} t_2) + ½ v(y_4; t_{1B,} t_2)$

$= ½ (5,1) + ½ (-1,9)$

$= (2,5)$

From the above table entries, it is easily seen that the entry (11,6) corresponds to the strategy pair (s_1^3, s_2^1) is a Bayesian Nash Equilibrium for the Mechanism M above because it satisfies the condition stated in Definition 5.6.3.

Definition 5.6.4. M = (A1 x A2 x… x A n, g) implements a function f :T ➔ Y

A Mechanism $M = (A_1 \times A_2 \times \ldots \times A_n, g)$ with a Bayesian Nash Equilibrium

$(s_1^*, s_2^* \ldots, s_n^*)$ implements a function f: $T_1 \times T_1$ to $T_1 \times T_2$ ➔ Y if the following condition is satisfied:

$g.(s_1^*(t_1), s_2^*(t_2), \ldots , S_n^*(t_n)) = f(t_1, t_1 \ldots ,t_n)$ for all $(t_1, t_1 \ldots ,t_n)$ in $T_1 \times T_2 \times \ldots T_n$. Equivalently, we can also say **the function f:T ➔ Y is implementable in Bayesian Nash Equilibrium.** The function f(.) is called the choice function.

(See Figure 5.1 how these functions f(.) and g(.) are related.)

Example 5.6.18: Given the mechanism M as in the last Example 5.6.4, and suppose we have a function f: $T_1 \times T_2$ ➔ Y defined by $f(t_{1A}, t_2) = y_3$, $f(t_{1B}, t_2) = y_1$. Then this Mechanism $M = (A_1 \times A_2, g)$ with the Bayesian Nash equilibrium (s_1^3, s_2^1) implements f(.). This is because

$g (s_1^3(t_{1A}), s_2^1(t_2)) = g(B, L)= y_3 = f(t_{1A}, t_2)$ and

$g (s_1^3(t_{1B}), s_2^1(t_2)) = g(T, L) = y_1 = f(t_{1B}, t_2)$.

Note that in this simple example, as there are 2 objects in the set $T_1 \times T_2$ and 4 objects in the set Y, so, there are 16 different functions f: $T_1 \times T_2$ ➔ Y in all. The reader can check out the other functions and look for other possible Bayesian Nash Equilibrium for implementing these other functions.

Now we know what it meant to implement a function f(.) by a Mechanism **M = (A1 x A2 x… x A n, g)**. There are two important special cases in this definition. The first special case is that when the function f happens to be the same function g in the mechanism M, and the second special case is that the type spaces and the action spaces are the same the M, in which case, the strategies $s_1^*, s_2^*\ldots, s_n^*$ will be just mappings from T_i to T_i, and there is an important strategy that corresponds to the identity mapping which we refer to as the truth-telling strategy. Combining these two special cases together, and if we also assume that the truth-telling strategies $s_1^*, s_2^*\ldots, s_n^*$ is also a Bayesian Nash Equilibrium, then this leads to the following important definition and concept used in the Revelation Principle to be described next.

Definition 5.6.5. A function f:T ➔ Y is truthfully implementable in Bayesian Nash Equilibrium.

A function f:T ➔ Y is truthfully implementable in Bayesian Nash Equilibrium if there is a mechanism $M = (T_1, T_2 \ldots, T_n, f(.))$ whose truth-telling strategy for each player is a Bayesian Nash Equilibrium.

Now the Revelation Principle relates the two concepts that a function is implementable and truthfully implementable together and can be stated as follows.

Proposition 5.6.1 The Revelation Principle in Bayesian Nash Equilibrium

Any choice function f:T ➔ Y is truthfully implementable in Bayesian Nash Equilibrium if and only if f:T ➔ Y is implementable in Bayesian Nash Equilibrium.

Before we go on to the proof of this principle, let us look at the contrapositive equivalent of this principle and see what the message that the original principle revealed. The principle tells us that if we cannot **truthfully implement** a choice function f( ), then we will not be able to **implement** f( ) by any means. In order words, if we cannot implement the choice function f:T ➔ Y with a direct mechanism M whose object function is f (.), that is, M= (T, f(.)) and that the truth-telling strategies of M is a Bayesian Nash Equilibrium, then there is no way we can find any mechanism M= (A, g) or game that can implement our choice function f(.). Thus, this principle is very powerful in the sense that it reduces the general problem of looking for games or mechanisms to implement a given choice function to the problem of checking out for a specific kind of mechanism associated with the intended choice function.

There are two parts to the proof of this principle.

First let us assume that **f:T ➔ Y** is truthfully implementable in Bayesian Nash Equilibrium, then this means that there is a direct mechanism $M =(T_1, T_2, \ldots, T_n, f(.))$ such that

$$E_{t_{-i}}[v_i(f(t_i,t_{-i}),t_i)\,|\,t_i] \geq E_{t_{-i}}[v_i(f(a_i,t_{-i}),t_i)\,|\,t_i] \quad where\ a_i \in A_i$$

Now we choose g(.) to be the function f(.) and choose the strategy profile $s_1^*, s_2^*\ldots, s_n^*$ to be the truth-telling strategies for each player, then the above conditions become

$$E_{t_{-i}}[v_i(g(s_i^*(t_i),s_{-i}^*(t_{-i})),t_i)\,|\,t_i] \geq E_{t_{-i}}[v_i(g(a_i,s_{-i}^*(t_{-i})),t_i)\,|\,t_i] \quad where\ a_i \in A_i$$

and $g(s^*(.)) = f(.)$ by the choice of g(.) and $s^*(.)$. The above condition means that f:T ➔ Y is implementable in Bayesian Nash Equilibrium.

To prove the converse, we assume that f:T ➔ Y is implementable in Bayesian Nash Equilibrium, this means that we can find a mechanism M = (A1 x A2 x… x A n , g) and a strategy profile s_1^*, s_2^*…., s_n^* such that $g(s^*(.)) = f(.)$ and

$$E_{t_{-i}}[v_i(g(s_i^*(t_i),s_{-i}^*(t_{-i})),t_i)\,|\,t_i] \geq E_{t_{-i}}[v_i(g(a_i,s_{-i}^*(t_{-i})),t_i)\,|\,t_i] \quad where\ a_i \in A_i$$

and all $t_i \in T_i$.

This means that player i's would not want to deviate to any other strategies from his or her current strategy $s_i^*(\,)$ with the specific type of t_i that corresponds to player i's action $s_i^*(t_i)$ in A_i which in turn give rise to his or hers largest expected value $E_{t_{-i}}[v_i(g(s_i^*(t_i),s_{-i}^*(t_{-i})),t_i)\,|\,t_i]$, all the other actions $a_i \in A_i$ are the result of $s_i^*(\,)$ on other types Θ_i different from t_i, that is $a_i = s_i^*(\Theta_i)$. Thus, the above condition becomes,

$$E_{t_{-i}}[v_i(g(s_i^*(t_i),s_{-i}^*(t_{-i})),t_i)\,|\,t_i] \geq E_{t_{-i}}[v_i(g(s_i^*(\theta_i),s_{-i}^*(t_{-i})),t_i)\,|\,t_i]\,,$$

But $g(s^*(.)) = f(.)$ from the chosen mechanism M= (A, g), so the above condition becomes,

$$E_{t_{-i}}[v_i(f(t_i,t_{-i}),t_i)\,|\,t_i] \geq E_{t_{-i}}[v_i(f(\theta_i,t_{-i}),t_i)\,|\,t_i]$$

, for all t_i, $\Theta_i \in T_i$

, which means that f (.) is truthfully implementable in Bayesian Nash Equilibrium by the mechanism M= (T, f(.)).

Before we show the many applications of the Revelation Principle in Section C, we first show a special case of Bayesian Nash Equilibrium called **Dominant Strategy Equilibrium** and apply the Revelation Principle to the underlying mechanism where the strategy profile consists of dominant strategies for each player. The Revelation Principle before then becomes

Proposition 5.6.2 The Revelation Principle in Dominant Strategy Equilibrium

If the choice function f :T➔Y is truthfully implementable in Dominant Strategy Equilibrium, then f :T➔Y is implementable in Bayesian Nash Equilibrium

This proposition follows trivially from Proposition 5.6.1 after we define what Dominant Strategy meant.

Definition 6.5.6 Dominant strategy and Dominant strategy Equilibrium

First, we use the following example to illustrate the definition of dominant strategy in a 2- players game. Suppose player 1 has n strategies and player 2 has m strategies and their payoff matrix is shown below:

Player 1 \ Player 2	s_2^1	s_2^2	.....	s_2^j		s_2^m
s_1^1	(a_{11}, b_{11})	(a_{12}, b_{12})		(a_{1j}, b_{1j})		(a_{1m}, b_{1m})
s_1^2	(a_{21}, b_{21})					
s^1i	(a_{i1}, b_{i1})					
s_1^n	(a_{n1}, b_{n1})					(a_{nm}, b_{nm})

The i th strategy for player 1, s_1^i, is the dominant strategy for player 1 if $a_{ij} \geq a_{kl}$ for all k from 1 to n, and for all j from 1 to m.

The j th strategy for player 2, s_2^j, is the dominant strategy for player 2 if $b_{ij} \geq b_{ik}$ for all k from 1 to m, and for all i from 1 to n.

The strategy pair (s_1^i, s_2^j) is called a dominant strategy pair. It is easily seen that a dominant strategy pair is also a Nash Equilibrium Strategy pair.

As another example below: the strategy pair (s_1^1, s_2^3) is a dominant strategy pair.

s_1^1 is the dominant strategy for player 1 because regardless of what strategies player 2 takes, the payoff values of player 1 is greater than or equal to that of all other player 1 strategies. Similarly, s_2^3 is the dominant strategy for player 2 because regardless of what strategies player 1 takes, the payoff values of player 2 is greater than or equal to that of all other player 2 strategies.

Player 1 \Player 2	s_2^1	s_2^2	s_2^3
s11	(5,0)	(2,1)	(7,2)
s12	(1,2)	(1,5)	(4,6)
s13	(4,3)	(0,4)	(6,5)

The strategy pair (s_1^1, s_2^3) is also a Nash Equilibrium strategy pair because if player 1's strategy s_1^1 remains unchanged, player 2 would not want to deviate herself from s_2^3 to the other two strategies because by doing so, player 2's payoff value could only get less and not more (ie. 2 is better than 1 or 0). Similarly, if player 2's strategy s_2^3 remains fixed, then player 1 would not want to deviate his current strategy s_1^1 to the other two strategies because, by doing so, he is not getting higher payoff values (i.e, 7 is better than 4 or 6).

We now extend this concept of dominant strategies equilibrium in a game to a mechanism in the following definition.

Definition 5.6.7 A strategy profile (s_1^*, s_2^*...., s_n^*) is a Dominant Strategy Equilibrium of a Mechanism M= (A, g(.)) if the following condition holds

$$v_i(g(s_i^*(t_i),a_{-i}),t_i) \geq v_i(g(a^1{}_i,a_{-i}),t_i) \quad where\ a^1{}_i \in A_i,\ a_{-i} \in A_{-i}$$

Note that the payoff value here of each player is associated with the outcome specified by the outcome function g(.). (Refer to Figure 5.1)

Now if we recall from Definition 5.6.3 of Bayesian Nash Equilibrium,

$$E_{t_{-i}}[v_i(g(s_i^*(t_i),s_{-i}^*)),t_i) \mid t_i] \geq E_{t_{-i}}[v_i(g(a_i,s_{-i}^*(t_{-i})),t_i) \mid t_i] \quad where\ a_i \in A_i$$

by comparing the two conditions, Dominant Strategy Equilibrium is a special case of Bayesian Nash Equilibrium and not the other way around, hence as a special case of Proposition 5.6.2, Proposition 5.6.3 below holds only for the if- part and not the only-if part in Proposition 5.6.2, and we stated explicitly the condition for f(.) to be truthfully implementable in dominant strategies.

Proposition 5.6.3 If the choice function f :T➔Y satisfies the following condition

$$v_i(f(t_i,t_{-i}),t_i) \geq v_i(f(\theta_i,t_{-i}),t_i) \quad where\ \theta_i \in T_i,\ t_{-i} \in T_{-i}$$

then f :T➔Y is implementable in Bayesian Nash Equilibrium

We will show the many applications of Proposition 5.6.3 in the following section.

5.6.C Applications of the Revelation Principle

1. VCG Mechanism and its applications
2. Vickrey Auction
3. Bilateral Trade
4. Public Project
5. Shortest Path in a Network

Thus far, we have only considered abstractly a general function **f :T→Y** called the choice function and looking for mechanism M=(A, g) to implement f(.), the set Y was called the set of outcomes, and the function g(.) was called the object function. These terms are meaningful names when we look at some real applications. In what follows, we will look at a very practical special scenario when Y is a product of the set X and R^n, where X is called the set of social choices or alternatives, and R^n, the set of payments to the individual players when the certain choice is made or certain alternative is chosen in X, this special scenario leads to a mechanism which we will call the VCG Mechanism. Before we give a formal definition of the VCG Mechanism and its applications, we first look at some actual problems where the term choice functions come up and see how the theory of Mechanism that we developed before is used to implement these choice functions.

Example 5.6.19. Choice of a player from a set X of players

Assume each player i has a private value of the good, say t_i and the problem is to allocate the private good to the player that values the good the most, that is to the player with the largest t_i. The set of choice X , in our example is , a collection of n-tuples x= (x_1, x_2, ….x_n), where each variable x_i is a binary variable of either 0 or 1, and $\sum_i x_i = 1$. In symbol, x = { (x_1, x_2, ….x_n) | x_i is either 0 or 1 and $\sum_i x_i = 1$}. Let

u_i (x, t_i) be player i 's value when his private value of the good is t_i.

The total utility is therefore $\sum_i u_i(x,t_i)$ In our example , u_i (x, t_i) = x_i t_i.

The choice function $f: T_1 \times T_2 \times ..... \times Tn \to X \times R^n$ chooses a player x^* in X such that x^* = (x_1, x_2, ….x_n) maximize the function $\sum_i u_i(x,t_i)$ over all x in X , which is same as maximizing $\sum_i x_i t_i$ over all x in X, , which means that when max {t_1, t_2, ….t_n}=t_k then x^*_k =1 and x^*_m = 0 for all other entries in x^*.

The choice function $f: T_1 \times T_2 \times ..... \times Tn \to X \times R^n$ also determines the payment for each player i, let (p_1, p_2, ……, p_n) be the payment for each player 1 to player n, then (p_1, p_2, ……, p_n) defines a point a tuple in R^n, so

$f: T_1 \times T_2 \times ..... \times Tn \to X \times R^n$ is given by f(t_1, t_2, ….t_n) = (x^* , p_1, p_2, ……, p_n)

For now, and in the next Example 5.6.7, we leave these functions

(p_1, p_2, ……, p_n) unspecified, but we will discuss them in detail in Section 5.6.C.I as they turn out to be the key component of the Vickrey-Clarke-Groves Mechanism.

Example 5.6.20 Choice of an alternative from a set X of social alternatives

Suppose we have a set X of social alternatives such as a government aerospace project, building of a citywide senior center, building of a public library, investment in statewide education, … etc. Suppose there are n citizens /players in supporting these projects. Our goal is to choose a single project to maximize the benefit of all players given the player's valuation on these projects. Let x be an alternative in X, let

u_i (x, t_i) be payoff value of alternative x when player i 's type is t_i , our object is to find an alternative x* in X such that $\underset{x \in X}{Max} \sum_i u_i(x,t_i)$

, in order words, $\sum_i u_i(x^*,t_i)$ is the largest amongst all other $\sum_i u_i(x,t_i)$ for all other x in X. Now with this x*, the function

$f: T_1 \times T_2 \times \ldots.. \times Tn \rightarrow X \times R^n$ is given by f(t_1, t_2, ….t_n) = (x^* , p_1, p_2, ……, p_n) is the choice function.

$\sum_i u_i(x^*,t_i)$ is the maximum social welfare.

5.6.C.I VCG Mechanism and its applications

As was shown in the last two examples, the alternative x* that the social function $f: T_1 \times T_2 \times \ldots.. \times Tn \rightarrow X \times R^n$ determines is done in two steps, first we formulate the expression for u_i (x, t_i) from the underlying problem , then we maximize the sum of the social welfare for each alternative x in X, that is $\underset{x \in X}{Max} \sum_i u_i(x,t_i)$.

In short,

$$x^* = \arg \underset{x \in X}{Max} \sum_i u_i(x,t_i)$$

The tuple (p_1, p_2, ……, p_n) was left unspecified as was stated in Examples 5.6.6 and 5.6.7. Now we are going to introduce the VCG Mechanism where the functions p_1, p_2, ……, p_n represent the amount of payment each individual player pays to the mechanism respectively, and these functions are given by specific expressions to be described below. Since our basic goal is to implement f(.) in Bayesian Nash Equilibrium, and so by Proposition 5.6.3, we only need to check if the condition below is satisfied,

$$v_i(f(t_i,t_{-i}),t_i) \geq v_i(f(\theta_i,t_{-i}),t_i) \quad where\ \theta_i \in T_i,\ t_{-i} \in T_{-i}$$

This suggests that we choose these functions (p_1, p_2, ……, p_n) in such a way to guarantee that the condition above could happen.

In the VCG Mechanism, the functions (p_1, p_2, ……, p_n) was as

$$p_i(t_1,t_2,...,t_n)=\underset{x}{Max}\sum_{j\neq i}u_j(x,\ t_j)-\sum_{j\neq i}u_j(x^*,\ t_j)$$

, where the term $\underset{x}{Max}\sum_{j\neq i}u_j(x,\ t_j)$ above was called the Clarke pivot rule, we denote it by $h_i(t_{-1})$, it is the social welfare for all players other than player i, thus we can say the payment functions p1, p2, ……, pn in the VCG Mechanism can be interpreted as the difference between the maximum social welfare if player i where absent and the social welfare of all other players when player I is present. Thus,

$$p_i(t_1,t_2,...,t_n)=h_i(t_{-i})-\sum_{j\neq i}u_j(x^*,\ t_j)\text{ , where } h_i(t_{-i})=\underset{x}{Max}\sum_{j\neq i}u_j(x,\ t_j)$$

There are several nice properties of this VCG Mechanism using this definition of the payment functions defined above. However, before we list and prove these properties, we first give two numerical examples to show how the payment functions and the Clarke pivot functions are computed below.

Example 5.6.21: (Two players and two alternatives)

Suppose there are two players N= { 1,2}, and there are two alternatives

X = { x_1, x_2}. The valuations of each player on the two alternatives are given in the following table:

Players\Alternatives	x_1	x_2
t_1	21	5
t_2	19	31

As shown, $u_1(x_1, t_1) = 21$, , $u_1(x_2, t_1) = 5$, , $u_2(x_1, t_2) = 19$, $u_2(x_2, t_2) = 31$. As $u_1(x_1, t_1) + u_2(x_1, t_2) = 21 + 19 = 40$ and $u_1(x_2, t_1) + u_2(x_2, t_2) = 5 + 31 = 36$ so by definition

$$x^* = \arg \underset{x\in X}{Max}\sum_i u_i(x,t_i)$$

$$= x_1$$

$$p_1(t_1,t_2)=\underset{x}{Max}\sum_{j\neq 1}u_j(x,\ t_j)-\sum_{j\neq 1}u_j(x_1,\ t_j)$$

$$= \underset{x\in X}{Max}\ \{u_2(x,t_2)\}-u_2(x_1,t_2)$$

$= Max\ \{19,31\} - 19$

$= 12$

$$h_i(t_{-i}) = \underset{x}{Max} \sum_{j \neq i} u_j(x,\ t_j)$$

So $h_1(t_2) = 31$.

Similarly,

$$p_2(t_1, t_2) = \underset{x}{Max} \sum_{j \neq 2} u_j(x,\ t_j) - \sum_{j \neq 2} u_j(x_1,\ t_j)$$

$= Max\ \{21,5\} - 21$

$= 0$

, and $h_2(t_1) = 21$.

Example 5.6.22 (Three players and three alternatives)

Suppose there are three players N= { 1,2,3}, and there are three alternatives

X= { x_1, x_2 x_3}. The valuations of each player on the three alternatives are given in the following table:

Players\Alternatives	x_1	x_2	x_3
t_1	10	5	12
t_2	9	15	10
t_2	4	2	3

As before, the optimal alternative x* from X is computed by

$$x^* = \arg \underset{x \in X}{Max} \sum_i u_i(x, t_i)$$

$= \arg \underset{x \in X}{Max} \{10+9+4,\ 5+15+2,\ 12+10+3\}$

$= \arg \underset{x \in X}{Max} \{23, 22, 25\}$

$= x_3$

$$p_1(t_1,t_2,t_3) = \underset{x}{Max} \sum_{j \neq 1} u_j(x,\ t_j) - \sum_{j \neq 1} u_j(x_3,\ t_j)$$

$= \underset{x \in X}{Max} \{u_2(x,t_2) + u_3(x,t_3)\} - \{u_2(x_3,t_2) + u_3(x_3\ ,t_3)\}$

$= \underset{x \in X}{Max} \{9+4, 15+2, 10+3\} - \{10+3\}$

$= 4$

$$h_i(t_{-i}) = \underset{x}{Max} \sum_{j \neq i} u_j(x,\ t_j)$$

So $h_1(t_{2,}\ t_3) =$ 17.

Similarly,

$$p_2(t_1,t_2,t_3) = \underset{x}{Max} \sum_{j \neq 2} u_j(x,\ t_j) - \sum_{j \neq 2} u_j(x_3,\ t_j)$$

$= \underset{x \in X}{Max} \{u_1(x,t_1) + u_3(x,t_3)\} - \{u_1(x_3,t_1) + u_3(x_3\ ,t_3)\}$

$= \underset{x \in X}{Max} \{10+4, 5+2, 12+3\} - \{12+3\}$

$= 0$

, and $h_2(t_1, t_3) = 15$.

$$p_3(t_1,t_2,t_3) = \underset{x}{Max} \sum_{j \neq 3} u_j(x,\ t_j) - \sum_{j \neq 3} u_j(x_3,\ t_j)$$

$= \underset{x \in X}{Max} \{u_1(x,t_1) + u_2(x,t_2)\} - \{u_1(x_3,t_1) + u_2(x_3\ ,t_2)\}$

$= \underset{x \in X}{Max} \{10+9, 5+15, 12+10\} - \{12+10\}$

$= 0$

,and h3(t1, t2) = 22.

Note that in the above computation of the functions $h_i(.)$, it only depends on the valuations $t_1, t_2, \dots, t_n$ and not on t_i .

Now we can state and prove the following nice properties of the VCG Mechanism below. But first we need to define what we mean by the utility v_i of a player i .

$v_i(t_1, t_2, \dots, t_n) = u_i(x^*, t_i) - p_i(t_1, t_2, \dots, t_n)$

Proposition 5.6.4 Individual Rationality (IR)

This proposition states that the utility of each player is positive, and no player can have negative utility , that is vi $(t_1, t_2 \dots, tn) \geq 0$. To prove this,

$v_i(t_1, t_2 \dots, t_n) = u_i(x^*, t_i) - p_i(t_1, t_2, \dots, t_n)$

$= u_i(x^*, t_i) - \{ \underset{x}{Max} \sum_{j \neq i} u_j(x, t_j) - \sum_{j \neq i} u_j(x^*, t_j) \}$

$= u_i(x^*, t_i) + \sum_{j \neq i} u_j(x^*, t_j) - \underset{x}{Max} \sum_{j \neq i} u_j(x, t_j)$

$= \sum_{j} u_j(x^*, t_j) - \underset{x}{Max} \sum_{j \neq i} u_j(x, t_j)$

$= \underset{x}{Max} \sum_{j} u_j(x, t_j) - \underset{x}{Max} \sum_{j \neq i} u_j(x, t_j)$

≥ 0

because the maximum of a group of respectively larger values is always bigger than the maximum of a group of smaller values.

Proposition 5.6.5 Incentive Compatible (IC)

This proposition states that the players will report their true valuation and no need to know how what other players report because the player's utility will be great if he reports the true valuation than reporting any other value. That is

$v_i(t_1, t_2, \ldots, t_i, \ldots, t_n) \geq v_i(t_1, t_2, \ldots, \theta_i, \ldots, t_n)$ where θ_i the value that player i misreports. To prove this, since from above derivation,

$$v_i(t_1, t_2, \ldots, t_i, \ldots, t_n) = \sum_j u_j(x^*, t_j) - \underset{x}{Max} \sum_{j \neq i} u_j(x, t_j)$$

$$v_i(t_1, t_2, \ldots, \theta_i, \ldots, t_n) = \sum_j u_j(x^*(t_1, t_2 \ldots \theta_i, \ldots t_n), t_j) - \underset{x}{Max} \sum_{j \neq i} u_j(x, t_j)$$

As the second term is common to both expressions above, we only need to compare

$\sum_j u_j(x^*, t_j)$ and $\sum_j u_j(x^*(t_1, t_2 \ldots \theta_i, \ldots t_n), t_j)$

But the first term above is the social welfare of the optimal solution x^*, the second term is the social welfare of only one allocation $x^*(t_1, t_2, \ldots, \theta_i, \ldots, t_n)$, and as such the first term is larger than the second term.

Proposition 5.6.6 No payment to the players from the VCG Mechanism

This proposition stays that the payment functions $p_i(t_1, t_2, \ldots, t_i, \ldots, t_n)$ are all positive, and by definition, there will be no payment to the players from the Mechanism. To prove this, since

$$p_i(t_1, t_2, \ldots, t_n) = \underset{x}{Max} \sum_{j \neq i} u_j(x, t_j) - \sum_{j \neq i} u_j(x^*, t_j)$$

The first term on the right-hand side of the above equation is the optimal social welfare without the ith player, and the second term is only the social welfare of a particular allocation x* without the ith player, and as such, the first term is larger than the second term.

Proposition 5.6.7 Choice of the alternative X* to maximize social welfare and individual utility

This proposition states that the choice of the alternative x* is crucial in the VCG Mechanism, recall x* is defined as

$$x^* = \arg \underset{x \in X}{Max} \sum_i u_i(x, t_i)$$

so that x* maximizes the social welfare,

and as shown in Proposition 5.6.4, the individual utility is expressed as

$$v_i(x^*(t_1, t \ldots, t_n)) = \sum_j u_j(x^*, t_j) - \underset{x}{Max} \sum_{j \neq i} u_j(x, t_j)$$

, since $\sum_{j} u_j(x^*, t_j)$ is same as $\underset{x}{Max} \sum_{j} u_j(x, t_j)$ and the second term

$\underset{x}{Max} \sum_{j \neq i} u_j(x, t_j)$ is independent of the valuations of the i th player, so x* maximizes the individual utility as well.

Proposition 5.6.8 Function implementable in dominant strategies

$f : T_1 \times T_2 \times ..... \times Tn \to X \times R^n$ defined before is implementable in dominant strategies with respect to the player's utilities v_i

Based on the definition of x* implemented in dominant strategy, we only need to show $v_i(x^*(t_1, t_2, ..., t_i, ..., t_n), t_i) \geq v_i(x^*(t_1, t_2, ..., \theta_i, ..., t_n), t_i)$,where θ_i is player i 's misreported valuation. But this follows from Proposition 5.6.5 on the Incentive Compatible property of v_i .

Now that we know what the VCG Mechanism is, we are in position to discuss some of its important applications including Vickrey Auction, Bilateral Trade, a Public Project, and the Shortest Path in a Network.

5.6.C.II Application to Vickrey Auction for an item

The Vickrey Auction is also shown as the second-price auction. It works as follows. Each player presents a bid and the player with the highest bid win the item in the auction, but the winner only pays an amount equal to what the second highest bidder had bid, and all other players pay nothing and receive nothing. We model the utility of the winner as the difference in the winner's value of the item and the amount they paid. In what follows , we are going to show that each player truthfully bidding exactly their own valuation is a Dominant Strategy Equilibrium.

Let t_i be bidder i 's true value of the item. Let b_i be bidder i 's bid for the item. The utility for bidder i is given by

$$u_i(t_1, t_2, ...t_n) = t_i - \underset{j \neq i}{Max}\, b_j$$

when $b_i \geq \underset{j \neq i}{Max}\, b_j$ else it is 0

We are going to prove that both the overbidding strategy and the underbidding are dominated by the truth telling strategy.

(a) To show the overbidding strategy ($b_i > t_i$) is dominated by truth telling strategy. There are three cases to consider,

Case 1. If $\underset{j\neq i}{Max}\, b_j < t_i$ then the bidder would win the item with both overbidding and truth telling strategy. The bidder's utility for both strategies are the same in this case.

Case 2. If $\underset{j\neq i}{Max}\, b_j > b_i$ then the bidder would lose the item in both strategies and have equal utility in this case.

Case 3. If $t_i < \underset{j\neq i}{Max}\, b_j < b_i$ then only the strategy of overbidding would win the auction. The utility value for bidder i would be negative from the definition given above, but the utility for the true-telling strategy would be zero. Hence in this case, the overbidding strategy is dominated by the true-telling strategy.

(b) To show the underbidding strategy ($b_i < t_i$) is dominated by truth truth-telling strategy. Again, there are three cases to consider.

Case 1. If $\underset{j\neq i}{Max}\, b_j > t_i$ then the bidder would lose the item with both truth telling and the underbidding strategy, so both strategies have equal utility in this case.

Case 2. If $\underset{j\neq i}{Max}\, b_j < b_i$ then the bidder would win the item in both strategies and have equal utility in this case.

Case 3. If $b_i < \underset{j\neq i}{Max}\, b_j < t_i$, then only the truth-telling strategy would win the auction. The utility for bidder i with the truth-telling strategy would be positive, but the utility for bidder i with the underbidding strategy would be 0. Thus, the underbidding strategy is dominated by the truth-telling strategy.

Thus by (A) and (B) above,the truth telling strategy dominates both underbidding and overbidding strategies, and thus it is an optimal strategy.

Now we are going to show that this second price auction can be formulated as a special case of the VCG Mechanism described before.

5.6.C.II.1 Vickrey Auction as a special case of the VCG Mechanism

Similar to Example 1 described before on choosing a player from a set X of players or choosing an alternative from a set of alternatives, the problem now is to allocate an object to one of the n players.

Assume that the value of owning the item for player i is t_i.

Let X = { ($x_1, x_2, \ldots x_n$) | x_i is either 0 or 1 and $\sum_i x_i = 1$}. Let $u_i(x, t_i) = x_i t_i$ be the value of player i owning the item, where $x = (x_1, x_2, \ldots x_n)$. Our job is to choose a player x^* in X such that $x^* = (x_1, x_2, \ldots x_n)$ maximize the function $\sum_i u_i(x, t_i)$ over all x in X , which is same as maximizing $\sum_i x_i t_i$ over all x in X, , which means that when max $\{t_1, t_2, \ldots t_n\} = t_k$ then $x^*_k = 1$ and $x^*_m = 0$ for all other entries in x^* with index different from k.

$$p_i(t_1,t_2,\ldots,t_n)=h_i(t_{-i})-\sum_{j\neq i}u_j(x^*,\ t_j)$$

Players\Alternatives	x_1	x_2	x_3		x_n
1	t_1	0			0
2	0	t_2	0		
3		0	.. t_3		
..		..	0		0
n	0	0			t_n

The payment function, as defined before,

$$p_i(t_1,t_2,\ldots,t_n)=\underset{x}{Max}\sum_{j\neq i}u_j(x,\ t_j)-\sum_{j\neq i}u_j(x^*,\ t_j)$$

becomes $p_i(t_1, t_2, \ldots\, t_n) = \text{Max}\ \{ t_1, t_2, \ldots\, t_n \text{ other than } t_i\} - 0.$

$$= \text{Max}\ \{ t_1, t_2, \ldots\, t_n \text{ other than } t_i\}$$

But this Max $\{ t_1, t_2, \ldots\, t_n$ other than $t_i\}$ is exactly the second largest valuation because t_I is the largest valuation as it is the winner, and all other payment functions $p_i(t_1, t_2, \ldots\, t_n)$ will be 0 if player i is not a winner. Note that we made the choice player x* to be the one that has the largest valuation amongst all the players, and the payment function does give us the second largest valuation amongst all the players; these two requirements are exactly those required for the second auction described before.

5.6.C.III Application to Bilateral Trade for a seller and a buyer

Consider a seller wants to sell an object (house, boat, business, etc.) and a buyer wants to buy if the price is right. Let t_s and t_b be the valuations of the object for the seller and the buyer respectively. To formulate this scenario by the VCG Mechanism, since the seller starts with owning the object, so if x_1 is the alternative that the seller has the object, and x_2 is the alternative that the buyer has the object, then the alternative x_1 means no trade has committed and the alternative x_2 means trade has committed . In the bilateral trade, there are only these two alternatives x_1 and x_2, and as before, let the set of alternatives be X= { x_1 (no trade) , x_2(trade)} . Note that the trade can happen, that is , the object goes to the buyer if $t_b > t_s$ and there would be no trade, that is, the seller keeps the object if $t_s > t_b$

. What the VCG Mechanism does is to take the inputs t_s, t_b from both buyer and seller then it determines if trading is possible and finally computes the payment function for each player.

We summarize the information of the VCG Mechanism in the following table:

Players\Alternatives	x_1 (no trade) $t_s \geq t_b$	x_2 (trade) $t_b > t_s$
Buyer t_b	$p_b = 0$ $h_s(t_b) = u_b(x_1, t_b)$	$u_b(x_2, t_b) = t_b$ $p_b = t_s$
Seller t_s	$p_s = 0$ $h_b(t_s) = u_s(x_1, t_s)$	$u_s(x_2, t_s) = - t_s$ $p_s = - t_b$

In the above table, for alternative x_1, since there will be no payments when there is no trade , so

$p_b(t_b, t_s) = 0$ and $p_s(t_b, t_s) = 0$

and when we apply the general payment formula shown before

$$p_i(t_1, t_2, \ldots, t_n) = h_i(t_{-i}) - \sum_{j \neq i} u_j(x^*, t_j) \qquad \text{Equation (1)}$$

$h_i(t_{-i}) = \underset{x}{Max} \sum_{j \neq i} u_j(x, t_j)$, we have

$p_b(t_b, t_s) = h_b(t_s) - u_s(x_1, t_s)$, $p_s(t_b, t_s) = h_s(t_b) - u_b(x_1, t_b)$. Since for the alternative x_1, $p_b(t_b, t_s) = 0$ and $p_s(t_b, t_s) = 0$, so the pivot terms are

$h_b(t_s) = u_s(x_1, t_s)$ and $h_s(t_b) = u_b(x_1, t_b)$. Now apply Equation (1) to alternative x_2, we have,

$p_b(t_b, t_s) = h_b(t_s) - u_s(x_2, t_s)$ and $p_s(t_b, t_s) = h_s(t_b) - u_b(x_2, t_b)$, and so

$p_b(t_b, t_s) = u_s(x_1, t_s) - u_s(x_2, t_s)$ and $p_s(t_b, t_s) = u_b(x_1, t_b) - u_b(x_2, t_b)$ ------(1)

Now we make the following interpretation, for the no-trade alternative x_1,

$u_s(x_1, t_s)=0$, $u_b(x_1, t_b) = 0$ and for the trade alternative x_2, the seller's valuation decreases by t_s and the buyer's valuation increases by t^b, and so

$u_b(x_2, t_b) = t_b$, $u_s(x_2, t_s) = -t_s$. Substitute these four utilities back to (1)

we get $p_b(t_b, t_s) = 0 - (-t_s) = t_s$ and $p_s(t_b, t_s) = 0 - t_b = -t_b$

The interpretation on the sign of the payment functions p_b and p_s is that the seller collects t_b from the Mechanism as it is negative, and the buyer pays the Mechanism t_s as it is positive. Since trade can only happen when $t_b > t_s$, so the Mechanism must subsidize the cost to make this condition true.

5.6.C.IV Application to Public Project

This application is on the use of VCG Mechanism to manage a public project on scenarios such as a private company is considering building a recreation center for the local community, a university is considering building a laboratory for the students, a department store is considering setting up an AI database system for the customers, …., etc. Here, we specifically look at the project of a private company that is building a recreation center for the local community. First the company needs to know how much the total cost is to build the recreation center. Let's say the cost is C. Next let us assume that there are 1, 2… n that many numbers of individuals are using the facility, . and everyone will value the facility at a cost c_i, for i from 1 to n. Note that these values c_i are private as it is known only to the individual using the facility and not others. Our objective is to use a VGC Mechanism to find the payment function for everyone using the facility how much this private company should charge to each user of the facility and to get each users' true valuation on the facility. There are two extreme cases to look at for these payment functions assuming payment and valuation are linearly related as in the case of the Single Auction described earlier. If there is no payment required for the facility users, then they will tend to overstate their valuation, and with payment required, they will tend to understate their valuation. The VCG Mechanism will give us the appropriate amount of payment as described below. There are only two alternatives in the Mechanism described below. Let x_1 be the alternative for the project not being undertaken, and x_2 be the alternative for the project being undertaken. The private company will undertake the project , that is to say , the company will choose alternative x_2 only if the condition is satisfied. Otherwise, the company will not undertake the project. As before, let the alternative set X be $X=\{x_1, x_2\}$. Let t_C be the valuation for the company, let $t_1, t_2, \ldots, t_n$ be the valuations for the n users respectively, then we have:

$u_C(x_1, t_C) = 0$, $u_C(x_2, t_C) = -C$.

$u_i(x_1, t_i) = 0$, $u_i(x_2, t_i) = c_i$.

Players\Alternatives	. x_1 Project not undertaken	. x_2 Project undertaken
t_1	0	c_1
t_2	0	c_2
t_3	0	c_3
t_i	$u_i(x_1, t_i) = 0$	$u_j(x_2, t_j) = c_i$
t_j	..	..
t_n	0	c_n
t_C	0	-C

Apply the payment function formula as before,

$$p_i(t_1, t_2, ..., t_n) = \underset{x}{Max} \sum_{j \neq i} u_j(x, t_j) - \sum_{j \neq i} u_j(x^*, t_j)$$

, since $x^* = x_2$, we get,

$$p_i(t_1, t_2, ... t_n, t_C) = Max\{\sum_{j \neq i} u_j(x_1, t_j), \sum_{j \neq i} u_j(x_2, t_j)\} - \sum_{j \neq i} u_j(x_2, t_j)$$

$$= Max\{0, \sum_{j \neq i} u_j(x_2, t_j) - C\} - \{\sum_{j \neq i} u_j(x_2, t_j) - C\}$$

$$= Max\{0, \sum_{j \neq i} c_j - C\} - \{\sum_{j \neq i} c_j - C\}$$

There are two conditions either one of which will cause $p_i(t_1, t_2, ... t_n, t_C)$ to become 0, the first one is the condition that when the project is not undertaken, that is $\sum_j c_j < C$ and the second condition that causes $p_i(t_1, t_2, ... t_n, t_C)$ to become 0 is when $\sum_{j \neq i} c_j - C \geq 0$ in the expression above, in which case $p_i(t_1, t_2, ... t_n, t_C)$ is the subtraction of two identical terms. Now we look at the condition under which $p_i(t_1, t_2, ... t_n, t_C)$ is not 0. Since the negation of the condition $\sum_j c_j < C$ or $\sum_{j \neq i} c_j - C \geq 0$ is the condition $\sum_j c_j \geq C$ and $\sum_{j \neq i} c_j < C$, under this condition, $p_i(t_1, t_2, ... t_n, t_C)$ is non-zero, and its exact value is computed as

$$p_i(t_1, t_2, ... t_n, t_C) = Max\{0, \sum_{j \neq i} c_j - C\} - \{\sum_{j \neq i} c_j - C\}$$

$$= 0 - \{\sum_{j \neq i} c_j - C\}$$

$$= C - \sum_{j \neq i} c_j$$

Note that in the above condition $\sum_j c_j \geq C$ and $\sum_{j \neq i} c_j < C$ means player i is called pivotal because its presence can change the choice of alternatives x_1 and x_2.

In summary, if a player is not pivotal, $p_i(t_1, t_2, \ldots t_n, t_C) = 0$, he has no incentive to pay any amount because his presence or not does not change the result, whereas if the player is pivotal, $p_i(t_1, t_2, \ldots t_n, t_C) = C - \sum$ he has the incentive to pay up to his valuation c_i. Note that $p_i(t_1, t_2, \ldots t_n, t_C) \leq c_i$. So, in general,

$\sum_i p_i \leq C$, and the project needs to be subsidized for it to start with the condition $\sum_j c_j \geq C$.

5.6.C.V Application on Computer Networks

Suppose a communication network is modeled by a directed graph G= (V, E), where V is a set of vertices and E is a set of edges. Assume there are two special nodes s and t , call them the start node and terminal node in G. We are going to formulate a VCG Mechanism to find the shortest path in the graph G from s to t. But first we need to describe the correspondence between the terminologies used in a graph and terminologies used in a VCG Mechanism. An edge e in G corresponds to a player in the Mechanism, the cost of the edge e corresponds to the player e's private valuation denoted by t_e . The set of all paths $p_1, p_2 \ldots, p_n$ from s to t corresponds to the set X of all alternatives. The objective function is the path's total cost. The problem of finding the shortest distance from s to t corresponds to choosing the problem of finding an optimal alternative x* from the alternative set X. Player e's valuation is set to 0 if it is not part of the chosen path, and an edge e's valuation is set to $-t_e$ if it is an edge in a chosen path p_i. We choose the negative sign because our objective of minimizing the total distance from s to t can is equivalent to maximizing the total negative distance and recall that the VCG is to maximiz e the social welfare. The following table shows the correspondences. $te_i, p_2, p_3, p_i \ldots\ldots p_n \cdot e_1, e_2, e_i, p_i \ldots\ldots e_n$

Players\Alternatives (edges\paths)	P_1	P_2	.. Each path starts from s to t e	..	P_n
e_1	$-te_1$				
e_2	- 0				
$e_{i,}$	- $-te_{i,}$				
e_n					

The usual payment functions formula in the VGC Mechanism

$$p_i(t_1,t_2,\ldots,t_n)=h_i(t_{-i})-\sum_{j\neq i}u_j(x^*,\ t_j)$$

. where the first term can be interpreted as the length of the shortest path which does not contain the edge e_i , and the second term is the length of the shortest path when the cost of edge e_i is assumed to be 0. In the context of shortest distance, let us define the following notations:

D(- e_i) to mean the total distance of the shortest path from s to t where the paths are not going through the edge e_i.

D (e_i =0) to mean the total distance of the shortest path from s to t going where the paths are going through the edge e_i but assuming the distance of the edge e_i is 0.

In the following examples, we show that p_i can be expressed as D(- e_i) - D (e_i =0) For edges not on the shortest path, D(- e_i) - D (e_i =0) .

Example 5.6.23 (For a network with 3 nodes 5 paths)

Consider the following network graph.

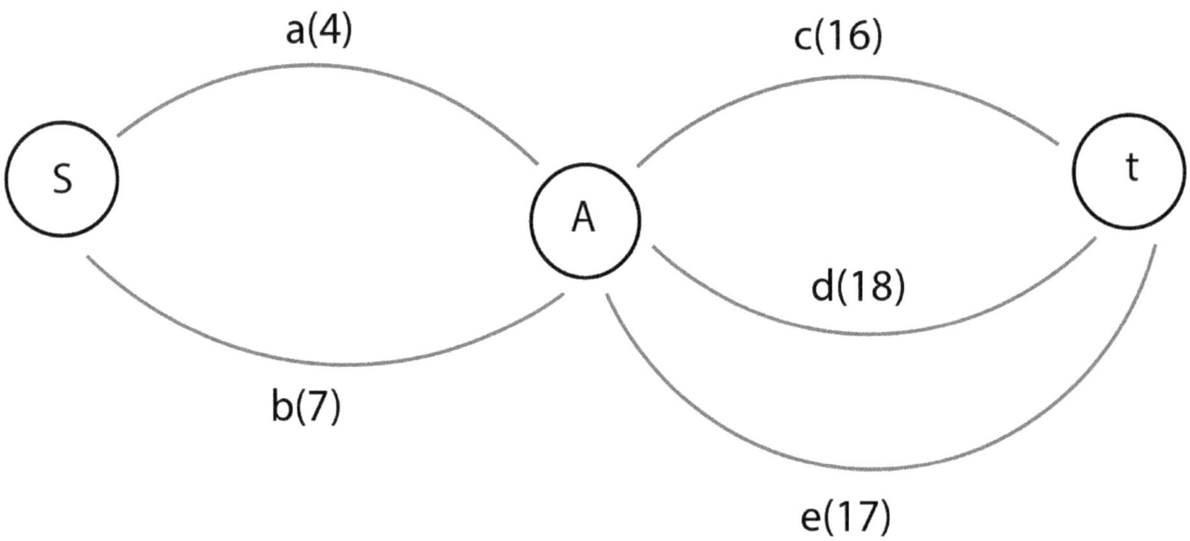

where s is the start node, t is the end node, and a, b, c, d, e are edges with corresponding cost (distance) shown. There are 6 paths to be considered and one of them is the shortest path. We put all these information into the framework of the VCG Mechanism.

Edges\paths	ac	ad	ae	bc	bd	be
a	-4	-4	-4	0	0	0
b	0	0	0	-7	-7	-7
c	-16	0	0	-16	0	0
d	0	-18	0	0	-18	0
e	0	0	-17	0	0	-17
	-20	-22	-21	-23	-25	-24

In this case, the set of alternatives X= { ac, ad, ae, bc, bd, be} , the shortest path is the path ac with shortest distance 20, so the optimal alternative x*= ac .

Using the payment function

$$p_i(t_1,t_2,\ldots,t_n) = \underset{x}{Max} \sum_{j \neq i} u_j(x,\ t_j) - \sum_{j \neq i} u_j(x^*,\ t_j)$$

, we find:

p_a= Max { -16, -18,-17, -23, -25, -24} – { -16} = -16 + 16 =0

p_b= Max { -20, -22,-21, -16, -18, -17} – { -20} = -16 + 20 =4

p_c= Max { - 4, - 22,-21, -7, -25, -24} – { -4} = -4 + 4=0

p_d= Max { - 20, - 4,-21, -23, -7, -24} – { -20} = -4 + 20= 16

p_e= Max { - 20, - 22,-4, -23, -25, -7} – { -20} = -4 + 20= 16

Using the expression D(- e_i) - D (e_i =0), we find:

p_a = 0

p_b = D(- b) - D (b =0) = 20 -16 =4

p_c = 0

p_d = D(- d) - D (d =0) = 20 - 4 =16

p_e= D (-e) – D (e=0) = 20 – 4 =16

Example 5.6.24 (For a network with 5 nodes 7 paths)

Consider the following network graph.

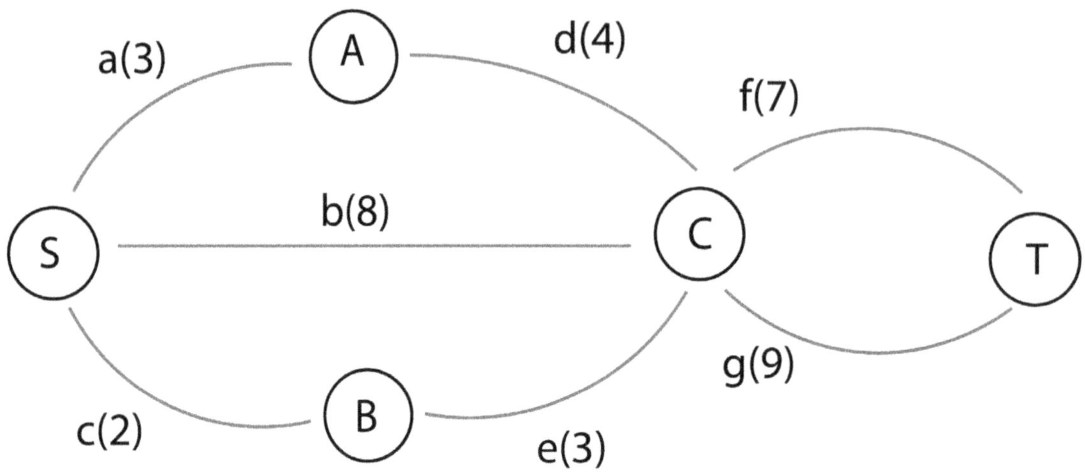

Seven edges are shown with corresponding distances in the graph above. There are 6 paths in all. The shortest path is cef with total distance 12.

Edges\Paths	adf	adg	bf	bg	cef	ceg
a	-3	-3	0	0	0	0
b	0	0	-8	-8	0	0
c	0	0	0	0	-2	-2
d	-4	-4	0	0	0	0
e	0	0	0	0	-3	-3
f	-7	0	-7	0	-7	0
g	0	-9	0	-9	0	-9
	-14	-16	-15	-17	-12	-14

Using the payment function

$$p_i(t_1,t_2,\ldots,t_n)=\underset{x}{Max}\sum_{j\neq i}u_j(x,\ t_j)-\sum_{j\neq i}u_j(x^*,\ t_j)$$

, we find:

p_a= Max { -11, -13,-15,-17,-12,-14} – { -12} = -11 + 12 =1

p_b= Max { -14,-16,-7,-9,-12,-14} – { -12} = -7 + 12 = 5

p_c= Max { -14,-16,-15,-17,-10,-12}- {-10} = -10 + 10 =0

p_d= Max { -10, -12,-15,-17,-12,-14} – { -12}=-10+12= 2

p_e= Max { - 14,-16,-15,-17,-9, -11} – { -9} = = -9+ 9 = 0

p_f= Max { -7,-16,-8,-17,-5, -14} – { -5} = = -5+ 5 = 0

p_g= Max { -14,-7,-15,-8,-12,-5} – { -12} = = -5+ 12 = 7

Using the expression D(-i) - D (i= 0), we find:

p_a = 12-11 =1

p_b = 12-7 = 5

p_c= 0

p_d= 12-10 =2

p_e = 0

p_f= 0

p_g= 14-7 =7

This shows that both methods give the same result.

5.7 Summary

This chapter gives a comprehensive description of the static Bayesian games and the associated concept of Bayesian Nash Equilibrium. A powerful tool established in the theory of Bayesian games is the Revelation Principle (summarized in Prepositions 5.6.1, 5.6.2, and 5.6.3) which allows us to solve many of the problems in practical situations (see 5.6.C). This celebrated principle basically states that any social choice function that can be implemented by a Mechanism can be implemented by a truthful direct mechanism. The concept of Direct Mechanism, Mechanism (Indirect Mechanism), function implemented by a Mechanism are explained in Section 5.6.A, Definition 5.6.2, Definition 5.6.3, Definition 5.6.4) , Since when a problem is formulated in terms of a choice function f: $T \rightarrow Y$, the Revelation principle is very powerful in the sense that we can reduce the effort in finding the solution to the problem by searching for the right Mechanism in a much and much smaller set of Mechanisms, called the direct Mechanisms, to implement the choice function f(.) .

Chapter 6

Subgame Perfect Nash Equilibrium

6.0. Introduction

This chapter gives a detailed exposition on dynamic games with complete, perfect, and imperfect information. Recalling that in Chapter 4, Section 4.3, we had informally introduced this type of dynamic game. A convenient data structure to represent this type of game is a tree structure (henceforth we refer to this as the game tree) and the use of an information set to store the proper information at each node in the tree.

Section 6.1 explains the concept of information set and shows the requirement of a subgame in the tree in terms of the information sets. Knowing what the subgames are would help us to identify the subgame perfect Nash Equilibriums SPE in the game which we will discuss in Section 6.4.

Section 6.2 gives the definition of pure strategies in a game tree and shows the common notation used to represent these pure strategies.

Section 6.3 , roughly speaking, puts the requirement on the players that each must be a rational player in the sense that when it is his(her) turn to move, he (she) must select the strategy (if there is choice) that would result in getting the optimal payoff value.

Section 6.4 gives the formal definition of subgame-perfect Nash equilibrium, and the notation used to represent these equilibriums.

Section 6.5 presents the two general approaches to searching for the subgame- perfect Equilibriums in a game tree, the Backwards and the Forwards Induction.

Section 6.6 provides three practical applications of subgame-perfect Equilibriums in the areas of Economics, business, Union and firm bargaining over wage.

Section 6.7 extends our key concept of subgame-perfect Equilibrium for a single game discussed before to multi-stage games. Using the concept of Award and Punishment introduced in a multi-stage game, we can create additional subgame-perfect Equilibriums other than the one guaranteed from Proposition 6.7.1 (Existence of subgame-perfect Equilibrium in a multi-stage game) .

Section 6.8. This last section focuses on Repeated Games, a special multi-stage game where all the stage component games are the same. In addition to the Award and Punishment concept introduced in the multi-stage game in 6.7 , a new concept called Grim Trigger strategy introduced here for the infinitively repeated games are used to create subgame-perfect Equilibrium in an infinitely repeated game. This section concludes with proof of the most celebrated result the Folk's Theorem in infinitely repeated games and an application of this theorem to economics.

6.1 What does it mean by a game with complete, perfect and imperfect information? What is a subgame in such a game?

In the previous chapters, we have only discussed static games (simultaneous games), we now introduce another kind of games that involve sequential interaction amongst the players, called the dynamic games (sequential games). We still restrict ourselves to games with **complete information** in which the players' payoff functions are common knowledge as the preceding chapters, but now since the players are interacting dynamically, there are two different situations to consider. The first situation is that at each move in the game the player with the move knows the full history of the play of the game thus far and we refer to this scenario as **game of perfect information**. The second situation is that if at some move the player with the move does not know the history of the game, then we refer to this scenario as **game of imperfect information**. Recall in static games, the game information is represented by a game matrix (**normal form or matrix representation**) , in dynamic games however, a convenient way we can capture the sequential interactions by means of a tree data structure called the game tree (**extensive form or tree structure representation**). For the same game, there are these two distinct representations by these two different data structures- matrices and trees. Each of these two representations can be converted from one form to another, and each form of the game representation has its own merit as we shall see in the following discussions. To continue, we first present two examples to illustrate the basic terminologies such as **information set, subgame, and strategy**.

Example 6.1.1 In this example as shown in Figure 6.1 and Figure 6.2, we show an example of a sequential game tree of two players 1 and 2 , and their corresponding action set {A,B} and { C, D} . Player 1 (she) has the move first and player 2 (he) move next.

Figure 6.1

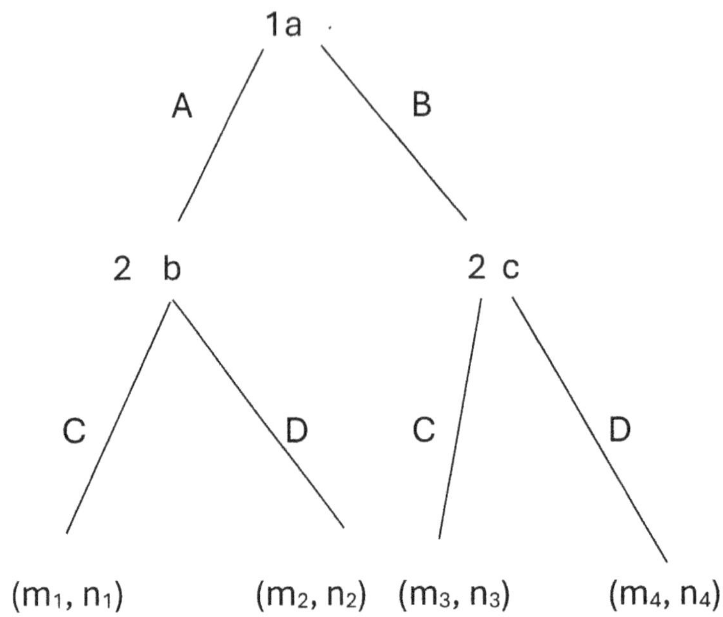

Player 1 has one decision node and player 2 has two decision nodes. We label each decision nodes by a, b, c. An information set of a player is a collection of decision nodes for that player. In this example, player 1 has only one decision node, so that is only one information set, denoted by

$I_1 = \{ a \}$ where the symbol I stands for information set and subscript 1 stands for player 1.

Figure 6.2

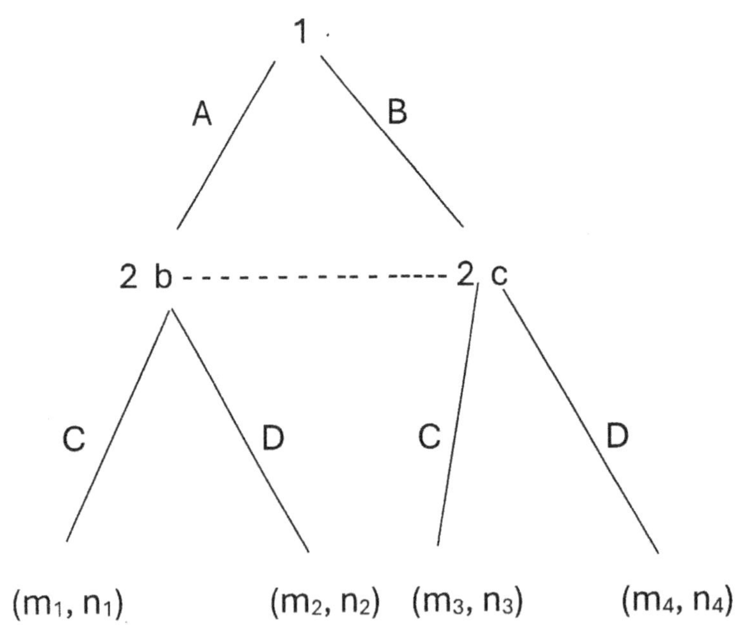

For player 2, there are two decision nodes b and c. So, there are three possible decision sets

$I_{2,1}$ = { b}, $I_{2,2}$ = {c } or $I_{2,3}$= { b, c } . The first subscript represents the player number, and the second subscript represents the number of possible information sets for that player. The important thing to note here is that the difference between a player has an information set that has a singleton decision node like I_1, I_2, , and a player has an information set that has more than one singleton decision node like $I_{2,3.}$. For the same game tree, different specifications of the information sets would give rise to different games. The game trees in Figure 1 and Figure 2 show two different games' sequential interactions. How so? In Figure 1, as there are two information sets $I_{2,1}$ = { b }, and $I_{2,2}$ = {c} for player 2, so at node b or node c, player 2 (he) can observed what player 1 (she) did in terms of taking which action A or action B, and so he can respond with a proper action from his action set { C, D}. In Figure 2 of the game tree however, as $I_{2,3}$= { b ,c}, $I_{2,3.}$ has two decision nodes, we have used a dotted line between node b and node c to signify that b and c are in the same information set. The interpretation of two decision nodes of the same player (player 2) in the same information set $I_{2,3}$ is that when player 2 gets the move, all he knows is that Player 1 has moved, and he does not know what player 1 (she) did in terms of which action A or action B she chooses.

Example 6.1.2 In this example as shown in Figure 6.3, we show an example of a sequential game tree of three players 1, 2 and 3. Player 1 move first, followed by player 2 then followed by player 3. Suppose the information set of this game is shown in Figure 6.3 below.

Figure 6.3

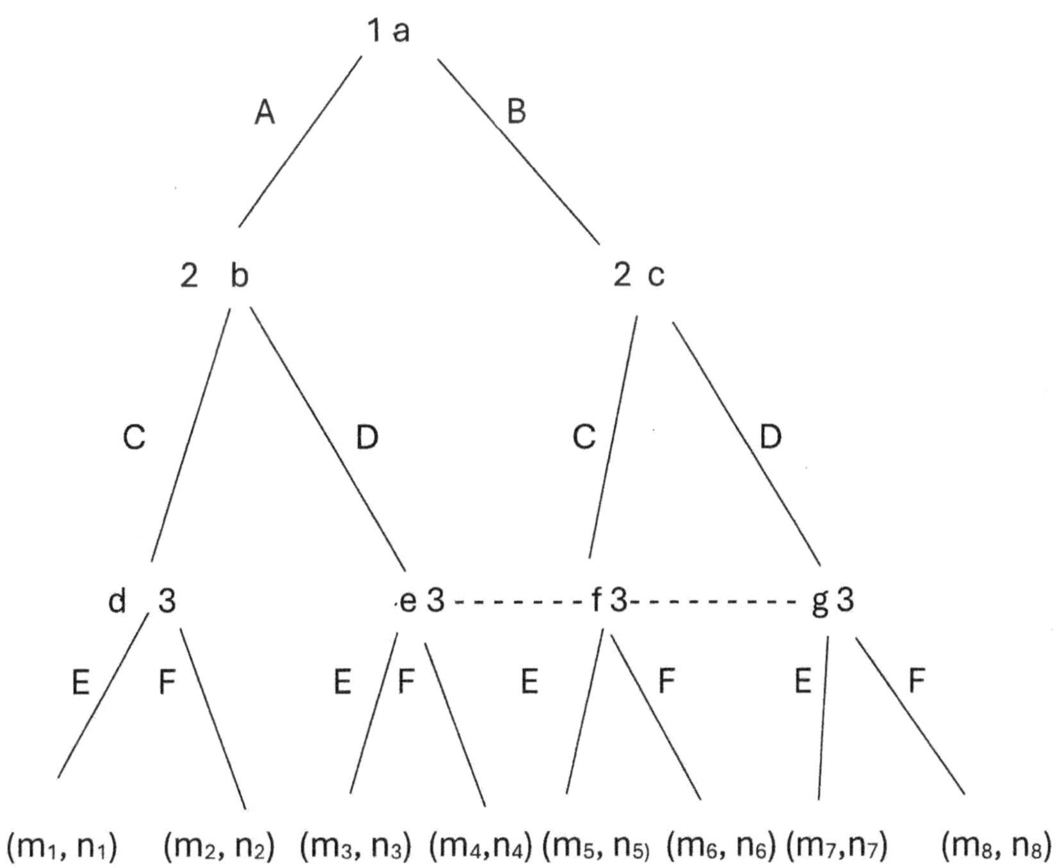

Using the information set notation described before,

$I_1 = \{ a\}$; $I_{2,1} = \{ b\}$,$I_{2,2} = \{ c\}$; ,$I_{3,1} = \{ d\}$, $I_{3,2} = \{ e, f, g \}$

Since each of player 2's information set is a singleton, so player 2 knows what player 1 did, or player 2 observes player 1 has chosen action A when he is at decision node b, and he observes player 1 has taken action B when he is at decision node c. Since $I_{3,1}$ has a singleton, so at decision node d, player 3 knows that player 1 has chosen action A and player 2 has chosen action C. Since $I_{3,2} = \{ e, f, g \}$ is not a singleton set, so at decision nodes e, f, or g, all player 3 knows is that he gets the move, but he does not know what player 1 and player 2 did, or player 3 does not know the full history of the game thus far. Note that in general, the information sets of each player disjointly covers all the decision nodes of that player, in other words, each decision node of a player belongs to one information set , if that information set is a singleton, then the player knows the history of the play up to player]s decision node in the game tree; on the other hand, if the information set is not a singleton, then the player does not know the full history of the play up to any of the player's decision nodes within the information set in the game tree.

After we defined and understood **the concept of information sets** in a game tree, we can now use it to define two related concepts, the **concept of a subgame** and the **concept of pure strategy of a player** and in a game tree.

We give a formal definition of a subgame in a game tree as follows.

Definition 6.1.1 ***A subgame of a game*** given in the extensive form representation is itself a game (we shall refer it as this game in the following definition) that satisfies the following three conditions.

C1: This game starts at a decision node, say node x, of the original game that is a singleton information set .

C2: This game includes all the decision and terminal nodes following node x in the original game, but it does any include any nodes that do not follow x in the original game.

C3: If node y is any other decision node follows x in the original game tree, then all other nodes z in the information set containing y must also follow x.

We examine the possible subgames in the game shown below:

Figure 6.4

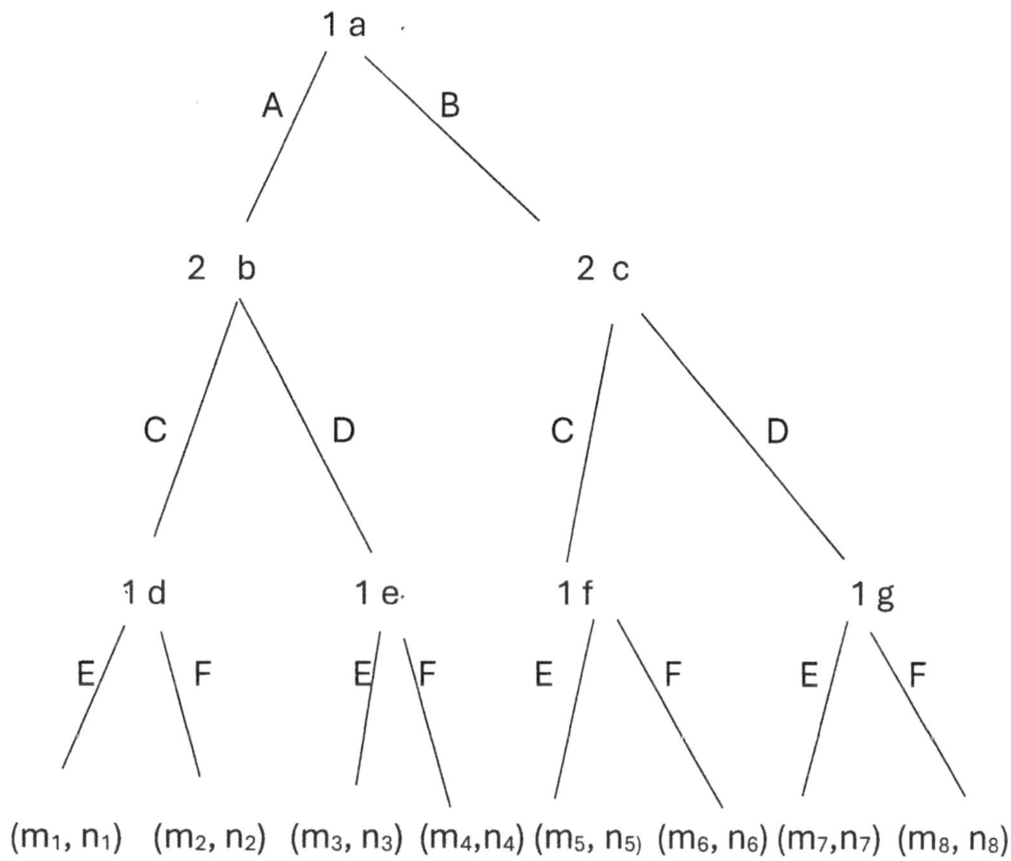

The four subgames are drawn in blue, starting at the decision nodes d, e, f, and g. The two subgames are those drawn in red, starting at the decision node b, and c. The original game can be considered as a subgame of the original game starting at decision node a. The conditions C1, C2, C3 can be easily checked and are satisfied if we let the node x in the subgame definition above to be the decision node at node d, e, f, and g respectively for the four subgames dawn in blue, let the node x be the decision node at b and c respectively for the two subgames dawn in red, and let the node x be the decision node at node a of the original game dawn in yellow. The following game however, not all these three conditions are satisfied.

Figure 6.5

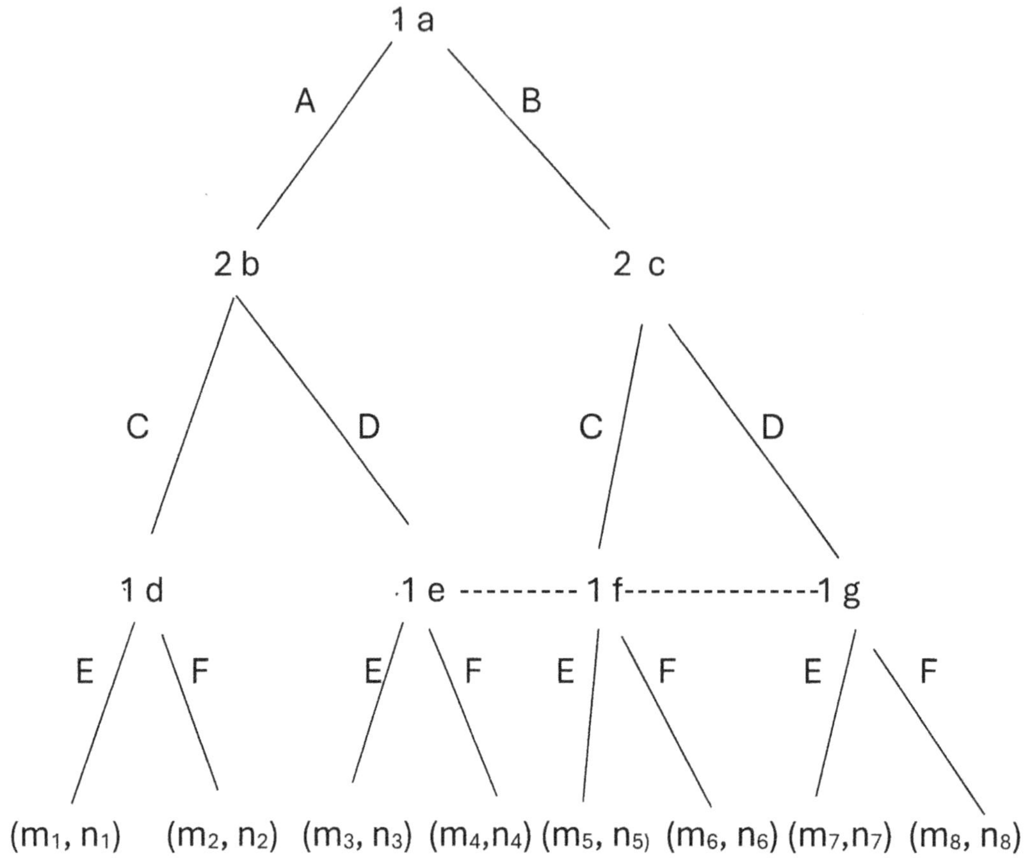

The main difference between this game and the game before is that now we an information set {e, f, g} that is not a singleton set, only the information set {d} and {a } are singleton information sets. The only subgames are the one shown in blue, and the one shown in yellow. These two subtrees are subgames because they both satisfies all the conditions C1, C2, and C3. But the other three subtrees shown in blue in Figure 6.5 are not subgames because they do not satisfy C1. The other two subtrees shown in red in Figure 6.5 are not subgames because they do not satisfy C2 and C3.

As we have just discussed, we saw how the concept of a subgame is defined in terms of the concept of information set, the following shows how a player's strategy is also defined in terms of the concept of the information set in a game.

6.2 What does it mean by player's strategies in a game and how do we represent these strategies?

First, we want to know what a pure strategy for a player is. Recall that in the simultaneous game, each player has a set of actions, these actions are played out by the player simultaneously with the other players, and there is no precedence amongst these actions for the player. But now since the game is played sequentially, we need to specify the order of these actions are played out for each player by means of the concept of information set discussed previously, we call this ordering of actions for the player to play a pure strategy for that player. And a follow up question is how many possible pure strategies a player can play. We show a simple combination formula for the total number of pure strategies possible in the following example.

Example 6.2.3

This example illustrates the definition of a pure strategy of a player.

Figure 6.6

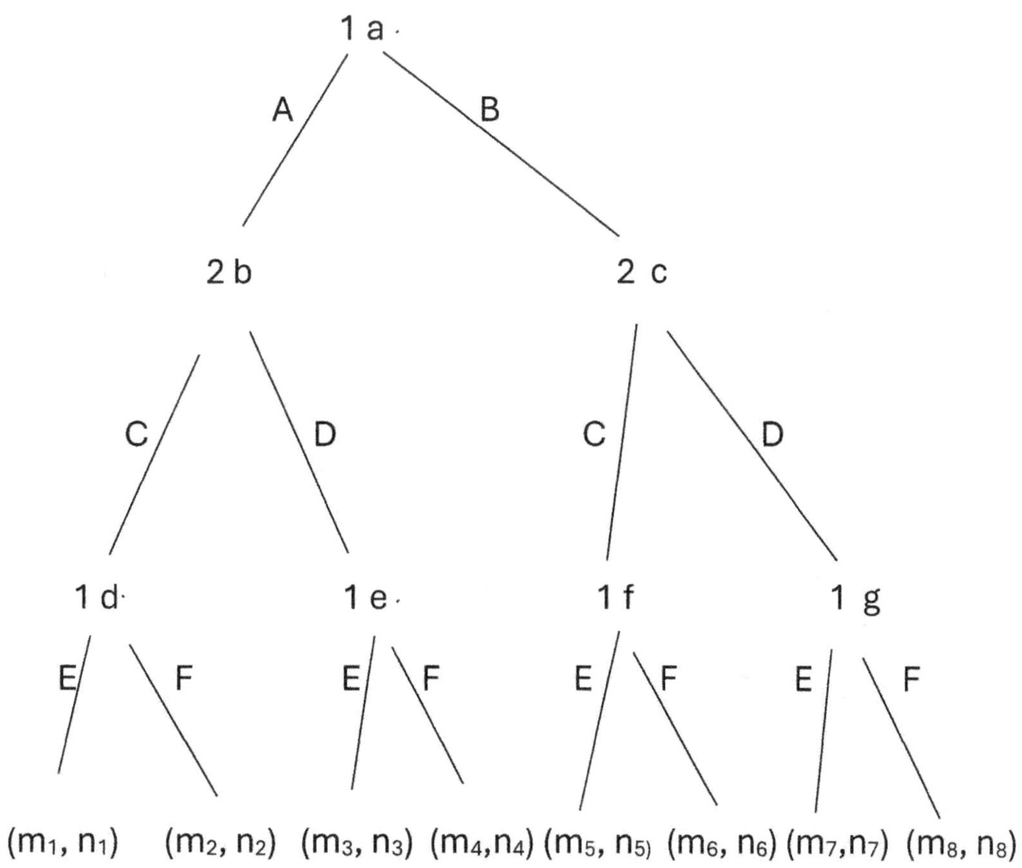

Definition 6.2.2. ***A pure strategy for player i*** is a complete plan of action that describes which action player i will choose at each of the player's information set. More precisely, by the definition of a function from one set to another , we can say that a pure strategy for player i is a function from the set of all the information sets of player- i to the set of all possible actions of player- i. We formulate this concept below and use Example 6.2.3 to Figure 6.6 as an illustration.

Player 1 has 5 information sets: {a}, {d}, {e}, {f}, {g}. Player 2 has two information sets: {b}, {c}. The action sets of player1 are { A, B} and {E,F}. The action set of player2 is {C,D}. Use the notations as before,

$I_{1,1}$ = {a}, $I_{1,2}$= {d}, $I_{1,3}$={e}, $I_{1,4}$= {f}, $I_{1,5}$={g} ; $I_{2,1}$= (b}, $I_{2,2}$={c}

Let C(i) be the collection of all the information sets for player i,

then C(1)= { $I_{1,1}$, $I_{1,2}$, $I_{1,3}$, $I_{1,4}$, $I_{1,5}$ }, C(2) = { $I_{2,1}$, $I_{2,2}$}.

Let A($I_{i,j}$) to denote the action set corresponds to the information set $I_{i,j}$, and let n(i) be the total number of information sets for player i, let A(i) to be the union of all these n(i) action sets , $A(i) = \bigcup_{j=1}^{j=n(i)} A(I_{i,j})$ for player i. Refer to the original definition stated above, a pure strategy for player i is a function s_i from C(i) to A(i) that assigns each information set in C(i) a possible action in A(i), we can compute the total number of such pure strategies in the above Example 6.6 as follows.

Let $k_{i,j}$ be the number of actions in $I_{i,j}$, then we can express the total number of pure strategies for player i, denoted by S_i , as

$$S_i = k_{i,1} \text{ x } k_{i,2} \text{ x } \ldots\ldots \text{ x } k_{i,n(i)} \qquad \text{(D 1)}$$

In this example Figure 6 above , A($I_{1,1}$) = {A, B} , A($I_{1,2}$) = {E,F},

A($I_{1,3}$) = {E,F}, A($I_{1,4}$) = {E,F}, A($I_{1,5}$) = {E,F}, A($I_{2,1}$) = { C, D}, A($I_{2,2}$) = { C, D}; n(1) = 5, n(2)=2; $k_{1,1}$= 2, $k_{1,2}$= 2, $k_{1,3}$= 2, $k_{1,4}$= 2, $k_{1,5}$= 2; $k_{2,1}$= 2 $k_{2,2}$= 2;

so , A(1) = { A, B, E,F}, and A(2) = {C, D}.

S_1= 2 x 2 x2 x 2 x2 =32 and S_2= 2 x 2 =4. Now that we know what pure strategies in a dynamic game are, next question is how these strategies are represented. The next three examples show the representation of each strategy using their corresponding action names.

Example 6.2.4. This example illustrates the definition of a pure strategy of a player

Figure 6.7

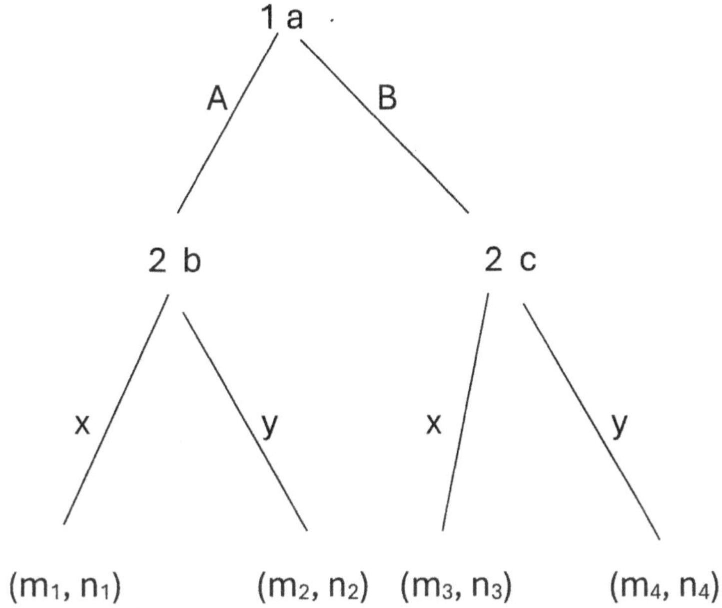

By Formula (D1) above, the number of strategies that player 1 has is $S_1 = 2$, the number of strategies that player 2 has is $S_2 = k_{2,1} x\, k_{2,2} = 2X2 = 4$. The strategy set for player 1 at the information set on a is { A, B}, and the strategy set for player 2 at the information set at b and c is { x,y } respectively. Thus, there are {xx, xy, yx, yy} strategy pairs for player 2. So, there are total 2X4=8 that many strategies that player 1 and player 2 can play out. But amongst all these 8 strategies, not all of them can be in Nash Equilibrium and we are going to examine which one is NE and which one is not. We can tabulate the strategies as follows

	x x	x y	y x	y y
A	(m_1, n_1)	(m_1, n_1)	(m_2, n_2)	(m_2, n_2)
B	(m_3, n_3)	(m_4, n_4)	(m_3, n_3)	(m_4, n_4)

The pure strategies of this game are represented as: < A, xx > , <A, xy >, < A, yx >, < A, yy>, < B,xx>, < B, xy>, < B, y x>, < B , yy>. The first component is one of the actions of player 1 at the information set ' a' can take, and the second component is one of the actions that player 2 at information sets b and c can take respectively. Any one of the entries in the above table can be a Nash Equilibrium depending on the playoff values (a_1, b_1). For example, <A, xx > is a NE if $m_1 \geq m_3$ and $n_1 \geq n_2$. We will discuss this in more detail in the next topic on Subgame Nash Equilibrium.

Example 6.2.5. Consider the following game tree.

Figure 6.8

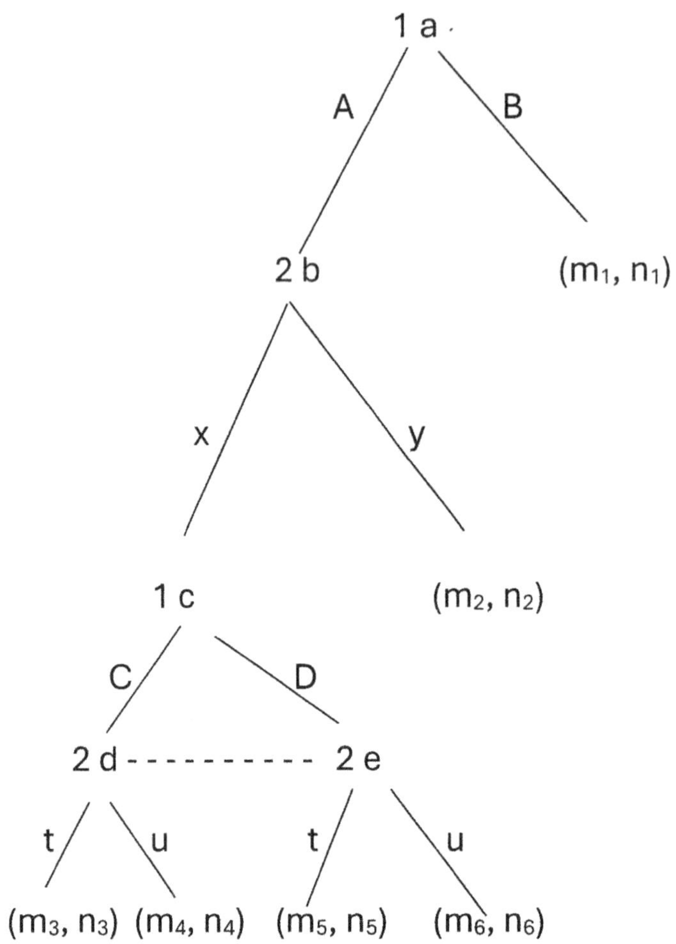

Player 1 has two information sets at 'a' and 'c' with actions {A, B} and {C,D} respectively, so, player 1's strategies are represented as AC, AD, BC, BD.

Player 2 has two information sets at 'b' and at 'd, e' with actions {x,y} and {t,u} respectively, so player 2's strategies are represented as { xt, xu, yt, yu} . Thus, all the strategies for both players in the game can be summarized in the following matrix and tracing out the strategies to fill out the playoff values.

	x t	x u	y t	y u
AC	(m_3, n_3)	(m_4, n_4)	(m_2, n_2)	(m_2, n_2)
AD	(m_5, n_5)	(m_6, n_6)	(m_2, n_2)	(m_2, n_2)
BC	(m_1, n_1)	(m_1, n_1)	(m_1, n_1)	(m_1, n_1)
BD	(m_1, n_1)	(m_1, n_1)	(m_1, n_1)	(m_1, n_1)

There are 16 pure strategy pairs, and they are represented as <AC, xt>, …., <AC, yu>, …… , <BD, xt>, <BD, yu> etc.

Again, any of the strategy pair for player 1 and player 2 can be a Nash Equilibrium pending on the values of the (m_1, n_1), we shall exploit these possible Nash Equilibrium in this game later. For now, we need to understand how these pure strategies come up in terms of the information sets of each player in the game.

Example 6.2.6 Consider the following game tree.

Figure 6.9

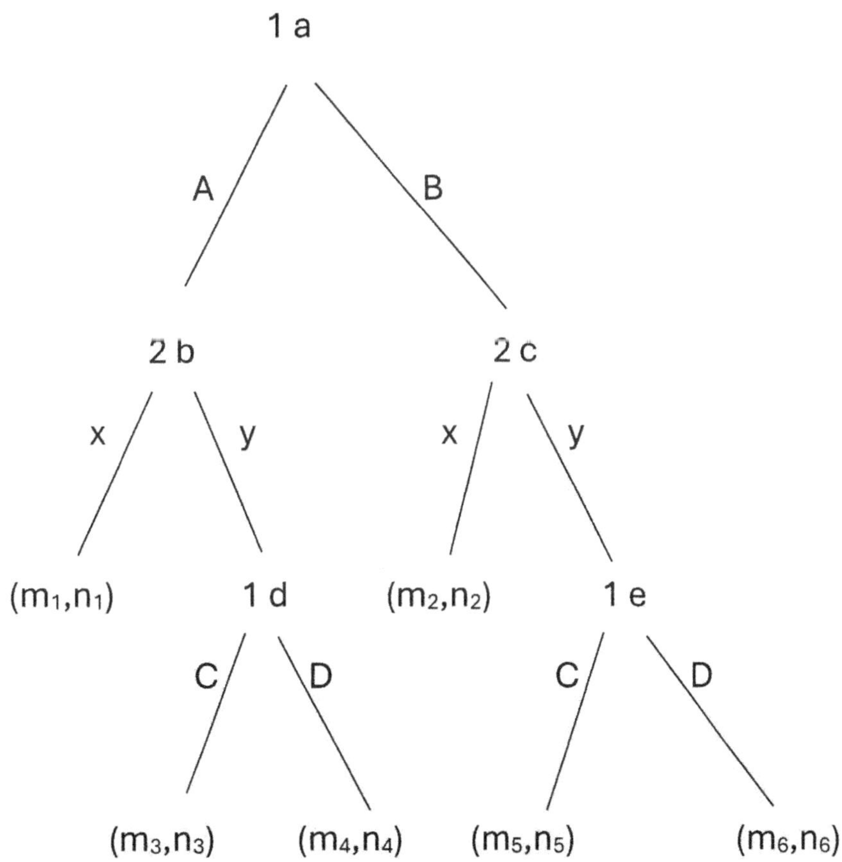

As there are two actions for player 1 at each of the information set 'a', 'd', and 'e' , the number of pure strategies for player 1 is 2 X 2 X2 =8 from Formula (D1) . Similarly, player 2 has two information sets at 'b' and 'c' and each information set has two actions, so the number of pure strategies for player 1 is 2x2 =4 by Formula 1 as well. Thus, we will end up with an 8X4=32 entries matrix of payoff values for each of each 32 pure strategy pairs.

The strategies for player 1 are { ACC, ACD, ADC, ADD, BCC, BCD, BDC, BDD} and for player 2 are { xx, xy, yx, yy}

Example 6.2.7. Consider the following game tree.

Figure 6.10

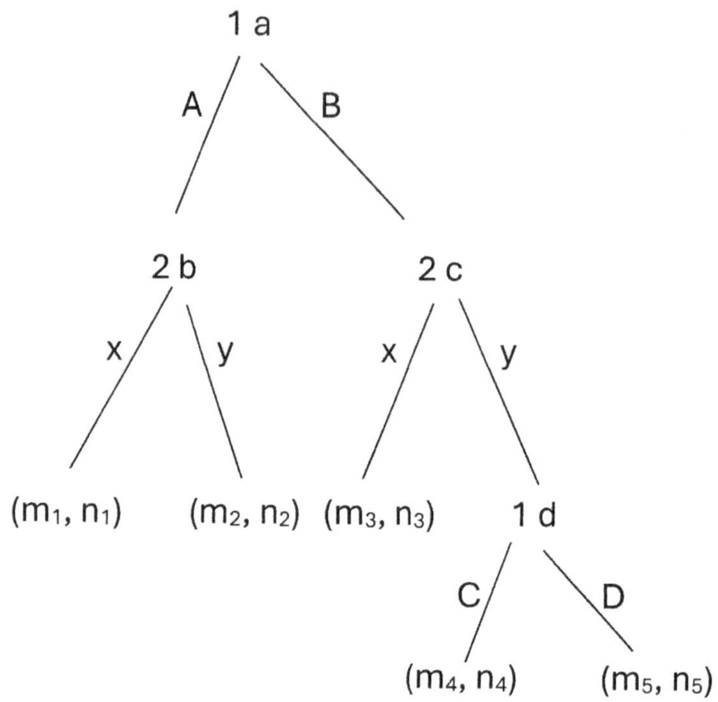

Player 1 has 2 information sets at 'a' and 'd', his pure strategy set is { AC, AD, BC, BD}. Player 2 has 2 information sets at 'b' and ' c', her pure strategy set is { xx, xy, yx, yy}. The matrix representation of all the strategies for the game is a 4 by 4 matrix.

	x x	x y	y x	y y
AC	(m_1, n_1)	(m_1, n_1)	(m_2, n_2)	(m_2, n_2)
AD	(m_1, n_1)	(m_1, n_1)	(m_2, n_2)	(m_2, n_2)
BC	(m_3, n_3)	(m_4, n_4)	(m_3, n_3)	(m_4, n_4)
BD	(m_3, n_3)	(m_5, n_5)	(m_3, n_3)	(m_5, n_5)

From this matrix representation of the strategy pairs, we can then determine which of the above strategy pairs will be a Nash Equilibrium from the basic definition of Nash Equilibrium.

6.3 What does it mean by sequential rationality and creditability of player's action in a dynamic game?

One of the basic assumptions in game theory is that the players are all rational, in terms of the game tree and the concept of information set, this means that the players must choose their optimal strategies at every information set. Since in a game tree, the order of movement for each player is specified in advance, when it is the player's turn to play, the player must choose the best action that is available to him at that moment that will give him (her) the highest payoff value. We refer to this tacit assumption as *the principle of sequential rationality*. As each player has finite set of actions he or she can take, if the action causes the other player a substantial loss in terms of that player ends up with a very low payoff value in the game tree when the game is done, we shall refer to the player's action as threat to the other player. We shall see in the following examples that not very threat is creditable. In other words, before playing the game, if player 1 says to player 2 that he (player 1) will play his threat action if she (player 2) does not respond to him (player 1) with an action that is in favor of him. There are two scenarios that the threat player 1 posts may not be creditable. The first scenario is that when it is player 1's turn to play where the threat action is available to him, another action is also available to him that will gives him a higher payoff value than the threat action does. He chooses the other action over the threat action because he follows the principle of sequential rationality. The other scenario is that the game tree dictates the order of movements for each player, as such , when the game begins playing and it follows a certain path and taking turns in executing the player's action in the game tree, the threat action may not be in the player's path. If , however, a player can make his threat creditable , it may well be to his advantage and putting his threat actions into good use. How so? We will give an example due to Thomas Schelling below.

We have seen from the previous examples how to set up the strategies payoff matrix from a game tree and from which we can find the Nash Equilibrium by the method we studied before. Amongst all the Nash Equilibriums, there is one that is of special interest to us, it is called the Subgame Nash Equilibrium, it is closely related to the concept of creditable threat we are discussing here. Before we give a formal definition of subgame perfect Nash Equilibrium in the next section, we will explore this concept along with creditable threat concept in the example below.

Example 6.3.8 This example illustrates the meaning of a non-creditable threat and Nash Equilibrium in a game tree.

Figure 6.11

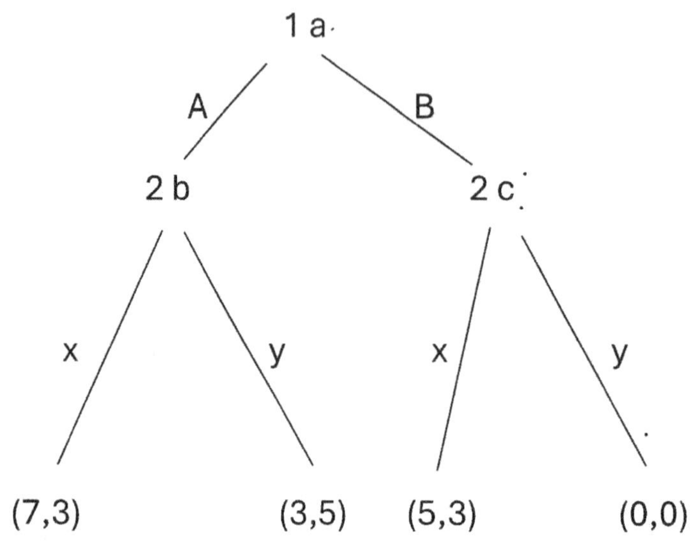

We obtain the game tree above from the tree in Example 4 by choosing $m_1 = 7$, $n_1= 3$, $m_2 = 3$, $n_2=5$, $m_3= 5$, $n_3=3$, $m_4=0$, $n_4= 0$

	x x	x y	y x	y y
A	(7,3)	(7,3)	(3,5)	(3,5)
B	(5,3)	(0,0)	(5,3)	(0,0)

If player 2 plays action y, then player 1 will result in getting the lowest payoff value of 0, so player 2 seemingly posts a threat to player 1 as "if you player 1 do not play action A, then I player 2 will take action "y". However, this threat "y" is not creditable, because when it is player 2's turn to choose an action, player 2 would chooses action x and gets a playoff of 3 rather than 0 by choosing y since she follows the principle of sequential rationality.

From the above payoff table, we can see the strategy pair <A, yy> in red with payoff value (3,5) is a Nash Equilibrium. This is because player 1 does not want to switch to B, and player 2 does not want to switch to yx, xy,and xx. Note that the threat action "y" is on the Equilibrium path.

By the same reason, we see that there is another strategy pair <B, yx> in blue is also a Nash Equilibrium but, in this case, the threat action "y "is not on the Nash Equilibrium path. In these two Nash Equilibriums, one contains the threat action, and one does not. In the next section, we shall give a more restricted form of Nash Equilibrium called the subgame perfect Nash Equilibrium and show its relationship with the concept of creditable threats.

Example 6.3.9 This example illustrates the meaning of a non-creditable favor and Nash Equilibrium in a game tree.

Figure 6.12

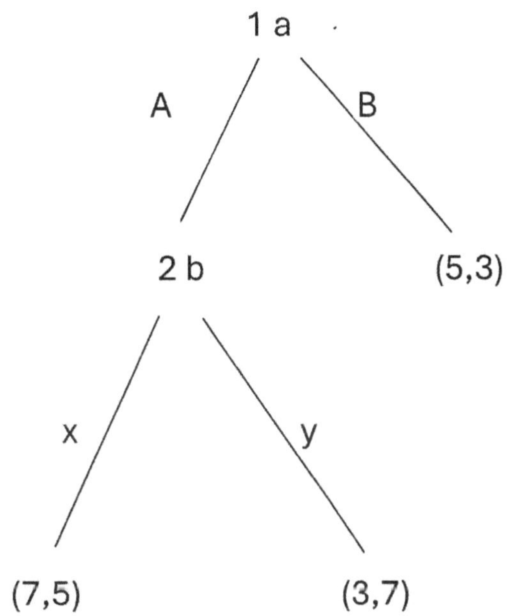

We obtain the game tree above from the tree in Example 4 by choosing $m_1 = 7$, $n_1 = 5$, $m_2 = 3$, $n_2 = 7$, $m_3 = 5$, $n_3 = 3$, $m_4 = 5$, $n_4 = 3$

	x x	x y	y x	y y
A	(7,5)	(7,5)	(3,7)	(3,7)
B	(5,3)	(5,3)	(5,3)	(5,3)

In this example, if player 2 plays action x, it is a favorable action to player 1. However, this favorable action x that player 2 posts to player 1 is non-creditable, because when it is player 2's turn to play action x, player 2 would rather play action y to achieve a higher payoff value of 7 rather than y. Note that in this game, the Nash Equilibrium < B, yx> is shown in red in the above table with payoff value (5,3), the favorable action x is not on the equilibrium path. Having seen the above two examples on non-creditable threat and non-creditable favor, we are now in position to formally introduce the definition of subgame perfect Nash Equilibrium (SPNE) and show how the SPNE can rule out these non-creditable threats and favors.

6.4 What is subgame - perfect Equilibrium SPE?

Definition 6.4.3 A subgame perfect Nash Equilibrium (SPNE) is a Nash Equilibrium in the original game tree such that when the strategies in the Nash Equilibrium are restricted to every subgame of the original game, the restricted strategies are also Nash Equilibriums in the subgames. We will show four examples of SPNE below, then we will describe a formal procedure called the backward induction to find such SPNEs.

Example 6.4.10 This example illustrates the difference between a regular Nash Equilibrium (NE) and a subgame perfect Nash Equilibrium (SPNE).

Figure 6.13

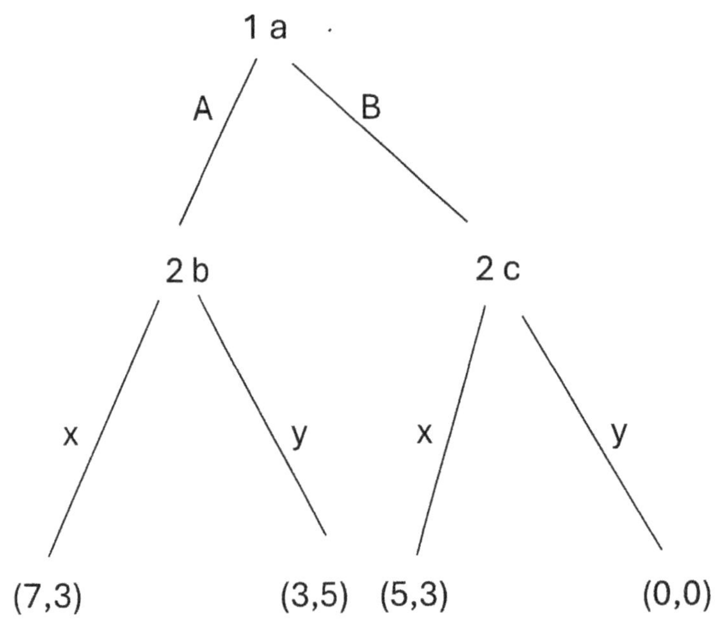

The above game tree is taken from Example 6.3.8, there we have shown that both the strategy pairs <A, yy> with payoff value (3, 5) in red and < B, yx> with payoff (5,3) in blue are Nash Equilibriums. <A,yy> is not a SPNE because when it is restricted to the subgame at the information set c , the strategy 'y' is not optimal because strategy x gives payoff value of 3nd y only gives 0 . The strategy pair <B, yx> is a SPNE because there are three subgames in all, one with root at 'b', one with root at 'c', and one with root at 'a'. The strategy y is optimal at 'b' as 5 is better than 3, the strategy x is optimal at 'c' as 3 is better than 0, and strategy B is optimal at 'a' because if A were chosen, player 1 only gets 3 as opposed to him getting 5.

Example 6.4.11 This example shows a SPNE in a game tree. Consider the game tree shown in Example 6.2.7 with

Figure 6.14

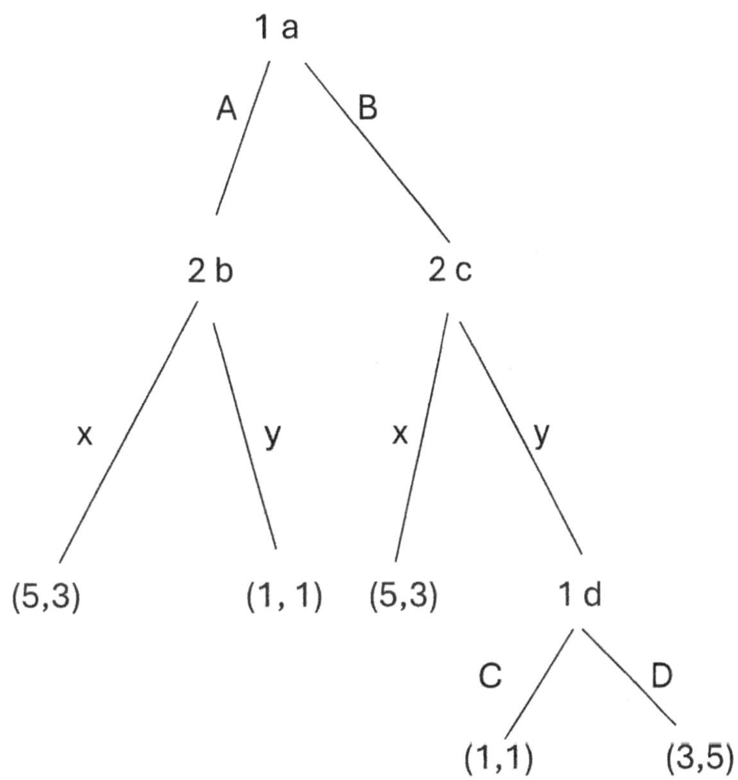

$m_1 = 5$, $n_1 = 3$, $m_2 = 1$, $n_2 = 1$, $m_3 = 5$, $n_3 = 3$, $m_4 = 1$, $n_4 = 1$, , $m_5 = 3$, $n_5 = 5$, we have the following table for the payoff values.

	x x	x y	y x	y y
AC	(5,3)	(5,3)	(1,1)	(1,1)
AD	(5,3)	(5,3)	(1,1)	(1,1)
BC	(5,3)	(1,1)	(5,3)	(1,1)
BD	(5,3)	(3,5)	(5,3)	(3,5)

From the above payoff tables, we can see that both <AD,xy> in red, and <AC, xy> in blue are both Nash Equilibriums, but only <AD, xy> is SPNE . As shown in red, there are 4 subgames with root at 'd', at 'b', at 'c', and at 'a'. At the subgame with root at 'd', D is the optimal strategy with payoff value 3 better than 1; at the subgame with root at 'b', x is the optimal strategy with payoff value 3 better than an 1; at the subgame with root at c, y is a strategy as good as x with payoff value 3; and at the subgame with root at a, A is a strategy with payoff value 5 same as the strategy B with payoff value 5. <AC,xy> is not SPNE because in the subgame with root at d, C is not optimal as its payoff is 1 less than 3.

So, the method of using the payoff matrix to find the Nash Equilibrium as shown in the examples above will not always give us a Nash Equilibrium (SPNE) that is also a Nash Equilibrium when it is restricted to all the subgames of the original game.

6.5 Backwards and Forwards Induction

There are two common approaches to finding the subgame-perfect Equilibriums in a game tree that we will discuss in detail below.

6.5.A Theory behind the Backwards Induction

The method of Backwards Induction described below allows us to find the subgame-perfect Nash Equilibrium. This method starts with the last player in the game tree, finds the player's optimal strategy, then finds the optimal strategy of the player before the last player, and continues to find the *optimal strategy of the first player in the tree.*

We illustrate this theory by a three players game three.

Figure 6.15

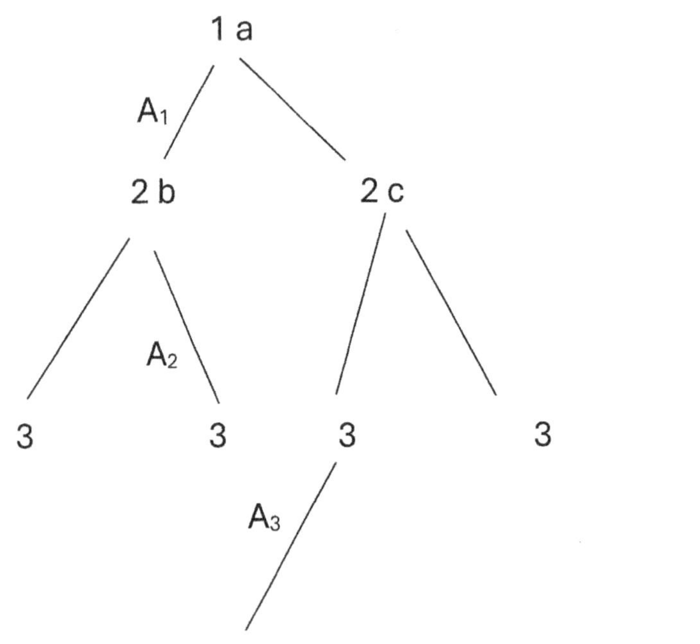

In the diagram above, A_1, A_2, A_3 are the possible actions that player's 1 , 2, and 3 can take from their finite action sets S_1 , S_2 , S_3 respectively. The functions l(*), m(*), n(*) are the payoff functions for players 1, 2, and 3 respectively.

The backwards induction starts from the last player, in this case, player 3 is looking for a best response strategy that maximize the payoff function n (*) given the actions A_1 and A_2 previously chosen by player 2 and player 1, denote player 3's best response strategy by $bR_3(A_1, A_2) = \max_{A_3 \in S_3} n(A_1, A_2, A_3)$. Now move backward to player 2, and similarly denote Player 2's best response strategy by $bR_2(A_1)$ then $bR_2(A_1) = \max_{A_3 \in S_2} m(A_1, A_2, bR_3(A_1, A_2))$. Finally, it is player 1's turn to move,

Player 1's best strategy to move is the strategy A_1^* that maximize the function l(*),

that is $A_1^* = \max_{A_1 \in S_1} l(A_1, bR_2(A_1), bR_3(A_1, A_2))$. After we find the best strategy for player 1, we can then move forward to find the best strategy A_2* player 2, then find the best strategy A_3* for player 3 as follows:

$$A_2^* = bR_2(A_1^*)$$

$$A_3^* = bR_3(A_1^*, A_2^*)$$

The strategy triple (A_1^*, A_2^*, A_3^*) is called the backwards induction outcome of the game. Because of the way the backward induction strategies are selected, the backwards induction outcome does not involve any non-credible threats. Player 1 anticipates what player 2 will respond optimally to any action A_1 that player 1 might choose, by playing $bR_2(A_1)$, player 1 gives no credence to threats by player 2 to respond in ways that will not be in player 2's best interest when it is player 2's turn to play. Similarly, player 2 gives no credence to threats by player 3. Player 3 will play the action $bR_3(A_1, A_2)$ regardless. In what follows, we are going to present six examples, Examples 6.5.12 to 6.5.17 to illustrate the concept of backwards induction and the notations used during the formulation of this process above.

Example 6.5.12 We use the game tree from Example 6.3.8

Figure 6.16

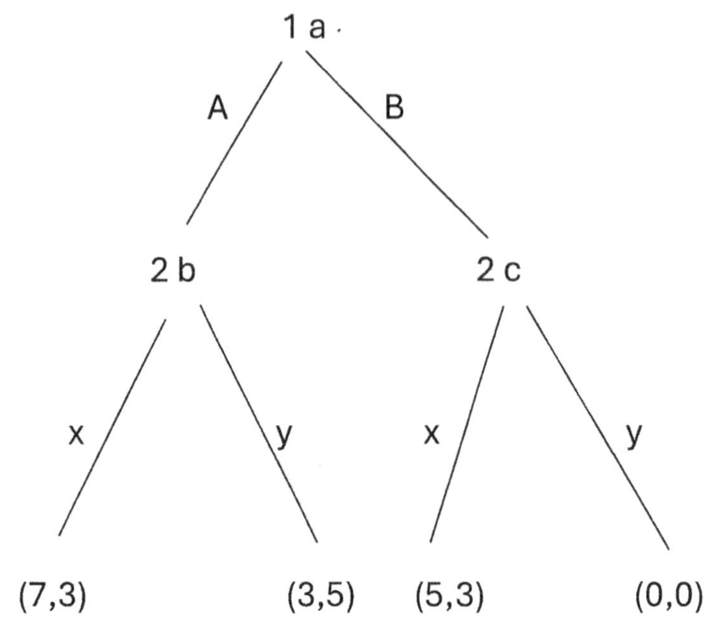

We start are the last player, player 2 whose payoff function is n (A_1, A_2)

$bR_2(A) = \max\limits_{\{x,y\}}\{n(A,x), n(A,y)\} = \max\limits_{\{x,y\}}\{3,5\} = y$ as the maximum is 5 which occurs at y'

Similarly, $b\,R_2(B) = \max\limits_{\{x,y\}}\{3,0\} = x$ as the maximum is 3 which occurs at x.

Thus player 1 will choose the action that would maximize player 1's payoff function m $(A_1, bR_2(A_1))$ which is in our case, $\max\limits_{\{A,B\}}\{m(A, R_2(A)),\ m(B, R_2(B))\ \}$

, which is $\max\limits_{\{A,B\}}\{m(A,y),\ m(B,x)\ \}$, which is, $\max\limits_{\{A,B\}}\{3,\ 5\ \}$, which is action B as 5 is the larger payoff value which occurs at action B. So, the backwards induction strategy is < B, yx> and it is shown in red in the game tree above.

Example 6.5.13. We refer to the game tree from Example 6.2.6.

Figure 6.17

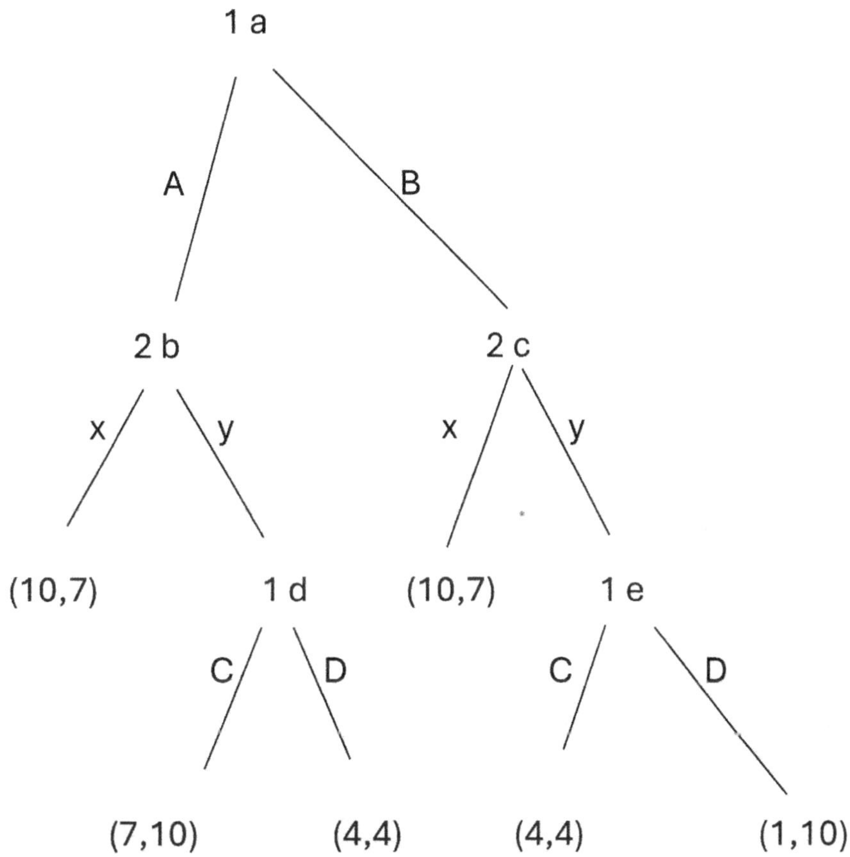

Let m (*) , n(*) be the payoff functions for player 1 and player 2 respectively. Starting from the last stage of the game which is player 1's turn to move. The best response action for player 1 is computed as follows.

$$bR_3(A,y) = \max_{\{C,D\}}\{m(A,y,C), m(A,y,D)\}$$

$$= \max_{\{C,D\}}\{7,4\}$$

= C, as the maximum is 7 which occurs at C

$$bR_3(B,y) = \max_{\{C,D\}}\{m(B,y,C), m(B,y,D)\}$$

$$= \max_{\{C,D\}}\{4,1\}$$

= C, as the maximum is 4 which occurs at C

At the next to the last stage, player 2's best response action is computed as:

$$bR_2(A) = \max_{\{x,y\}}\{n(A,x), n(A,y,C)\}$$

$$= \max_{\{x,y\}}\{7,10\}$$

$$= y$$

$$bR_2(B) = \max_{\{x,y\}}\{n(B,x), n(B,y,C)\}$$

$$= \max_{\{x,y\}}\{7,4\}$$

$$= x$$

Now back at the first stage, player 1's best strategy is to choose strategy A_1 such that m ($A_{1,}$ $bR_2(A_1)$, $bR_3(A_{1,}$ $bR_2(A_1)$)) is maximum.

When A_1 is A, the payoff value of player 1 is m(A, y, C) = 7 and when A_1 is B, the payoff value of player 1 is m (B, x, C)= 10, and 10 is larger than 7, so player 1's best strategy is to choose B . Thus the backwards induction shown in red in the above game tree is represented as < (B, C, C), (y,x) >.

We have just shown in the above two examples the mechanics of how to find the subgame perfect solution of a game when it is represented in a game tree. Now we ask this question, does the backwards induction solution always give us a meaningful solution to the underlying competitive scenarios when these scenarios are represented by game trees? We will give three examples below to demonstrate that the answers are not always yes.

Example 6.5.14 We refer to the game tree below, this is a short version of the classical Centipede Game.

Figure 6.18

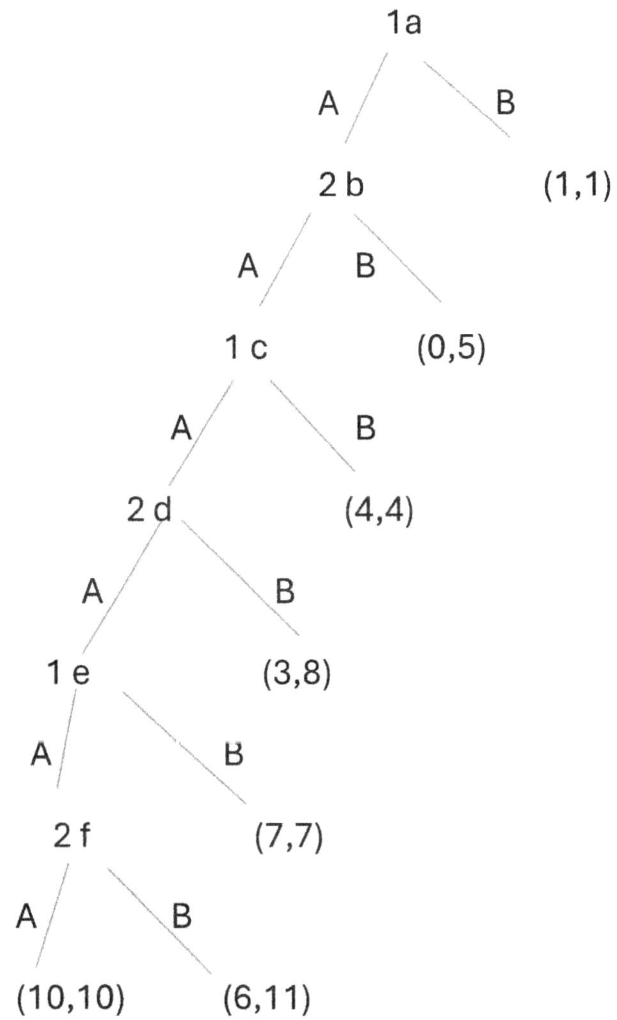

As shown, this is a 2-player game, each player has two choices of actions, A and B. At the beginning of the game, each player puts 1 dollar into a pot. Either player can choose to continue the game by putting 3 dollars into the pot or choose to stop the game by splitting the total amount of money in the pot according to the way described below. So, action A in the game tree is "add \$ 3 into the pot, except if the player wants to end the game, then just split the amount in the pot ", and action B is "splitting the money in the pot". If the total amount in the pot is an even number, then each player gets half. Otherwise, the player does the splitting gets to take \$ 5 first and then split the rest of the amount each player get half. The pot starts with 2, hence after player 1 at the information set 'a' decides to end the game, each will get (1,1) as payoff by splitting the 2, and player 2 at the information set 'b' gets (0,5) as the total amount at the pot is only 5 as there is nothing left to split after player 2 takes the first 5. The subsequent payoffs (4,4), (3,8), (7,7), (6,11), (10,10) are generated accordingly. Looking at the actual scenario of this game shown in the game tree, the optimal solution would have been for both players to finish playing the game, by then each player would achieve payoff (10,10). However, as we shall see below, by backwards induction, the SPE solution is to end the game from the start, the payoff should be (1,1) instead of (10,10). This shows that the backwards induction solution is not always meaningful.

To demonstrate the backwards induction in the game tree above, we start at player 2's information set "f", as 11 is greater than 10, player 2 chooses B; moving back at player 1's information set "e", player 1 would choose B because that will give him 7, and choosing A would give him 6; moving back at 2's information set "d", player 2 would choose B because that will give her 8, and choosing A would give her 7; similarly , at 1's information set 'c', player 1 would have chosen B; at player 2's information set 'b', 2 would have chosen B, and finally, at 1's information set 'a', player 1 would have chosen B. The entire backward induction process is shown in red on the game tree.

Example 6.5.15 Our next example will show that the backwards induction solution does not gives the optimal solution to the chain store problem, what we are describing below is a simplify version of the classical Chain Store Paradox.

Suppose a monopolist, player M, has branches in 10 cities, and M has 10 potential competitors C one in each city. Each competitor C has the choice of two actions IN and OUT. IN means C will compete with M, and OUT means C will stay out the competition with M. Monopolist M has two actions Cooperative (c) or Aggressive (a) when a competitor C choose the IN action. Suppose there are no collaboration amongst the 10 competitors, each one decides independently and sequentially one at a time as to choose IN or OUT. Suppose the payoff values of a typical competitor and the monopolist interaction are shown in the following game tree:

Figure 6.19

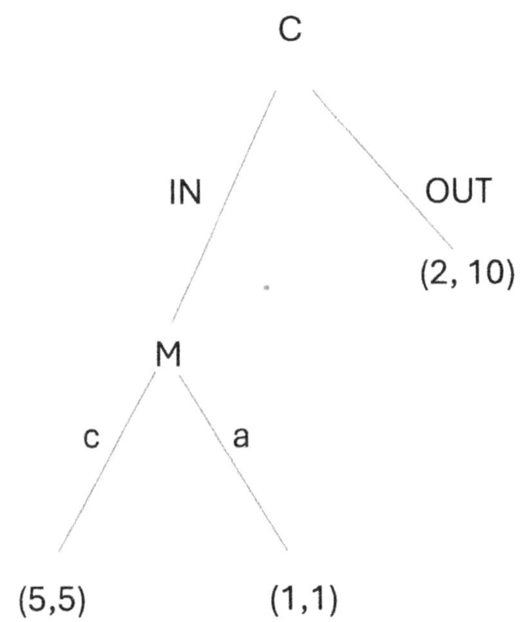

Apply backwards induction to the above game tree, the monopolist M would choose the cooperative strategy "c" to achieve the better payoff 5 rather than choosing the aggressive strategy "a" ending up in a lower payoff value of 1. The competitor C would choose the strategy IN to achieve the payoff 5 rather than a lower payoff of 2. The solution is shown in red. Thus, when the chain stores and the competitors each employ the backward induction solution to their business, the total payoff for the monopolist M from the 10 cities is 10 X 5 =50, whereas each competitor from each city would be 5.

But what about the total payoff if the monopolist M employs other strategies rather than the backwards induction? Suppose M plays the aggressive strategy "a" to deter the competitors from joining the competition. Suppose 6 out of 10 are deterred, then the monopolist would receive a payoff of 6 X 10 +4 X 5= 80, this is a much better outcome than the backwards induction payoff outcome of 50. Suppose even there is only 3 competitors got deterred and choose OUT, the remaining 7 enters, then the monopolist's payoff would be 5 x 5 + 2X 2 + 3 X10 = 59, which is still better payoff than the backwards induction solution of 50.

Example 6.5.16 Our last example involves many players, and it will also demonstrate that the backwards induction solution does give the optimal solution under a big assumption. The game tree below involves 4 players, each player has a choice of two actions A and B. A means to contribute and B means to take. If all the 4 players in a row take the same action A to contribute, then each will earn S 100 at the end of the game. However, if any player chooses action B to take, then the player will earn $1 and the game ends.

Figure 6.20

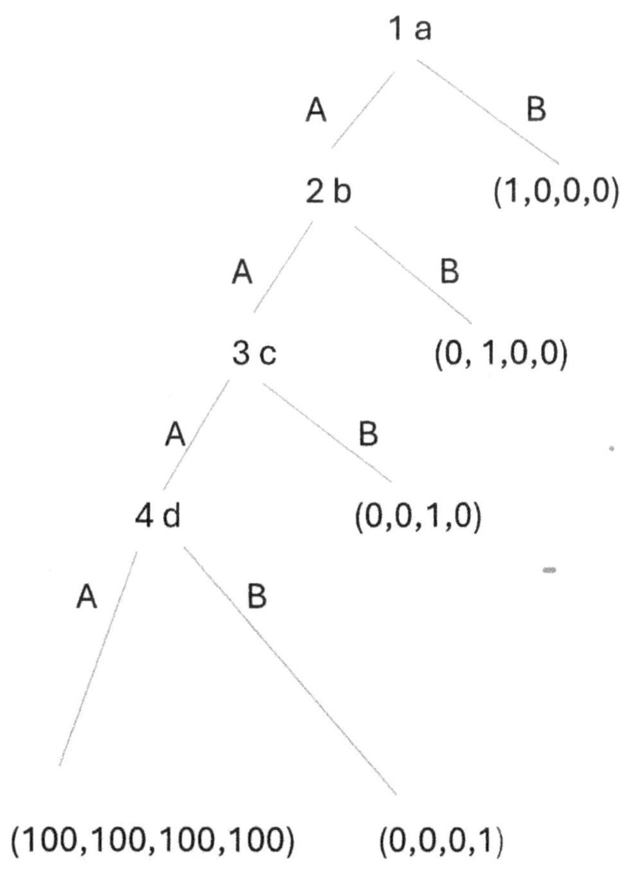

Apply backwards induction at player 4's information set "d", action A is better than B as 100 is better than 1. Moving backwards to player 3's information set c, player 3 will choose A because he will get 100 rather than 1 by choosing B, and his choice of A is based on the assumption that player 4 will choose A. By similar reasoning, player 2 at information set "b" will choose A based on the assumption that player 3 will choose A, and player 1 at the information set "a" will choose A based on the assumption that player 2 will choose A. The backwards induction solution is shown in red in the game tree above. This example shows that an optimal solution in the backwards induction, in game scenarios that involved many players (not just 4 as shown above for demo), could happen only if all the players are all rational, and this is one of our big assumptions in game theory.

We have seen from all the previous examples that the backwards induction process applies only to games with perfect information, that is, to games that have only singleton information set. But in games with imperfect information that involves at least one non-singleton information set, we see that the formulation for backwards induction cannot be directly applied. In such games, there are two general basic approaches. The first approach involves using the concept of forwards induction to be described in the next section. Second approach involves using the concept of Bayesian equilibrium to be described in great details in the next Chapter.

For now, we give an example to demonstrate a slight modification to the existing backwards induction process to a simple game tree that involves an element of imperfect information.

Example 6.5.17 Consider the following game tree we have seen from previous Example 6.2.5 before.

Figure 6.21

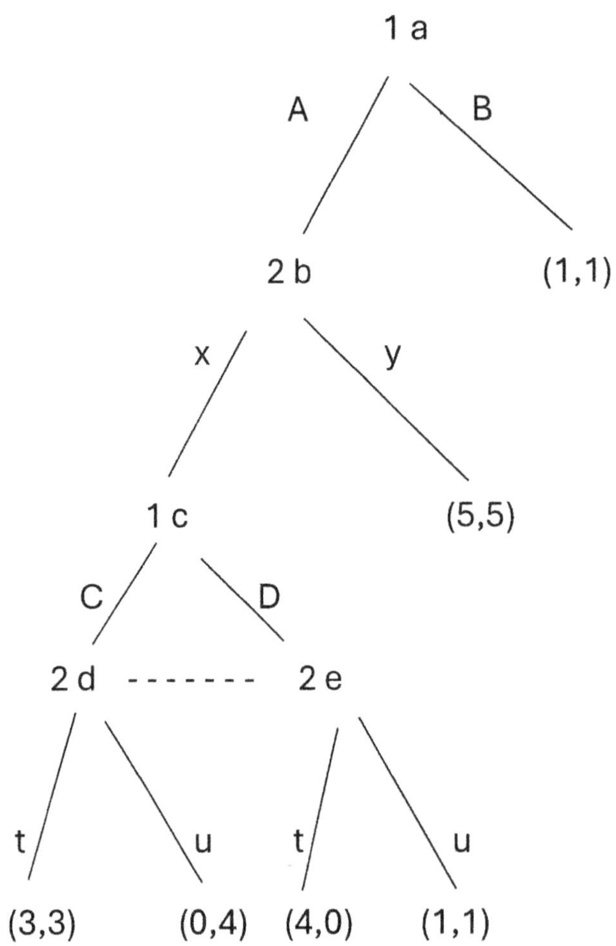

We cannot apply backwards induction directly at player 2's information set at "d" or at "e" because they are not subgame per our definition we defined before. But the game rooted at player 1's information set at "c" is a subgame per our previous definition. In fact, this subgame is just a simultaneous game described by the following payoff matrix:

	t	u
C	(3,3)	(0,4)
D	(4,0)	(1,1)

This is the well-known Prisoner's Dilemma Payoff Matrix we studied before, and the Nash Equilibrium for this subgame is known to be (D,u) with payoff (1,1). The slight modification to the game tree before we can apply the backwards induction is to replace the entire subgame with the payoff (1,1) and keep in mind that the Nash Equilibrium happens at (D,u). Thus, the modified game tree is:

Figure 6.21A

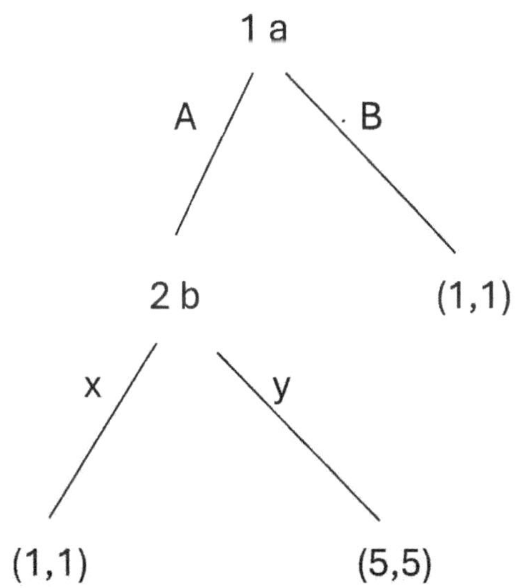

Now apply backwards induction at "2b", player 2 will choose "y", and moving backwards at information set 1 a, player 1 will choose "A" as 5 is greater than 1. So, (A,y) is a subgame perfect Nash Equilibrium of the modified game above, and so (AD,yu) is a subgame perfect Nash Equilibrium to the original game.

If we were to find the rest of the Nash equilibriums, we can first obtain the payoff matrix representation for the entire game by tracing out each path from the game tree above and then determine the Nash equilibrium from the matrix accordingly.

	xt	xu	yt	yu
AC	(3,3)	(0,4)	(5,5)	(5,5)
AD	(4,0)	((1,1)	(5,5)	(5,5)
BC	(1,1)	(1,1)	(1,1)	(1,1)
BD	(1,1)	(1,1)	(1,1)	(1,1)

These strategy pairs circled in reds are possible Nash equilibriums, but only the pair (AD, yu) is also subgame perfect Nash Equilibrium.

6.5.B Theory behind the Forwards induction

As we have seen from the previous examples, Backwards Induction assumes that all future play will be rational, and this method of predicting how a game will be played cannot be applied to games with imperfect information. Forward induction is yet another approach but assumes all past play was rational, however, unlike Backwards induction that follows a rather fixed procedure, the players in a game tend to make logical inferences from other players. We will give three examples below to demonstrate this inferences process.

Example 6.5.18 Consider the following game tree. This is a familiar simultaneous matrix game represented in a game tree. Note however, player 1 has an additional choice of action C other than A and B. It is this additional action C that player 1 has, that allows player 2 to make logical inference about her history when she is at information set 'b' or 'c'. Without the existence of action C it player 1, player 2 would not know what action to take when it is her turn to play. Here are the details.

First, when player 1 is asked to move, he is confronted with three choices of actions, A, B or C. If he chooses A, regardless of which actions x,y that player 2

Figure 6.22

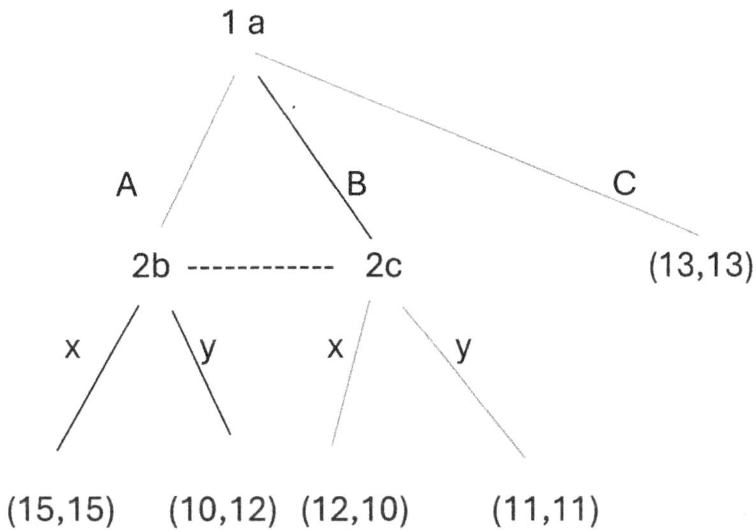

takes, he will get 15 or 10 depending on what player 2's future move is; if he chooses B, regardless of which actions x, y that player 2 takes, he will get 12 or 11; if he chooses C, player 1 surely will get 13. As 13 surely more than 12 or 11, so amongst the three choices A, B, C, player 1 is surely not to play B. At this point in time, player 1 can only choose between A or C, but not sure which one to choose. To determine which is the right choice A or C, player 1 now turn to look at player 2' choice of actions between x and y. As forwards induction assumes all the past plays are rational, so player 2 can infer that player 1 will not play B but to play A because B is surely worse than C, and thus the rational choice for B is to choose x rather than y because 15 is more than 12. Thus < A, x> is the equilibrium solution to this game and when player 2 is at the information set "c", she will choose x as well. The matrix representation of the game is shown below:

	x	y
A	(15,15)	(10,12)
B	(12,10)	(11,11)
C	(13,13)	(13,13)

Note in this example, the existence of action C in player 1 signals to player 2 that he will take A given that player 2 is a rational player, and thus player 1 can achieve his highest payoff value 15 given the mere existence of his other action C.

Example 6.5.19 This example is like Example 16.5.8, there are however, two payoff value pairs (1,1), and (2,2) that are relatively low as compared to the rest of the payoff values in the tree. These low values occur at the action pairs <A,x> and <C,y> respectively, and we can think of these actions as relatively threat to either player.

Figure 6.23

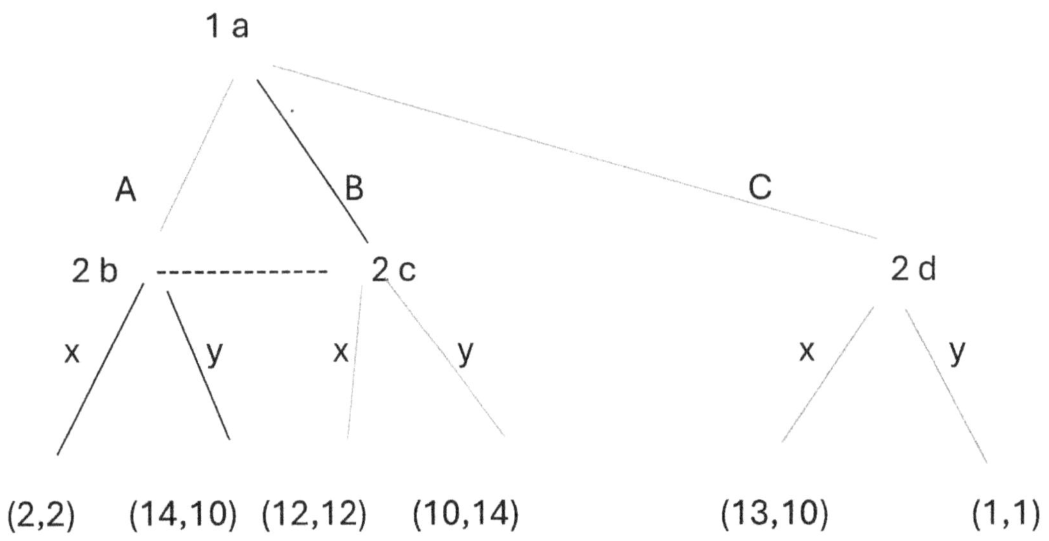

The matrix representation for the tree above is shown below:

	x	y
A	(2,2)	(14,10)
B	(12,12)	(10,14)
C	(13,10)	(1,1)

As before, player 2's uncertainty about which action between A and B that player 1 has taken is when she is at information set 'b' and 'c'; at information set 'd', player 2 surely knows that player 1 has chosen action C. Apply forwards induction to player 1 at the beginning of the game, player 1 is sure that he will get 13 if he plays C, and if he plays B, the highest he can get is 12 regardless of what player 2's choice is, as 12 is less than 13, so player 1 would choose between A or C. Now player 2 infers that player 1 would not choose B, and hence player 2 would choose action y to get 10 rather than to choose x to get 2. Thus our forwards induction equilibrium is <A,y> , and if player 1 were to choose action B, player 2 will still respond with action y giving the payoff (10,14).

Note that in this example, the existence of the action C in player 1 allows player 2 to infer that player 1 will choose action A, and as a result player 2 can respond appropriately by choosing the right action y to avoid the harmful payoff (2,2) for both players if player 2 were choosing x. In a way, the existence of C in player 1 signals to player 2 that A is a credible threat to player 2 if she does not respond appropriately.

Example 6.5.20 This example adds a bit of complication to Example 6.5.19 in that player 2 has four information sets shown in the tree diagram below:

Figure 6.24

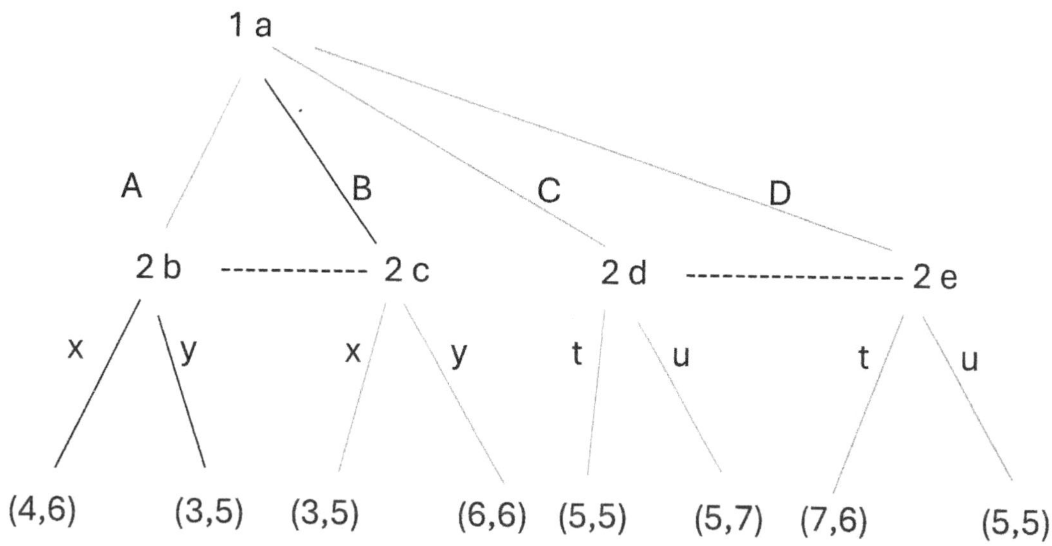

First note that there are four different information sets for player 2 at 'b', 'c', 'd', or 'e'. When Player 2 is asked to move, she can only tell if player 1 has taken actions A or B, or if player 1 has taken action C or D. She can take rational move only if she knows or infer what action player 1 has taken. She has two possible actions x or y. The rational action she made will be the same at information sets ' b' and 'c', or same at information set 'd' and 'e'. When forwards induction starts, player 1 is asked to decide which action A,B,C,D to choose. Regardless of what player 2 will do, A will have the maximum payoff of 4, and this maximum payoff 4 is still less than the payoff values 5,6,7 if player 1 were choosing actions C or action D regardless of what player 2's choices (refer to the payoff values in the above figure). Thus, as a rational player, player 1 will not choose action A. Knowing this, when player 2 is at the information set 'c', she will infer that player 1 has selected action B because action A is not rational strategy as compared to C or D, and as a rational player, she will choose y rather than x as 6 is more than 5. And when player 2 is at information set 'b', y will be chosen.

Next, we need to decide which strategy C or D player 1 will choose. As we just established, if strategy B were chosen, player 1 gets 6, but if strategy C is chosen, player 1 gets 5 regardless of what player 2 chooses, and 5 is less than 6, so strategy C is not a rational strategy between B and C. Thus, player 2 can infer that at information set 'e', player 1 must have chosen strategy D, and being a rational player, she will choose to get 6 as opposed to choose u to get 5. Thus, at information set 'd', player 2 would have chosen t to get 5 as well. (Refer to the green lines shown in the figure above). Both the red and green lines above show the unique equilibrium solution to this game tree using the forwards induction inferences.

6.6 Applications of Backwards Induction to Economics and Business

Application 6.6.1 The Cournot and Stackelberg Model

Recall that we first used the Cournot Model in Chapter 3 before to illustrate the concept of Nash Equilibrium in a static game, here we use it again to demonstrate its application as a dynamic game. This application we present here has to do with two competing firms each produces certain number of homogeneous products and sells in the market to obtain respectively profits. The question is: what is the optimal number of products should each firm produce to gain the individual firm's maximum profit. The underlying scenario can be described by the *Cournot Model of Duopoly* due to A. Cournot in 1838.

This model stated that, if x_1 and x_2 are the quantities of products produced by firms 1 and 2 respectively, then the market -clearing price 'p' of the product is modeled as $p = A - (x_1 + x_2)$ for some constant A, and the unit cost of producing the product in both firms is the same, say, another constant C. There are no fixed costs, and the marginal cost is constant at C, and we also assume $C < A$.

Based on the Cournot Model stated above, we can provide answers to the above question in two different situations. First situation is that when both firms choose to produce their quantities simultaneously, and the second situation is that each firm chooses to produce its quantities sequentially.

First Scenario:

Suppose firm 1 chooses to produce x_1 and firm 2 chooses to produce x_2 that many units simultaneously, let P_1 and P_2 be the profit of firm 1 and firm 2 respectively, then

$P_1 = (p - C) * x_1$, where the difference (p - C) is unit product profit gain,

so, $$P_1 = (A - (x_1 + x_2) - C) * x_1, \qquad (1)$$

and similarly,

$P_2 = (p - C) * x_2$, where the difference (p - C) is unit product profit gain,

so, $$P_2 = (A - (x_1 + x_2) - C) * x_2 \qquad (2)$$

In the above expressions for P_1 and P_2 are functions of x_1 and of x_2, and the problem is to simultaneously choose a fixed value of x_1^* and of x_2^* such that

$P_1(x_1^*, x_2^*) \geq P_1(x_1, x_2^*)$ for all possible x_1 (3)

and

$P_2(x_1^*, x_2^*) \geq$ and $P_2(x_1^*, x_2)$ for all possible x_2 (4)

Inequality (3) is same as finding x_1^* that maximizes the function $P_1(x_1, x_2^*)$, thus, using simple Calculus by differentiating $P_1(x_1, x_2^*)$ with respect to x_1 using the product rule, and setting the derivative to 0 to solve for x_1, we have:

$$\frac{dP_1}{dx_1} = (A - (x_1 + x_2^*) - C) + (-1) * x_1 = 0 \text{ which gives}$$

$$x_1^* = 1/2 \{A - C - x_2^*\} \quad (5)$$

Similarly, from (4), we differentiate $P_2(x_1^*, x_2)$ with respect to x_2,

$$\frac{dP_2}{dx_2} = (A - (x_1^* + x_2) - C) + (-1) * x_2, \text{ setting it to 0 to solve for } x_2 \text{ gives}$$

$$x_2^* = 1/2 \{A - C - x_1^*\} \quad (6)$$

Thus, solving the two simultaneous equations (5) and (6) together algebraically gives:

$$x_1^* = x_2^* = 1/3 (A - C) \quad (7)$$

Note that this is an optimization problem for two players, two firms in this case, the strategy for each player is to select the optimal number of product units x_1 and x_2 so that the payoff value for each player $P_1(x_1, x_2)$ and $P_2(x_1, x_2)$ is maximum respectively.

Thus, this economic scenario can be viewed as a two players simultaneous game with the payoff matrix shown below:

Table 6.1

… … x_2^* … …

:							
:							
x_1^*			$(P_1(x_1, x_2), P_2(x_1, x_2))$				
:							
:							

($x1^*$, $x2^*$) in the above payoff matrix is a Nash Equilibrium pair because $P1(x1^*, x2^*) >= p1(x1, x2^*)$ and $p2(x1^*, x2^*) > p2(x1^*, x2)$ for all x1 and x2 from the way we computed $x1^*$ and $x2^*$.

Second Scenario:

Instead of both firms select the production quantity simultaneously, another dynamic model due to Stackelbery in 1934 known as the *Stackelbery Model of Duopoly. In this model, a dominant firm moves first, we call it firm 1, and a subordinate firm moves second, we call it firm 2.* To formulate the backwards induction for this sequential scenario, we use the same notations as the simultaneous case before,

Diagram 6.1

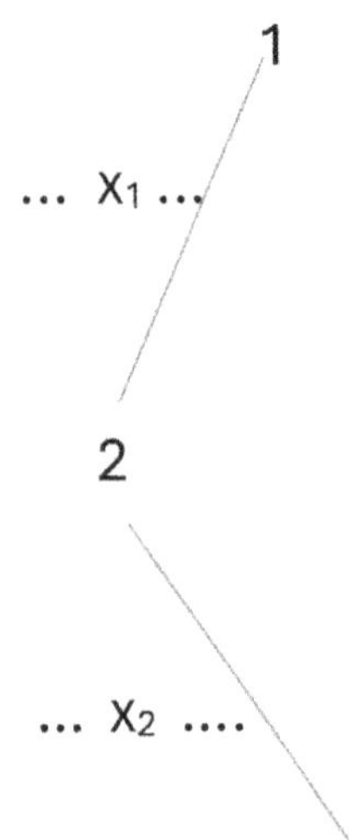

Backwards induction begins from player 2, player 2 needs to find out what is her best response given that player 1 has chosen x_1, this means player 2 needs to maximize her profit function $P_2(x_1\ \ x_2)$ in (7) by choosing a proper x_2, call it the best response of x_1 denoted by $bR_2(x_1)$, so we differentiate $P_2(x_1\ \ x_2)$, and set the derivative to zero, and to solve for x_2:

$$\frac{dP_2}{dx_2} = A\text{-}C\text{-}\ x_1 - 2\,x_2 \quad = 0,$$

so, $bR_2(x_1) = (A\text{-}C\text{-}x_1)/2$ (8)

Now, backwards induction moves back one step to player 1, he needs to find x_1 that will maximize his profit function $P_1(x_1\ \ x_2)$ in (1), where x_2 will be replaced by $bR_2(x_1)$. Simple calculus tells us to differentiate function $P_1(x_1, bR_2(x_1))$ with respect to x_1, and set the derivative to 0 to solve for x_1:

$$P_1(x_1\ \ x_2)= \{ A\text{-}(x_1 + (A\text{-}C\text{-}x_1)/2)\text{-}\ C\} * x_1$$

$$=\{(A\text{-}\ x_1\ \text{-}C)/2\} *\ x_1$$

$$\frac{dP_1}{dx_1} = ½*\{ A\text{-}2x_1\ \text{-}C\} =0$$

So, $x_1^* =1/2\ (A\text{-}C)$ (9)

Putting (9) back into (8), we find:

$x_2^* = bR_2(x_1^*)$, which is,

$= \{ A\text{-}\ C\text{-}\ 1/2\ (A\text{-}C)\}/2$, so,

$$x_2^* = ¼\ (A\text{-}\ C) \qquad (10)$$

Note that in the above formulation for the backwards induction, the expressions (1) and (2) for payoff functions P_1 and P_2 are symmetric with respect to the players 1 and 2, thus in the game above, if player 2 were to move first, then player 1 's best response to player 2 would be $bR_1(x_2)$, and (8) become

$$bR_1(x_2) = (A\text{-}C\text{-}x_2)/2 \qquad (11)$$

Now we are going to do a comparison of these two models, to see for each player which model gives them better payoff values. We can compute the payoff values for both players in the two model. Results are shown in the table below.

Cournot Model (simultaneous scenario)	*Stackelberg Model (sequential scenario)*
	$P_1 = (A - (x_1 + x_2)\text{-}C) * x_{1,}$ (1) $P_2 = (A - (x_1 + x_2)\text{-}C) * x_2$ (2) $bR_2(x_1) = (A\text{-}C\text{-}x_1)/2$ (8) $bR_1(x_2) = (A\text{-}C\text{-}x_2)/2$ (11)
$x_1^* = 1/3\ (A\text{-}C)$ (7)	$x_1^* = 1/2\ (A\text{-}C)$ (9)
$x_2^* = 1/3\ (A\text{-}C)$ (7)	$x_2^* = ¼\ (A\text{-}C)$ (10)
$x_1^* + x_2^* = 2/3(A\text{-}C)$ The expression $A\text{-}(x_1^* + x_2^*)$ is called the economic clearing price, it can be shown to be (A+2C)/3 Since A>C by assumption, (A+2C)/3 > (A+3C)/4, So, the economic clearing price in the Cournot Model is larger than that of the Stackelberg Model.	$x_1^* + x_2^* = ¾(A\text{-}C)$ $A\text{-}(x_1^* + x_2^*)$ is shown to be (A+3C)/4
$P_1(x_{1*}, x_{2*}) = 1/9\ (A\text{-}C)^2$ $P_2(x_{1*}, x_{2*}) = 1/9\ (A\text{-}C)^2$	$P_1(x_{1*}, x_{2*}) = 1/8\ (A\text{-}C)^2$ $P_2(x_{1*}, x_{2*}) = 1/16\ (A\text{-}C)^2$

Thus, we see from the table above, both players in the Cournot Model have the same payoff outcomes, player 1 has better payoff outcome in the Stackelberg Model than that of player 2 in the Stackelberg Model. Player 1's payoff in the Cournot Model is worse than that of the Stackelberg Model, player 2 in the Cournot Model is however, better than that of the Stackelberg Model.

Before we leave the Stackelberg Model, we observe that in the factor

$(A - (x_1 + x_2) - C)$ is common in both the payoff expressions for players 1 and 2:

$$P_1 = (A - (x_1 + x_2) - C) * x_1 \tag{1}$$

$$P_2 = (A - (x_1 + x_2) - C) * x_2 \tag{2}$$

so, the bigger the player's choice of production units x_1 or x_2, the bigger the player's payoff value P_1 or P_2. Now we apply this simple observation to the following situation in the Stackelberg Model.

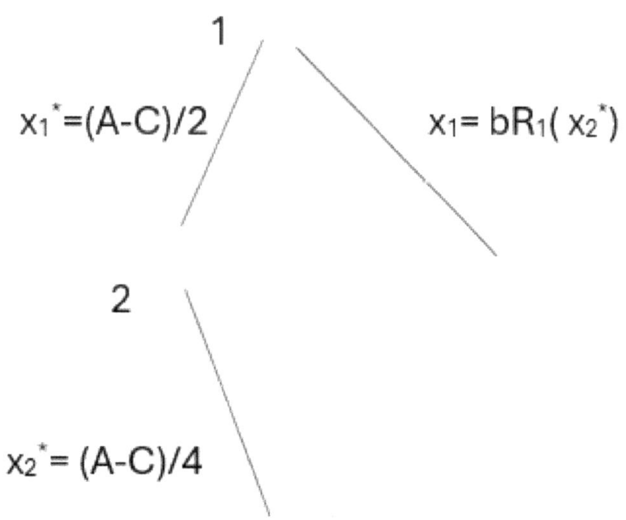

Recall in the formulation of the Stackelberg Model, we have assumed that player2 selects $x_2^* = (A\text{-}C)/4$ only after she observes that player 1 has chosen $x_1^* = (A\text{-}C)/2$. But suppose player 1 assumes player 2 always responds with $x_2^* = (A\text{-}C)/4$ and he goes ahead to play the best response $x_1 = bR_1(x_2^*)$,

then $x_1 = bR_1(x_2^*)$

$= (A\text{-}C\text{-}x_2^*)/2$

$= 3/8\,(A\text{-}C)$

$<= 1/2(A\text{-}C) = x_1^*$

So, the payoff for player 1 is getting worse if he is choosing $bR_1(x_2^*)$ than choosing the original x_1^* by the observation before.

Application 6.6.2 Graphical illustration of The Leontief Model (1946)

This application we are going to present has to do with formulating a competing scenario between the firm and the union by the Leontief Model. The union is considered as the monopoly seller of labor to the firm, it has exclusive control over the wages, and the firm has exclusive control over the employment. To describe and specify this game scenario, lets call the union our player 1, and the firm our player 2, and define the following notations.

Let w be the wage of an employee demanded by the union

Let n be the number of employees controlled by the firm.

Let the union's payoffs be described by the set of utility functions U (w,n) = k, for some constants k, each value of k corresponds to a indifference curve of level k. Assume U(n,k) increases in both n and w. These typical indifference curves are shown below. Note that, for the union's indifference curves, when n is fixed, the union does better when w is higher, higher indifference curves means higher utility level for the union.

Figure 6.25A

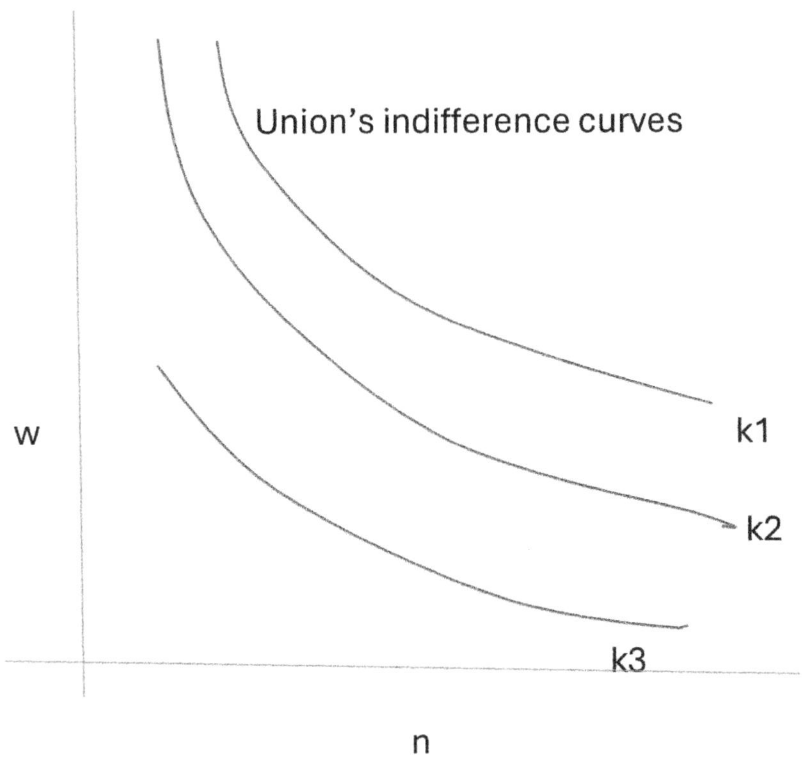

Let the firm's profit function be P(w,n). It is reasonable to assume the firm's profit function takes the form P(w,n)= R(n)- w*n, where R(n) is the firm's revenue function only depends on the number of employees n , and it is an increasing function with n and concave as shown below:

Figure 6.25B

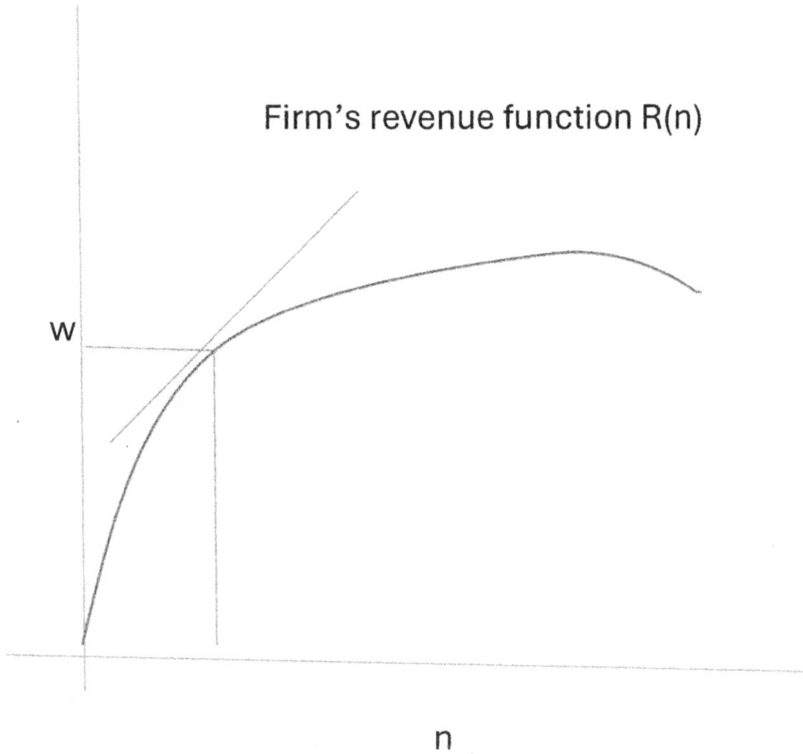

The firm's isoprofit curves will be given by the equation P(w,n)= R(n)- w*n = f, for some constants f. So, a typical firm's isoprofit curve is :

R(n)- n*w =f or, w = (R(n) – f)/n,

So, the slope of the isoprofit curve at level f is :

$\frac{dw}{dn}$ = { n* R'(n)- (R(n)-f) }/n^2 by the quotient rule for differentiation,

= { n* R'(n)- n*w} /n^2

= { R'(n)- w}/ n

So the zero slope of the isoprofit curve at level f is located at the point (n,w) where R'(n) = w. Thus, for a given w demanded by the union, P(w,n) is only a function in n, so $\frac{dP}{dn}$ = R'(n)- w, and the second derivative $\frac{d^2P}{dn^2}$ = R''(n) <0 because R(n) concave down, so (w,n) maximize P(w,n) at the point (w,n).

Now if we plot all pairs (w,n) on the isoprofit curves P(w,n) = f with R′(n) = w for different constants f1, f2, f3….etc, we will have the following isoprofit curves for the firm and the trajectory of the pairs (w,n) shown in red :

Figure 6.25C

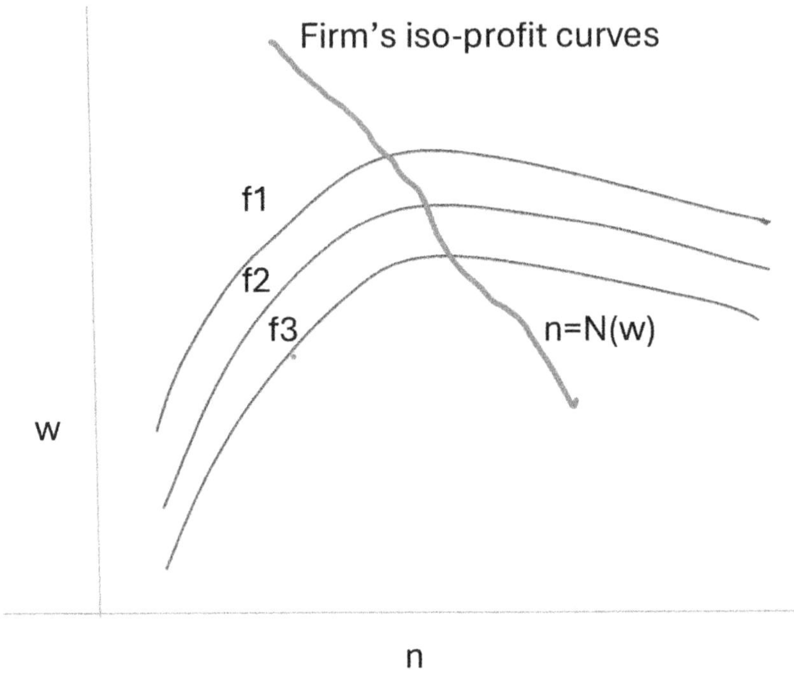

As P(w,n)= R(n)- w*n, when n is fixed, the firm does better when w is lower, so when the isoprofit curve is lower (that is when the horizontal line w is lower), it represents higher profit level P(w,n) .

To recap, the curve shown in red above contains the correspondences between w and n. Given a fixed w demanded from the union, the firm comes up with a n that maximizes the its isoprofit function P(w,n). Denote this correspondence by a functional expression n =N(w).

After we have described these functions U(w,n), P(w,n), N(w), we are now going to describe this union and firm competitive scenario as a sequential game, and show how the backwards induction procedure, which is usually solving the problem algebraically, can be done graphically. Suppose the sequence of the game is that the union moves first by making a wage demand w to the firm, then the firm observes w and chooses an employment level n. As shown below:

Figure 6.25D

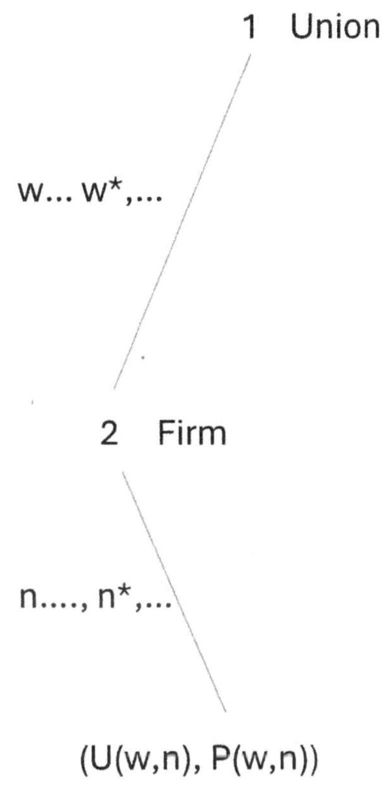

Recall that the backwards induction starts with player 2, the firm must maximize its profit function P(w,n) by choosing a suitable n for every w the union demands. If we were given the specific functional form for P(w,n) and R(n) , we could have solved this optimization problem algebraically similar to Application 1 shown previously. However, we have done this graphically by showing the correspondence between w and n in Figure 3, by the function n =N(w). As the backwards induction moves back to player 1, player 1 must choose the right w to maximize his utility function U(w, N(w)) . As before, we could solve this optimization algebraically when the specific functional form of U(w,n) is given, instead Figure 4 below shows how this optimal w is determined graphically.

We now claim and show that curve n=N(w) shown in red together with all the union indifference curves all plotted in the same w-n plan will give us the backwards induction outcome of this game.

Figure 6.25E

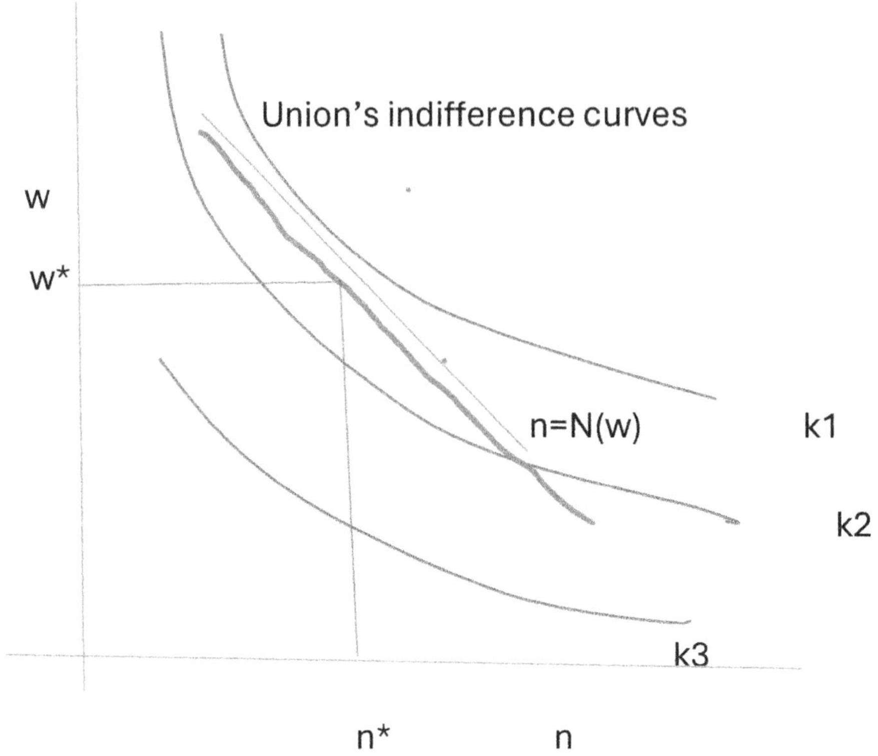

Since in the first stage of the backwards induction, the union must choose a w that would maximize its utility function U (w, N(w)), this means that graphically in Figure 25E, we must choose an indifference curve as high as possible and at the same time tangent to the firm's information graph n =N(w) shown in red above. The tangent point shown at (w*, n*) then gives the solution to the union-firm problem.

Application 6.6.3 Application to tariff and production competition between countries

In this application we are going to extend Application 6.6.1 to the situation when the two firms each is locating in two different countries, each firm is selling its products at its local country as well as selling its products in the other foreign country where the other firm is located. As such, there will be tariffs imposed by the government of the local country on products produced by the firm in the foreign country.

To simplify the problem modelling, assume the model parameters A and C are the same as those in Application 1, and the parameter C is same in both countries.

Let t_1 be the tariff imposed by country 1 on country 2 's product.

Let t_2 be the tariff imposed by country 2 on country 1's product.

Recall that the profit functions (1) and (2) for the two firms in Application 1 :

$$P_1 = (A - (x_1 + x_2) - C) * x_{1,} \quad (1)$$

$$P_2 = (A - (x_1 + x_2) - C) * x_2 \quad (2)$$

These functions need to be modified in our present situation as shown below:

Let l_1 be the number of products produced by firm 1 selling in the local market.

Let f_1 be the number of products produced by firm 1 selling in the foreign market.

Let l_2 be the number of products produced by firm 2 selling in the local market.

Let f_2 be the number of products produced by firm 2 selling in the foreign market.

In the present case, (1) is modified as:

$$P_1 = (A - (l_1 + f_2) - C) * l_1 + (A - (f_1 + l_2) - C) * f_1 - t_2 * f_1 \quad (3)$$

This means that firm 1's total payoff is firm 1's profit on country 1, the term shown in red, which is a function on l_1 and f_2 , and firm 1's profit on country 2, the term shown in blue, which is a function on f_{1}, l_{2}, t_2 . Since $l_{1}, f_{1,}$ and $l_{2,}$ f_2 are independent, so maximizing (3) is same as maximizing the term in blue and the term in red. Thus differentiate the term in red with respect to $l_{1,}$ set it to zero, and solve for l_1, we find

$$l_1 = ½ (A - C - f_2) \quad (4)$$

And differentiate the term in blue with respect to f_1, set it to zero, and solve for f_1, we find

$$f_1 = ½ (A - C - l_2 - t_2) \quad (5)$$

Similarly, (2) is modified as

$$P_2 = (A - (l_2 + f_1) - C) * l_2 + (A - (f_2 + l_1) - C) * f_2 - t_1 * f_2 \quad (6)$$

Just as what we did to Firm 1, we find from (6),

$$l_2 = ½(A - C - f_1) \tag{7}$$

$$f_2 = ½(A - C - l_1 - t_1) \tag{8}$$

By combining (4) and (8) we find:

$$l_1 = 1/3\,(A - C + t_1) \tag{9}$$

By combining (5) and (7) we find:

$$f_1 = 1/3\,(A - C - 2t_2) \tag{10}$$

By combining (7) and (10), we find:

$$l_2 = 1/3\,(A - C + t_2) \tag{11}$$

By combining (8) and (9), we find:

$$f_2 = 1/3\,(A - C - 2t_1) \tag{12}$$

Thus form (9), (10), (11), and (12), we know, for a fixed tax rates t_1, t_2 between the two governments G1 and G2 , what are the optimal units of productions l_1, f_1 and l_2, f_2 should be. In fact, these values give the Nash Equilibrium for the two firms.

We are going to model the present scenario as four players, two stage game depicted in Figure 26 below:

Figure 6.26

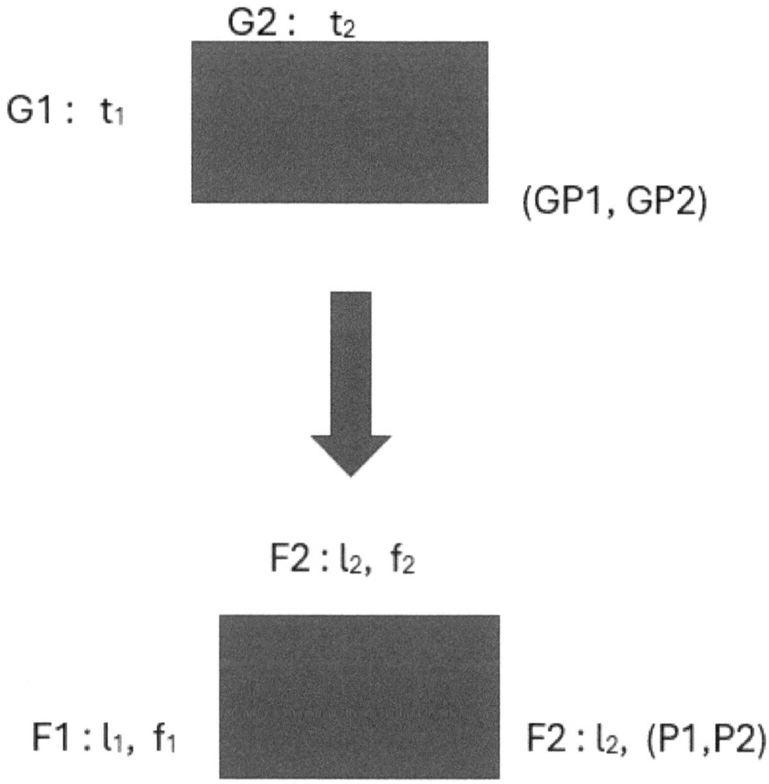

The total payoff for each (GP1, GP2) government G consists of:

1. the consumer surplus (CS1, CS2)
2. profit from the firm located in that country (P1, P2)
3. Tariff revenue collected from the other country (TR1, TR2)

As the consumer surplus is ½ of the square of the number of units sold given the demand curve in the Stackelberg Model, we have :

$$CS1 = \tfrac{1}{2}(l_1 + f_2)^2 \quad (13)$$

$$CS2 = \tfrac{1}{2}(l_2 + f_1)^2 \quad (14)$$

$$TR1 = t_1 * f_2 \quad (15)$$

$$TR2 = t_2 * f_1 \quad (16)$$

$$GP1 = CS1 + P1 + TR1$$

$$= ½(l_1 + f_2)^2 + (A – (l1 + f2)-C) * l1 + (A – (f1 + l2)-C) * f1 - t2 * f1 + t_1 *f_2 \quad (17)$$

GP2 = CS2 + P2 + TR2

$$= ½(l_2 + f_1)^2 + (A – (l_2 + f_1) - C) * l_2 + (A – (f_2 + l_1) - C) * f_2 - t_1 * f_2 + t_2 *f_1 \quad (18)$$

Having found the Nash Equilibrium between the two players F1 and F2 in the second stage game with the solution given by (9), (10), (11), (12), using the concept of backwards induction described earlier, we proceed to find the Nash Equilibrium between G1 and G2 in the first stage. But first, we need to express the payoff values GP1 and GP2 of the two players G1 and G2, in terms of t_1 and t_2.

Putting (9), (10), (11), (12) into (17) and (18) we find:

$$GP1= 1/18 * (2(A – C) – t_1)^2 + 1/9* (A – C + t_1)^2 +1/9* (A – C - 2t_2)^2 + 1/3* t_1 (A – C - 2t_1)^2 \quad (19)$$

$$GP2= 1/18 * (2(A – C) – t_2)^2 + 1/9* (A – C + t_2)^2 +1/9* (A – C - 2t_1)^2 + 1/3* t_2 (A – C - 2t_2)^2 \quad (20)$$

In (19) above, holding t_2 fixed, and differentiate GP1 with respect to t_1, and set it to zero, and solve for t_1, we find
$t_1 = 1/3(A – C)$ (21)

Similarly, in (20) above, holding t_1 fixed, and differentiating GP2 with respect to t_2, and set it to zero, and solve for t_2, we find

$$t_2 = 1/3(A – C) \quad (22)$$

By putting (21),(22) into (9), (10), (11),(12), we finally obtain the Nash Equilibrium for the entire two stage game:

$$l_1 = 4/9 (A – C), \quad f_1 = 1/9 (A – C), \quad (23)$$

$$l_2 = 4/9 (A – C), \quad f_2 = 1/9 (A – C), \quad (24)$$

$$t_1 = 1/3(A – C), \quad t_2 = 1/3(A – C) \quad (21), (22)$$

It is interesting to note that , in the Nash Equilibrium solution given above, the aggregate quantity on each market $l_1 + f_2$ and $l_2 + f_1$ is 5/9 (A- C) , but if both governments had chosen zero tariffs, then (9),(10), (11), (12) would imply that the aggregate quantity on each market is 2/3 (A- C) , which is given by the Cournot Model, and which will give larger consumers surplus for the country with zero tariff than with tariff given by the Nash Equilibrium, this is because 2/3 (A -C) is larger than 5/9 (A-C) and consumers surplus is ½ the square of the aggregate quantity.

6.7 *Multi-stage games*

A multi-stage game is a composite game consist of a finite sequence of normal form component stage-games, each component game is an independent, well-defined game of complete and imperfect information such as simultaneous -move games). These component games are played sequentially by the same players, and the total payoffs of the composite game are evaluated using the sequence of outcomes in the component games that are played. In this section, we will

1. show how the total payoffs are computed using the concept called discounted sum with discount factor d
2. show how the play out of the strategies between the component games are arranged to make the composite game to achieve an expected goal.
3. explain the concept of subgame perfect Nash Equilibrium in the Multi-stage composite games.

We will use the following example to illustrate the various concept mentioned above.

Example 6.7.21. Suppose the component games G1, G2, and G3 with the payoff values as shown:

Figure 6.27

G1:	a	b
A	1,1	5,0
B	0,5	4,4

G2 :	c	d
C	5,5	1,4
D	4,1	2,2

G3:	e	f
E	2,2	6,1
F	1,6	5,5

Definition 6.7.4 (Total payoff in multi-stage game)

If there are T stages in a multi-stage game, the discounted sum (total payoff) of player i from playing the sequence of games with payoffs

$v_i^1, v_i^2, \ldots. v_i^T$ is defined as:

$$v_i = v_i^1 + dv_i^2 + d^2 v_i^3 + \ldots. + d^{T-1} v_i^T = \sum_{t=1}^{T} d^{t-1} v^t{}_i$$, where d is called the discount factor.

Thus, if the sequence of play is (A,b), (C, c). and (E, f) in the composite game shown in Figure 6.27, then

$V_1 = 5 + 5d + 6d^2$, $V_2 = 0 + 5d + 1.d^2$

6.7.1 Representation of strategies in multi-stage games

We will show how to represent the strategies between the component games in a muti-stage game, to visualize the need of this representation, we redraw the composite game in normal form shown in Figure 6.27 by the same game in extensive form shown in Figure 6.28 below:

Figure 6.28

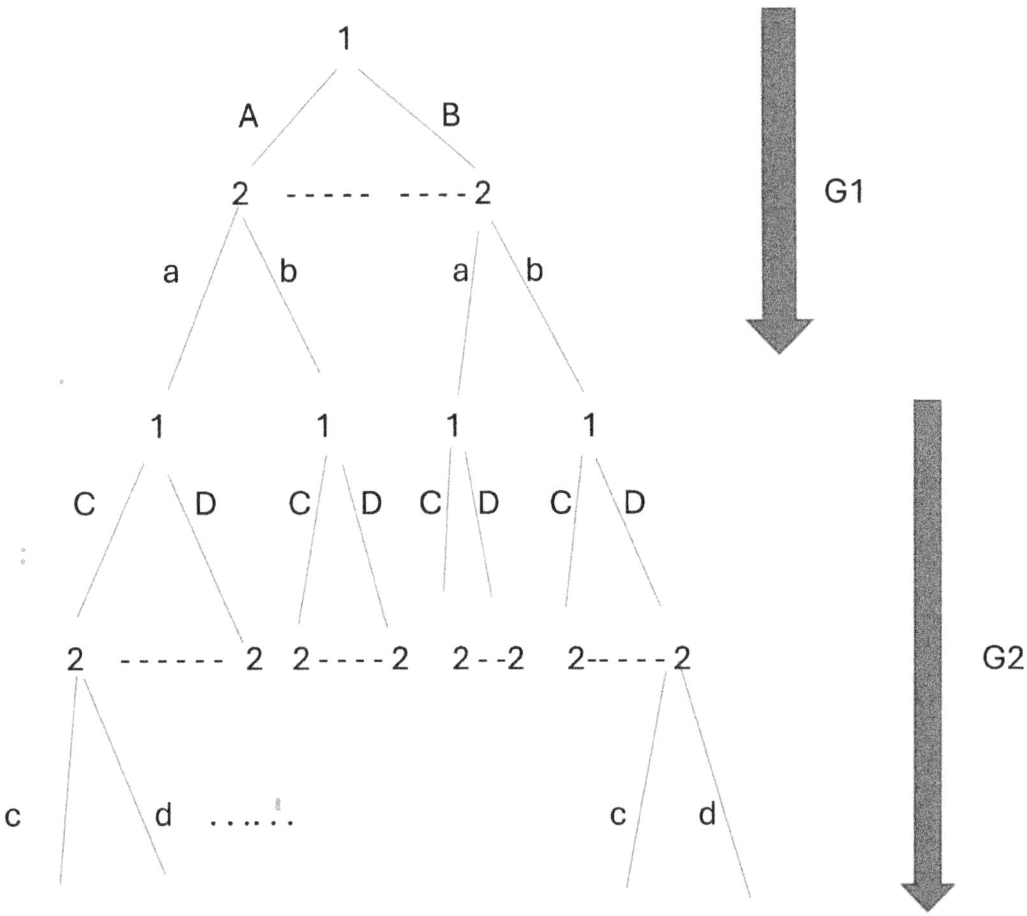

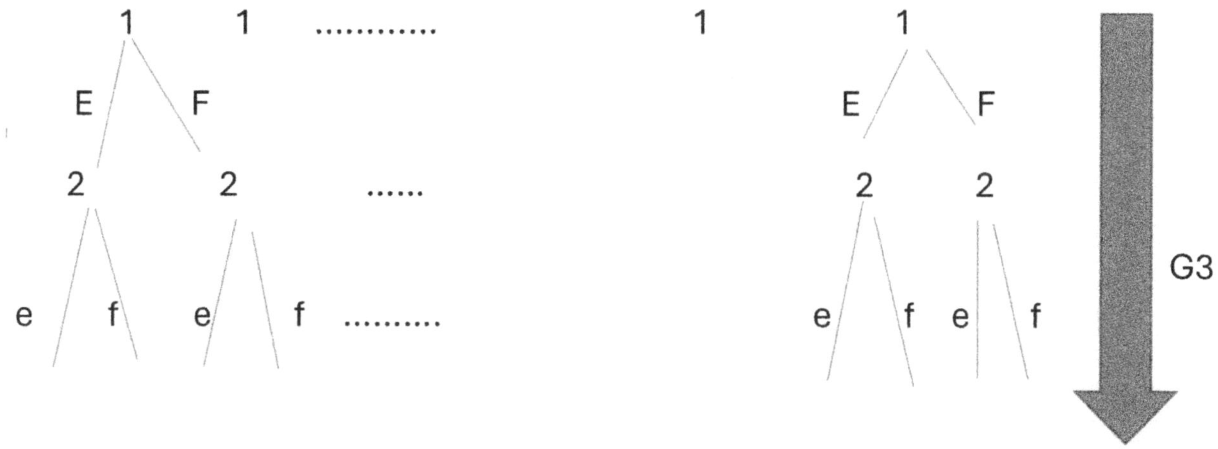

Below we are going to describe a procedure and give an idea how to represent the strategies of the component games in this simple multi-stage composite game shown in Figure 28 above. This idea can be applied to multi-stage games in general.

We adopt the notation $s_i^1, s_i^2, \ldots . s_i^T$ to represent the strategies of player i from stages 1, 2....., T in the subsequent discussion.

At stage 1, T=1: the strategies for player 1 are simply any action from the action set {A,B}, and the strategies for player 2 are simply any action from the action set {a,b},

So $s_1^1 \in$ {A,B}, $s_2^1 \in$ { a, b}

At stage 2, T=2: the strategies s_i^2 are from the set {C,D} for player 1 and {c,d} for player 2. Unlike in stage 1, this time, s_i^2 also depends on what was played in stage 1 before, there were 4 possible playouts by both players then, they are (A a, Ab, Ba, Bb), so we need to specify what were these actions s_i^2 (Aa), s_i^2 (Ab), s_i^2 (Ba), s_i^2 (Bb). They are in {C,D} or {c,d} pending on if i is player 1 or player 2. To simplify the notation, we denote,

h_1= {Aa, Ab, Ba, Bb} to mean the history of the 4 possible outcomes at stage 1, so it makes sense to define a strategy for player i in this 2 -stage game as,

$$s_i = \{ s_i^1, s_i^2(Aa), s_i^2(Ab), s_i^2(Ba), s_i^2(Bb)\}$$

$$s_i = \{ s_i^1, s_i^2(h_1)\}$$

At stage 3, T=3, $s_i = \{ s_i^1, s_i^2(h_1), s_i^3(h_2)\}$, where h_2 is the history of all the outcomes up to stage 3, In the present case, h_2 are the 16 outcomes:

{Aa, Ab, Ba, Bb} x {Cc, Cd, Dc, Dd} = { AaCc, AaCd.... } in Figure 28, $s_i^3(h_2)$ is in the action set {E,F} or {e,f} pending on if i is player 1 or player 2.

In general, if the multi-stage game as T stages, a strategy for player i is defined as follows:

$$s_i = \{s_i^1, s_i^2(h_1), ..., s_i^t(h_{t-1}), ..., s_i^T(h_{T-1})\}$$

Where $s_i^t(h_{t-1})$ is an action for player i from the tth stage game, and h_{t-1} is a particular history of events that occurred up to stage t.

Using these notations defined above, we can define the Nash Equilibrium strategies of each component game in Figure 28.

6.7.2 Nash Equilibrium and Subgame Perfect Nash Equilibrium in multi-stage games

In Figure 28 above, G1 has the Nash Equilibrium (A,a), so it is represented as

$s_i = (\ s_i^1)$, where s_i^1 = A, s_2^1 = a. This Equilibrium path is shown in red in Figure 28.

For G2, its Nash Equilibrium is (C,c) or (D,d), so for (C,c) it is represented as,

$s_1 = \{\ s_1^1, s_1^2(Aa), s_1^2(Ab), s_1^2(Ba), s_1^2(Bb)\}$, and regardless of what happened in stage 1, player 1 will take the action C, so

S_1 = {*A, C, C, C, C*}. Similarly,

$s_2 = \{\ s_2^1, s_2^2(Aa), s_2^2(Ab), s_2^2(Ba), s_2^2(Bb)\}$, and regardless of what happened in stage 1, player 2 will play c, so S_2 = {*a, c, c, c*}. The equilibrium path is shown in blue.

For G3, the Nash Equilibrium is (E,e). To represent this Nash Equilibrium in the 3-stage game, we follow the notation above,

$s_i = \{\ s_i^1, s_i^2(h_1), s_i^3(h_2)\}$, where $s_1^3(h_2)$ = E and $s_2^3(h_2)$ = e regardless all the history of h_2,

S_1= (A, C,C,C,C, E,.....E) , S_2= (a, c ,c, c, c, e,.....e). The equilibrium path is shown in dark blue in Figure 28. The path consists of the concatenation of the red, blue, and dark blue paths is the Nash Equilibrium path for the composite multi-stage game.

Follow this simple example in Figure 28 above and the way we define the conditional strategies between the component games in a multi-stage game, we can make the following general observation:

Proposition 6.7.1 (Existence of the subgame-perfect Equilibrium in multi-stage games)

If G1, G2, ... GT are games in stage 1, 2...., T, and each has a Nash Equilibrium $S_*^1, S_*^2, ..S_*^t, ..., S_*^T$, then there exist a subgame perfect Nash Equilibrium in the multi-stage game (G1,G2,...., GT) where the equilibrium path is given by the concatenation of each individual equilibrium path from each component game

$$S_*^1, S_*^2, ..S_*^t, ..., S_*^T$$

Proof. We demonstrate the general proof using the 3-stage game shown in Figure 28. By assumption, (E,e) is a Nash Equilibrium in the stage 3 game, this is the path shown by dark blue in Figure 28. Back at stage 2, the total payoffs are the sum of the second stage game shown earlier, repeated below, together with a fixed payoff from stage 1 , say (1,1) out of the 4 pairs, and the payoff from stage 3, (2,2)

Figure 6.29

G2:	c	d
C	8,8	4,7
D	7,4	5,5

And the Nash Equilibrium for each game in stage 2 is still (C,c)

Now back at stage 1, the total payoffs from plays in stage 1 is the sum of the payoffs from G1 plus the expected payoff from (C,d) which is (5,5) and (E,f) which is (2,2)

		a	b
G1:	A	1,1	5,0
		0,5	4,4

		a	b
G1:	A	8,8	12,7
		7,12	11,11

And (A, a) is still the Nash Equilibrium in the stage 1 game.

6.7.3 The concept of award and punishment in multi-stage games

Proposition 6.7.1 stated that if S_*^1, S_*^2 , ..S_*^t ,...,S_*^T are Nash Equilibriums in each of stage game, then the composite path must be a Subgame Perfect Nash Equilibrium in the composite multi-stage game. Using the concept of award and punishment, we can show that not every S_*^1, S_*^2 , ..S_*^t ,...,S_*^T is required to be Nash Equilibrium, and yet the composite path can still be a Subgame Perfect Nash Equilibrium in the composite game. We could demonstrate how this is done using the 3-stage composite game in Figure 29, but for simplicity, we will restrict ourselves to only the 2-stage game.

To recap, the component game G1 and G2 are repeated here,

Figure 6.29A

G1:

	a	b
A	1,1	5,0
B	0,5	4,4

G2:

	c	d
C	5,5	1,4
D	4,1	2,2

In G1, (B,b) with payoff (4,4) is not a Nash Equilibrium, but we can show that (B,b) followed by the Nash Equilibrium (C,c) with payoff (5,5) in G2, the composite path will be a Subgame Nash Equilibrium in the 2-stage composite game. This is done by properly choosing the strategies for player 1 and player 2 in this two-stage game. Note that in G2, there are two Nash Equilibriums, one has better payoff (5,5), the other has lower payoff (2,2), we want to choose a sequential strategy between G1 and G2 as follows:

- If both players end up with (4,4) in G1, then we want both players get the rewards by giving them with the better payoff (5,5) in G2.
- If both players end up with any other payoff other than (4,4) in G1, then we want both players get the punishment by giving them with the lower payoff (2,2) in G2.

These two sequential conditions translate into the following notations:

$$s_1 = \{\ s_1^{\,1}, s_1^{\,2}(Aa), s_1^{\,2}(Ab), s_1^{\,2}(Ba), s_1^{\,2}(Bb)\}$$

$$s_1 = \{B, D, D, D, C\}$$

$$s_2 = \{\ s_2^{\,1}, s_2^{\,2}(Aa), s_2^{\,2}(Ab), s_2^{\,2}(Ba), s_2^{\,2}(Bb)\}$$

$$s_2 = \{b, d, d, d, c\}$$

We want to check that if (s_1, s_2) forms a subgame perfect equilibrium. Since at stage 2, both players are playing either Nash Equilibrium (C,c) or (D,d), each of the subgames are playing a Nash equilibrium, so we need only check at stage 1, if Player 1 wants to change from B to A, or player 2 wants to change from b to a. But in either case, we have $v_1(B, s_2) = 4 + 5d$, $v_1(A, s_2) = 5+ 2d$, this means that B is the best choice over A if d> 1/3.

Thus (B, b) followed by (C, c) is a subgame perfect Nash equilibrium in the 2-stage game if d>1/3.

As another example of the award and punishment concept as described above, we can show below that (B,a) in the first stage followed by (C,c) in the second stage is a subgame perfect Nash Equilibrium. We award (B,a) in the first stage with the better payoff (5,5) in the second stage, and anything else in the first stage with the lower payoff (2,2).

$$s_1 = \{\ s_1^1, s_1^2(Aa), s_1^2(Ab), s_1^2(Ba), s_1^2(Bb)\}$$

$$s_1 = \{B, D, D, C, D\}$$

$$s_2 = \{\ s_2^1, s_2^2(Aa), s_2^2(Ab), s_2^2(Ba), s_2^2(Bb)\}$$

$$s_2 = \{a, d, d, c, d\}$$

As before, we only need to check if player 1 wants to change from B to A, or player 2 wants deviate from a to b. Since a is a dominating strategy for player 2, so we only need to check if player 1 wants to change B to A.

$v_1(B, s_2) = 0 + 5d$, $v_1(A, s_2) = 1+ 2d$, so B is the best response if d >1/3.

6.8 Repeated games

Now we are going to turn our attention to the repeated games. Repeat game is a special case of multi-stage game in which each stage game is the same game. Repeated games, especially infinitely repeated games, can be used to model many of the real-life scenarios and applications. We are going to state and formulate a celebrated theorem in an infinitely repeated game called the Folk Theorem. But to understand this theorem, first we need to understand the concept of award and punishment in a finitely repeated game and the concept of Grim Trigger strategy played by the players which we are now going to describe.

6.8.1 The concept of award and punishment in repeated games

We are going to illustrate this concept by three of our examples shown below.

Example 6.8.22

We use the following familiar Prisoners Dilemma as the stage game and consider the two stages repeated game as shown below. We want to know what strategy to be taken in each stage so that the concatenated strategy in the 2-stage repeated game is a Subgame Perfect Nash Equilibrium.

Figure 6.30

G1:	a	b
A	1,1	10,0
B	0,10	5,5

⇓

G1:	a	b
A	1,1	10,0
B	0,10	5,5

Note in the stage game, there is only one Nash Equilibrium (1.1). If both players play (A,a) in stage 1, and both play (A,a) again in stage 2, then this sequence of strategies played by both players, denoted by $(A,a) \Rightarrow (A,a)$ is a Subgame Perfect Nash Equilibrium. This can be easily explained as follows. Regardless which of the 4 strategies players play in stage 1, the players in stage 2 must play the Nash Equilibrium (A,a) which corresponds to the payoff (1,1), so the payoff for the first stage in the two stage repeated game is the same as the one-shot game with the payoff matrix is obtained by adding the payoff (1,1) to each entry of the original one stage payoff matrix.

G1:	a	b
A	2,2	11,1
B	1,11	6,6

And obviously, players playing (A,a) with payoff (2,2) is a Nash Equilibrium in the above payoff matrix. Thus (A,a) followed by (A,a) is a Subgame perfect Nash Equilibrium in the 2-stage repeated game.

To recap, the one-shot payoff matrix in stage 1 above is obtained by the sequential interaction or negotiation between stage 1 and stage 2 strategies and they are stated as follows:

- IF players play (A,a) in stage 1, THEN the players play (A,a) in stage 2 (this amounts to adding (1,1) to the entry (1,1) in the stage payoff matrix)
- IF players play (A,b) in stage 1, THEN the players play (A,a) in stage 2 (this amounts to adding (1,1) to the entry (10,0) in the stage payoff matrix)
- IF players play (B,a) in stage 1, THEN the players play (A,a) in stage 2 (this amounts to adding (1,1) to the entry (0,10) in the stage payoff matrix)
- IF players play (B,b) in stage 1, THEN the players play (A,a) in stage 2 (this amounts to adding (1,1) to the entry (5,5) in the stage payoff matrix.

The example above does not involve the reward and punishment concept because at each stage, the players are playing the Nash Equilibrium strategy, and the concatenated strategy is shown to be the Subgame Perfect Nash Equilibrium in the 2-stage repeated game. If, however, when players do not play the Nash Equilibrium in the first stage, then the concatenated strategy can still be a Subgame Perfect Nash Equilibrium by using the reward and punishment concept illustrated in the two examples below.

Example 6.8.23

Suppose now the stage game in the 2-stage repeated game is:

	a	b	c
G1: A	1,1	6,0	0,0
B	0,6	5,5	0,0
C	0,0	0,0	3,3

Now the stage game has two Nash Equilibriums (1,1) and (3,3) corresponds to the strategy (A,a) and (C,c). Note that (C,c) has a higher payoff (3,3) than that of (A,a) with payoff (1,1). Also note that (B,b) is not a Nash Equilibrium strategy. We are going to show that the concatenated strategy (B,b) $\Rightarrow$ (A,a) is a Subgame Perfect Nash Equilibrium in the 2-stage game by choosing a proper sequential negotiation (reward and punishment) strategy between the players in the two stages as follows:

- IF both players play (B,b) in stage 1, THEN both players will play (C,c) in stage 2. (This is to reward both players by adding the higher payoff (3,3) to the entry (B,b) in the payoff matrix above)
- IF both players play any of the 9 strategies other than (B,b) in stage 1, THEN both players will play (A,a) in stage 2. (This is to punish both players by adding a lower payoff (1,1) to each of the 9 entries in the payoff matrix above)
- Thus, the players' first stage in the 2-stage repeated game is same as the one shot game with the payoff shown below:

	a	b	c
G1: A	2,2	7,1	1,1
B	1,7	8,8	1,1
C	1,1	1,1	4,4

And obviously, (B,b) with payoff (8,8) is a Nash Equilibrium. And so (B,b) followed by (A,a) is a Subgame perfect Nash Equilibrium in the 2-stage repeated game. This example shows that we could have Subgame perfect Nash Equilibrium in a 2-stage repeated game even through the strategy used in the first stage is not a Nash Equilibrium.

Example 6.8.24

Suppose now the stage game in the 2-stage repeated game is:

	a	b	c	d	e
G1: A	1,1	6,0			
B	0,6	5,5			
C			3,3		
D				5,1	
E					1,5

where the empty entries are (0,0).

First note that there are 4 Nash Equilibriums in the stage game, (A,a), (C,c), (D,d), and (E,e), but note that (B,b) is not a Nash Equilibrium. Again, we want to demonstrate through proper negotiation (reward and punishment) between strategies players play in stage 1 and stage 2 to be described below, the concatenated strategy (B,b) $\Rightarrow$ (C,c) is a Subgame Perfect Nash Equilibrium.

- IF both players play (B,b) in stage 1, THEN both players will play (C,c) in stage 2. (This is to reward both players by adding the higher payoff (3,3) to the entry (B,b) in the payoff matrix above)
- IF the row player plays 'B', and the column player plays any of the other 4 strategies other than 'b' in stage 1, THEN both players will play (D,d) in stage 2. (This is to reward player 1 and punish player 2 by adding a payoff (5,1) to each of the 4 entries in row B other than (B,b) in the payoff matrix above)
- IF the column player plays 'b', and the row player plays any of the other 4 strategies other than 'B' in stage 1, THEN both players will play (E,e) in stage 2. (This is to reward player 2 and punish player 1 by adding a payoff (1,5) to each of the 4 entries in column b other than (B,b) in the payoff matrix above)

- IF the players play any other strategies other than player 1 playing B or player 2 playing b (there are 16 such strategy pairs in the payoff matrix above) in stage 1, THEN both players will play (C,c)

Thus, the players' interactions in the first stage is the same as the following one-shot game with the modified payoff shown below:

By adding the entry (3,3) to entry (5,5) we get:

Figure 6.31

a	b	c	d	e
1,1	6,0			
0,6	8,8			
		3,3		
			5,1	
				1,5

By adding entry (5,1) to the four entries in Row B we get,

1,1	6,0			
5,7	8,8	5,1	5,1	5,1
		3,3		
			5,1	
				1,5

By adding (1,5) to the 4 entries in column b we get:

1,1	7,5			
5,7	8,8	5,1	5,1	5,1
	1,5	3,3		
	1,5		5,1	
	1,5			1,5

By adding (3,3) to the remaining entries we get:

	a	b	c	d	e
A	4,4	7,5	3,3	3,3	3,3
B	5,7	8,8	5,1	5,1	5,1
C	3,3	1,5	6,6	3,3	3,3
D	3,3	1,5	3,3	8,4	3,3
E	3,3	1,5	3,3	3,3	4,8

From the above payoff matrix, (B,b) is obviously a Nash Equilibrium , and hence (B,b) followed by (C,c) is a Subgame perfect Nash Equilibrium in the 2-stage repeated game as claimed.

6.8.2 The concept of Grim Trigger Strategy in an infinitely repeated games

We have seen from examples in a 2-stage repeated game above, Subgame perfect Nash Equilibrium can be achieved without having a Nash Equilibrium strategy in the first stage if there is more than one Nash Equilibrium in the stage game. We will show in this section, Subgame perfect Nash Equilibrium can be achieved in an infinitely repeated game even though there is only one Nash Equilibrium in the stage game. To do this, we need to introduce the concept of a Grim Trigger strategy in a infinitely repeated game. In general, when we repeat a game infinitely many times and there is sequential interaction between strategies from consecutive stages, it can be imagined that the task of managing these strategies will be very complex and beyond control. However, the Grim Trigger strategy that the players play makes the task and the book keepings simplified. We will define Grim Trigger strategy below and illustrated it with an example.

So, what is grim trigger strategy in a repeated game?

Definition 6.8.5 (Grim Trigger Strategy)

Roughly speaking, it is a set of pre-selected strategies each player repeatedly plays without ending until any player fails to play his or her pre-selected strategy by playing any other strategies, if that happens, then all the players will play another set of pre-selected strategies repeatedly, again without ending in the infinitely repeated game. We will give an example of illustration for two players.

Example 6.8.25

The following stage game is going to repeated infinitely many times:

G1:	a	b
A	1,1	10,0
B	0,10	5,5

As noted, in this stage game, there is only one Nash Equilibrium (1,1) ,and note that (5,5) is not, it is clear then , from Proposition 6.7.1 stated before, that

(A,a) $\Rightarrow$ (A,a) $\Rightarrow$.... infinitely is a Subgame perfect Nash Equilibrium. The point in this example is to show that we can start with playing a non-Nash Equilibrium strategy (B,b), and it could also give us a Subgame perfect Nash Equilibrium in an infinitely repeated game when a proper grim trigger strategy is used by the players.

The grim trigger strategy for player 1 (row player) and player 2 (column player) are stated as follows:

Player Row:

Play B in the first stage. In the nth stage.

IF both players have been playing (B,b),(B,b).... (B,b) in the previous (n-1) stages, THEN play B

ELSE play A

Player Column:

Play b in the first stage. In the nth stage.

IF both players have been playing (B,b),(B,b).... (B,b) in the previous (n-1) stages, THEN play b

ELSE play a

Using these trigger strategies defined above, we are going to show that:

(A). These two strategies is a Nash Equilibrium in the infinitely repeated game

(B). These two strategies is a Subgame perfect Nash Equilibrium in the infinitely repeated game

Proof of (A):

This is equivalent to proving the following parts (A1) and (A2)

(A1) IF player 1 has adopted the trigger strategy, THEN we want to prove that the best response for player 2 is to adopt the trigger strategy also.

(A2) IF player 2 has adopted the trigger strategy, THEN we want to prove that the best response for player 1 is to adopt the trigger strategy also.

Now focus on the proof of (A1):

Since player 1 has adopted the trigger strategy, this means that, Player 1 will play strategy A forever once one stage outcome differs from (B,b) , player 2's best response is to play strategy 'a' because 1 is better than 0 from the payoffs matrix shown in Figure 1. It remains to look at what happen when there is no change to the outcome (B,b) in all the previous stages we want to determine what would player 2's best response then?

If player 2 plays ‘a’, she will get payoff of 10 at the present stage but will trigger the payoff of (1,1) forever, so that the payoff in every future stage will be 1. Therefore, the present value of player 2 when she plays ‘a’ is:

$$10 + 1^* d + 1^* d^2 + \ldots.. = 10 + 1/(1-d)$$

If player 2 plays ‘b’, she will get a payoff of 5 in the present stage and it will lead to the same choice between ‘a’ and ‘b’ in the next stage, therefore, the present value of player 2 when she plays ‘b’ is:

$$5 + 5^*d + 5^*d^2 + \ldots \quad = 5/(1-d)$$

By comparing the two present values above, player 2 will choose ‘b’ when

$$5/(1-d) \geq 10 + 1/(1-d), \text{ or}$$

$$d \geq 3/5$$

Thus, given that player 1 has adopted the trigger strategy, in the first stage, and in any stage such that all the preceding outcomes have been (B,b) player 2 best response is to play ‘b’ when $d \geq 3/5$. This shows that player 2 is best responding with the trigger strategy as defined above.

Proof of (A2) is similar and is left as an exercise to the reader. (A1) and (A2) together proves (A).

Now the proof of (B). We need to recall the definition of subgames in an infinitely repeated game and keep in mind that each subgame beginning at state (t+1) is identical to the original game. We want to show the Trigger strategy defined above is a Nash Equilibrium on every subgame of the infinitely repeated game. As shown in Figure 6.32 below, only first 3 stages out of the infinitely many stages are shown, there are as many subgames that begin at stage (t+1) as there are possible histories of play through stage t which is an infinite number. However, the good news is that many subgames can be grouped into only two classes:

(B1) Subgames in which all the outcomes of earlier stages have been (B,b). One of such subgames is shown in blue in Figure 6.32.

(B2) Subgames in which all the outcomes of at least one earlier stage is not (B,b). One of such subgames is shown in red in Figure 6.32.

, and in each of these two cases, we claim that the grim trigger strategy that both players play through out the whole game is a Nash Equilibrium. But (B1) has been proven in (A). In case (B2), the strategies that the players play are simply the repetition of the stage-game Nash Equilibrium (A,a) forever, which is also a Nash Equilibrium by Proposition 1 stated before.

Figure 6.32

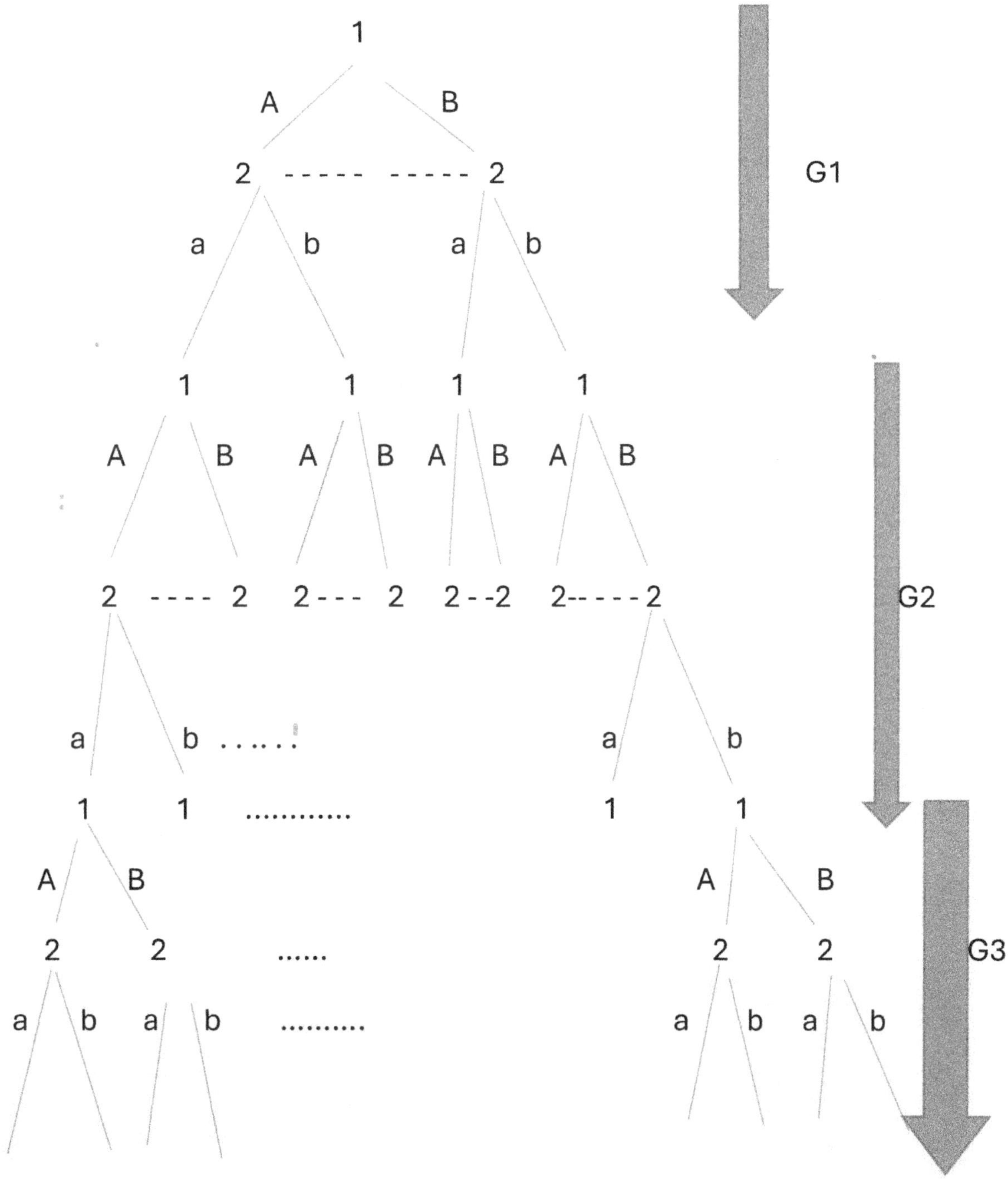

Now that we know what a grim trigger strategy is in a infinitely repeated game, we are almost in a position to state and formulae the celebrated Folk's Theorem in the next section, but in order to formulate the theorem in a neatly way, we need to introduce what is known as the average payoff of a player defined below rather than to use the total payoff that we defined earlier in Definition 6.7.1

Definition 6.8.6 The average payoff of player i with an infinite sequence of payoffs .. $v_i^1, v_i^2, \ldots . v_i^T$ from playing the sequence of games is defined as:

$$A(v_i) = (1\text{-d})\{\ v_i^1 + dv_i^2 + d^2v_i^3 + \ldots . + d^{T-1}v_i^T + \ldots\ \}$$

, where d is called the discount factor, and T approach ∞. In other words, the average payoff is defined as the constant factor (1-d) times the total payoff.

We call the expression $A(v_i)$ defined above the average payoff this is because when T is finite, and when all the $v_i^1, v_i^2, \ldots . v_i^T$ are the same payoff 'v' in each stage , then A(v) is indeed v as expected. This can be easily shown as follows:

$$A(v) = (1\text{-d})\ \{\ v + dv\ + d^2v\ + \ldots . + d^{T-1}v\ \}$$

$$= (1\text{-d})\ v\ .\ \frac{1-d^T}{1-d} = v$$ as T goes to ∞ and d^T goes to 0.

Now both Grim Trigger Strategy and Average Payoffs in an infinitely repeated game are defined, we can state the Folk Theorem as follows:[2]

6.9 The Folk's Theorem: (Friedman 1971):

First, we need to state and make clear the following assumptions and notations used in the theorem below.

IF the stage game G is a finite, static complete information game of n-players, where A_1, A_2……, A_n are the action set for the n -players.

Let (a_1 , a_2 , a_n) be a set of actions, each a_i in A_i.

Let $p_1(a_1, a_2, a_n)$, $p_2(a_1, a_2, a_n)$,… $p_n(a_1, a_2, a_n)$ be the payoff-functions for each player.

Denote (ae_1 ,ae_2 , ae_n) be a set of actions, where each ae_i in A_i, such that it is a Nash Equilibrium of G and the corresponding payoffs is denoted by (e_1 ,e_2 , e_n). We use the prefix ' a' to signify that it is an action. That is to say,

$e_1 = p_1(ae_1, ae_2, ae_n)$,

$e_2 = p_2(ae_1, ae_2, ae_n)$,

.. (1)

..

$e_n = p_n(ae_1 , ae_2 , \quad ae_n)$

Let $(x_1 , x_2 , \quad x_n)$ be any feasible payoffs of G with the restriction that $x_i > e_i$ for each player. Let $(ax_1 , ax_2 , \quad ax_n)$ be the corresponding set of actions to the payoffs $(x_1 , x_2 , \quad x_n)$. That is to say,

$x_1 = p_1(ax_1 , ax_2 , \quad ax_n)$,

$x_2 = p_2(ax_1 , ax_2 , \quad ax_n)$,

.. (2)

..

$x_n = p_n(ax_1 , ax_2 , \quad ax_n)$

Assume d is the discount factor close to 1.

THEN there exist a Subgame perfect Nash Equilibrium in the infinitely repeated game of G such that the average payoff is $(x_1 , x_2 , \quad x_n)$.

Proof: Before proofing this theorem, a few words about the significance of this theorem is in order. In general, we may think of existence of Nash Equilibrium in a stage game G game can provide some sort of useful information about the game, and usually there are not that many Nash Equilibriums, however, the Folk Theorem stated that, if we repeated the stage game many times, then there will be many subgame perfect Nash Equilibriums with payoff $(x_1 , x_2 , \quad x_n)$ as long as $x_i > e_i$. This explains that many real-life applications such as negotiations amongst players involve repetition of playing the stage game.

The proof of the theorem is like Example 6.8.25 we have shown before. Again, we will show that if each player is playing the Grim Tigger Strategy, then these strategies will give rise to a Subgame perfect Nash Equilibrium in the infinitely repeated state game G.

Precisely, for each play i in the stage game G, we consider the grim trigger strategy S(i):

In the first stage, player i plays ax_i

In the kth stage,

IF the outcome of all the previous (k-1) stages has been (ax_1 , ax_2 , ax_n) THEN player i plays ax_i ELSE play ae_i .

We will show that (a) if each of the player plays this strategy S(i), then together they form a Nash Equilibrium and (b) furthermore they form a Subgame perfect Nash Equilibrium. The logic of proofing (a) and (b) are like that shown in the previous example of two players' Prisoners Dilemma.

To prove (a):

Assume all the players other than player i have adopted the grim trigger strategy S(i) as described above, we will show that the best response that player i plays is in fact the trigger strategy also.

To do this, we first find a strategy a in the action set A_i that maximizes the payoff function pi(ax_1 , ax_2 , …, a,.. ,ax_n). Call this action ad_i.

$$\underset{a \in A_i}{Max}\ p_i\,(ax_1, ax_2, \ldots.\ , a, \ldots\ , ax_n) \qquad (3)$$

Thus, $d_i = p_i\,(ax_1, ax_2, \ldots.\ , ad_i, \ldots\ , ax_n) \geq x_i = p_i(ax_1\ , ax_2\ , \ldots\ , ax_i, \ldots\ , ax_n)$ by the choice of the action ad_i and $x_i = p_i(ax_1\ , ax_2\ , \ldots\ , ax_i, \ldots\ , ax_n) > e_i = p_i(ae_1\ , ae_2\ , \ldots\ , ae_n)$ since (ae_1 , ae_2 ... ,ae_n) is a Nash Equilibrium by assumption, so $d_i \geq x_i > e_i$.

If player i plays ad_i it will yield a maximal payoff of d_i at the present stage but then it will trigger all the other players to play (ae_1 , ae_2 ,.. $ae_{i-1,}$ $ae_{i+1,}$… ,ae_n), and so the best response to player i is to play ae_i as well and thus the payoff in all future stage will be

$p_i(ae_1\ , ae_2\ , \ldots\ ae_{i-1},\ ae_i,\ ae_{i+1},\ \ldots\ , ae_n)$ which is e_i . The present value PV of the sequence of payoffs is therefore,

$$PV(d_i) = d_i + d\,e_i + d^2\,e_i + \ldots. = d_i + d\,\{1 + d + d^2 + \ldots..\}\,e_i = d_i + \frac{d}{1-d}e_i \qquad (4)$$

If player i plays ax_i, then it will yield a payoff of x_i at the present stage, and it will lead to the same choice between adi and ax_i in the next stage. The present value PV of playing ax_i is,

$$PV(ax_i) = x_i + d\,x_i + d^2\,x_i + \ldots. = \frac{x_i}{1-d} \qquad (5)$$

Therefore, player i will play ax_i rather than ad_i if and only if $PV(ax_i) \geq PV(ad_i)$, or

$$\frac{x_i}{1-d} \geq d_i + \frac{d}{1-d}e_i \qquad (6)$$

Solving d gives $d \geq = \frac{d_i - x_i}{d_i - e_i}$. Thus, given that all others are playing the trigger strategy, and that $d \geq \frac{d_i - x_i}{d_i - e_i}$, the best response for player i is also the grim trigger strategy. Recall that the discount factor d above was selected for a particular player i, so if we select d to be

$d \geq \underset{i}{Max} \frac{d_i - x_i}{d_i - e_i}$ for all the players as the discount factor, then playing (ax_1, ax_2,… ,ax_i,.. ,ax_n) with grim trigger strategy for each player is a Nash Equilibrium. Note that $\frac{d_i - x_i}{d_i - e_i} < 1$ since $d_i \geq x_i > e_i$.

To prove (b):

We need to show if every players plays the grim trigger strategy, then it will also be a Nash Equilibrium in every subgame of the infinitely repeated game of the stage game G that has the Nash Equilibrium (ae_1, ae_2,.. ae_{i-1}, ae_{i+1},… ,ae_n)

Note that every subgame of the infinitely repeated game is identical to the infinitely repeated game itself. Since each player is only playing the grim trigger strategy, all the subgames can be grouped into two different groups.

(G1) Subgames in which all the outcomes of earlier stages have been

(ax_1, ax_2, ...,ax_i,.. ,ax_n)

(G2) Subgames in which all the outcome of at least one earlier stage differs from

(ax_1, ax_2, ...,ax_i,.. ,ax_n)

(G1) follows from the proof of (a) because the players are all playing the grim trigger strategies. (G2) follows from Proposition 67.7.1 stated earlier because the subgames in this group are simply repeatedly playing the Nash Equilibrium

(ae_1, ae_2, ae_n) of the stage game G.

(G1) and (G2) together show that the grim trigger strategies that the players play is sub-game perfect Nash Equilibrium in the infinitely repeated game.

Example 6.9.26 Impact on the choice of the discount factor "d"

Recall that the choice of the discount factor d has a big impact on the proof of the Folk's Theorem. To reiterate Example 6.9.25 with slight input data changes, the following simple example - a classical two-player Prisoner's Dilemma, shows that for any feasible and rational payoff to be sustained as a credible equilibrium it depends on the proper choice of d. Assume the one-shot game has the following payoff matrix:

	Player 2\ Player 1	Cooperate (C)	
Defect (D)	(C)	(3,3)	(0,5)
	(D)	(5,0)	(1,1)

The only equilibrium in this payoff matrix is (1,1). Suppose that both players agree to cooperate forever with the more desirable payoff (3,3) by both playing (C) above , to maintain this equilibrium, either player playing D must not be profitable in the long run. Let's now compute the total payoff (discounted sum) by 6.7.1 for each of these two scenarios. Let S(C) be the discounted payoff for cooperating forever, then

S(C) = 3 + 3d + 3 d2 + 3 d3 + …
= 3 [1 + d + d2 + d3 + …]
= 3 [1/ (1-d)]

Let S(D) be the discounted payoff for playing D once and gets punished forever, then

S(D) = 5 + 1d + 1 d2 + 1 d3 + …..
= 5 + d [1 + d + d2 + d3 + …]
= 5 + d [1/ (1-d)]
= (5-4d)/ (1-d)

Thus for (3,3) to be sustainable as Nash equilibrium, S(C) must be larger than or equal to S (D), which gives

3/(1-d) ≥ (5-4d)/ (1-d) , which gives d ≥ 0.5

Example 6.9.27 A practical economic application

We are going to give an application of the Folk Theorem to an economic scenario to show that cooperation between two firms could be achieved in an infinitely repeated game where the underlying stage game is described by the Cournot and Stackelberg Model (Application 6.6.1). To be precise, recall that if x_1 and x_2 be the number of products firm 1 and firm 2 produced respectively, then the profit functions are given by $P_1 = (A – (x_1 + x_2)-C) * x_1$, and

$P_2 = (A – (x_1 + x_2)-C) * x_1$ respectively. We find from (7) , page 195 that the Nash Equilibrium pair (x_1^*, x_2^*) is given by $x_1^* = x_2^* = 1/3$ (A-C). From the individual firm's perspective, it is desirable for each of them to produce the monopoly quantity x_1^m for firm 1 and x_2^m for firm 2. As x_1^m maximize P_1 when x_2 is 0, so

by setting $\frac{dP_1}{dx_1} = \frac{d}{dx_1}(A-C-x_1)x_1$ to 0, so A - C – 2 x_1 =0, so x_1^m = ½(A-C). Similarly,

$x_2^m = ½(A\text{-}C)$. If both firms cooperate and each produces half of this monopoly quantity $½\, x_1^m$ or $½\, x_2^m$ or $¼(A\text{-}C)$, then the profit for both firms would be

$P_1 = P_2 = 1/8\,(A\text{-}C)^2$. This monopoly profit is larger than that at the Nash Equilibrium $x_1^* = x_2^* = 1/3\,(A\text{-}C)$, which is $P_1 = P_2 = 1/9\,(A\text{-}C)^2$. This motivates both firms to cooperate and to achieve this higher profit in an infinitely repeated game via the Folk Theorem.

Now we can think of the underlying stage game is represented by the following payoff matrix:

	$½\, x_2^m$	x_2^*
$½ x_1^m$	(1/8, 1/8) M	(5/48.5/36) M
x_1*	(5/36,5/48) M	(1/9 , 1/9) M

where $M = (A\text{-}C)^2$.

Now we are going to repeat the above stage game infinitely many times and firm 1 and firm 2 each play the grim trigger strategy stated below: Keep in mind, $x_1^* = x_2^*$ and $x_1^m = x_2^m$.

For firm 1:

Produce $½\, x_1^m$ units in the first period. In the tth stage, keep producing $½\, x_1^m$ units if both firms have produced $½\, x_1^m$ units in each of the previous (t-1) stages, otherwise produce x_1^* units

For firm 2:

Produce $½\, x_2^m$ units in the first period. In the tth stage, keep producing $½\, x_2^m$ units if both firms have produced $½\, x_2^m$ units in each of the previous (t-1) stages, otherwise produce x_2^* units.

Recall the general expression (6) from the proof of the Folk Theorem,

$$\frac{x_i}{1-d} \geq d_i + \frac{d}{1-d} e_i. \tag{6}$$

Here x_i is the profit when firm i =1, 2 produces $½\, x_i^m$, which was shown to be $1/8(A\text{-}C)^2$, e_i is the profit when firm i produces x_i^*, which was shown to be $1/9(A\text{-}C)^2$, and d_i is computed as follows:

Recall that the definition of ad_i and d_i shown in expression (3) of the Folk Theorem, here d_i is the maximum of the profit function P_i of firm i when the other firm j produces $½\, x_j^m$. So, $P_i = (A\text{-}C\text{-}\, x_i - ¼(A\text{-}C))x_i$, by setting the first derivative of P_i to 0 gives $x_i = 3/8(A\text{-}C) = ad_i$, and the corresponding profit $P_i = 9/64(A\text{-}C)^2 = d_i$.

Thus (6) becomes,

$$\frac{1}{1-d}\ 1/8(A\text{-}C)^2 \geq 9/64(A\text{-}C)^2 + \frac{d}{1-d}\ 1/9(A\text{-}C)^2,$$

Solving d, we found d $\geq$ 9/17. Thus, by choosing the discount factor d larger than 9/17. Folk Theorem tells us that the two firms can achieve the payoff (1/8(A-C)^2, 1/8(A-C)^2) as a subgame perfect Nash Equilibrium with the grim trigger strategies defined above.

6.10 Summary

This chapter focuses on dynamic games with complete, perfect and imperfect information. The usual concept of the Nash Equilibrium in static games is extended to the concept of subgame-perfect Equilibrium (SPE) in dynamic games (Definition 6.4.3) . Two common approaches to finding the SPE are the Backwards and Forwards Induction (6.5) . The concept of subgame-perfect Equilibrium in a single game can also be extended to the Multi-stage Games (6.7) and Repeat Games (6.8) using the concept of Award and Punishment (6.7.3) and the Grim Trigger Strategy (Definition 6.8.5). The chapter ends with the celebrated Folk's Theorem (6.9) on infinitely repeated games with a practical application (Example 6.9.27) of the theorem to economics. We have shown many examples to facilitate the understanding of various concepts discussed in this chapter.

Chapter 7

Perfect Bayesian Nash Equilibrium and Signaling games

7.0 Introduction

This chapter introduces the fourth and final equilibrium concept discussed in this book—perfect Bayesian equilibrium—as applied to dynamic games of incomplete information.

Previously, we examined three equilibrium concepts:

1. Nash equilibrium in static games of complete information
2. Bayesian Nash equilibrium in static games of incomplete information
3. Subgame-perfect Nash equilibrium in dynamic games of complete information

It may seem that each equilibrium concept was specifically developed for the corresponding class of games. As we shall see, however, perfect Bayesian equilibrium is particularly well-suited for Signaling Games, a fundamental type of dynamic game with incomplete information.

All four equilibrium concepts are closely related. Just as subgame-perfect Nash equilibrium refines the basic concept of Nash equilibrium, the perfect Bayesian equilibrium concept refines Bayesian Nash equilibrium, making it a more precise tool for analyzing strategic interactions with sequential decision-making and uncertainty.

Chapter Structure:

Section 7.1 shows the logical dependence of the previous three equilibrium concepts to Perfect Bayesian Equilibrium

Section 7.2 defines perfect Bayesian equilibrium by outlining its four key requirements.

Section 7.4 introduces Signaling Games, describing their three fundamental requirements and providing a formal definition.

Section 7.4.5 presents two practical examples of Signaling Games. Example 7.4.11 illustrates the concept of limit pricing and entrance deterrence, Example 7.4.12 shows an application of the Spence Model on using education signaling in job markets.

Section 7.5 extends the discussion by introducing Requirements 5 and 6, which account for more complex and nuanced Signaling Games.

Section 7.6 highlights the versatility of the Spence Model discussed earlier. It demonstrates the existence of Separating, Pooling, and Hybrid Equilibria that fulfill Requirements 5 and 6, which are not achievable in other models.

By the end of this chapter, readers will have a clear understanding of how perfect Bayesian equilibrium refines strategic reasoning in dynamic games with incomplete information.

7.1 Perfect Bayesian Nash Equilibrium and its relationship with the three other equilibriums

Previously, we introduced the basic concept of Nash Equilibrium for static games with complete information in Chapter 4, then we extended this basic concept to the concept of Bayesian Nash Equilibrium to static games with incomplete information in Chapter 5. Independently, we extended the basic concept of Nash Equilibrium in Chapter 4 to the concept of Subgame -perfect Nash Equilibrium to dynamic games of complete information in Chapter 6. The Perfect Bayesian Nash Equilibrium concept shown in this Chapter7 can be thought of as an extension of the previous concepts to dynamic games with incomplete information.

The logical connection on the complexity of the four kinds of Nash Equilibrium is shown in the following Figure 7.0:

Figure 7.0

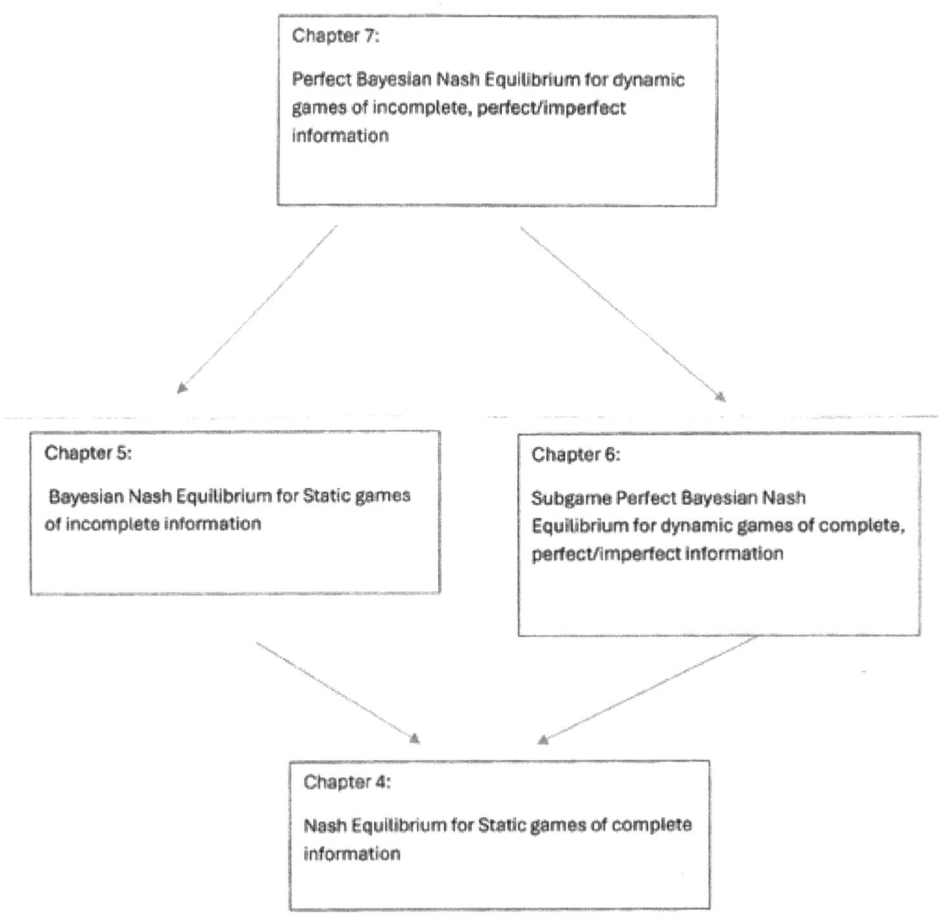

7.2 Definition and 4 Requirements of Perfect Bayesian Nash Equilibrium

A perfect Bayesian Nash Equilibrium in a game tree consists of two pieces of information: the first piece is the strategy that each player choses and the second piece is the beliefs for each information set in the game tree the player chooses to believe. To be precise, a belief is a probability distribution over the nodes in the information set of each player. If the information set is a singleton, then the player's belief is 1, and if the information set is non-singleton, then the beliefs for each node in the information set must be non-zero probabilities.

Note that it is this additional second piece of information, the beliefs for each information set, that differentiates the perfect Bayesian Nash Equilibrium from the other three types of Nash Equilibrium that discussed before in the previous chapters. We shall incrementally describe below four requirements on these beliefs and define that a perfect Bayesian Nash Equilibrium in a game tree is a set of strategies and beliefs satisfying these four requirements.

7.2.1 Requirement 1

At each information set in a game tree, the player with the move must have a belief (probability) about which node in the information set has been reached.

7.2.2 Requirement 2

Given the players' beliefs on their information sets by Requirement 1, the players' strategies must be sequential rational. By sequential rational, we mean that each information set the action taken by the player with the move must be optimal given the player's belief at that information set and the other players' subsequent strategies.

Note that Requirements 1 and 2 imposed on the beliefs thus far only says that these beliefs are any probability distribution on the information set for each player and each player acts optimally given these beliefs, it does not care above the actual value of the beliefs. Requirement 3 below imposes that these beliefs must be determined by Bayes' conditional probability rule. To formulate this requirement precisely, we need to define what is meant by an information set is on the equilibrium path and off the equilibrium path.

Definition 7.2.1 (An information set is on or off the equilibrium path)

An information set is on the equilibrium path it is reached with positive probability if the game is played according to the equilibrium strategies, and it is off the equilibrium path if it is not reached if the game is played according to the equilibrium strategies.

7.2.3 Requirement 3

For each player, at each information set on the equilibrium path in the game tree, the beliefs are determined by the Bayes' rule and the players' equilibrium strategies.

7.2.4 Requirement 4

For each player, the belief of an information set that is off the equilibrium path in the game tree is assigned with arbitrary value if it cannot be determined by the Bayes rule and the players' equilibrium strategies. However, the belief of an information set that is off the equilibrium path is always determined by the Bayes' rule and the players' equilibrium strategies if it is possible to do so. Example 7.3.3 below shows the scenarios on the applicability of this requirement.

Definition 7.2.2 (Perfect Bayesian Nash Equilibrium)

If the strategies for the players and the beliefs for the information sets are satisfying Requirement 1 to 4 ,then they are Perfect Bayesian Nash Equilibrium strategies.

7.3 Simple examples of Perfect Bayesian Equilibrium

The next two examples, Example 7.3.1 and Example 7.3.2 below are examples of perfect Bayesian Nash Equilibrium that satisfy all the four requirements stated above. The game tree in Example 7.3.1 represents a scenario that the game information is incomplete and perfect, and the game tree in Example 7.3.2 represents a scenario that the game information is complete and imperfect.

Example 7.3.1

In this example, player 1 has two types, t1 and t2. Assume nature chooses t1 with probability p, and chooses t2 with probability 1-p. Player 1 has two actions A and B. Player 2, has actions C and D. As there are two types for player 1, player 1 has the possible strategies {AA, AB, BA, BB} , whereas player 2 has strategies {C, D}. The game tree with incomplete information is shown below:

Figure 7.1

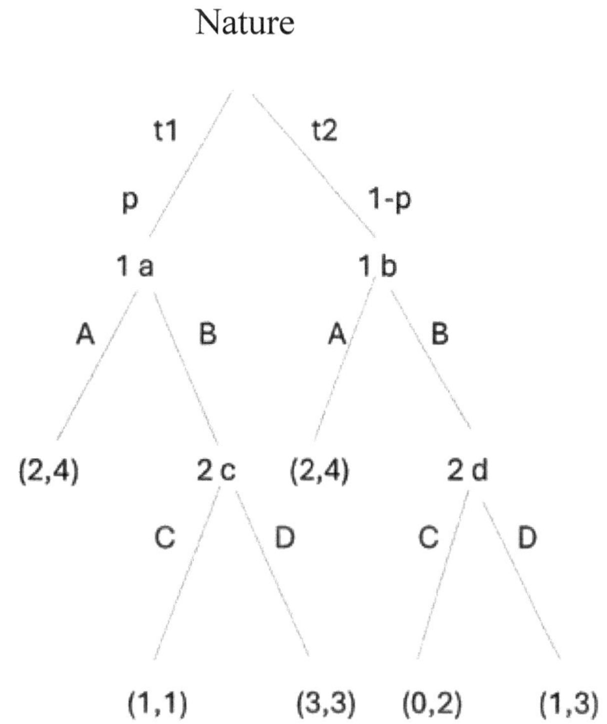

As shown, let a, b be the information set for player 1, let c, d be the information set for player 2. Let u(a), u(b), u(c), u(d) be the belief for the information sets a, b, c, d respectively. We will determine these beliefs for a perfect Bayesian equilibrium in the extensive game tree shown above. First, we need to find out all the possible Bayesian Nash equilibriums, and then select the one that is qualified for being a perfect Bayesian equilibrium. To do so, we need to compute the expected payoffs of the strategies pairs for player 1 and player 2 as shown in the matrix below:

Player 1 has two types t1 and t2, so the strategy space is {AA, AB, BA, BB},

Player 2 is a single type, so the strategy space is same as the action space {C, D}

The matrix for the expected value of strategy pairs is shown below:

Player 1 \ Player 2	C	D
AA	(2, 4)	(2, 4)
AB	(2p, 2p + 2)	(p + 1, p + 1)
BA	(-p + 2, -3p + 4)	(p + 2, -p + 4)
BB	(p, -p + 2)	(2p + 1, 3)

It is easily verified that the strategy pairs (AA, C) and (BA, D) are Bayesian Nash equilibriums as for the pair (AA, C), 2 is the maximum amongst {2, 2p, -p+2, p} for any value of p, and 4 is greater or equal to 4, and for the pair (BA,D), p+2 is the maximum amongst {2, p+1, p+2, 2p+1}, and -p+4 is greater than -3p+4 for all p.

Now we are going to examine which of the Bayesian Nash equilibriums (AA, C) and (BA, D) are candidates for perfect Bayesian Nash equilibrium by checking each against each of the four requirements stated before.

For the pair (AA, C), it does not satisfy Requirement 2 of sequential rationality, this is because, regardless of the value of the degree of belief u(c) is, when it is player 2's turn to play, player 2 will choose to play D instead of C because

$EV2(C) = u(c)\,(1) + (1-u(c))\,(2) = 2-u(c)$

$EV2(D) = u(c)(3) + (1-u(c))\,(3) = 3$, and so $EV2(D) > EV2(C)$

It remains only to check the pair (BA, D) for perfect Bayesian Equilibrium. The pair (BA, D) satisfies Requirement 2 for sequential rationality because.

$EV2(D) = u(c)\,(3) + (1-u(c))\,(3) = 3$,

$EV2(C) = u(c)\,(1) + (1-u(c))\,(2) = 2-u(c)$, and

EV2(D) > EV2(C).

To check if (BA, D) satisfies Requirement 1 to 3, note from Figure 1 shown above, if (BA, D) were the equilibrium path,

u(a) = 1,

u(b) = 0,

u(c) is player 2's belief that player 1 is of type t1 given the player 1 has chosen action B, by Baye's Theorem, it can be expressed as

u(c) = P(t1 | B), and by Bayes' Theorem, this is

= P(B | t1) P(t1)/{P(B | t1) P(t1) + P(B | t2) P(t2)}

= (1)(p)/{(1)(p) + (0) (1-p)}

= 1, because P(B | t1) =1, P(B | t2) = 0 as (BA, D) is the equilibrium path.

Thus, u(d) = 1- u(c) = 0, u(a) = 1, u(b) = 0 together with the equilibrium path (BA, D) satisfies Requirements 1 and 3. Now Requirement 4 is trivially satisfied because there is no information set off this equilibrium path as player 2 has only one information set {c, d}.

Before we leave this example, we note that both Nash equilibriums (AA, C) and (BA, D) are also trivially sub-game perfect Nash equilibriums in this example because there is no proper subgame in this game of incomplete information.

Example 7.3.2

In this example, both players are single type, Player 1 has three actions A, B, C and player 2 has two actions D, E.

Figure 7.2

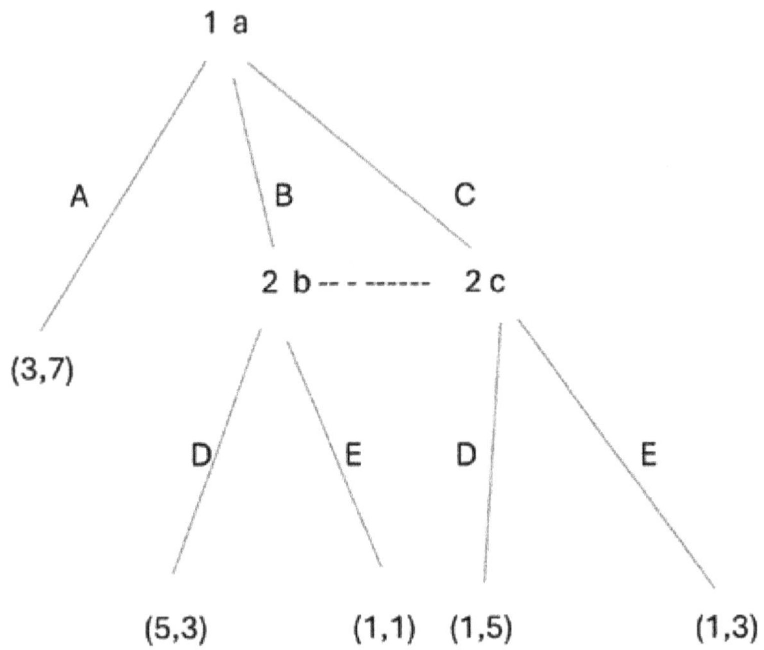

The dynamic game tree shown in Figure 7.2 above is an example of complete but imperfect information. Imperfect because when it is player 2's turn to move, he does not know which action B or C player 1 had chosen, all he knows is Player 1 did not choose A. To find the perfect Bayesian equilibrium in this game, first we write down the payoff matrix from the game tree:

		DD	DE	ED	EE
	A	(3,7)	(3,7)	(3,7)	(3,7)
Player 1	B	(5,3)	(5,3)	(1,1)	(1,1)
	C	(1,5)	(1,3)	1,5)	(1,3)

We check that both pairs (A, EE) and (B, DD) are Nash Equilibriums, and since there is no proper subgame, both pairs are subgame perfect Nash equilibrium as well, but the pair (A, EE) depends on a noncredible threat. This is because when player 2 is given the chance to move, Player 2 would choose to play D rather than to play E as D dominates E, and so E is a noncredible threat to player 1, as such player 1 does not have to choose A to counter act that, in fact the pair (B, DD) gives player 1 better payoff value of 5 rather than 3.

To check Requirement 1, refer to Figure 7.2, let u(a) be the degree of belief for Player 1's information set, let u(b), u(c) be the beliefs for Player 2's information set, u(c) is Player 2's belief that Player 1 has played B, note that u(c) = 1-u(b). To check Requirement 2, we compare the expected value of player 2 playing D, EV2 (D), to the expected value of 2 playing E, EV2(E)

EV2(D) = 3 u(b) + 5(1-u(b)) = 5 - 2u(b)

EV2(E) = 1 u(b) + 3(1-u(b)) = 3 - 2u(b)

As EV2(D) is greater than EV2(E) for any belief u(b), therefore by Requirement 2 on Sequential Rationality, the pair (B, DD) is selected over (A, EE).

Requirement 2 only insists that each decision point in an information set must have a belief, however it does not require specifying an exact value. Requirement 3, however, demands that the belief must be computed from the Bayes Theorem. In our example, the equilibrium path under consideration is (B, DD), this means that P(D | B) = 1, P(D | C) = 1, and u(b), player 2's belief that Player 1 has chosen B can be computed from the Bayes Theorem as:

u(b) = P(B | D)

= P(D | B) P(B) / {P(D | B) P(B) + P(D | C) P(C)}

= 1*1 / {1*1 + 1* 0}

= 1

Thus, with u(a) = 1 and u(b) = 1, Requirement 3 is satisfied.

Requirement 4 is trivially satisfied because for the equilibrium (B, DD), there is no information set off the equilibrium path. Therefore (B, DD) is a perfect Bayesian equilibrium.

Before we leave this Example 2, we want to illustrate a more general use of the Bayes Theorem with regards to Requirement 3 in a perfect Bayesian equilibrium. Suppose in this example, Player 1 is playing a mixed strategy with probabilities p_A, p_B, p_C playing A, B ,C respectively, then for the same equilibrium path (B, DD), player 2's belief that player 1 has chosen A would have been:

u(b) = P(B | D)

= P(D | B) P(B) / {P(D | B) P(B) + P(D | C) P(C)}

= 1 * p_B / {1 * p_B + 1*1 * p_C}

= p_B / {p_B + p_C}

Example 7.3.3

We use this example to illustrate Requirement 4, same game tree structure as in Example 7.3.2 but different payoff values for both players as shown in Figure 7.3 below.

Figure 7.3

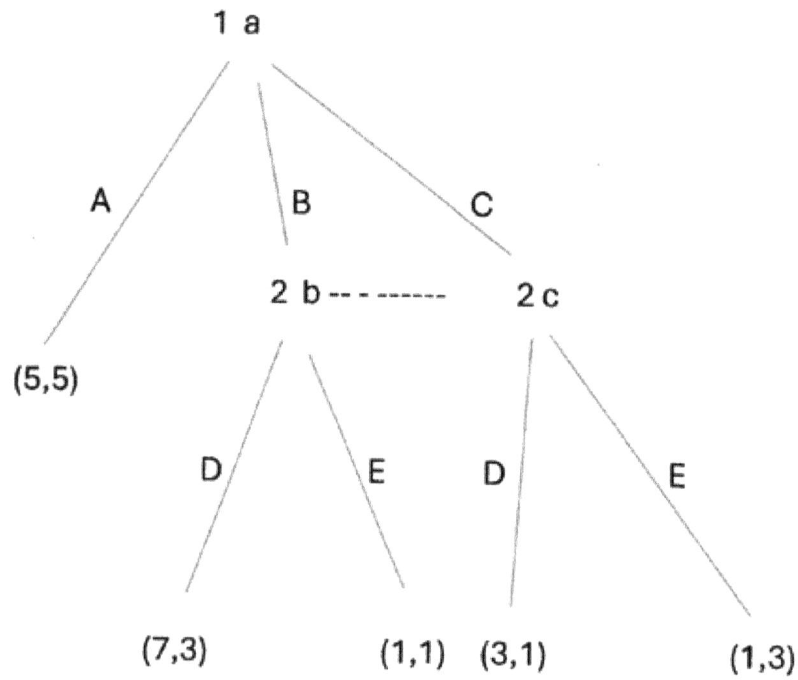

The payoff matrix for both players is shown below:

		Player 2			
		DD	DE	ED	EE
	A	(5,5)	(5,5)	(5,5)	(5,5)
Player 1	B	(7,3)	(7,3)	(1,1)	(1,1)
	C	(3,1)	(1,3)	(3,1)	(1,3)

It is easily seen that (B, DD) and (A, EE) are two Nash equilibria from the matrix above, and (B, DD) with u(b) = 1, and (A, EE) with u(b) < ½ can be shown below to be two perfect Bayesian equilibria as there are no subtree in this game. For the equilibrium (B, DD), both information sets b and c are on the equilibrium path, so Bayes Theorem in Requirement 3 applies and u(b) = 1 as before. As for the equilibrium (A, EE), the information set b is off the equilibrium path and so Requirement 4 puts no restriction on u(b) as Bayes Theorem cannot be applied. But Requirement 2 insists that player 2's strategy must be rational to play E rather than to play D, so the expected value of player 2 playing E is greater than that of D. So, EV2(E) > EV2(D) implies u(b) < ½ as

EV2(E) = 1* u(b) + 3* (1-u(b)) = 3 - 2u(b)

EV2(D) = 3 * u(b) + 1* (1-u(b)) = 1 + 2u(b)

from above. Thus, in this example, both equilibria satisfy Requirement 1 to 4 trivially. So, they are perfect Bayesian Nash Equilibria. Next example shows scenario that a Nash Equilibria does not satisfies Requirement 4.

Example 7.3.4

We consider a three-person game with the payoff values of the three players as shown.

Figure 7.4

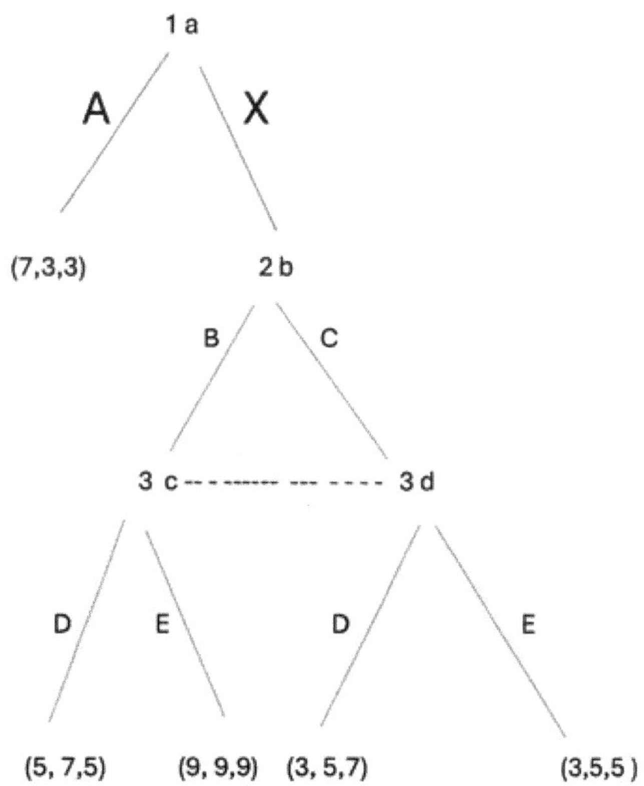

The game between player 2 and player 3 in the tree-diagram above is the only subgame of the entire game. Let u(a), u(b), u(c), u(d) be the belief for the information sets a, b, c, d respectively. It is easily seen that (B, EE) is with u(c) = 1 is the only perfect Nash equilibrium in this subgame from examples before. The strategies (X, B, EE) is Nash Equilibrium as no player wants to deviate unilaterally which we can verify as follows:

As shown in Figure 7.4, the first player gets payoff 9 when selecting X but only gets 7 when deviates to A. The second player gets payoff 9 when selecting B but only gets 5 when deviates to C. The third player gets payoff 9 when selecting E but only gets 5 when deviates to D.

Thus (X, B, EE) with u(c) = 1 satisfies Requirement 4 trivially because there is no information set is off the equilibrium path. Thus (X, B, DD) with the degree of beliefs at the information sets is a perfect Nash Equilibrium.

Example 7.3.5

In this example, we show that the strategies (A, B, DD) with belief u(c) = 0 in Figure 7.4 above cannot satisfy Requirement 4. First note that (A, B, DD) when u(c) = 0 is a Nash Equilibrium as no player wants to deviate. Again, this can be verified as follows. Player 1 will get 5 instead of 7 if deviates from A to X. Player 2 will get 5 instead of 7 if deviated from B to C, and Player 3 will get 5 instead of 5 if deviated from D to E. Now we are going to show that the equilibrium (A, B, DD) and the belief u(c) = 0 cannot satisfy the 4 requirements simultaneously. First note that player 3's belief at c can be anything because node c is not on the equilibrium path (A, B, DD) as such its belief is not governed by Requirement 1, 2, and 3. However, Requirement 4 does determine player 3's belief on node c based on player 2's strategy of choosing B or choosing C. If player 2's strategy is B, the player 3's belief at c, u(c) = 1; and if player 2's strategy is C, then player 3's belief at c, u(c) = 0 as u(d) = 1. To determine player 3's strategy based on player 3's belief at c, as before, we compute the expected value for strategy D, EV3(D) and the expected value for strategy E, EV3(E):

EV3(D) = u(c) 5 + (1-u(c))7 = 1 - 2 u(c)

EV(E) = u(c) 9 + (1-u(c))5 = 1 + 4 u(c)

Thus, for u(c) = 1, EV3(E) = 5 > EV3(D) = -1, so the equilibrium is (A, B, EE) rather than (A, B, DD). And for the equilibrium (A, B, DD), u(c) cannot be 0 by Requirement 4. So, the strategies (A, B, DD) with u(c) = 0 cannot simultaneously satisfy Requirement 1 to Requirement 4.

Example 7.3.6

In this example, we show a scenario where Requirement 4 does not apply in which the beliefs cannot be determined by the Bayes' rule and the players' equilibrium strategies. Consider the game in Figure 7.4 above is modified as follows: player 2 has an additional strategy Y that ends the game (See Figure 7.5 below)." Now if the equilibrium were (A, Y, DD), then Requirement 4 does not apply to put restrictions on player 3's belief because the information set c or d is not on the equilibrium path.

Figure 7.5

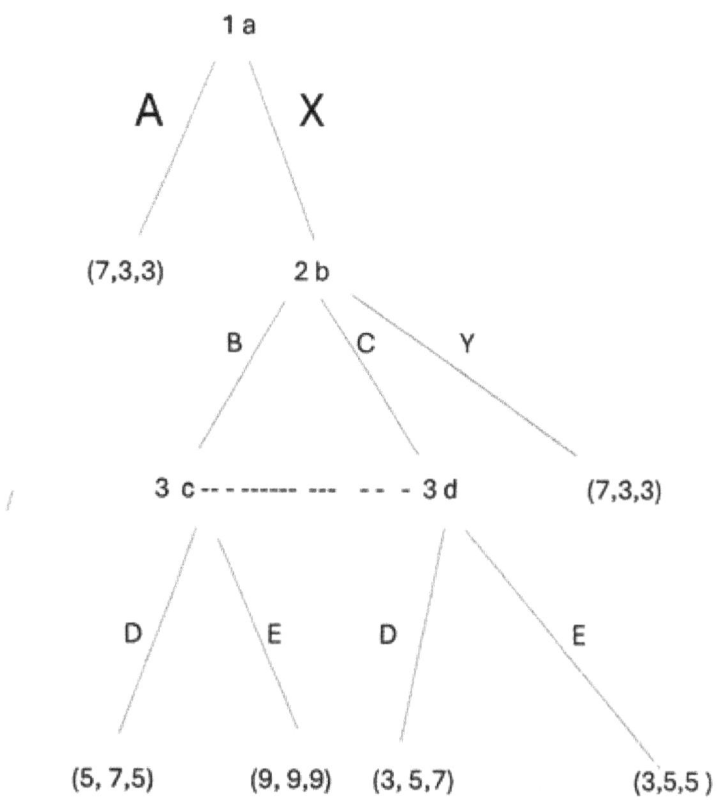

7.4 What are Signaling Games ?

In this section and the rest of the chapter, we will discuss a class of dynamic game of incomplete information involving two players called the Sender (S) and the Receiver (R) called Signaling Games. The general structure of these games is described below:

7.4.0 General structure of the Signaling Games (notations and terminologies refer to Figure 7.6 below)

(1) Nature chooses a type from a finite set of types $T = \{t_1, t_2 .... t_I\}$ for player 1 (Sender) according to a known probability distribution , say $p(t_i)$, where $p(t_i) > 0$ for each type i, and that $p(t_1) + p(t_2) + ... + p(t_I) = 1$.

(2) First player 1 (Sender) chooses an action A_i from a list of possible actions in an action set $A = \{A_1, A_2 ... A_I\}$, then player 2 (Receiver) responds with another action B_k from another action set $B = \{B_1, B_2, ..., B_K\}$ after observing which action A_i that player 1 took.

(3) Player 1 certainly knows which type t_i from T that he belongs to, Player 2, on the other hand, only knows about player 1's probability distribution $p(t_i)$ of player 1's type but not the type itself. Player 2's initially makes a belief about player 1's strategy, namely $p(A_j | t_i)$, then updates her belief after observing player 1's

choice of the strategy A_j, namely $u(t_i \mid A_j)$. Furthermore, based on her updated beliefs, Player 2 then makes her best choice B_k of action from B.

(4) Payoff values of both player 1 (Sender) and player 2 (Receiver) are given by the known functions $V1(t_i, A_j, B_k)$ and $V2(t_i, A_j, B_k)$

A schematic diagram for the signaling game is shown below:

Figure 7.6

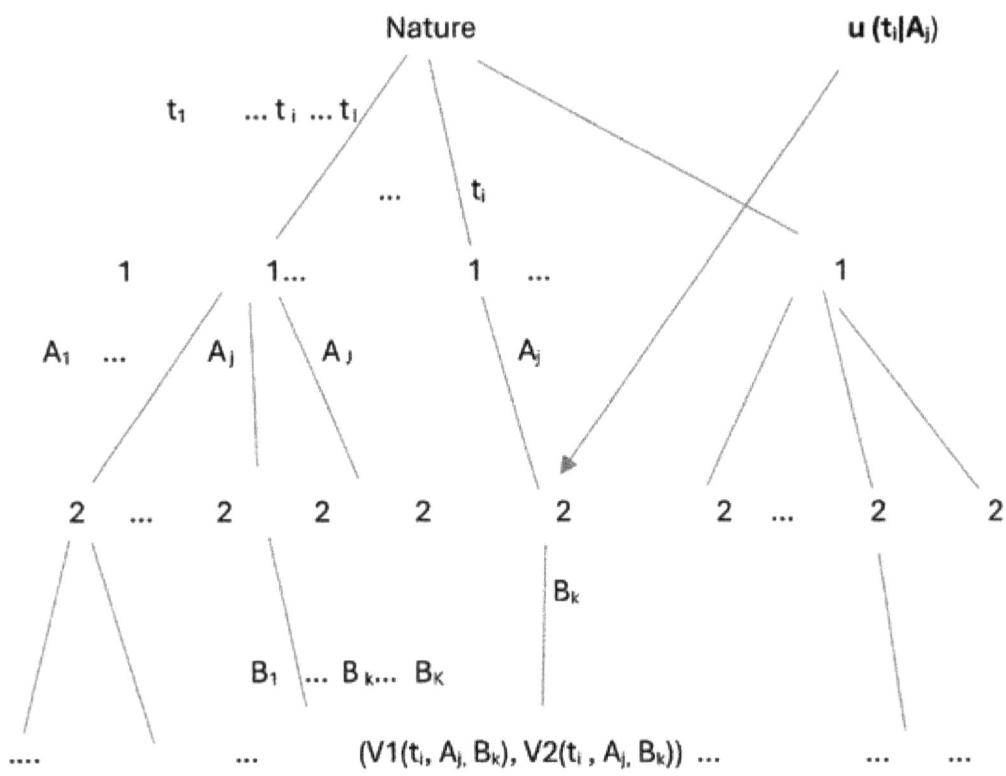

The four requirements for a perfect Bayesian equilibrium of a dynamic game defined in Section 7.2 before can be translated into the following three requirements for Signaling Games:

7.4 .1 Signaling Game Requirement 1

For each action (message) A_j in A, player 2 has the beliefs $u(t_1 \mid A_j)$, $u(t_2 \mid A_j)$.....$u(t_I \mid A_j)$ that it comes from the types $t_1, t_2 \dots t_I$ respectively, then

$u(t_1 \mid A_j) + u(t_2 \mid A_j) + \dots u(t_I \mid A_j) = 1$

7.4.2R Signaling Game Requirement 2 for player 2 (Receiver)

For each action (message) A_j in A that player 1 plays, player 2 will respond with the particular action B_k that maximizes the following expected value over all the B_k'.

Sum over t_i of the expression u(ti | A_j) V2(t_i, A_j, B_k).

Denote this particular action in B by B*(A_j)

7.4.2S Signaling Game Requirement 2 for player 1 (Sender)

For each type t_i in T, player 1 will choose the particular action (message) A_j in A that maximizes player 1's (Sender's) payoff value V1(t_i, $A_{j,}$ B*(A_j)). Denote this particular action in A by A*(t_i). In other words, A*(t_i) maximize the expression V1(t_i, A_j ,B*(A_j)) over all the message A_j in A.

7.4.3 Signaling Game Requirement 3

For each player 1's message A_j in A, let T_j be the subset of T that contains all the types t_i such that A*(t_i) = A_j. This is possible because of Requirement 2 above. Now we impose that player 2's belief at the information set corresponding to A_j namely, u(t_i | A_j) must satisfy Bayes' rule and Player 1's strategy, that is to say,

u(t_i | A_j) = p(t_i) / Summation of p(t_i) over all t_i in T_j

Note that Requirement 4 for Perfect Equilibrium for general dynamic games does not have a corresponding requirement for the Signaling games. This is because the general structure of the Signaling game shown in Figure 7.6 before and the typical scenario of Example 7.3.6 discussed earlier that player 1's strategy may not be able to determine player 2's belief given the game structure as shown in Figure 7.6.

Thus, we give the following definition for perfect Bayesian equilibrium for a signaling game as follows.

Definition 7.4.3 (perfect Bayesian Nash equilibrium in Signaling games)

For each t_i in T, a pure perfect Bayesian equilibrium strategy in a Signaling Game is a pair of strategies A*(t_i) and B*(A_j) and a belief u(t_i | A_j) satisfying the three Signal games requirement stated above.

Definition 7.4.4 (pooling, separating , hybrid strategies)

In the above definition, if all the types t_i are sending the same message A*(t_i), that is to say, A*(t_1) = A*(t_2) =.... A*(t_I), then both players are playing a pooling strategy. If different types are sending to different messages, that is to say, A*(t_i) <> A*(t_j) then it is called separating strategies. If some types are sending to the same message but other types are randomly sending to other messages, then it is called hybrid strategies.

7.4.4 Simple examples of Perfect Bayesian Equilibrium in Signaling games

In the following section, we will show you some simple examples of perfect Bayesian equilibrium in Signaling game when there are only two types t_1 and t_2, and that player 1 has only two actions A_1, A_2 and player 2 has only two actions B_1, B_2. Refer to Figure7.6, $T = \{t_1, t_2\}$, $A = \{A_1, A_2\}$, $B = \{B_1, B_2\}$. From the discussion above, there are only four possible candidates of perfect Bayesian equilibriums. Assume that the payoffs for the sender and receiver are the same in each of the four candidates.

(1) Type t_1 is sending to message A_1 and t_2 is sending to message A_1. (Pooling)
(2) Type t_1 is sending to message A_2 and t_2 is sending to message A_1. (Separating)
(3) Type t_1 is sending to message A_2 and t_2 is sending to message A_2. (Pooling)
(4) Type t_1 is sending to message A_1 and t_2 is sending to message A_2. (Separating)

In each of these candidates, we need to know player 2 responds to the player 1's action A_1 or A_2 by choosing the proper action B_1 or B_2 so that the resulting pair of the player 1 and player 2 strategies can satisfy the four requirements of a perfect Bayesian equilibrium given in Definition 7.2.2

The next four examples below will discuss the above candidates (1), (2), (3), and (4) respectively.

Example 7.4.7 (Pooling strategies)

This is the first case where $A(t_1) = A_1$ and $A(t_2) = A_1$ as shown by the red lines in Figure 7.7 below. This is an example of pooling strategies on player 1's message A_1.

Figure 7.7

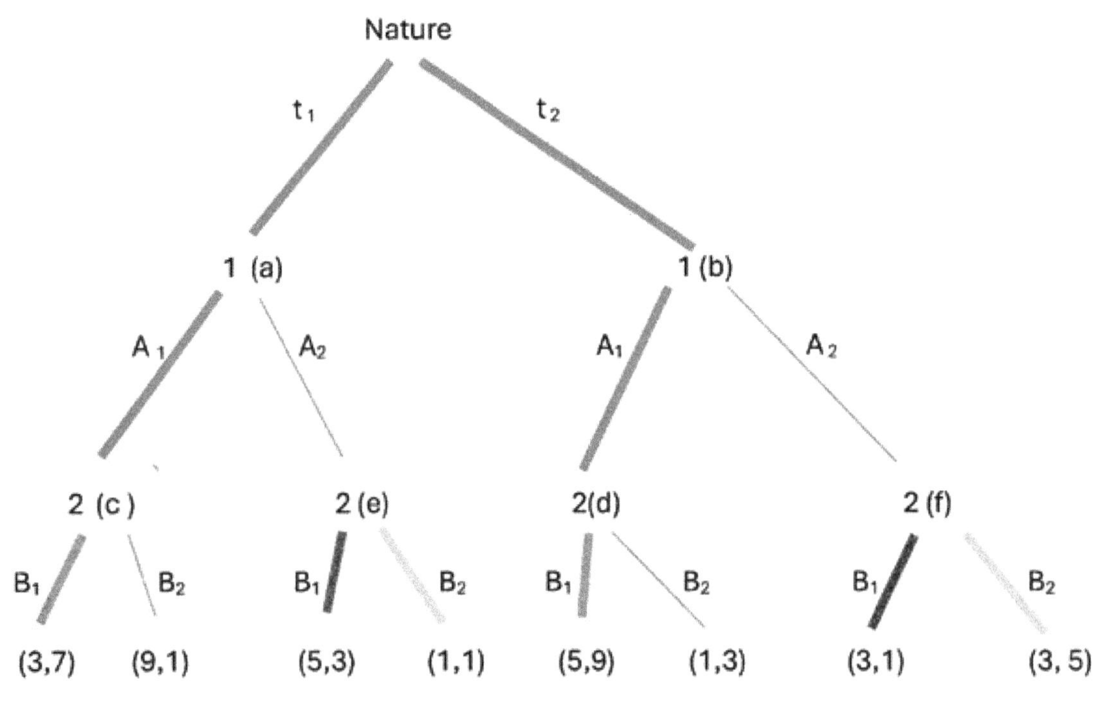

In Figure 7.7 above, we need to determine player 2's strategy at the information set c and d given that her belief is $u(t_1 \mid A_1)$ and $1 - u(t_1 \mid A_1)$ respectively. we thus compute the expected value of B_1 and B_2 as follows: To simplify the notation, let $u(t_1 \mid A_1) = p$, and let E2(B, p) to denote the expected value of player 2 for strategy B when her degree of belief is p, then

$E2(B_1, p) = p(7) + (1 - p)(9) = 9 - 2p$

$E2(B_2, p) = p(1) + (1 - p)(3) = 3 - 2p$

Thus, $E2(B_1, p) > E2(B_2, p)$ for all p, thus B_1 is selected as shown by the two green lines in Figure 7.7,

Similarly, let $u(t_1 \mid A_2)$ and $1 - u(t_1 \mid A_2)$ be the degree of belief of player 2 at information set e and f respectively. Let $q = u(t_1 \mid A_2)$, then

$E2(B_1, q) = q(3) + (1 - q)(1) = 2q + 1$

$E2(B_2, q) = q(1) + (1 - q)(5) = 5 - 4q$

Now we assume $E2(B_2, q) > E2(B_1, q)$, which is equivalent to $5 - 4q > 2q + 1$ or $2/3 > q$, which is equivalent to player 2 choosing the strategy B_2 as shown in the two yellow lines in Figure 7. Had we assumed $E2(B_1, q) < E2(B_2, q)$ or $2/3 < q$, then player 2 would have chosen strategy B_1 as indicated by the two purple lines in the Figure 7.7. We must now verify that the pair (A_1, B_1) with type t_1, and the pair (A_2, B_2) with type t_2 are both Nash Equilibrium. This means that both players already have their best payoff values, and no player wants to change strategy unilaterally in an attempt to get a better payoff value. Indeed, player 1 has the best payoff value 3 and 5 respectively by playing strategy (A_1, A_1), and note that player 2 could not have chosen B_1 as indicated by the two purple lines in the figure, because then player 1's payoff value would have been 5 and 3 respectively, and this would force player 1's strategy changes from A_1 to A_2, and this in turn violates the basis assumption that $A(t_1) = A_1$ and $A(t_2) = A_1$.

Now the strategy the pair (A_1, B_1) with type t_1, and the pair (A_2, B_2) with type t_2 together with the degree of belief at p and q at the information set c and e respectively is a perfect Bayesian Nash Equilibrium for this game. Why? This is because it satisfies all the three requirements of the Signaling Game stated above. Requirement 1 and Requirement 3 are satisfied because $u(a) = p(t_1)$, $u(b) = p(t_2)$, and for p,

$$u(c) = p = u(t_1 \mid A_1) = P(A_1 \mid t_1)\, p(t_1) / \{P(A_1 \mid t_1)\, p(t_1) + P(A_1 \mid t_2)\, p(t_2)\}$$

$$= 1p(t_1) / \{p(t_1) + p(t_2)\} = p(t_1)$$

$u(d) = 1 - u(c) = p(t_2)$. Well, what about q? q was chosen above to be $2/3 > q$ so that Requirement 2 is satisfied.

Example 7.4.8 (Separating strategies)

This is the second case $A(t_1) = A_2$ and $A(t_2) = A_1$ as shown by the red lines in Figure 8 below. Player 1 (sender) plays the separation strategy A_2 when the type is t_1 and play A_1 when the type is t_2. As before, $u(a) = p(t_1)$ and $u(b) = p(t_2)$.

$u(t_1 \mid A_2) = p(A_2 \mid t_1)\, p(t_1) / \{p(A_2 \mid t_1)\, p(t_1) + p(A_2 \mid t_2)\, p(t_2)\} = p(t_1) / \{p(t_1) + 0\} = 1$

Similarly $u(t_2 \mid A_1) = 1$. To simplify the notations, let $p = u\,(t_2 \mid A_1)$, let $q = u\,(t_1 \mid A_2)$, then $u(c) = p = 0$, $u(d) = 1\text{-}p = 1$, $u(e) = q = 1$, and $u(f) = 1\text{-}q = 0$. Now player 2 can decide which strategy to choose between B_1 and B_2 by computing the expected value at the information set c and d respectively:

$E2(B_1, p) = p(7) + (1\text{-}p)\,(9) = 1\text{-}2p$

$EV2(B_2, p) = p(1) + (1\text{-}p)\,(3) = 1\text{-}2p$

$EV2(B_1, q) = q(3) + (1\text{-}q)\,1 = 2q+1$

$EV2(B_2, q) = q\,(1) + (1\text{-}q)\,5 = 5\text{-}4q$

As $E2(B_1, q) > E2(B_2, q)$ when $q = 1$ and $E\,2(B_1, p) > E2(B_2, p)$ when $p = 0$, so player 2 would choose B_1 at information set c and d as shown by the green lines in Figure 7.8, and again choose B_1 at information set e and f as shown by the yellow lines in the Figure. As before, the strategies (A_2, B_1) for t_1 and (A_1, B_1) for t_2 are Nash equilibriums because no player wants to deviate because both achieve the best payoff values (5, 3) and (5, 9) respectively.

Figure 7.8

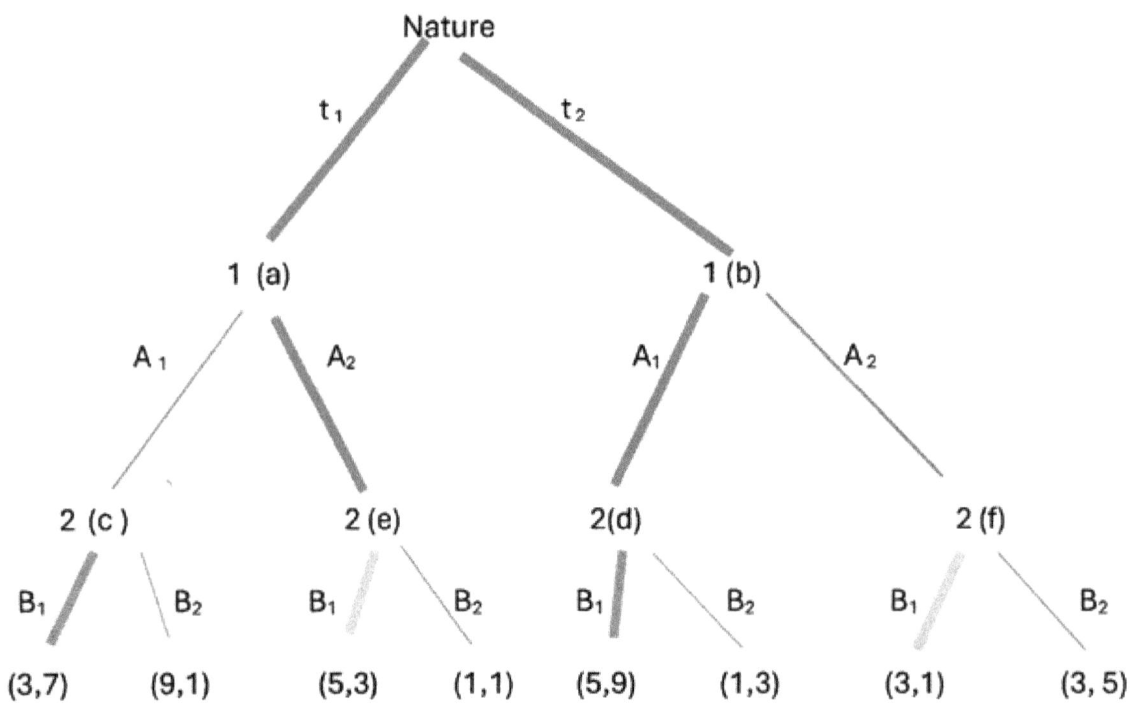

Example 7.4.9 (Perfect Bayesian Equilibrium cannot exist)

This is the case $A(t_1) = A_2$ and $A(t_2) = A_2$ as shown by the red lines in Figure 7.9 below. Player 1 (sender) plays the pooling strategy A_2 for both type t_1 and t_2. We want to show that there is no perfect Bayesian equilibrium in this case when $p(t_1) = p(t_2) = 1/2$. As before,

$u(a) = p(t_1)$ and $u(b) = p(t_2) = ½$, $u(c) = p(t_1 \mid A_1) = p$, $u(e) = q$, $u(f) = 1\text{-}u(e)$

$q = p(t_1 \mid A_2)$

$= p\,(A_2 \mid t_1)\,p(t_1) / \{p(A_2 \mid t_1)\,p(t_2) + p(A_2 \mid t_2)\,p(t_2)\}$

$= p(t_1) / \{p(A_2 \mid t_1)\,p(t_1) + p(A_2 \mid t_2)\,p(t_2)\}$

$= p(t_1) / \{p(t_1) + p(t_2)\}$

$= p(t_1) = 1/2$

Figure 7.9

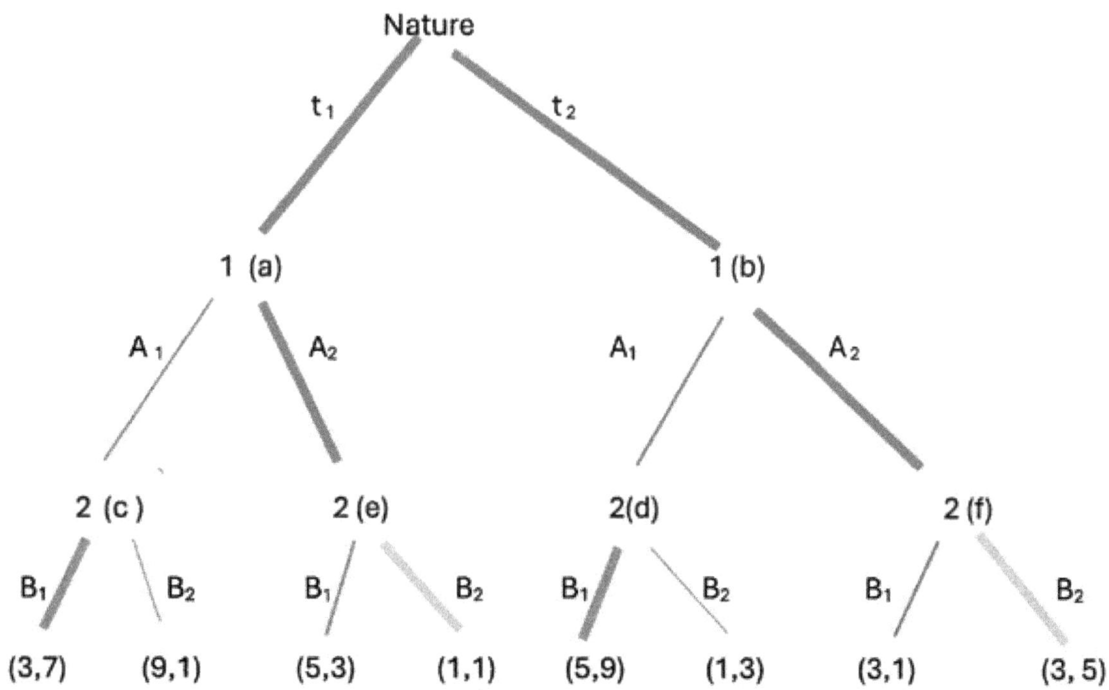

To determine player 2's strategy, we compute the expected value of B_1 and B_2 at the information set {c, d}, and {e, f} respectively:

$EV2(B_1, q) = q(3) + (1\text{-}q)\,(1) = 1 + 2q = 2$ when $q = ½$

$EV2(B_2, q) = q(1) + (1\text{-}q)\,(5) = 5 - 4q = 3$

$EV2(B_1, p) = p(7) + (1\text{-}p)(9) = 9 - 2p$

$EV2(B_2, p) = 1(p) + (1\text{-}p)(3) = 3 - 2p$

As $EV2(B_1, p) > EV\ 2(B_2, p)$ for all p, and $EV2(B_2, q) > EV2(B_1, q)$. So, the possible perfect Bayesian Nash Equilibrium could be (A_2, B_2) for t_1 and (A_2, B_2) for t_2 as shown in the green lines and yellow lines. But the pair (A_2, B_2) is not Nash Equilibrium because player 1 can deviate from A_2 to A_1 and improve his payoff value from 3 to 5 for t_2.

Example 7.4.10 (Perfect Bayesian Equilibrium cannot exist)

This is the case $A(t_1) = A_1$ and $A(t_2) = A_2$ as shown by the red lines in Figure 7.10 below. Player 1 is playing separation strategy for the two types. In this case, both player 2's information set {c, d} and {e, f} are on the equilibrium path. We want to show that there can be a perfect Bayesian Nash Equilibrium under this condition.

As before, $u(a) = p(t_1)$ and $u(b) = p(t_2)$.

Figure 7.10

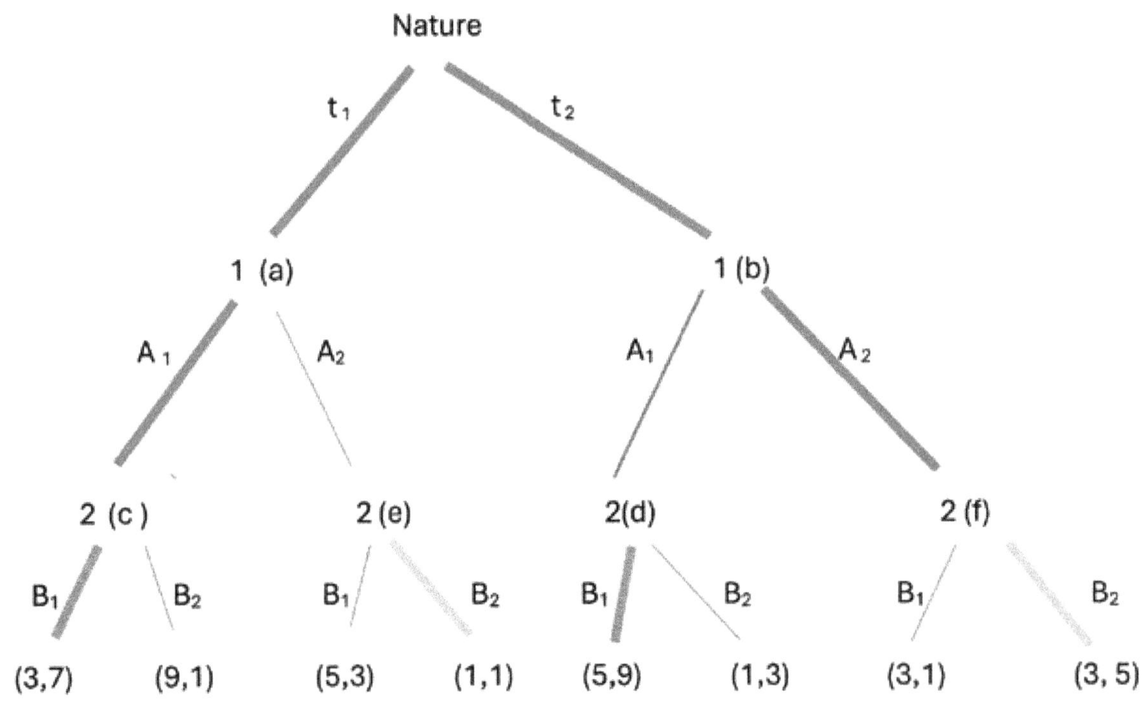

As in Example 7.4.8, $u(c) = 1$, $u(d) = 1 - u(c) = 0$, $u(f) = 1$, and $u(e) = 1\text{-}u(f) = 0$.

Simplify the notation, $u(c) = p = 1$, $u(e) = q = 0$. As before, computing the expected values at the information sets, we find $EV2(B_1, p) = p(7) + (1\text{-}p)(9) = 7$, $EV2(B_2, p) = p(1) + (1\text{-}p)(3) = 1$, so B_1 is chosen as shown by the green lines in the figure above.

Similarly, $EV2(B_1, q) = q (3) + (1-q) (1) = 1$, $EV2(B_2, q) = q (1) + (1-q) (5) = 5$, so B_2 is chosen as shown by the yellow lines. But now the pair (A_1, B_1) for t_1 and the pair for (A_2, B_2) for t_2, are not Nash Equilibrium because player 2 can deviate from A_2 to A_1 to improve her payoff value from 3 to 5.

7.4.5 Two applications of Perfect Bayesian equilibrium

Now we are ready to present two applications for the signaling games to real life scenarios. Example 7.4.11 below describes limit pricing and entry deterrence and Example 7.4.12 that follows describes how the employees signal their education level to the employer to get better payment.

Example 7.4.11 (limit pricing and entry deterrence)

The basic idea of limit pricing and entry deterrence can be described by the following simple scenario. Imagine in a market there is an established company, denoted by player 1, selling a homogeneous good and a potential entrant company, denoted by player 2. Initially, each company knows its own unit cost, but it does not know the other company's cost level. To simply the formulation in what follows, we assume there are only two periods in this game. In period 1, the established company is selling a certain number of its product q_1 at cost c_1, company 2 is just an observant. In period 2, company 2 observes what the established company did and decides whether to enter the market to produce q_2 that many units each at a cost c_2 or to stay out of the market. If company 2 (player 2) decided to stay out of the market, then company 1(player 1) will be the monopolist and if company 2 decided to enter then both companies will agree to follow the Cournot economic model. In general, if company 2 knows that its unit cost is higher than that of company 1 in advance, then company 2 will be deterred from entering; however, this is not guaranteed in advance as each company does not know about the other company's cost level as mentioned before and that is what complicates the matters.

7.4.5.1 Cournot Model Revisited

To formulate the above economic situation described, we summarize the basic facts and profit formulae for the Cournot model that will be needed.

1. We assume that the market demand model is given by,

   $$p = a-b*q \qquad (1)$$

   where p is the price of the homogenous product, q is the number of product produced, a, b are model parameters.

2. If c_1 is the unit cost of the product for Company 1, and q_1 is the number of units produce by Company 1, then the total profit Company 1 is:

   $$P_1 = (a-b*q_1 - c_1)\, q_1 \qquad (2)$$

   We denote :

   $$P_1(c_1, q_1) = (a -b*q_1 - c_1)\, q_1 \qquad (2A)$$

   for the profit for Company 1 when it produces q_1 and its cost type is c_1.

3. In the above, the value of q_1 that maximizes P_1 is called the monopoly output $m(c_1)$ and the corresponding maximum profit is called the monopoly profit denoted by $M(c_1)$.

   By simple calculus, we can find $m(c_1)$ and $M(c_1)$ as follows:

   Differentiate P_1 with respect to q_1, setting it to zero, and solving for q_1 gives,

   $$m(c_1) = (a\text{-}c_1)/2b, \qquad (3)$$

   and substitute $m(c_1)$ back to P_1 gives,

   $$M(c_1) = (a\text{-}c_1)^2/4b \qquad (4)$$

4. If Company 2 enters the market and produce q_2 that many units, then the Cournot Competition Model dictates that the common price of each unit in the market is

   $$p_c = a\text{-}bq_1 - bq_2, \qquad (5)$$

   so, the Cournot profit for Company 1 and 2 are respectively,

   $$P_1 = (a\text{-}bq_1 - bq_2 - c_1)q_1 \text{ and } P_2 = (a\text{-}bq_1 - bq_2 - c_2)q_2 \qquad (6), (7)$$

   As we need to choose q_1 and q_2 simultaneously maximize both P_1 and P_2 above, so simple calculus tells us to take partial derivatives of P_1 with respect to q_1 and of P_2 with respect to q_2, setting both derivates to zero, and solving q_1 and q_2, we find:

   $$q_1 = (a - bq_2 - c_1) / 2b \qquad (8)$$

   and

   $$q_2 = (a - bq_1 - c_2) / 2b. \qquad (9)$$

   The above two equations (8) and (9) can be thought of as the best response of Company 1 (the optimal number of units Company will produce given the number of units that Company 2 has produced) to Company 2, and the best response of Company 2 to Company 1 respectively. Solving (8), (9) simultaneously, gives

   $$q_1 + q_2 = (2a\text{-}c_1\text{-}c_2)/3b$$

   $$q_1 = (a + c_2 - 2c_1) / 3b \qquad (10)$$

   $$q_2 = (a + c_1 - 2c_2) / 3b \qquad (11)$$

   For later reference, we use the notations $c(q_1)$ and $c(q_2)$ for the above two expressions.

   $$c(q_1) = (a + c_2 - 2c_1) / 3b \qquad (10a)$$

   $$c(q_2) = (a + c_1 - 2c_2) / 3b \qquad (11a)$$

   The qualifiers 'c' signify that they are from the Cournot competition.

   Now substitute (10) and (11) back into (6) and (7), we find the maximum profits for each company called the Cournot profit C_1 and C_2 of Company 1 and Company 2 respectively,

   $$C_1(c_1, c_2) = (a + c_2\text{-}2c_1)^2/ 9b \qquad (12)$$

   $$C_2(c_1, c_2) = (a + c_1\text{-}2c_2)^2/ 9b \qquad (13)$$

5. The difference between the Monopoly profit of Company 1, $M(c_1)$, and the Cournot profit of Company 1, $C_1(c_1, c_2)$ denoted by $R_1(c_1, c_2)$ can be thought of as the reward to Company 1 for deterring Company 2 to enter the market, denote this reward to Company 1 by $R_1(c_1, c_2)$,

$$R_1(c_1, c_2) = M(c_1) - C_1(c_1, c_2) \tag{14}$$

After describing all these preliminary formulas above, we are in position to present two scenarios for Example 7.4.11, call it scenarios A and B below, each scenario represents a slightly different input condition. In both scenarios, we will show step by step, how the Separating and Polling equilibriums are computed.

Scenario A:

Suppose there are two airline manufacturing companies, and they are competitors, and we assume the market follows the Cournot competition model so that all the formulas derived in the previous section apply. Suppose the market demand model is given by,

$p = 150 - 2q$, where the price is in millions of dollars. Assume also Company 1 is the established company, a monopolist, and Company 2 is a potential entrant company. Initially, each company only knows its own unit cost, both companies have two types of cost levels (High and Low denoted by H and L) but it does not know the other company's cost level. Specifically, $c_{1L} = 60$ and $c_{1H} = 100$, and $c_{2L} = 75$ and $c_{2H} = 90$. Also assume the two probability functions one for each company on the two cost levels are known. Specifically,

let p1 = Probability that Company 1 will select the cost level c_{1H}, and

let p2 = Probability that Company 2 will select the cost level c_{2H}.

As Company 1 is a monopolist in the beginning, it must pick a quantity q_1 to produce, Company 2 then decide to enter the market or stay out given its observation knowing the value of c_2 but not the value of c_1. If Company 2 enters, it will incur an entry cost of C_e, say $C_e = 240$, and then the two companies operate to follow the Cournot duopoly model described before. If Company 2 does not enter, Company 1 will stay on as monopoly henceforth.

To simply formulation, we view the underlying scenario as having two periods. In period 1, Company 1 (player 1) must produce, and the strategy for Company 1 is to produce an optimal number q, and Company 2 (player 2) is just an observant. In period 2, Company 2 has two strategies, one to stay out, denoted by O, and the other strategy is to enter the market, denoted by I. If it stays out, Company 2 produces 0 units, and if Company 2 enters, then both companies will produce the Cournot quantities determined by the formulas described in the previous section.

The scenario above can be described by the following diagram:

Figure 7.11

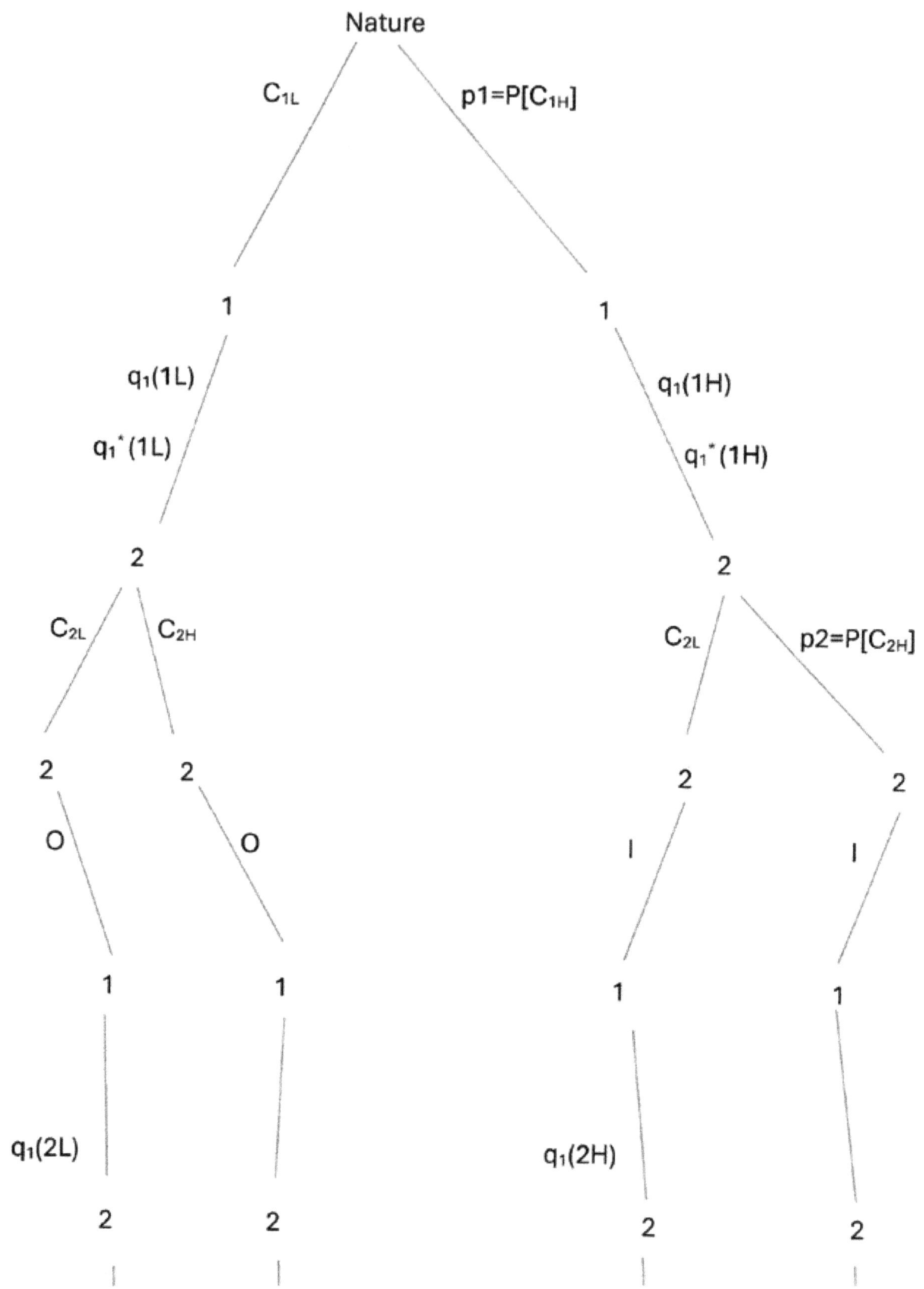
Nature
C1L
p1=P[C1H]
1
1
q1(1L)
q1*(1L)
q1(1H)
q1*(1H)
2
2
C2L
C2H
C2L
p2=P[C2H]
2
2
2
2
O
O
I
I
1
1
1
1
q1(2L)
q1(2H)
2
2
2
2

Before we proceed to find the Bayesian Nash Equilibrium for the above game tree, we need to define the following notations shown in the diagram above.

$q_1(1L)$ is the strategy of Company 1 in the first period when its cost level is low,

$q_1(1H)$ is the strategy of Company 1 in the first period when its cost level is high,

$q_1(2L)$ is the strategy of Company 1 in second period when its cost level is low,

$q_1(2H)$ is the strategy of Company 1 in the second period when its cost level is high.

Note that the strategy of Company 2 has only two choices, to enter or not to enter, depending on its cost level and the number of units that Company 1 has produced. $q_2(2L)$, $q_2(2H)$ have similar meaning for Company 2.

Now we apply the formulas (1) to (14) described in the last section to compute the monopoly profit for Company 1, Cournot profits for both companies, and rewards for Company 1, all these data will be needed to determine the Separating and Polling equilibrium for this competition problem.

From formula (1), a = 150, b = 2 and with $c_{1L} = 60$, $c_{1H} = 100$, $c_{2L} = 75$, $c_{2H} = 90$ given for the problem.

From (3), $m(c_{1L}) = (150 - 60)/4 = 22.5$, $m(c_{1H}) = (150- 100)/4 = 12.5$

From (4), $M(c_{1L}) = (150 - 60)^2/8 = 1012.5$, $M(c_{1H}) = (150 - 100)^2/8 = 312.5$

From (12) and (13) we compute the Cournot profit for Company 2:

$C_2(60,75) = (150 + 60 - 150)^2/ 18 = 200$

$C_2(60,90) = (150 + 60 - 180)^2/ 18 = 50$

$C_2(100,75) = (150 + 100- 150)^2/ 18 = 555.56$

$C_2(100,90) = (150 + 100- 180)^2/ 18 = 272.22$

As the entry cost for Company 2 is Ce = 240, the net payoffs for Company 2 in the above four situations are,

$C_2(60,75) - 240 = -40$ (negative!)

$C_2(60,90) - 240 = -190$ (negative!)

$C_2(100,75) - 240 = 315.56$

$C_2(100,90) - 240= 32.22$

Thus, we see that if Company 1's low-cost $C_{1L} = 60$ were known to Company 2, then Company 2 would not want to enter to the market as this would incur negative profit, but if Company 1's cost $C_{1H} = 100$ were known to Company 2, then Company 2 would enter regardless of its cost levels.

As we define earlier p1 is the probability that Company 1 would play it cost level C_{1H} and (1-p1) is the probability that Company 1 would play C_{1L}, therefore, Company 2 will enter or not depends on its expected value of entry defined as follows:

$EV(entry) = p1C_2(c_{1H}, c_2) + (1\text{-}p1)\, C_2(c_{1L}, c_2) - C_e$ is positive or negative.

If this quantity EV (entry) is positive, then Company 2 will enter otherwise will not. (C_2 above could be C_{2L} or C_{2H})

In our case,

$p1C_2(100, 75) + (1\text{-}p1)\, C_2(60, 75) - 240 > 0$ implies $p1 > 0.1125$ and

$p1C_2(100, 90) + (1\text{-}p1)\, C_2(60, 90) - 240 > 0$ implies $p1 > 0.8550$.

Thus, within the interval of p1 between (0.1125, 0.8550), Company 2 with the low-cost level will enter, but Company 2 with the high cost will not enter. If p1 > 0.8550, Company 2 with high and low cost will enter, and if p1< 0.1125, Company 2 will not enter regardless of its cost level. In what follows we also need to compute the rewards to Company 1 from deterring Company 2 to enter, but first we need to compute Company 1's Cournot profit by formula (13) in the last section:

$C_1(c_{1L}, c_{2L}) = C_1\ (60, 75) = (150 + 75 - 120)^2/18 = 612.5$

$C_1(c_{1L}, c_{2H}) = C_1\ (60, 90) = (150 + 90 - 120)^2/18 = 800$

$C_1(c_{1H}, c_{2L}) = C_1\ (100, 75) = (150+ 75 - 200)^2/18 = 34.72$

$C_1(c_{1H}, c_{2H}) = C_1\ (100, 90) = (150 + 90 - 200)^2/18 = 88.89$

Thus by (14) , we find the rewards of Company 1 from deterring Company 2 to enter the market in various cases:

$R_1(c_{1L}, c_{2L}) = M(c_{1L}) - C_1\ (c_{1L}, c_{2L}) = 1012.5 - 612.5 = 400$

$R_1(c_{1L}, c_{2H}) = M(c_{1L}) - C_1\ (c_{1L}, c_{2H}) = 1012.5 - 800 = 212.5$

$R_1(c_{1H}, c_{2L}) = M(c_{1H}) - C_1\ (c_{1H}, c_{2L}) = 312.5 - 34.72 = 277.78$

$R_1(c_{1H}, c_{2H}) = M(c_{1H}) - C_1\ (c_{1H}, c_{2H}) = 312.5 - 88.89 = 223.61$

With the analysis above, and by the definition of separating equilibrium, we claim that the following pair of strategies for Company 1 and Company 2 is a separating Bayesian Nash equilibrium:

Strategy for Company 1:

$q_1^*(1L) = 24.29$, is the strategy of Company 1 in the first period when its cost level is low. (Note we will explain why 24.29 below)

$q_1^*(1H) = m(c_{1H}) = 12.5$, is the strategy of Company 1 in the first period when its cost level is high,

Strategy for Company 2 :

The strategy for Company 2 is Company 2 to decide if it should enter or not to enter the market. This decision rule is described by a binary entry function $E_2(C_2, q_1)$, which depends on the cost level of Company 2 itself, and on the number of units q_1 has produced by Company 1. This binary function is defined as follows:

$$E_2(C_2, q) = \left\{ \begin{matrix} 1 & if\ q < 24.29 \\ 0 & if\ q >= 24.29 \end{matrix} \right\}$$

where 1 to mean Company 2 shall enter if $q < 24.29$ and 0 to mean Company 2 shall not enter to market if $q >= 24.29$, and where C_2, the cost-level of Company 2 could be C_{2L} or C_{2H}.

To show that the two strategies constitute a Separating Nash Equilibrium, we need to show

(A) Company 2's strategy is best response to Company 1's strategy.

(B) Company 1's strategy is best response to Company 2's strategy.

For the proof of (A), this follows from the expected value of entry for Company 2 defined earlier : EV(entry) = p1 $C_2(c_{1H}, c_2) + (1\text{-p1})\, C_2(c_{1L}, c_2) - C_e$ we computed earlier.

When p1 = 1, $C_2(c_{1H}, c_2) - 240$ is 315.56 or 32.22 pending on if C_2 is C_{2L} or C_{2H}. When p1 = 0, $C_2(c_{1L}, c_2) - 240$ is -40 or -190 pending on if C_2 is C_{2L} or C_{2H}.

Thus Company 2 will stay out regardless of its cost-level given the low-cost Company 1's strategy (q = 24.29) and Company 2 will enter given the high-cost Company 1's strategy (q = 12.5) as shown in the figure above.

For the proof of (B), we assume that Company 2 is taking the strategy given by the binary entry function $E_2(C_2, q)$ defined above. We want to show that the strategy taken by Company 1, $q_1^*(1L) = 24.29$, and $q_1^*(1H) = m(c_{1H}) = 12.5$ is optimal. This means that we need to show:

(C) The high-cost Company 1 would not want to deviate from its current value of $q_1^* = 12.5$ to $q_1^* = 24.29$

(D) The low-cost Company 1 would not want to deviate from its current value of $q_1^* = 24.29$ to other values

Now for the proof of (C), we need to demonstrate that if the high-cost Company 1 were to deviate to $q_1^* = 24.29$, then Company 1's payoff would still be less than the high-cost Company 1's payoff when it is producing its monopoly value $q_1^* = 12.5$.

Recall from previous computation, when the monopoly unit $m(c_{1H}) = 12.5$, the payoff was $M(c_{1H}) = 312.5$.

Now Company 1's expected payoff when producing $q_1^* = 24.29$ is given by:

Company 1's first period profit when producing $q_1^* = 24.29$ plus the expected rewards to Company 1 from deterring entry of Company 2,

which is,

$(150\text{-}2^*\ 24.29\text{-}100)^*24.29\ + p2\ R_1(c_{1H}, c_{2H}) + (1\text{-}p2)\ R_1(c_{1H}, c_{2L})$,

where the first term in the above expression is given by formula 2, which is,

34.49 + p2 (223.61) + (1-p2)(277.78), which is,

312.27 - p2(54.17) , which is less than 312.5 for all values of p2.

Now for the proof of (D), the low-cost Company 1 would not produce more than $q_1^* = 24.29$ because then Company 1's expected payoff given by the expression

$(150\text{-}2^*\ q_1^*\text{-}100)q_1^* + p2\ R_1(c_{1H}, c_{2H}) + (1\text{-}p2)\ R_1(c_{1H}, c_{2L})$ would be less. If Company 1 produced less than 24.29, it must produce the $m(c_{1L}) = 22.5$ as we computed before, and Company 2 will enter the market by its entry strategy $E_2(C_2, q_1)$. To show that $q_1^* = 24.29$ is optimal for low-cost Company 2, we must show that the expected payoff for low-cost Company 1 when producing $q_1^* = 24.29$ is at least the monopoly payoff $M(c_{1L}) = 1012.5$ when producing $m(c_{1L}) = 22.5$ as we computed before.

Now Company 1's expected payoff when producing $q_1^* = 24.29$ is the Company 1's first period profit plus the expected rewards to Company 1 from deterring entry of Company 2, which is,

$(150\text{-}2^*\ 24.29\text{-}60)^*24.29 + p2\ R_1(c_{1L}, c_{2H}) + (1\text{-}p2)\ R_1(c_{1L}, c_{2L})$,

1006.09 + p2(212.5) + (1 - p2) (400), which is,

1406.09 – 187.5 p2, which is greater than 1012.5 for all p2.

We have just verified that the two companies' strategy pair defined above is a Separating Bayesian Nash Equilibrium. Now we must explain in the strategy taken by Company 1, how did we come up with $q_1^*(1L) = 24.29$?

This is because $q_1^*(1L)$ was chosen such that if high-cost Company 1 were to produce it, the profit of Company 1 from choosing it in the first period plus the expected profit of Company 1 to deter Company 2 to enter, this total profit must less than the monopoly profit $M(c_{1H}) = 312.5$ of Company 1 when producing the unit $m(c_{1H}) = 12.5$. In other words, $q_1^*(1L) = q_1^*$ is chosen such that the following inequality is satisfied:

$(150\text{-}2\ q_1^*\ \text{-}100)\ q_1^* + p2(223.61) + (1\text{-}p2)\ (277.78) <= 312.5,$

and solving the above inequality gives $q_1^*(1L) = 24.29$ be a possible candidate.

Similarly, *we claim that the following strategies for Company 1 and Company 2 constitute a Pooling equilibrium.*

Strategy for Company 1:

$q_1^*(1L) = q_1^*(1H) = m(c_{1L}) = 22.5$, is the strategy of Company 1 in the first period when its cost level is both low and high.

Strategy for Company 2:

The strategy for Company 2 is Company 2 to decide if it should enter or not enter the market depending on its cost level. It is described by the entry function described below.

$$E_2(C_{2H}, q) = \begin{cases} 1 & \text{if } q < 22.5 \\ 0 & \text{if } q >= 22.5 \end{cases} \Big\} ,$$

for the low-cost level, $E_2(C_{2L}, q) = 1$ regardless the value of q.

Proof of the above claim that the two strategies is a Pooling Equilibrium is like that of Separating Equilibrium. From earlier discussion, we know that when p1 is in the interval (0.1125, 0.8550), Company 2 with the low-cost level will enter, but Company 2 with the high-cost level will not enter. This shows that strategy for Company 2 is best response to the strategy for Company 1 when p1 is within the interval (0.1125, 0.8550). Now to show the converse, given Company 2's strategy, we need to show $q_1^*(1L) = q_1^*(1H) = m(c_{1L}) = 22.5$ is optimal when (a) Company 1 is low-cost type (b) Company 2 is high-cost type.

For the proof of (a), when $C_1 = C_{1L}$, any increase of 22.5 the Company 1 chooses would not deter Company 2 from entering, and any decrease of 22.5 would induce Company 2 to enter and reduce Company 1's first period profit.

For the proof of (b), when $C_1 = C_{1H},$ Company 1 would have no incentive to produce more than 22.5. But Company 1 may produce less than 22.5 because this gives the monopoly first period return but would induce Company 2 to enter, and the total profit of this situation is given by:

$P_1(C_{1H}, 22.5) + p2\ R_1(C_{1H}, C_{2H})$ which is ,

(150 – 2* 22.5 - 100)* 22.5 + p2(223.61) which is,

112.5 + 223.61 p2, which will be greater than $M(C_{1H}) = 312.5$ if p2 > 0.894

Thus, when p1 lies in the interval (0.1125, 0.8550), and p2 > 0.894 , $q_1^*(1L) = q_1^*(1H) = m(c_{1L}) = 22.5$ as Company 1's strategy, and the entry function $E_2(C_2, q)$ defined before as Company 2's strategy, then they constitute a pooling equilibrium.

Scenario B:

In this second scenario, the basic background setting and data are just like Scenario A. The only difference is that for Company 2, there is only one cost-level $C_2 = C_{2H} = C_{2L} = 100$, and cost-level for Company 1 is as before, $C_{1L} = 70$, $C_{1H} = 100$. For numerical illustration, we assume that the probability function $p1(C_{1H} = 100) = 1/2$, and $p2\ (C_2 = 100) = 1$, the entry cost for Company 2 is now Ce = 25, the Cournot Model p = 150 - 2q and all other notations used in this example are similarly defined as in Scenario A. We also take this example to illustrate the concept of Incentive Compatibility Constraints for finding the Separating and Pooling Bayesian Nash Equilibrium in Signaling Games (Books [17], Papers [11]). Roughly speaking, these incentives impose that when in equilibrium, each type has an incentive to choose its own prescribed strategy and not choose the prescribed strategy for the other types.

Before we proceed to find the Bayesian Nash Equilibrium, as we did in Scenario A, we are going to compute the following monopoly and Cournot profits for Company 1 and Company 2 using the formulas discussed before in the previous section.

If Company 2 stays out, Company 1 is a monopoly, and we can compute

$$m(C_{1L}) = 20,\ M(C_{1L}) = 800,\ m(C_{1H}) = 12.5\ ,\ M(C_{1H}) = 312.5,$$

If Company 2 enters, the Cournot profits for Company 2 are computed as:

$C_2(C_{1H,}\ C_2) = 138.89$ with $c(q_{2H}) = 8.33$, $C_2(C_{1L},\ C_2) = 5.56$ with $c(q_{2L}) = 1.67$. So, if Company 1 is a high-cost type, Company 2 will enter with a net positive profit of 138.89 - 25.0 = 113.89, if Company 1 is a low-cost type, Company 2 will enter with a net profit of 5.56 - 25.0 = -19.44 (negative). Thus Company 2 will not enter if Company 1 were a low-cost type, and it will enter if Company 1 were a high-cost type.

Cournot profits for the high-cost Company 1 are: $C_1(C_{1H,}\ C_2) = 138.89$ with $c(\ q_{1H}) = 8.33$, and the Cournot profit for the low-cost Company 1 are: $C_1(C_{1L,}\ C_2) = 672.22$ with $c(\ q_{1L}) = 20$ (formulas 11a, 11b in the previous section).

Before we proceed to find the separating and pooling Bayesian Nash Equilibrium, we need to define clearly what the terminology on the degree of belief meant in the present problem. Refer to Figure 7.12 shown below:

The degree of belief in the present problem is defined as the probability that Company 2 thinks that Company 1 is of high-cost type after Company 1 has produced q_1. Remember that Company 2 only knows about its own cost-level but has no idea about the cost types of Company 1, so it can only guess Company 1 's cost type with certain probability based on what Company 1 has done. Company 2's guess is given by the conditional probability shown below:

$$\mu(q_1) = P\,(C_1 = C_{1H} \mid q_1)$$

Refer to Figure 7.12 shown below and the definition of the notations defined in Scenario A, if $q_1(1L)$ and q1(1H) were part of the perfect Separate Equilibrium, then from the Cournot profits of Company 2 computed before, Company 2 will stay out if Company 1 choses $q_1(1L)$ and Company 2 will enter if Company 1 choses q1(1H) as shown in the figure. Therefore, $\mu(q_1(1L)) = 0$ and, $\mu(q_1(1H)) = 1$ for the beliefs of Company 2 on the equilibrium path. But what about the beliefs of Company 2 for all other quantities q_1 different from q1(1L) and q1(1H), that is those quantities that are off the equilibrium path ? This is the tricky part, and it is also the simplest way to keep Company 1 on the equilibrium path. This is to define that Company 2 will surely believe that Company 1 is of high-cost type if Company 1 chooses any quantity q_1 different from $q_1(1L)$. Precisely,

$$u(q_1) = \left\{ \begin{array}{ll} 0 & \textit{if } q_1 = q_1(1L) \\ 1 & \textit{if } q_1 <> q_1(1L) \end{array} \right\}$$

Figure 7.12

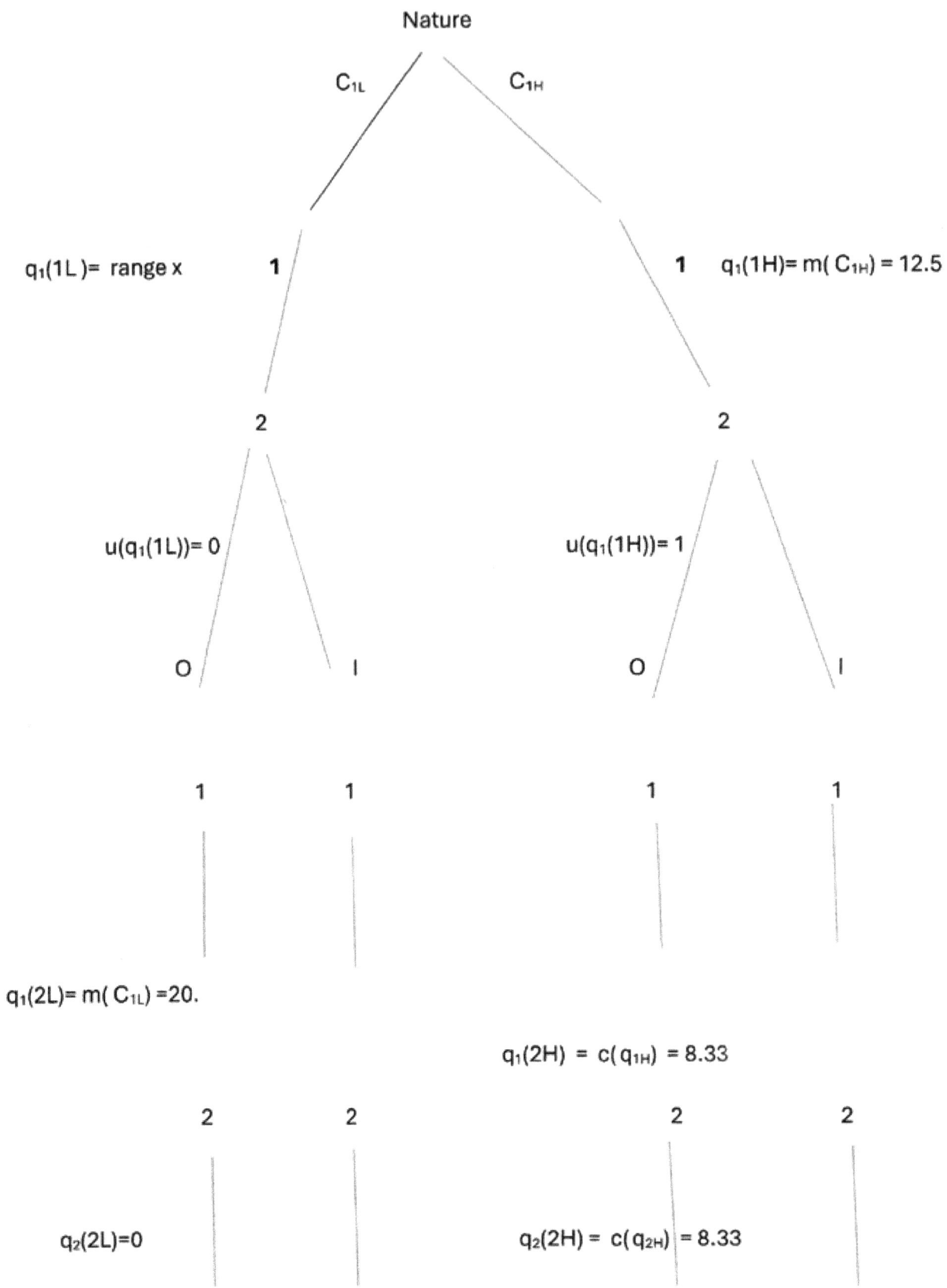

Nature
C_{1L}
C_{1H}
$q_1(1L)$= range x
1
1
$q_1(1H)$= m(C_{1H}) = 12.5
2
2
$u(q_1(1L))$= 0
$u(q_1(1H))$= 1
O
I
O
I
1
1
1
1
$q_1(2L)$= m(C_{1L}) =20.
$q_1(2H)$ = c(q_{1H}) = 8.33
2
2
2
2
$q_2(2L)$=0
$q_2(2H)$ = c(q_{2H}) = 8.33

From the above analysis and the Figure 7.12, it seems that $q_1(1H) = m(C_{1H}) = 12.5$ is a likely candidate for a separating Bayesian equilibrium for the high-cost Company 1. As we now explain, it turns out that it is the only candidate for Company 1 of the high-cost type. From Company 2's degree of belief defined earlier, even Company 1 deviates from $q_1(1H)$, Company 2 will still enter and plays the Cournot model and that gives both companies the Cournot profit of 138.89. Company 1 produces q_1 , which is the number that gives the maximum monopoly profit $M(C_{1H}) = 312.5$, so Company 1 will not deviate from $q_1(1H)$ to any other q_1 values in order to maximize the total monopoly and the Cournot profit. Thus, $q_1(1H)$ is the best that the high-cost Company 1 would prefer over any other q_1 that induces Company 2 entry. This led us to ask the similar question of what should $q_1(1L)$ be? Intuitively, like the choice of $q_1(1H)$, $q_1(1L)$ could be chosen to be $m(C_{1L}) = 20$ and gives the profit $M(C_{1L}) = 800$ in the first period, but it turns out this cannot be the case because of the two Incentive Compatibility Constraints that we are going to explain. In what follows, we name these two constraints ICH and ICL, one for the Company 1 high-cost type and the other for the Company 1 low-cost type respectively.

[ICL]: When Company 1 is a low-cost type, it prefers to choose $q_1(1L)$ over $q_1(1H)$ and other quantity.

[ICH]: When Company 1 is a high-cost type, it prefers to choose $q_1(1H)$ over $q_1(1L)$

For the explanation of [ICL]:

As explained, Company 1 chooses $q_1(1L) = m(C_{1L}) = 20$ gives the profit of $M(C_{1L}) = 800$ in the first period, and if this is part of a separating perfect Bayesian equilibrium, it will induce Company 2 to stay out, and thus Company 2 will again get the highest monopoly profit of $M(C_{1L}) = 800$ in the second period. Now if Company 2 chooses any other quantity or $q_1(1H)$. This is bad for Company 1 because in period 1, it will get lower than the monopoly profit and it will induce Company 2 to enter, and thus Company 1 will also get lower than its monopoly profit in the second period.

Now for the explanation of [ICH]:

For the Company 1 high-cost type, choose of $q_1(1H) = m(C_{1H}) = 12.5$ gives rise to the profit of $M(C_{1H}) = 312.5$ in the first period, and from the definition of Company 2's degree of belief, Company 1 will induce Company 2 to enter the market, and Company 1 will get the Cournot profit $C_1(C_{1H,} C_2) = 138.89$. Therefore Company 1 of high-cost type will get a maximum profit of $M(C_{1H})$ plus $C_1 (C_{1H,} C_2)$, which is, 451.39. For later reference, we call this maximum profit of Company 1's high-cost type MHC1, that is MHC1 = 451.39. And similarly, we will define and compute the minimum profit of Company 1's low-cost type MLC1 later.

Again, from the definition on the degree of belief of Company 2, Company 1 with the high-cost type would not deviate to any other quantity that induces Company 2 's entry because that would lower Company 1's first period profit without deterring Company 2's entry. So, it remains to show that the high-cost Company 1 type, $q_1(1H)$ would not want to deviate from $m(C_{1H}) = 12.5$ to $m(C_{1L}) = 20$.

Suppose on the contrary, the high-cost Company 1 type deviates from $q_1(1H) = 12.5$ to $q_1(1L) = 20$, then by the definition of the degree of belief of Company 2, Company 1 will deter entry of Company 2 and so Company 1 can obtain the monopoly profit in the second period by producing $q_1(1H)$. Therefore, the total profit for Company 1 of high-cost type would be Company 1's profit in the first period, which is $P_1(C_{1H}, q_1(1L))$ plus Company 1's profit in the second period, which is $P_1(C_{1H}, q_1(1H))$. By formula (2A) before,

$$P_1(C_{1H}, q_1(1L)) = (150 - 2*20 - 100)\,(20) = 200$$

$$P_1(C_{1H}, q_1(1H)) = (150 - 2*12.5 - 100)\,(12.5) = 312.5$$

Thus, the total profit is 512.5 (200 + 312.5). But then this total profit is larger than MHC1 = 451.39, which is a violation to ICH.

Therefore, for the ICH to hold, we must choose $q_1(1L)$ to be other values say x, other than the original value of $m(C_{1L}) = 20$ and to enforce the following condition to satisfied:

$$P_1(C_{1H}, q_1(1L)) + P_1(C_{1H}, q_1(1H)) <= 451.39$$

$$\text{Or } (150 - 2x - 100)x + (150 - 2*12.5 - 100)\ 12.5 <= 451.39$$

which is,

$$(50 - 2x)x <= 138.89$$

Or $x \geq 21.82$. This gives a lower bound for $q_1(1L)$. To obtain the upper bound for $q_1(1L)$, we need the ICL to be explained below.

Now for the explanation of [ICL]:

As mentioned in the explanation for ICH above, we need to find the optimal profit of Company 1's low-cost type LMC1. Referring to the Figure 7.12 shown above, both Company 1 high-cost type and Company 2 would produce the same output $q_1(2H) = q_2(2H) = 8.33$ with the same Cournot profit of 138.89. So, for Company 1 low-cost type, the best deviation for $q_1(1L) = 20$ would be it deviates to the best response of $q_2 = 8.33$, this best response for Company 1 is computed by formula (8) shown below:

$$q_1 = (150 - 2*\ 8.33 - 70)/\ 4 = 15.84,$$

and thus, the profit for Company 1 in second period corresponds to this output level 15.84 is by formula 6,

$$\begin{aligned} P1 &= (a - bq_1 - bq_2 - c_{1L})\, q_1 \\ &= (150 - 2*\ 15.84 - 2*\ 8.33 - 70) *\ 15.84 \\ &= 501.49 \end{aligned}$$

Because the monopoly profit made by the low-cost type of Company 1 is $M(C_{1L}) = 800$ in the first period. This means that the best deviation for Company 1, when it is the low-cost type, would yield a total profit of 800 + 501.49 = 1301.49. As mentioned above, we call this total profit, for the low-cost Company 1 type MLC1, which equals 1301.49. With MLC1 defined, we can compute an upper bound for $q_1(1L)$ below. Let x be the upper bound of $q_1(1L)$., then the following inequality must hold:

$P_1(C_{1L}, x)) + M(C_{1L}) \geq MLC1$, which is,

$(150 - 2x - 70)x + 800 \geq 1301.49$, or,

$x \leq 32.21$

Combining the previous result, we must have $21.82 \leq q_1(1L) \leq 32.21$.

In summary, we can state the Separating Perfect Bayesian Nash Equilibrium for this problem. Refer to Figure [ ]:

In period one, $q_1(1H) = 12.5$, and $21.82 \leq q_1(1L) \leq 32.21$.

In period two, $q_1(2H) = 8.33$, $q_1(2L) = 20$

The degree of belief for Company 2 that Company 1 has type C_{1H} is given by the binary function below :

$$u(q_1) = \begin{cases} 0 & if\ q_1 = q_1(1L) \\ 1 & if\ q_1 <> q_1(1L) \end{cases} \Bigg\}$$

Where 1 means Company 2 to enter, and 0 means Company 2 to stay out.

Company 2's output $q_2 = q_2(2L) = q_2(2H)$, depends on Company 1's output $q_{1,}$ is given by the function below:

$$q_2 = \begin{cases} 0 & if\ q_1 = q_1(1L) \\ 8.33 & if\ q_1 <> q_1(1L) \end{cases} \Bigg\}$$

At last, knowing the range x for $q_1(1L)$ in Separating Equilibrium as shown above g, we are going to find the Pooling Equilibrium in the following Figure 7.13 shown below:

Figure 7.13

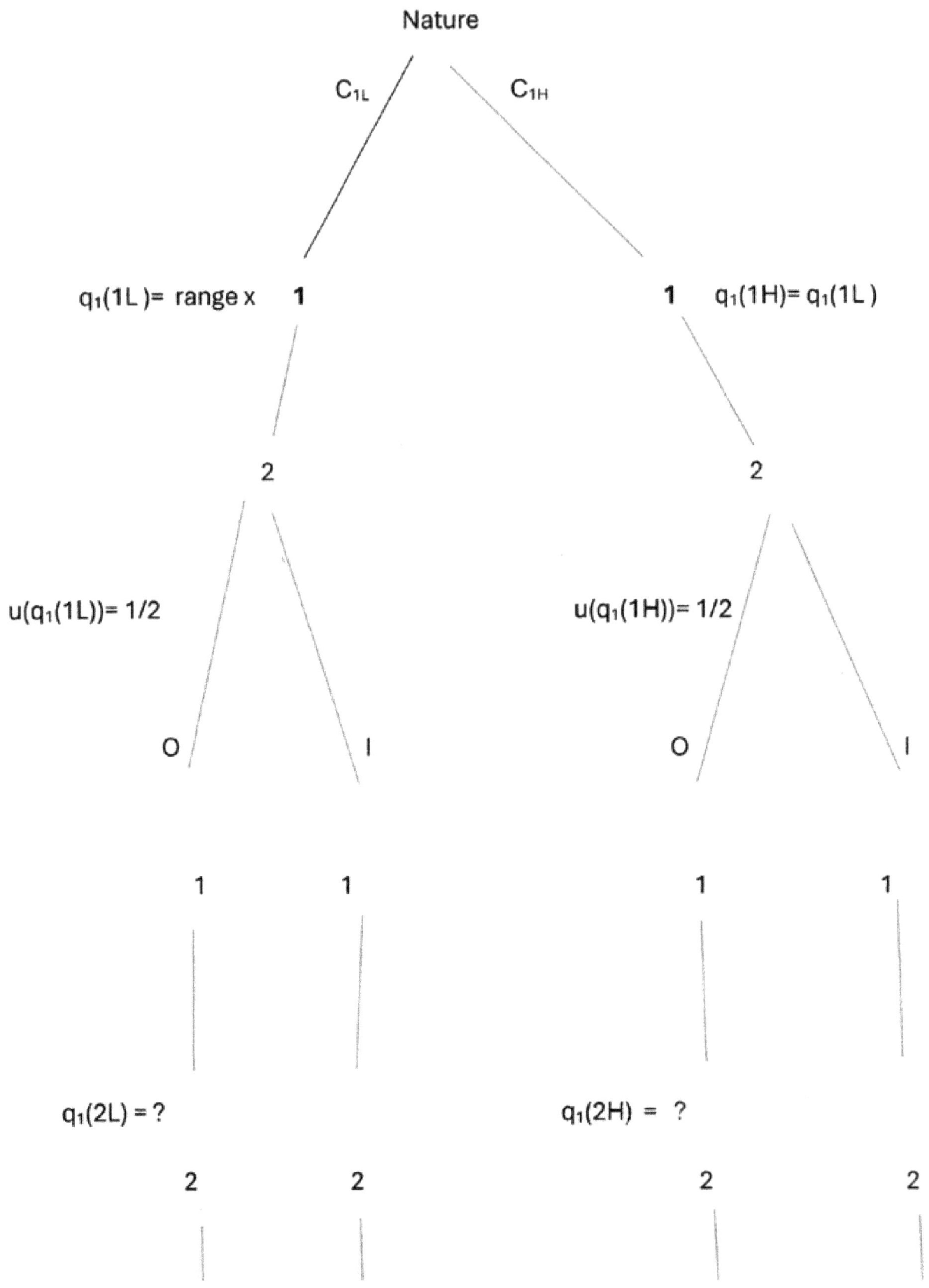
Nature
C1L
C1H
q1(1L)= range x
1
1
q1(1H)= q1(1L)
2
2
u(q1(1L))= 1/2
u(q1(1H))= 1/2
O
I
O
I
1
1
1
1
q1(2L) = ?
q1(2H) = ?
2
2
2
2

As shown in Figure 7.13 above, a Pooling Equilibrium would consist of the following information $\{q_1(1L) = q_1(1H) = q_1(1^*), q_1(2L), q_1(2H), q_2(2^*), u(q_1)\}$, we are going to find $q_1(2L)$, $q_1(2H)$, $q_2(2^*)$ first, then the range for $q_1(1^*)$ using the values MLC1 and MHC1 as we did in the Separating Equilibrium before.

Because Company 1 is playing $q_1(2L)$, $q_1(2H)$ with Company 2 playing $q_2(2^*)$ following the Cournot Model in the second period, so Company 1 would want to maximize its Cournot profit follow Formula 6, 7:

$$(150 - 2\,q_1(2L) - 2q_2(2^*) - C_{1L})\,q_1(2L) \qquad (1)$$

$$(150 - 2\,q_1(2H) - 2q_2(2^*) - C_{1H})\,q_1(2H) \qquad (2)$$

Note that $q_2(2^*)$ depends on $q_1(1^*)$ and Company 2 would want to maximize its expected Cournot profit with the high-cost Company 1 and low-cost Company 1, this is given by:

$$u\,(q_{1L})\,\{150 - 2q_1(2L) - 2q_2(2^*) - C_2\}q_2(2^*) +$$

$$u\,(q_{1H})\,\{150 - 2q_1(2H\,) - 2q_2(2^*) - C_2)\}q_2(2^*) - C_e, \qquad (3)$$

where $C_e = 25$, $C_{1L} = 70$, $C_{1H} = 100$, $C_2 = 100$, and $u\,(q_{1L})$, and $u\,(q_{1H})$ are the conditional probability that Company 2 will believe that Company 1 is of high-cost type given that Company 1 has chosen output q_{1L} and q_{1H} respectively, which are same as the probability that Nature chooses Company 1's type to be C_{1L} and C_{1H} respectively when we are setting $q_1(1L) = q_1(1H) = q_1(1^*)$ in finding pooling equilibrium, and both probabilities equal to 1/2 in our case.

The expressions (1), (2), (3) above involves three unknowns $q_1(2L)$, $q_1(2H)$, and $q_2(2^*)$, each represents the Cournot profit of Company 1's low-cost type, Company 1's high-cost type, and Company 2's expected Cournot profit. We need to determine these three quantities simultaneously. Again, from simple Calculus application, a standard approach is to take partial derivatives of each expression with respect to the specific variable, setting the derivatives to zero, and solve for these variables from the resulting algebraic equations.

After taking partial derivative in (1) with respect to $q_1(2L)$, setting it to zero, we find

$$q_1(2L) = (40 - q_2(2^*))/2 \qquad (4)$$

Similarly, taking partial derivative in (2) with respect to $q_1(2H\,)$, setting it to zero, we find

$$q_1(2H) = (25 - q_2(2^*))/2 \qquad (5)$$

And finally, taking the derivative in (3) with respect to $q_2(2^*)$, setting it to zero, we find,

$$q_2(2^*) = (50 - q_1(2L) - q_1(2H))/4 \qquad (6)$$

Equations (4), (5), (6) above are three equations in three unknowns, solving these simultaneous equations we find,

$q_2(2^*) = 35/6$, $q_1(2L) = 205/12$, $q_1(2H) = 115/12$ (7)

Putting these values back to (1), (2), (3), we find the profits for Company 1's low-cost type, Company 1's high-cost type, and the expected return of Company 2 are respectively $\frac{1}{2}(205/6)^2$, $\frac{1}{2}(115/6)^2$, $(35/3)^2 - 25$.

Note that $q_2(2^*) = 35/6$ computed above depends on q_1 that is on the equilibrium path, that is to say, $q_1 = q_1(1^*)$. The best response of q_2 when q_1 is off the equilibrium path when $q_1 \neq q_1(1^*)$ is still 8.33 as was computed in the case of Separating Equilibrium before. Our question is then, what should $q_1(1^*)$ be ?

As was done in the Separating Equilibrium, $q_1(1^*)$ must be chosen so that it is better than Company 1's optimal deviation MHC1, MLC1 for the low- cost and high-cost type respectively.

In our case, MHC1 = 451.39, MLC1 = 1301.49 were computed from before. Thus, for Company 1's high-cost type,

$$(150 - 2^* q_1(1^*) - 100)\, q_1(1^*) + \tfrac{1}{2}(115/6)^2 \leq 451.3 \quad (8)$$

And for Company 1's low-cost type,

$$(150 - 2^* q_1(1^*) - 70)\, q_1(1^*) + \tfrac{1}{2}(205/6)^2 \geq 1301.49 \quad (9)$$

Equation (8) implies that the range of $q_1(1^*)$ lies in [7.76, 17.24], and Equation (9) gives $q_1(1^*)$ lies in [13.59, 26.41]. Because $q_1(1^*)$ must satisfy both Incentive Constraints ICH and ICL, so $q_1(1^*)$ must be in the common overlapping interval [13.59, 17.24]. Now we can summary the following formation that constitutes a Pooling Equilibrium for Company 1 and Company 2.

Refer to Figure 7.13 above, Company 1' s strategy is:

$q_1(1^*)$ lies in [13.59, 17.24], $q_1(2L) = 205/12$, $q_1(2H) = 115/12$,

Company 2 responds as:

$$u(q_1) = \begin{cases} 1/2 & \text{if } q_1 = q_1(1^*) \\ 1 & \text{if } q_1 <> q_1(1^*) \end{cases}$$

and

$$q_2(2^*) = \begin{cases} 35/6 & \text{if } q_1 = q_1(1^*) \\ 8.33 & \text{if } q_1 <> q_1(1^*) \end{cases}$$

Example 7.4.12 (Discrete and continuous version of the Spence Model)

The next Example 7.4.12 on perfect Bayesian equilibrium is from the Nobel Prize -winning contribution of Spence's Model that was developed in his Ph.D. thesis 1973 (Papers [19]). The original model was presented in the form of continuous input and output data described by continuous functions. Recall that in the last Example 7.4.11, inputs to Company 1 and Company 2, $q_1(1^*)$ and $q_2(2^*)$ are continuous ranges, in this example, a slight complexity is added on, the inputs and outputs are also continuous functions.

To describe the underlying concept of Spence's Model, we shall first present it using discrete data by a simple example much like those we had presented in examples on this chapter before, followed by the presentation of the original model with continuous data.

Example 7.4.12A

Spencer's model is simplified in this example. There are two players, the employee and the employer in this extended form game shown below in Figure 7.14.

Terms and Notations:

Player 1, the employee, is a worker in a company of two possible types, a skillful type with high productivity at work, we call it type H, and a not so skillful type at work with low productivity, we call it type L. Player 1 could have two educational degree levels (denoted by B if he has a basic degree and by A if he has an advanced degree). B is the level that he already has, and A is an advanced degree, or a higher education level which he can achieve at the expense of getting extra training with a fixed cost of say, 4 for the H type and 6 for the L type in this example. Denote these two cost levels by C_L and by C_H respectively. Thus ,we assume $C_L = 5$, and $C_H = 2$. We emphasis here that $C_L > C_H$ because for the skillful type H, it is easier to get trained and hence at a lower cost. Denote the two strategies that player 1 can take by {B, A}. Obviously, which of the two strategies player 1 will take will depend on the net payoff he will be getting to be determined below.

Player 2 is an employer in a company, his strategy is to assign player 1 one of the two possible positions, the position titled Worker (denoted by W) and the position entitled Manager (denoted by M). Obviously, Player 2's strategy of assigning which position to player 1 will depend on the payoff he is getting given by the input data below and on which position he assigned to Player 1. Assume player 1 gets a wage of 21 if he gets the Manager position and gets a wage of 17 if he gets the Worker position, we denote these wages by $W_m = 21$ and $W_w = 17$. Player 2's payoff, on the other hand, depends on how skillful he assigns the different skill levels H and L of the workers, to the two different positions W and M available to the workers, these payoffs are shown in table below:

Position assignment to Player 1

	M	W
H	21	22
L	11	12

The game tree and the payoff in each of the eight possible scenarios is shown below. We assume that the nature player is picking up randomly the skillful worker with probability p = 1/3, and the non-skillful worker with probability 2/3.

Figure 7.14

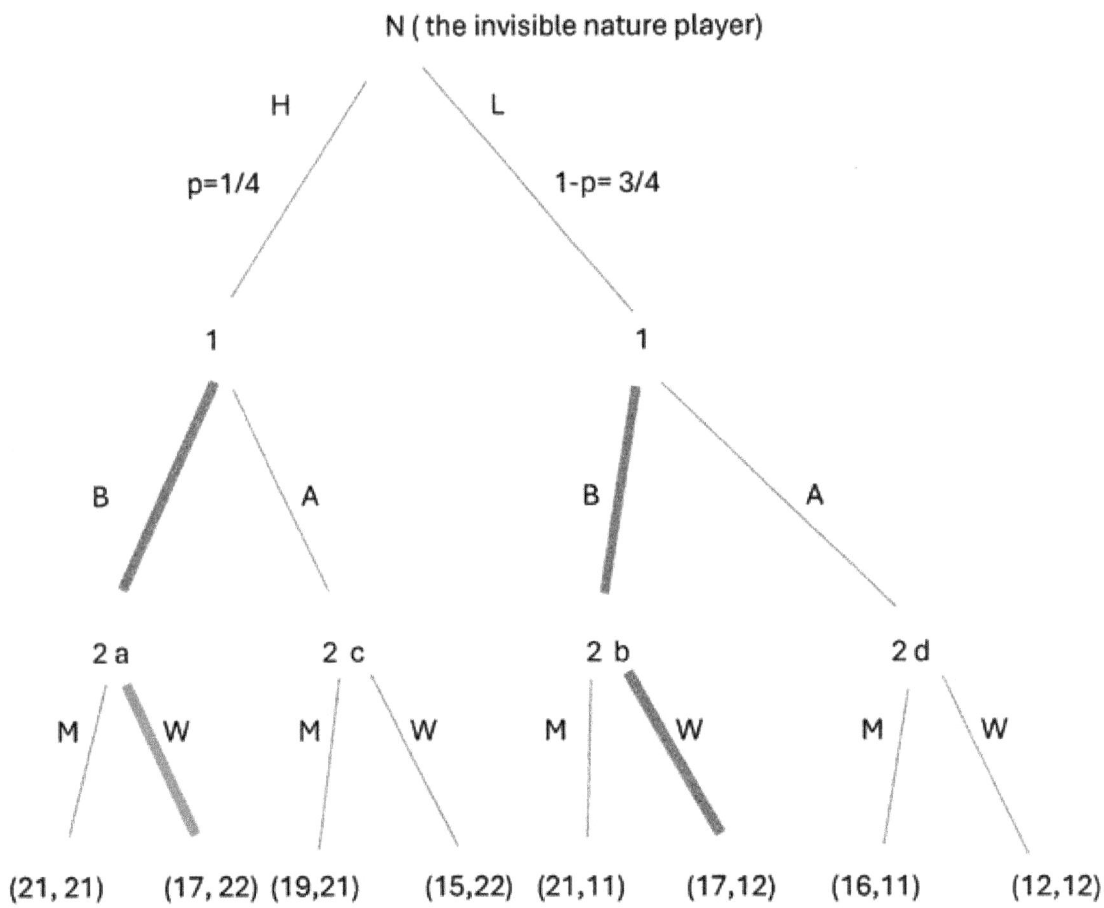

By the convention we used before, the first component of each of the payoff pairs is the wage of the employee and, the second component is the wage of the employer. We shall explain how did we come up with the payoffs (21, 21), (15, 22), (12, 12) in the above figure, and the rest are done similarly.

For the pair (15, 21), this is under the assumption Nature is picking up the skillful employee of type H with a basic degree and the employer is assignment the employee with the Manager position, thus from the assignment table, the employer would get a payoff of 21, since the employee is getting a Manager position, he would get a payoff of 21 by assumption $W_m = 21$, and since he already has a basic degree B, he does not need to pay for this education cost, so his net payoff is still 17, thus we have the pair (21, 21).

For the pair (15, 22), this is under the assumption that the employee is of type H having an advance degree A and the employer is assigning the employee the position Worker, thus from the assignment table, the employer would get a payoff of 22, since the employee is getting a Worker position, his payoff is 17 by assumption, but since is having an advance degree A, so he must pay for his education cost of 2 as he is high ability that $C_H = 2$, thus resulting a net payoff of 17 - 2 = 15, thus we have the pair (15, 22)

For the pair (12, 12), this is under the assumption that nature is picking up an employee with low ability with an advance degree level A and the employer is assigning the employee the position of Worker, thus from the assignment table, the employee is getting the payoff 12, since the employee is getting the Worker position, his payoff is 17 by assumption, and he has to payback is education cost for an advance degree of 5, so his net payoff is 17 - 5 = 12, thus we have the pair (12, 12).

To find the Perfect Bayesian Nash Equilibrium, we first find out all the possible Nash Equilibriums. The four possible strategy pairs for player 1(Employee) are {BB, BA, AB, AA} and that of the player 2(Employer)are {MM, MW, WM, WW} and the two players expected values are shown in the table below:

	Player 2 (Employer)		Player 1 (Employee)	
	MM	MW	WM	WW
BB	(21, 13.5)	(18, 14.25)	(20, 13.75)	(17, 14.5)
BA	(16.75, 13.5)	(13.75, 14.25)	(15.75, 13.75)	(12.75, 14.5)
AB	(17.25, 13.5)	(14.25, 14.25)	(16.25, 13.75)	(13.25, 14.5)
AA	(20.5, 13.5)	(17.5, 14.25)	(19.5, 13.75)	(16.5, 14.5)

To demonstrate the computation of a couple of the expected value of the strategy pair (BB, MW) and that of the pair (AB, MW)

EV (BB, WW) = ¼(17, 22) + ¾(17, 12) = (17, 14.5)

EV(AB, MW) = ¼(19, 21) + ¾(17, 12) = (17.5, 14.25)

The rest of the expected values for each remaining strategy pair in the table are similarly computed. Thus, by definition, we see that the pair (BB,WW) is a possible pooling Bayesian Nash Equilibrium, and to see if it is also a perfect Bayesian Nash Equilibrium, all we need to confirm we can assign proper degree of belief on each of the information set {a, b} and {c, d} and the strategy for Player 2, the Employer in this case, is rational.

Let u(a) and u(b) be the degree of belief at the information set {a, b}

By Bayes' Theorem,

u(a) = P(H | B) by definition,

= P(B | H) P(H) / {P(B | H) P(H) + P(B | L) P(L)]

= 1*p /{1*p + 1*(1-p) }

= p = ¼

This means the degree of belief that Player 2 (the employer) believes that Player 1 (the employee) 1 is of type H after observing that Player 1's action of playing B, is the same as the common knowledge prior probability p.

So, u(b) = 1 - u(a) = 1 - ¼ = 4/3

Based on Player 2's belief u(a) and u(b) above, it is rational for Player 2 to choose strategy W over M because, the expected value of W and M based on these believes are:

EV (W) = ¼ (22) + ¾ (12) = 14.5

EV(M) = ¼ (21) + ¾ (11) = 13.50,

and EV(W) > EV(M).

As for Player 2's belief on the information set {c, d}. we can choose u(c) to be any value between 0 and 1 if the expected value of W is larger than the expected value of M to ensure that Player 2 will choose W than M.

EV(M) = u(c)* 21 + (1 - u(c)) * 11 = 10u(c) + 11

EV(W) = u(c)* 22 + (1 - u(c)) * 12 = 10u(c) + 12

As EV(W) is always larger than EV(M) regardless of any u(c) between 0 and 1, so there are beliefs that support Player 2 to choose strategy W in the information set {c, d} as well. Thus, the pair (BB, WW) is a pooling perfect Bayesian Equilibrium with the degree of beliefs u(a), u(b), u(c), u(d) shown before.

Following the idea of computing the table of all possible expected value pairs between two players to identify possible Bayesian Nash Equilibrium as shown in Figure 7.14 before, we show an example of the Separating Bayesian Equilibrium below with a slightly different data set from the one used in Example 7.4.12 A before.

Example 7.4.12B

In this example, we choose P(H) = ¾, $C_L = 6$, and $C_H = 4$, by $W_m = 17$ and $W_w = 12$, and the assignment table shown below:

	M	W
H	21	16
L	11	14

The game tree and payoffs now becomes:

Figure 7.15

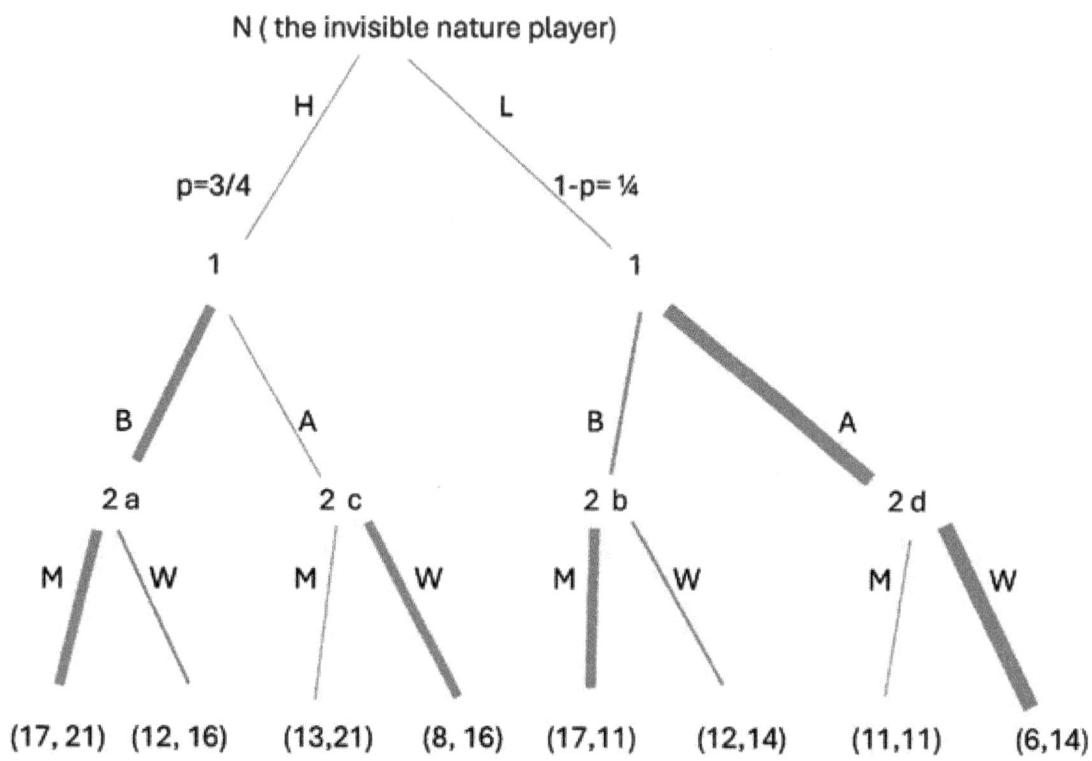

The table of the expected value pairs before now becomes:

	MM	MW	WM	WW
BB	(17, 18.50)	(15.75, 19.25)	(13.25, 14.75)	(12,15.50)
BA	(15.50, 18.50)	(16.75, 19.25)	(11.75, 14.75)	(10.5, 15.5)
AB	(13.25, 21)	(12.75, 19.25)	(10.25, 14.75)	(9, 15.5)
AA	(11.75, 18.50)	(11.25, 19.25)	(8.75, 14.75)	(7.5, 15.50)

Thus, the pair (BA, MW) is a possible pair of Perfect Separating Bayesian Nash Equilibrium. Now we try to assign rational degree of beliefs to the information set {a, b} and {c, d}.

$$u(a) = P(H \mid B) = P(B \mid H)\,P(H) / \{ P(B \mid H)\,P(H) + P(B \mid L)\,P(L) \}$$
$$= 1 * p / \{ 1 * p + 0 * (1 - p) \} = 1,$$

and similarly, $u(d) = 1$. Thus $u(b) = 0$, $u(c) = 0$. With these degree of beliefs for Player 2, the pair (BA, MW) is likely a Perfect Separating Bayesian Nash Equilibrium. However, given the strategies of Player 2 as shown in Figure [] above, the strategy BA for Player 1 is not optimal because for the L type, Player 1 can improve its payoff from 6 to 17 by changing its strategy from A to B. Thus (BA, MW) cannot be a Perfect Separating Bayesian Nash Equilibrium.

Example 7.4.13 (Continuous Version of the Spence Model)

We are now already to extend the Spence Model from the discrete case as shown in the previous Example 7.4.12 A and Example 7.4.12 B to a continuous version. Recall that in the two previous examples, the education levels (basic and advance) is discrete, the costs of education C_L, C_H for the low and high ability employee are discrete values, the playoffs W_m and W_w for the employee are discreate values pending on if the employee is a manager position or a worker position, and the payoff W of the employer is also discrete value pending on the employee's ability type and the employee's position. In the present extension of the Spence Model, all these quantities described before are continuous functions. To be precise, we make the following assumptions to this model and define the following notations and terminologies. Our presentation of the Spence Model in this book primarily follows the graphical approach outlined in reference (Books [8] by Gibbons). However, we have also endeavored to elaborate further for clearer understanding. To this end, we have incorporated algebraic interpretations alongside the graphical representations throughout the discussion, wherever possible.

1. The education level is denoted by "e", it is now a continuous variable, it is no longer just ' basic' and 'advance' as we did in the last two examples.
2. For simplicity, we assume that the employee's ability has only two types, L and H as before. We assume that the ability of the employee is private information, which means that the employer has no knowledge about the ability of the employee's ability prior to observing the employee's action, the action in this case is the employee's education level "e" to be revealed to the employer. We also assume that Nature picks on the high-ability employee with probability p, P(H) = p, and P(L) = 1-p
3. Let $C_L(e)$ and $C_H(e)$ be the cost functions of the high level and low level of a particular employee e.
4. Let $Y_L(e)$ and $Y_H(e)$ be the wages that the employee of type L and type H gets.
5. For each type H and L of the employees, we identify the optimal education level e(L) and e(H) and the optimal wage w(L) and w(H) for them.

   $$e(H) = \underset{e}{Max}\ Y_H(e) - C_H(e)$$
   $$e(L) = \underset{e}{Max}\ Y_L(e) - C_L(e)$$
   $$w(H) = Y_H(e(H))$$
   $$w(L) = Y_L(e(L))$$

6. For each ability type of the employee, we assume that there are indifference cures I_L and I_H passing through the points (e(L), w(L)) and (e(H), w(e(H)) respectively. The points (e, w) in each of the indifference curves shown satisfies the following relationship.

   $w(L) - C_L(e(L)) = w - C_L(e)$ for the I_L curve,

   $w(H) - C_H(e(H)) = w - C_H(e)$ for the I_H curve.

   Thus, we see that all the point on the I_L curve consists of all point (e, w) such that

   $w = w(L) - C_L(e(L)) + C_L(e)$ and similarly, all the points on the I_H curve consists of all points (e, w) such that $w = w\ (H) - C_H(e(H)) + C_H(e)$

7. To reflect that the low -ability employees cost more than the high-ability employees to gain additional education e, we assume that the slope of the I_L curve is higher than the slope of the I_H curve, or the low-ability indifference curve is steeper than the high-ability indifference curve.
8. The Spence Model consist of all the preceding information

Based on these assumptions above, we are going to show that the Spence Model can have many Pooling, Separating, and hybrid perfect Bayesian Nash Equilibriums. To begin, let's use the following figure as a mental note for the Spence Model in all the following examples:

Figure 7.16

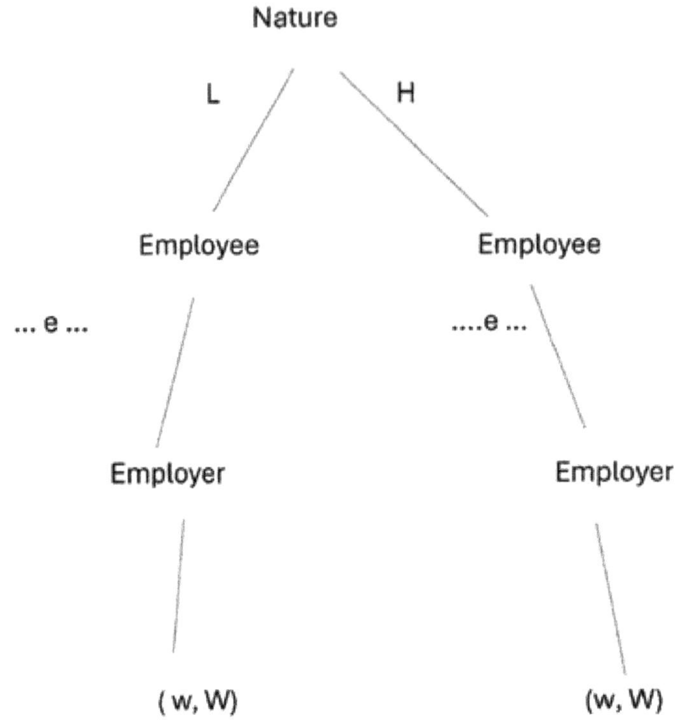

For the high and low ability employee, the education level e(H) and e(L) are well defined before. There are also education levels e(S) and e(P) that we need to demonstrate the existence of Separating and Polling equilibriums in the Spence Model. First, we show these distinct education levels and the indifference curves graphically for the low and high ability types in a progressive manner.

Figure 7.17

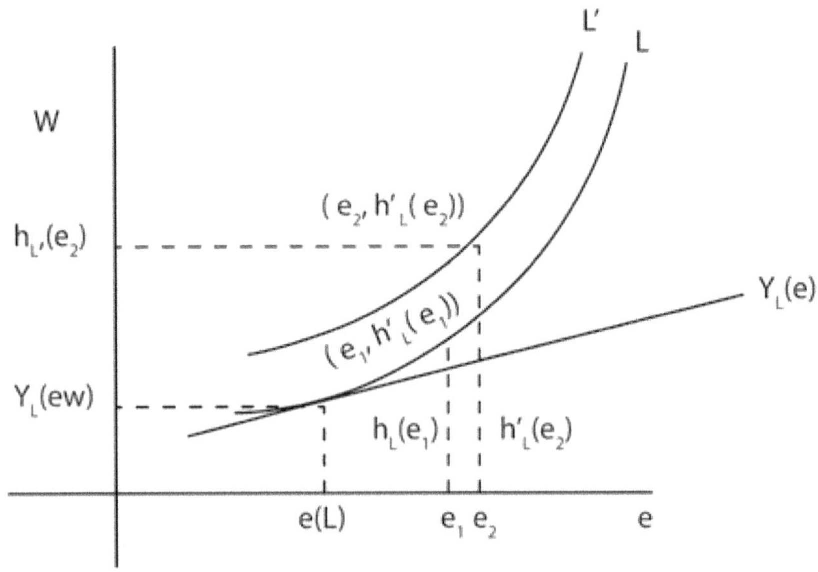

Figure 7.18

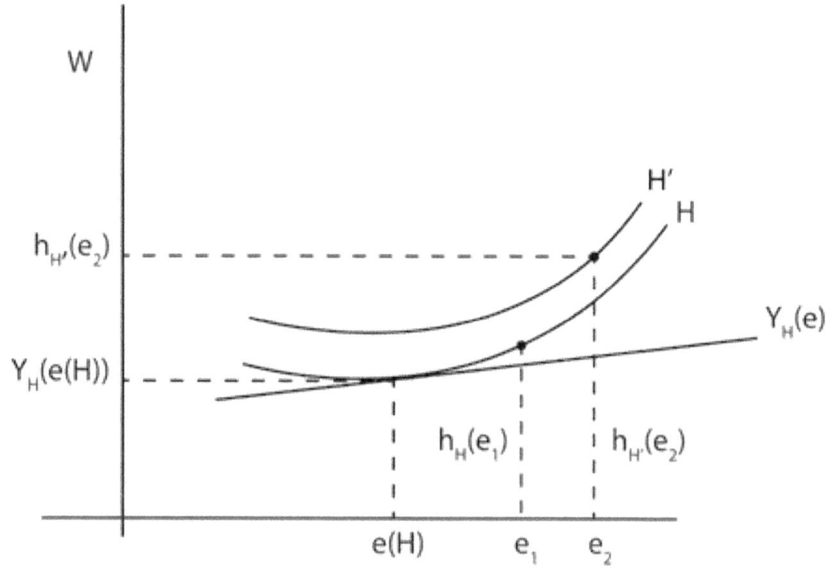

Figure 7.17, Figure 7.18 shows the low-ability and high-ability indifference curve respectively, $Y_L(e)$, $Y_H(e)$ are the respectively linear payment functions, e(L), e(H) are the respective optimal education level for each ability types, the indifference curves L and L' shown in Figure 1 are passing thorough the point (e(L), Y_L (e(L)) and (e_2, $h_{L'}(e_2)$) respectively, and similarly for Figure 2 for the H, H' indifference curves. Note that each indifference curve of a given type is determined by a single point that it passes through.

There are many possible Separating, Polling , and hybrid equilibriums existed in the Spencer Model. For our purpose, however, we are going to show the existence of only three representative equilibriums, one for Separating Equilibrium, one for Polling Equilibrium, and one for Hybrid Equilibrium respectively in the following sections A, B, and C.

Section A: Existence of a Separating Equilibrium

For the existence of a Separating Equilibrium, we are going to look at a special case and a general case. The special case assumes that the following condition hold true between the lower ability and high ability types of employees:

Figure 7.19

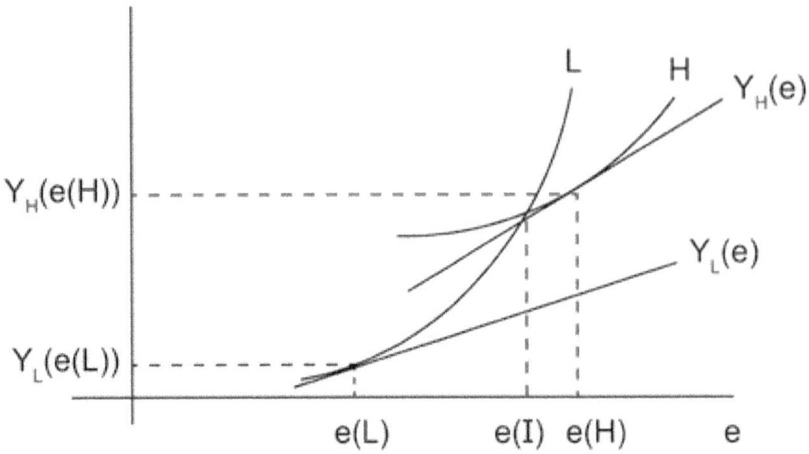

Condition X : $Y_L(e(L)) - C_L(e(L)) > Y_H(e(H)) - C_L(e(H))$

This condition basically assumes that the cost C_L (e(H)) of the low ability employee to obtain the optimal educational level e(H) for the high ability employee is very high compared to what the payment he (she) would get Y_H(e(H)) if he(she) was treated as a high ability employee, he(she) will be better off to be a low ability employee and get pay as a low ability employee with the optimal education level e(L). This assumption may or may not be feasible pending on the given functions Y_L(e), Y_H(e), C_L(e), henceforth, we refer to this condition as Condition X. Figure 7.21 below shows a scenario where condition X does not satisfy.

Figure 7.20

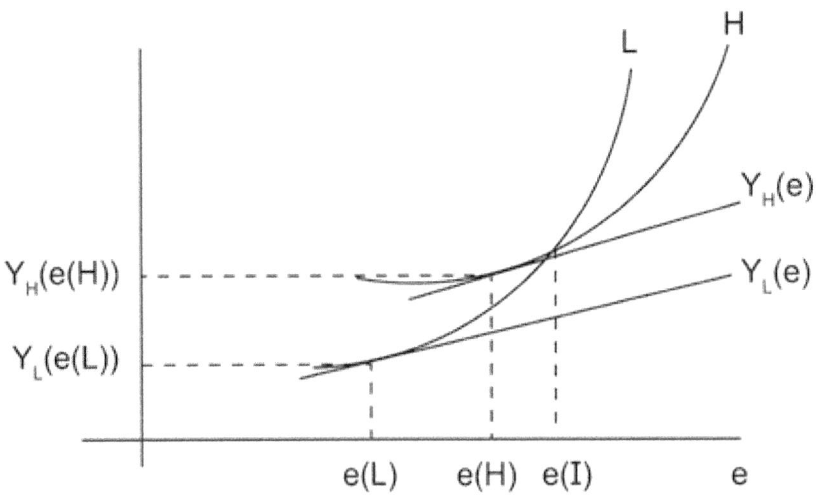

Condition X : NOT SATISFIED

Existence of Separating Equilibrium in this Special Case:

We claim that, if the employee and the employer are choosing the following strategies, then they are Separating Equilibrium.

Refer to Figure 7.21, for the low ability employee, choose e = e(L), and for the high ability employee, choose e = e(H) to be on equilibrium path respectively, this means that from the Bayes Theorem, the degree of beliefs of the employer in believing that the employee is of high ability is :

u(H | e(L)) = 0 and u(H | e(H)) = 1. Next, we choose the degree of beliefs for those e that are out of equilibrium path as follows:

u (H | e) = 1 if $e \geq e(H)$ and 0 if $e < e(H)$.

The wage that the employer assigns to the employee is given by the general expression:

$W(e) = u(H \mid e)\, Y_H(e) + (1 - u(H \mid e))\, Y_L(e)$,

which based on the employer's degree of beliefs that the employee is of high ability, reduces to the following simplified employer strategy:

$W(e) = Y_H(e)$ if $e \geq e(H)$ and $Y_L(e)$ if $e < e(H)$

Figure 7.21

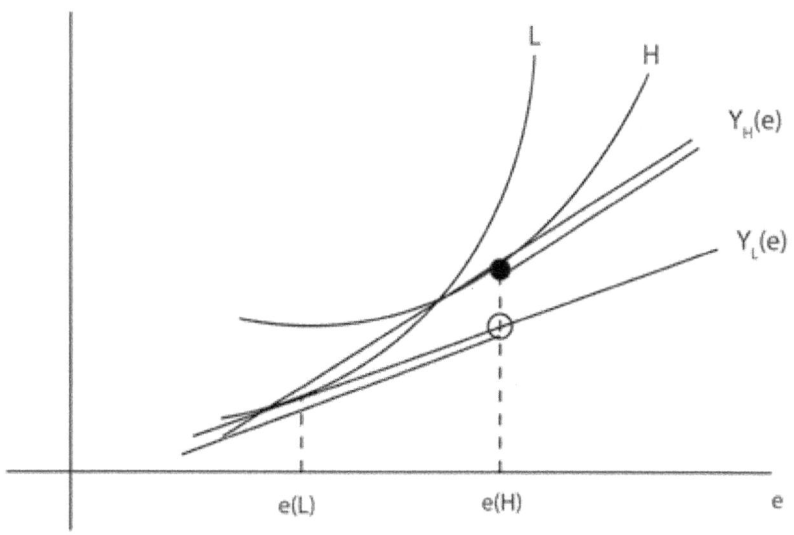

To show that the above employee and employer strategy pair is a Separating Equilibrium, it remains for us to show:

(1) For the low ability employee, e = e(L) is the employee's best response to the employer's wage function W(e) for all education level e

(2) For the high ability employee, e = e(H) is the best response to the employer's net wage function for all education level e

Proof of (1):

This means that we need to show that the net wage to the employee, refer to Figure 0 above, $W(e) - C_L(e)$ is maximum at $e = e(L)$ for all education level e. We break down e into two intervals $e < e(H)$ and $e \geq e(H)$.

For $e < e(H)$, $W(e) - C_L(e) = Y_L(e) - C_L(e)$, which is maximum when $e = e(L)$ by the definition of e(L).

For $e \geq e(H)$), $W(e) - C_L(e) = Y_H(e) - C_L(e)$, to show $e = e(L)$ is also a maximum for the employee's net wage function, first we need to establish the observation that

$Y_H(e) - C_L(e)$ is bound above by $Y_H(e(H)) - C_L(e(H)$ by the following Lemma 7.1.

Lemma 7.1

a. $C_L(e) - C_L(e(H)) > C_H(e) - C_H(e(H))$ for $e \geq e(H)$
b. $Y_H(e) - C_L(e) < Y_H(e(H)) - C_L(e(H))$ for all $e \geq e(H)$

Proof of (a): Since the low-ability curve is steeper than the high-ability indifference curve, so

$$h_H(e) - h_H(e(H)) < h_L(e) - h_L(e(H)) \quad (1)$$

Now $h_H(e) = h_H(e(H)) - C_H(e(H)) + C_H(e)$ (2)

and

$$h_L(e)) = h_L(e(L)) - C_L(e(L)) + C_L(e) \quad (3)$$

$$h_L(e(H)) = h_L(e(L)) - C_L(e(L)) + C_L(e(H)) \quad (4)$$

Substitute (2), (3), and (4) back to (1), we get,

$C_L(e) - C_L(e(H)) > C_H(e) - C_H(e(H))$

Proof of (b):

Since $Y_H(e)$ is less than $h_H(e)$, it suffices to show that:

$h_H(e) - C_L(e) < Y_H(e(H)) - C_L(e(H))$ for all $e \geq e(H)$.

Or, to show that :

$$Y_H(e(H)) - C_L(e(H)) - \{h_H(e) - C_L(e)\} > 0 \quad (5)$$

But $h_H(e) = h_H(e(H)) -- C_H(e(H)) + C_H(e)$ (6)

Putting (6) into (5), we need to show that,

$C_L(e) - C_L(e (H)) - \{C_H(e) - C_H(e(H))\} > 0$, which is true by (a).

Now we can go back and continue with the proof of (1):

$Y_H(e) - C_L(e) = Y_H(e) - C_L(e)$

$\leq Y_H(e(H)) - C_L(e(H))$ by Lemma 7.1(b)

$= Y_L(e(L)) - C_L(e(L))$ by Assumption X

$< Y_H(e(L)) - C_L(e(L))$

$= W(e(L)) - C_L(e(L))$

So, e(L) is maximum for the low ability employee's net wage function. This completes the proof of (1).

For the proof of (2):

This means that we need to show $W(e) - C_H(e)$ is maximum when $e = e(H)$ for all education level e. Again, we break down the range of e into two intervals $e \geq e(H)$ and $e < e(H)$.

For $e \geq e(H)$: $W(e) = Y_H(e)$, $Y_H(e) - C_H(e)$ is maximum when $e = e(H)$ by definition of e(H).

For $e < e(H)$, $W(e) = Y_L(e)$, as $Y_L(e) - C_H(e) < Y_H(e) - C_H(e)$ and e(H) is maximum for $Y_H(e) - C_H(e)$ by the definition of e(H), so e(H) is also maximum for $Y_L(e) - C_H(e)$.

Existence of Separating Equilibrium in the general case:

Figure 7.22

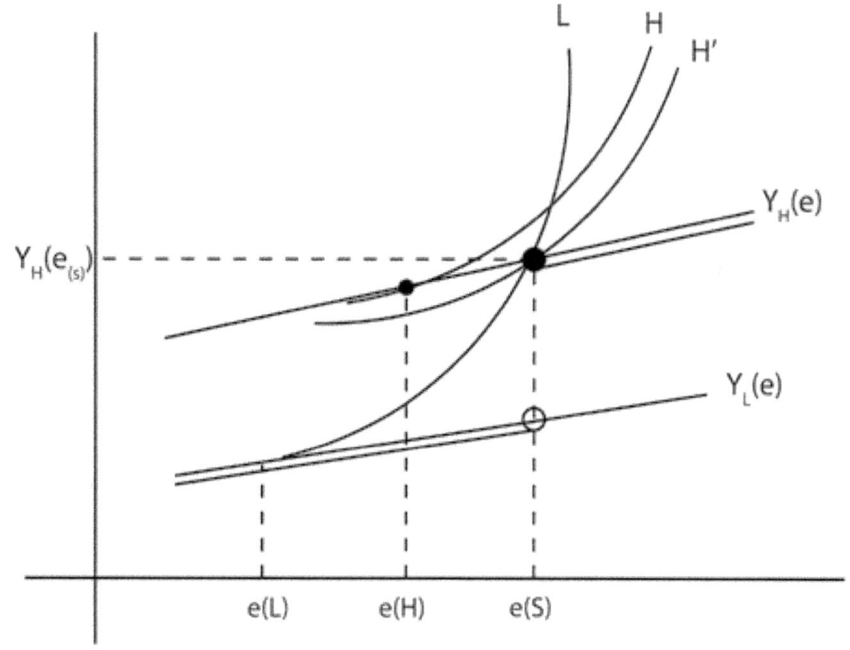

When condition X is not satisfied, the above approach for the low ability employee to choose e = e(L) and the high ability employee to choose e = e(H) would not work. Instead, as shown in Figure 9.6 above, the high ability employee will choose an education level e(S) that is higher than e(H) and such that his (hers) payment matches the optimal payment of the low ability employee, ie. e(S) is the education level that the indifference curve of the low ability employee intersects with the high ability employee's payment function $Y_H(e)$. e(S) is chosen by the high ability employee to be larger than e(H). Formally, the strategy chosen by the employee and the employer is as follows. As before, refer to Figure 7.22, for the low ability employee, e = e(L), and for the high ability employee e = e(S), the degree of belief $u(H \mid e(S)) = 1$, $u(H \mid e(L)) = 0$ on the equilibrium path, and

$u(H \mid e) = 1$ for $e \geq e(S)$ else 0 for $e < e(S)$ off the equilibrium path.

For the employer, the wage function $W(e) = u(H \mid e)\, Y_H(e) + (1 - u(H \mid e))\, Y_L(e)$ then reduces to

$W(e) = Y_H(e)$ for $e \geq e(S)$ and $Y_L(e)$ for $e < e(S)$

As in the special case, for the above employee and employer strategy pair to be a Separating Equilibrium, it remains for us to show:

(1) For the low ability employee, e = e(L) is the best response when e < e(S) and e = e(S) is the best response when e ≥ e(S)
(2) For the high ability employee, e = e(S) is the best response for all education level e

Proof of (1): For low ability employee:

When $e < e(s)$, $W(e) - C_L(e) < W(e(L)) - C_L(e(L))$ because

$Y_L(e) - C_L(e) < Y_L(e(L)) - C_L(e(L))$ by the definition of e(L)

When $e \geq e(S)$, we need to show $W(e) - C_L(e) < W(e(S)) - C_L(e(S))$, which is equivalent to show $Y_H(e) - C_L(e) < Y_H(e(S)) - C_L(e(S))$.

Now $Y_H(e) - C_L(e) \leq h_{H'}(e) - C_L(e)$ because $h_{H'}(e) > Y_H(e)$ for $e \geq e(S)$,

$h_{H'}(e) - C_L(e) = h_{H'}(e(S)) - C_H(e(S)) + C_H(e) - C_L(e)$

As $h_{H'}(e(S)) = Y_H(e(S))$, so by Lemma 7.1(a),

$Y_H(e) - C_L(e) \leq Y_H(e(S)) - C_L(e(S))$ and this completes the proof of (1).

Proof of (2): For high ability employee:

(a) For $e \geq e(S)$, since the indifference curve H' is above $Y_{H,}$

$$Y_H(e) - C_H(e) \leq h_{H'}(e) - C_H(e)$$
$$= h_{H'}(e(S)) - C_H(e(S)) + C_{H'}(e) - C_H(e)$$

$= Y_H(e(S)) - C_H(e(S))$

and so, $W(e) - C_H(e) \leq W(e(S)) - C_H(e(S))$

(b) For $e < e(S)$, we need to show $Y_L(e) - C_H(e) \leq Y_L(e(S)) - C_H(e(S))$ or equivalently $Y_L(e) - Y_L(e(S)) \leq C_H(e) - C_H(e(S))$. Since $Y_L(e) - Y_L(e(S)) < h_L(e) - h_L(e(S))$, it suffices to show $h_L(e) - h_L(e(S)) \leq C_H(e) - C_H(e(S))$.

But $\quad h_L(e) - h_L(e(S))$

$$= h_L(e(S)) - C_L(e(S)) + C_L(e) - h_L(e(S))$$

$$= C_L(e) - C_L(e(S))$$

$$\leq C_H(e) - C_H(e(S))$$

by Lemma 7.1 (a) and the direction of inequality is reversed because $e < e(S)$.

Section B: Existence of a Polling Equilibrium

Figure 7.23

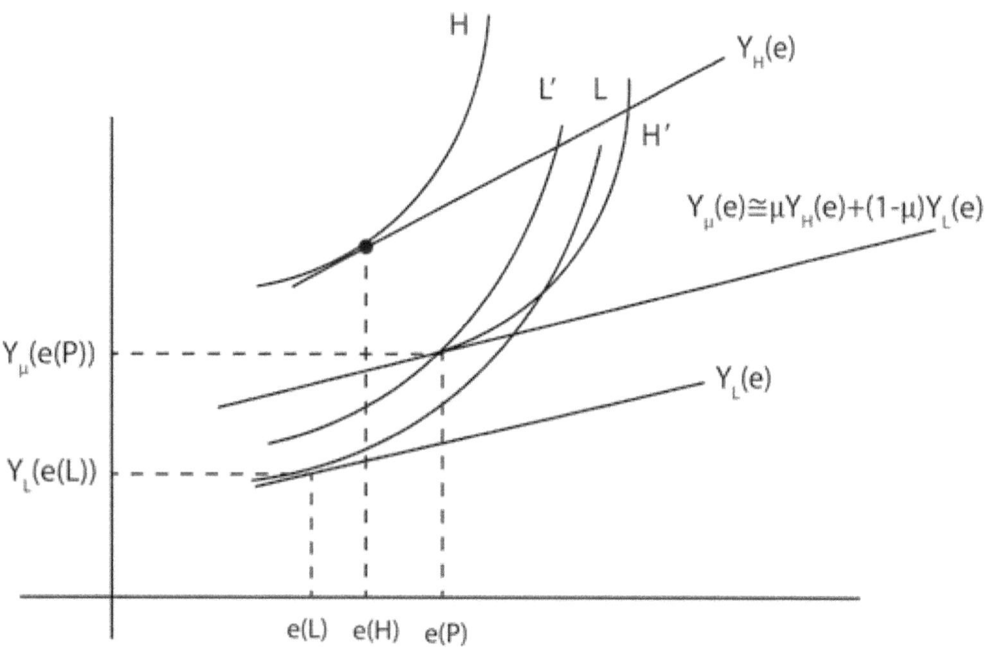

Figure 7.23 above depicts another education level e(P), different from e(L), e(H), and e(S) we discussed before. For any value u between 0 and 1, we consider a hypothetical wage function Y_u offered by the employer defined as follows.

$Y_u(e) = u\, Y_H(e) + (1 - u)\, Y_H(e)$. We then select an education level e(P) such that the point $(e(P), Y_u(e(P)))$ is above the indifference curve L of the low ability employee, and that the two indifference curves L' and H' passing through the point (e(P), Y_u(e(P)) are that L' is above L and H' is above Y_L(e) respectively. We are going to show that the following employee and employer strategies form a Polling Equilibrium.

For both the low and high ability employee, select e = e(P) on the equilibrium path, refer to Figure 7.23 before. For the employer, the degree of belief that he(she) believes that the employee is of high ability is:

u(H | e) = u when e = e(P) and 0 when e ≠ e(P). From the employer's wage function

$Y_u(e) = u\,Y_H(e) + (1 - u)\,Y_H(e)$ and let u be the employer's degree of belief u(H | e), the employer's strategy becomes:

$W(e) = Y_u(e(P))$ for e = e(P) and $Y_L(e)$ when e ≠ e(P)

We are going to show that the above employee and employer strategies form a Polling Equilibrium. As before, it remains to show:

(1) For the low ability employee, e = e(P) is the best response for all education level e
(2) For the high ability employee, e = e(P) is the best response for all education level e

Proof of (1):

(a) For e < e(P), $W(e) = Y_L(e)$, we need to show that :

$$W(e) – C_L(e) \le W(e(P)) - C_L(e(P))$$

$Y_L(e) - C_L(e) \le Y_L(e(L)) - C_L(e(L))$ by the definition of e(L),

$$= h_L(e(P)) - C_L(e(P))$$
$$\le h_{H'}(e(P)) - C_L(e(P))$$
$$= W(e(P)) - C_L(e(P))$$

(b) For e > e(P), $W(e) = Y_L(e)$,

$$Y_L(e) - C_L(e) \le h_L(e) - C_L(e)$$

$\le h_{L'}(e) - C_{L'}(e)$ because L' is above L

$$= h_{L'}(e(P)) - C_{L'}(e(P))$$
$$= W(e(P)) - C_L(e(P))$$

Thus, $W(e) - C_L(e) \le W(e(P)) - C_L(e(P))$ for all e.

Proof of (2):

$W(e) = Y_L(e)$ for all e ≠ e(P)

We need to show $W(e) - C_H(e) \le W(e(P)) - C_H(e(P))$ for all e.

$Y_L(e) - C_H(e) \le h_{H'}(e) – C_H(e)$ since H' is above the line $Y_{L'}$,

$= h_{H'}(e(P)) – C_{H'}(e(P))$

$= W(e(P)) - C_H(e(P))$

Section C: Existence of a Hybrid Equilibrium

Figure 7.24

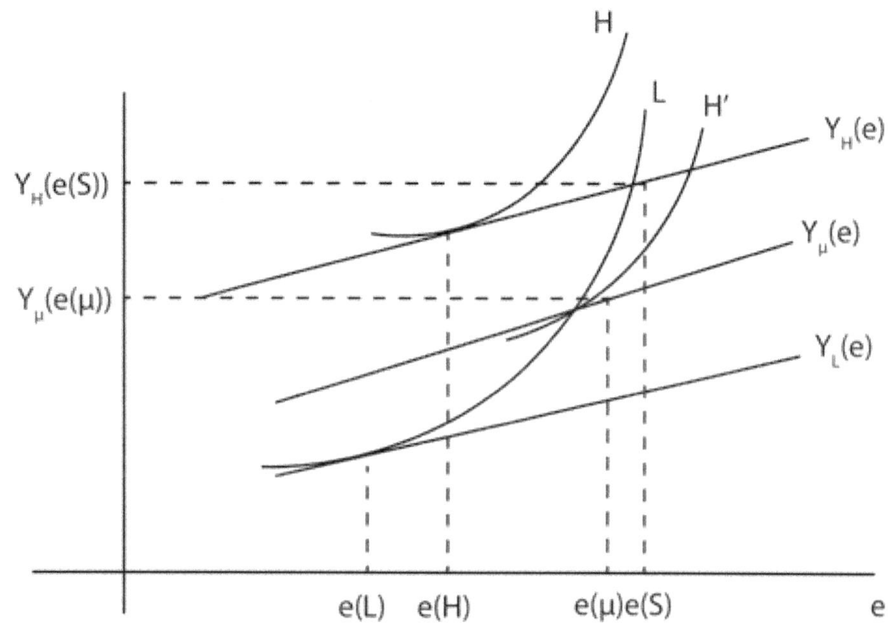

In this last section, we are going to explain the concept of Hybrid Equilibrium in the Spence Model.

First, we explain hybrid education level $e(H_h)$ for the high ability employee, and hybrid education level $e(L_h)$ for the low ability employee. The subscript h here to mean it is a hybrid education level.

By definition, $e(H_h)$ is the education level that the high ability employee will choose with certainly, but it will be chosen by the low ability employee at random between $e(H_h)$ with certain probability $p(H_h)$ and the education level e(L) with probability $(1 - p(H_h))$. In short, we refer to this process the low-ability employee does the randomizing in our later discussions. The hybrid education level $e(L_h)$ has similar meaning.

By definition, $e(L_h)$ is the education level that the low ability employee will choose with certainly, but it will be chosen by the high ability employee at random between $e(L_h)$ with certain probability $p(L_h)$ and the education level e(H) with probability $(1 - p(L_h))$. In short, we refer to this process the high-ability employee does the randomizing in our later discussions.

Second, discussion of the existence of these education levels $e(H_h)$ and $e(L_h)$ and the proper choices of these probabilities $p(H_h)$ and $p(L_h)$ and their involvement in coming up with the Hybrid Equilibrium between the employee and employer strategies is our next objective. But due to the similarities in the discussion of $e(H_h)$ and $e(L_h)$, we focus on the formulation $e(H_h)$ and $p(H_h)$ left that of $e(L_h)$ and $p(L_h)$ as an exercise. In other words, we focus for now on the formulation of $e(H_h)$ and $p(H_h)$ when the low-ability employee does the randomizing.

Refer to Figure 7.25 below:

Figure 7.25

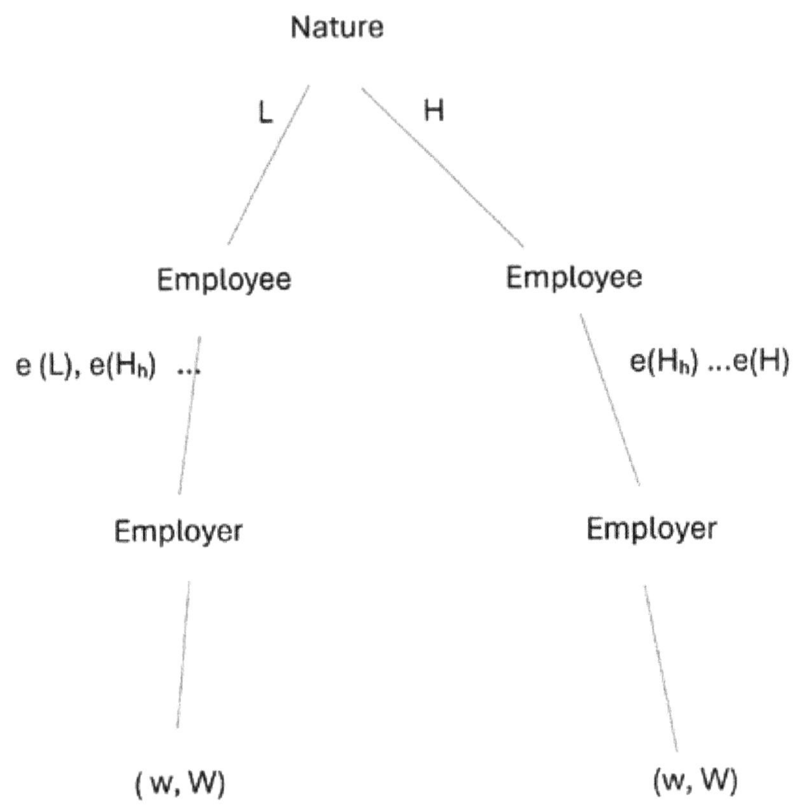

Let p_H be the probability of Nature picking up the high ability employee,

$p_H = P(H)$, and $p_L = P(L)$ be the probability that Nature picks up the low ability employee, then Bayes Theorem gives the following conditional probability that can be interpreted as the employer's degree of belief that the employee is high ability H given that he(she) observed the employee's hybrid education level $e(H_k)$

$$P(H \mid e(H_h)) = \frac{P(e(H_h) \mid H)P(H)}{P(e(H_h) \mid H)P(H) + P(e(H_h) \mid L)P(L)} \qquad (1)$$

Now $P(e(H_h) \mid H) = 1$ since high ability employee chooses $e(H_h)$ with certainty,

$P(e(H_h) \mid L) = p(H_h)$ because the low ability employee will choose $e(H_h)$ with probability $p(H_h)$.

Thus (1) becomes, $$P(H \mid e(H_h)) = \frac{p_H}{p_H + (1 - p_H)p(H_h)} \qquad (2)$$

Note the probability above, $P(H \mid e(H_h)$ is also the employer's degree of belief $u(H \| e(H_h))$ that the employee is high ability given that the employer observed the hybrid equilibrium education level $e(H_h)$. Because the high-ability employee chooses $e(H_h)$ at all time, and the low-ability employee randomizes between $e(H_h)$ and $e(L)$, so $P(H \mid e(H_h))$ is greater than $P(H)$, or, $u(H \mid e(H_h)) > p_H$.

In what follows, we proceed to determine the education level $e(H_h)$ and $p(H_h)$

Recall that the wage W offered by the employer after observing the employee's education level e with the degree of belief $u(H \mid e)$ that he(she) is of high ability H type is given by :

$$W(e) = u(H \mid e)\, Y_H(e) + (1 - u(H \mid e))\, Y_L(e) \tag{3}$$

Thus, when $e = e(H_h)$, the wage is

$$W(e(H_h)) = \frac{p_H}{p_H + (1-p_H)p(H_h)} Y_H(e(H_h)) + \frac{(1-p_H)p(H_h)}{p_H + (1-p_H)p(H_h)} Y_L(e(H_h)) \tag{4}$$

Since the low ability employee randomizes $e(H_h)$ and $e(L)$, this means that two points $B(e(L), Y_L(e(L))$ and the point $A(e(H_h)), W(e(H_h))$ must be on the same low ability employee' s indifference curve L as shown in Figure 9.8. Therefore, we must have,

$$\begin{aligned} &Y_L(e(L)) - C_L(e(L)) = W(e(H_h)) - C_L(e(H_h)), \text{ or} \\ &W(e(H_h)) = Y_L(e(L)) - C_L(e(L)) + C_L(e(H_h)) \end{aligned} \tag{5}$$

From (4) and (5), we can check that if any given education level $e(H_h)$ chosen by the high-level employee H is a hybrid education level by performing the following test:

1. Compute $Y_L(e(H_h))$, $Y_H(e(H_h))$, $C_L(e(H_h))$, $Y_L(e(L))$, $C_L(e(L))$
2. From (5) compute $W(e(H_h))$
3. From (4), $W(e(H_h))$ is a weighted average of $Y_H(e(H_h))$ and $Y_L(e(H_h))$, therefore, we must check if it is in between the range between $Y_H(e(H_h))$ and $Y_L(e(H_h))$ when $W(e(H_h))$ is computed from (5). If it is in between the range, then there must be a proper weighting factor α between 0 and 1 such that $W(e(H_h)) = \alpha\, Y_H(e(H_h)) + (1 - \alpha)\, Y_L(e(H_h))$. In fact,

Let $A = Y_H(e(H_h))$, $B = Y_L(e(H_h))$, and $C = W(e(H_h))$, then we find $\alpha = \frac{C-B}{A-B}$ and we can compute the probability $p(H_h)$ from the equation:

$$\alpha = \frac{p_H}{p_H + (1-p_H)p(H_h)} \tag{6}$$

Thus, we completed checking the hybrid education level H_h with the associated probability $p(H_h)$:

$$p(H_h) = \frac{p_H(1-\alpha)}{(1-p_H)\alpha} \tag{7}$$

4. If, however, $W(e(H_h))$ computed from (5) is not in range, then such an education level $e(H_h)$ cannot be a hybrid education level, and we start looking for another one.

Having defined what is meant by the hybrid education level $e(H_h)$, we are in position to state the hybrid equilibrium strategies between the employee and the employer.

For the high ability employee, choose $e = e(H_h)$ with certainty and for the low ability employee, choose $e = e(H_h)$ with probability $p(H_h)$ given by (7) above and choose $e = e(L)$ with probability $(1 - p(H_h)$. For the employer, the degree of belief $u(H \mid e) = \alpha$ for all $e \geq e(H_h)$ else 0 for all $e < e(H_h)$. The employer's wage function from (3) is thus simplified to:

$W(e) = \alpha Y_H(e) + (1 - \alpha)\, Y_L(e)$ for all $e \geq e(H_h)$ else $Y_L(e)$ for $e < e(H_h)$.

Like the Separating and Pooling examples in Section A, B, we can easily verify that for the high ability employee, $W(e) - C_H(e)$ is maximum at $e = e(H_h)$, and for the low ability employee, $W(e) - C_L(e)$ is optimal at $e = e(H_h)$ when $e \geq e(H_h)$ but it is optimal at $e = e(L)$ when $e < e\,(H_h)$.

7.5 Requirements 5 & 6 for Signaling Games

We have seen many examples of equilibrium in signaling games like those discussed in the previous discussions. These examples basically satisfy the Signal Requirements 1 to 4 discussed before. However, these four requirements 1 to 4 do not cover all aspects of signaling games that occur in practice.(Papers [5], [8]). The Spence Model provides scenarios, as we shall see shortly, that make the additional requirements, Requirement 5, and Requirement 6, reasonable. In this section, we first state Requirement 5 and 6 then followed by the description of these scenarios. But before that, we must understand what it means to say a message is dominated for a particular type and a message is equilibrium dominated for a particular type in a perfect Bayesian equilibrium. We will see shortly that if a message is equilibrium dominated for a particular type, then it is dominated for that type, but not vice versa.

Definition 7.5.5 (A message is dominated for a particular type)

In a signaling game with type set T, message set M, and action set A, a message m in M is dominated for a particular type t in T if there exists another message m' in M such that,

$$\min_{a \in A} U_s(t, m', a) > \max_{a \in A} U_s(t, m, a)$$

Definition 7.5.6 (Equilibrium payoff of a particular type)

The equilibrium payoff of a particular type t in T, denoted by $U_s(t)$, is the payoff $U_s(t, m, a)$ where (m, a) is on the equilibrium path for the type t.

In what follows we will give two examples to illustrate the above two definitions.

Example 7.5.14

In Figure 7.26 below, the sender's strategy responded to the types t_1 and t_2 are the strategy A and A respectively, and the receiver's strategy responded to the sender's strategy A, B are X, Y respectively.

Figure 7.26

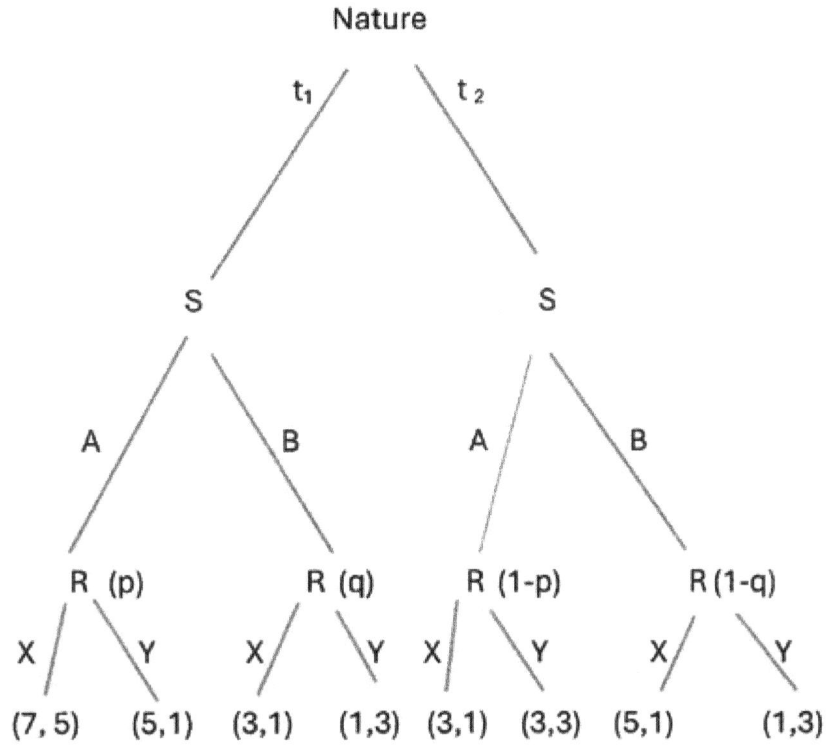

Probability p is the receiver's belief that the sender is type t_1 after he(she) has observed that the sender has played the strategy 'A', similarly,

Probability q is the receiver's belief that the sender is type t_1 after he(she) has observed that the sender has played the strategy 'B'. From the payoffs shown in the figure, it is easily checked that p = 0, 5 and q could be any value of belief, the strategy pair (A, A), (X, Y) is a Perfect Bayesian Nash equilibrium.

In this example, the message B is dominated for type t_1 because, min $\{U_s(t_1, A, X), U_s(t_1, A, Y) > \max (t_1, B, X), U_s(t_1, B, Y)\}$ as $\min\{7, 5\} > \max \{3, 1\}$ from the figure. But message B is not dominated type t_2 because, min $\{U_s(t_2, A, X), U_s(t_2, A, Y) < \max (t_2, B, X), U_s(t_1, B, Y)\}$as $\min\{3, 3\} < \max \{5, 1\}$ As for t_1, AX is the equilibrium path, so $U_s(t_1) = 7$, $U_s(t_2) = 3$

We are going to introduce the next example to illustrate a similar but slightly different concept. We are going to define what it means by a message is equilibrium-dominated for a given type t, this is a concept needed to impose the new requirement 6 on Bayesian Nash equilibrium.

Definition 7.5.7 (A message is equilibrium-dominated for a particular type)

In a signaling game with type set T, message set M, and action set A, a message m in M is dominated for a particular type t in T if the following condition holds:

$$U_s(t) > \max_{a \in A} U_s(t, m, a)$$

The following example illustrates this concept.

Example 7.5.15

Figure 7.27

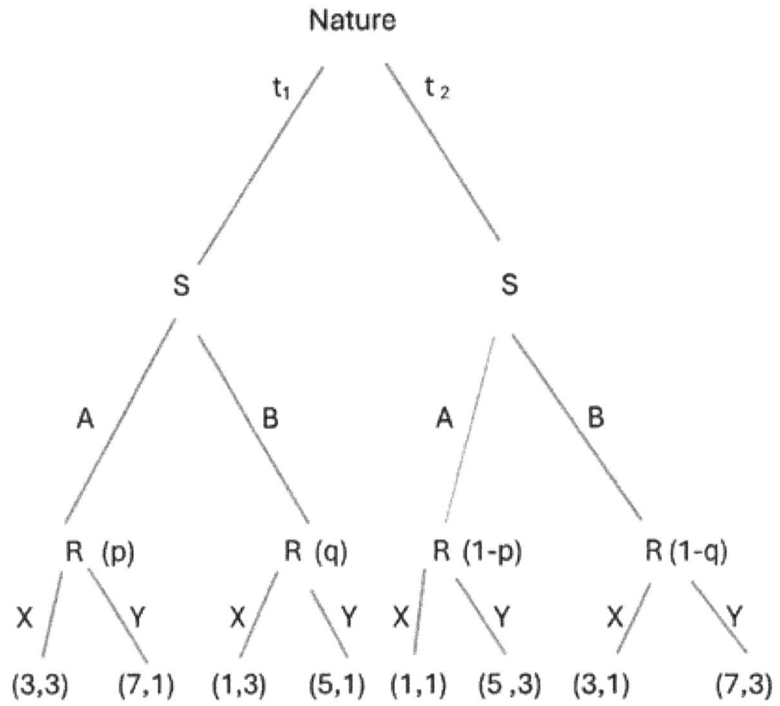

In this example, message 'B' is equilibrium-dominated for t_1, this is because $U_s(t_1) = 7$ as AX is the equilibrium path for t_1, $U_s(t_1) > \max\{U_s(t_1, B, X), U_s(t_1, B, Y)\} = \max\{1, 5\} = 5$. But the same message 'B' is not equilibrium-dominated for t_2 because $U_s(t_2) = 5$ which is not larger than $\max\{U_s(t_2, B, X), U_s(t_2, B, Y)\} = \max\{3, 7\} = 7$.

From the direct definitions of a message is dominated for a type t, and a message is equilibrium dominated for type t, we can see that if a message is equilibrium dominated for type t, then the same message is dominated for the same type t. But the converse it not true. This observation can be seen from the above example. Here message B is equilibrium dominated for t_1, but B is not dominated for t_1 because min $\{3, 7\}$ is not larger than max $\{1, 5\}$.

With these two concepts squared away, we can formally state the Signaling Requirement 5 and 6 as follows.

7.5.1 Signaling Requirement 5

Given a perfect Bayesian equilibrium game, if a message m in M is off the equilibrium path, and if m is also dominated for a particular type t in T defined above, then by setting the receiver's degree of belief of the message m is of type t to zero, u(t | m) = 0, will not contradict with any of the Signaling Requirements 1 to 4 for original perfect Bayesian equilibrium.

We illustrate this requirement 5 by Example 7.5.14 earlier. Here, as shown in Figure 7.27, the degree of belief $q = u(t_1 \mid B)$ can be any value without contradicting any of the perfect Nash Equilibrium requirements from 1 to 4, for example, the receiver's response to the sender's message A is the action X rather than the action Y and the receiver's response to the sender's message B is the action Y rather than the action X. These decisions are determined by the following expected value calculations:

EV(p, A, X) = p(5) + (1 - p) (1) = 1 + 4p = 3

EV(p, A, Y) = p(1) + (1 - p) (3) = 3 - 2p = 2 when p = 1/2.

As 3 > 2, action X is selected. Similarly,

EV(q, B, X) = q(1) + (1 - q) (1) = 1

EV(q, B, Y) = q(3) + (1 - q) (3) = 3

As 3 > 1 for all q, so the receiver selected action Y in response to the sender's message B. Thus, q can be set to zero without disturbing the perfect Bayesian Nash equilibrium. So, this example satisfies Signaling Requirement 5. Now we can state Signaling Requirement 6 below.

7.5.2 Signaling Requirement 6

Given a perfect Bayesian equilibrium game, if a message m in M is off the equilibrium path, and if m is also equilibrium-dominated for a particular type t in T defined above, then by setting the receiver's degree of belief of the message m is of type t to zero, u(t | m) = 0, will not contradict with any of the Signaling Requirements 1 to 4 for the original perfect Bayesian equilibrium.

We illustrate this requirement 6 by Example 7.5.15 shown earlier. Here, as shown in Figure 7.27, $P(t_1) = 0.1$, [(A, A), (Y, X), p = 0.9, q > 1/2} is a Polling perfect Bayesian Nash equilibrium for any value q. Here, although message B is equilibrium-dominated for type t_1, we cannot set q to 0 because that will violate Signaling Requirements 1 - 4 as the following computations show.

Since $P(t_1 = 0.1)$, for the Sender to choose strategy (A, A) as polling equilibrium, the receiver's degree of belief that $u(t_1 \mid A)$ must also be 0.1 as the Bayes' Theorem shows:

$$p = u(t_1 \mid A) = \frac{P(A\mid t_1)P(t_1)}{P(A\mid t_1)P(t_1)+P(A\mid t_2)P(t_2)} = \frac{1(0.1)}{1(0.1)+1(0.9)} = 0.1$$

Now the receiver can decide which action to take in response to the sender's message A from the following expected values:

EV(A, X, p) = p(3) + (1 - p) (1) = 1.2

EV(A, Y, p) = p(1) + (1 - p) (3) = 2.8

Action Y was selected in response to message A as 2.8 > 1.2.

Similarly, the receiver can decide which action to take in response to the sender's message B from the following expected values:

EV(B, X, q) = q(3) + (1 - q) (1) = 1 + 2q

EV (B, Y, q) = q(1) + (1 - q) (3) = 3 - 2q

As the receiver selected action X than action Y, so EV(B, X, q) > EV(B, Y, q), so it would not be a perfect Bayesian Nash equilibrium. Thus, in this Example 7.5.15, the equilibrium [(A, A), (Y, X), p = 0.1, q > ½] does not satisfy Requirement 6 although the message B is equilibrium-dominated for type t_1 but $q = u(t_1 \mid B)$ cannot be set to 0 to maintain the equilibrium Requirements 1 to 4.

As we have noted before, if a message B is equilibrium-dominated for type t, then it is dominated for type t, but not the other way around. Thus, if a perfect Bayesian Nash equilibrium satisfies Requirement 6, that it must satisfy Requirement 5.

7.6 Spence Model Revisited

After we understood the distinction of these two concepts just described, we are returning to the Spencer Model to demonstrate the richness of this model in the sense that there exist many perfect Bayesian Nash equilibriums that can satisfy Requirement 5 and Requirement 6 in this model although these equilibriums can occur in other applications. In this section, we will establish the following results:

7.6.1 Existence of Separating equilibrium satisfying Requirement 5/6

7.6.2 Existence of Pooling equilibria satisfying Requirement 5

7.6.3 Existence of the Hybrid equilibria satisfying Requirement 5

7.6.4 Equilibria that cannot satisfy Requirement 5

7.6.1 There exists a unique Separating Nash equilibrium that can satisfy Signaling Requirement 5 and Requirement 6

Figure 7.28

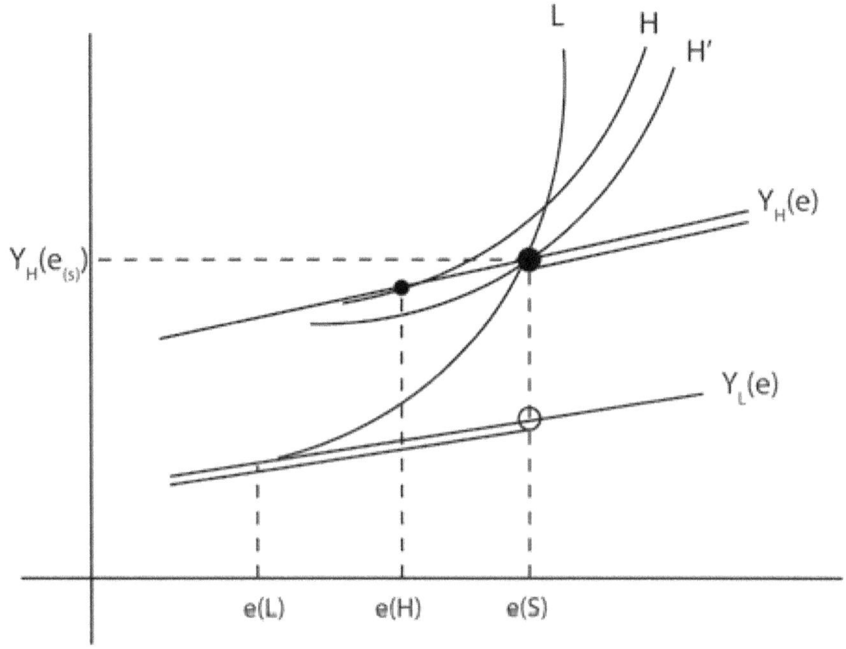

To see that this is true, refer to Figure 7.22 in Section 7.4. we reiterate the main points here. We showed that when we introduce a new education level e(S) determined by intersecting the low ability employee's indifference curve L with the high ability employee's payment function $Y_H(e)$, then if the low ability employee chooses e = e(L), the high ability employee choses e = e(S), the equilibrium beliefs

u(H | e(L)) = 0, u(H | e(S)) = 1, and if the employer chooses equilibrium wage

$w(e(L)) = Y_L(e(L))$, $w(e(S)) = Y_H(e(s))$,

and the beliefs off the equilibrium path are:

u(H | e) = 1 for e ≥ e(S) and 0 for e < e(S), so that the employee's wage function is

$w(e) = Y_L(e)$ for e < e(S) else $Y_H(e\ )$ for e ≥ e(S), then the employee and employer strategy just described is a Separating equilibrium. The education level e(S) represents the high ability employee's minimal education level such that if the low ability employee was able to achieve e(S), then there is no difference between the low-ability employee and the high-ability employee as far as the payment is concern.

Now we look at those education level e > e(S) in Figure 7.28 above.

We claim that these education e are dominated for the low-ability employee.

To show this, first recall that $w(e) = u(H \mid e)\, Y_H(e) + (1 - u(H \mid e))Y_L(e)$ is the wage function that the employer will give to the employee after the employer believe that the employee has high ability is $u(H \mid e)$. Note $w(e)$, is at least $Y_L(e)$ no matter what the employer's belief $u(H \mid e)$ is after observing e. Therefore, the low-ability employee's utility $W(e)$ for those $e > e(S)$,

$$W(e) = Y_L(e) - C_L(e) \leq w(e) - C_L(e)$$

$$= w(e(L)) - C_L(e(L)),$$

this is because the points $A(e, h_L(e))$ and $B(e(L), Y_L(e(L))$ are on the L- indifference curve,

$$= Y_L(e(L)) - C_L(e(L))$$

$$= W(e(L)),$$

which is the low-ability employee's utility at the education level e(L). Thus, by definition, the existing Separating Nash equilibrium described before can satisfy Signaling Requirement 5 because we can set $u(L \mid e)$ to 0 or $u(H \mid e) = 1$ without contradicting the requirement, namely $u(H \mid e) = 1$, of the existing Separating Nash equilibrium.

Note from above discussion, for $e > e(S)$, the utility for the low-ability employee is bounded above by $Y_L(e(L)) - C_L(e(L))$ for the Separating equilibrium to satisfy Requirement 5. Similarly, we can show that for $e > e(S)$, the utility for the high-ability employee is bounded below by $Y_H(e(S)) - C_H(e(S))$. Or we can state in another way, if Requirement 5 is satisfied, then for $e > e(S)$, e is not dominated for the high-ability employee, and $u(H \mid e) = 1$.

Thus, we have shown the existence of the Separating Nash equilibrium that satisfies Requirement 5 above. Now we show that such a Separating equilibrium that also satisfies Requirement 5 must be unique. Suppose there exists e(S)' such that e(S)' > e(S) and e(S)' is also a Separating equilibrium and Requirement 5 is also satisfied with respect to this new e(S)'. Existence of such e(S)' will lead to an impossible situation because if e is any education level between e(S) and e(S)', then, on one hand, $u(H \mid e) = 1$ since e(S) satisfies Requirement 5, but on the other hand, $u(H \mid e) = 0$ because $u(H \mid e)$ is the degree of belief of the employer for the new Separating equilibrium e(S)', which was set to 0 in the definition of e(S)'.

Now we know that the Separating Nash equilibrium shown before is the only one Separating Nash equilibrium satisfying the Requirement 5. But does it also satisfy Requirement 6? The answer is yes because the education level $e > e(S)$ was shown to be dominated by the low-ability employee, and it can be shown that is also equilibrium dominated by the low-ability employee.

Recall $W(e) = Y_L(e) - C_L(e) \leq Y_L(e(L)) - C_L(e(L)) = W(e(L))$ and note that e(L) is on the equilibrium path for the low-ability employee, so W(e(L)) is the equilibrium payoff for the low-ability employee, so Requirement 6 is also satisfied for this Separating Nash equilibrium.

7.6.2 There exist many Pooling equilibria that can satisfy Signaling Requirement 5

Recall from Figure 7.29, for the pair (e, w(e)) to be a Pooling equilibrium, w(e) must be the employer's wage function w such that,

$w = q\,Y_H(e) + (1 - q)\,Y_H(e)$, where q is the probability of the high-ability employee, the low-ability indifference curve going through the point (e, w(e)) must be on the top of the low-ability indifference curve through (e(L), Y_H(e(L)), and the high-ability indifference curve going through the point (e, w(e)) must be above the low-ability wage function Y_L(e).

Figure 7.29

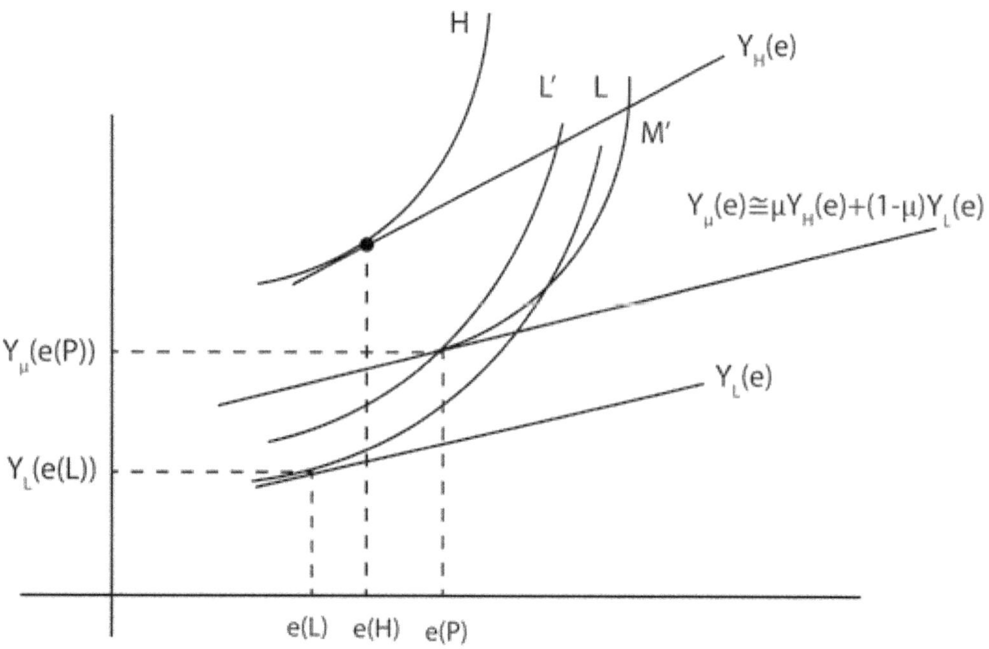

Now let's look at the following Figure 7.30

Figure 7.30

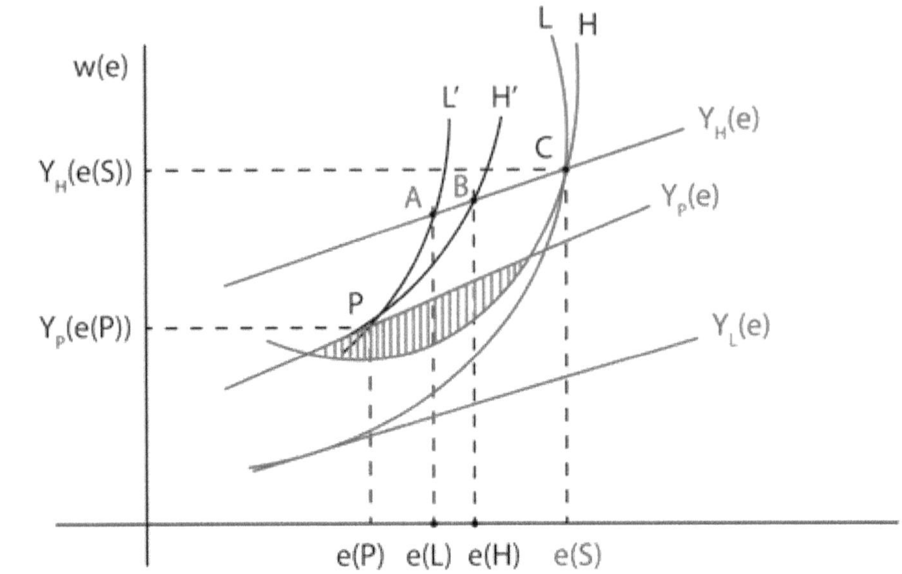

This figure depicts a typical scenario where the wage function $Y_q(e)$, where $Y_q(e) = w = q\, Y_H(e) + (1 - q)\, Y_H(e)$ bounds a region with the high-ability employee's indifference curve going through the point $C(e(S), Y_H(e(S))$ as shown in the shaded area. We see any point P on the wage function $Y_q(e)$ can be a Polling equilibrium satisfying Requirement 5. This is because the low-ability and high-ability indifference curves L' and H' passing through the point P satisfies the conditions of a Polling equilibrium as shown in Figure 9.7, and this Polling equilibrium also satisfies Requirement 5 because all the education level $e > e(P)$ is dominated for type H and $u(H \mid e) = 0$ was given in the Polling equilibrium of $P(e(P), Y_q(e(P))$ and the high-level employee's payoff at e(P) is larger than that employee's payoff at e. Thus any point corresponds to a Polling equilibrium with the shaded region in the above figure can satisfy Requirement 5 as long as there is a region bounded by the H-indifference curve passing through C and the line given by the wage function $Y_q(e) = w = q\, Y_H(e) + (1 - q)\, Y_H(e)$ for all values of degree of belief $q = u(H \mid e(P))$. After we show that the Polling equilibrium described before satisfies Requirement 5, it remains for us to ask : Can the Pooling equilibrium P described above also satisfy Requirement 6? The answer is NO. We explain this as follows. We draw the low-ability and high ability indifference curves through P, each intersects with the high-ability wage function $Y_H(e)$ at points A and B respectively with the corresponding education levels e_L and e_H as shown in the figure. We now focus our attention on the education levels e in the interval I between e_L and e_H, that is $I = (e_L, e_H)$, Then

(a) All e in I is equilibrium-dominated for the low-ability employee.

This is because the low-ability indifference curve passing through P is higher than the wage of any e in I, including the wage of the high-ability employee $Y_H(e)$

(b) All e in I is not equilibrium-dominated for the high-ability employee.

This is, because if it were, say some e* convinces the employer that the employee has high ability, then the employer will offer the employee $Y_H(e^*)$, which will make the high-ability employee better off than in the given pooling equilibrium at P, where the payoff for both the low-ability and high-ability employees are equal.

Thus, in interval I, if Requirement 6 were satisfied, then u(H | e) = 1 by the definition of Requirement 6, but since P is a Polling equilibrium , u(H | e) = 0 from the description of the degree of belief for a Polling equilibrium described in Section B of Example 7.4.13 earlier. Thus, this contradiction established that Requirement 6 cannot be satisfied in the Pooling equilibrium scenario described above.

7.6.3 Existence of the Hybrid equilibria satisfying Requirement 5

Refer to Figure 7.30 above, the hybrid equilibria in which the high-ability employee does the randomizing can satisfy Requirement 5. In the figure, suppose now the point P lies on the high-ability employee indifference curve passing through the point C(e(S), Y_H(e(S)), then it is Hybrid equilibrium, because by definition of the high-ability employee does the randomizing, the Polling equilibrium must be on the high-ability H-indifference curve, and the Polling equilibrium satisfies Requirement 5 follows from (b). Thus, any point, which is a (education level, employee payoff) pair in the shaded region is a Hybrid equilibrium satisfying Requirement 5.

7.6.4 Conditions under which Polling equilibria , and Hybrid equilibria where the low-ability employee does the randomizing cannot satisfy Requirement 5

Figure 7.31 below depicts situations under which the Polling equilibria, and Hybrid equilibria where the low-ability employee does the randomizing cannot satisfy Requirement 5.

Figure 7.31

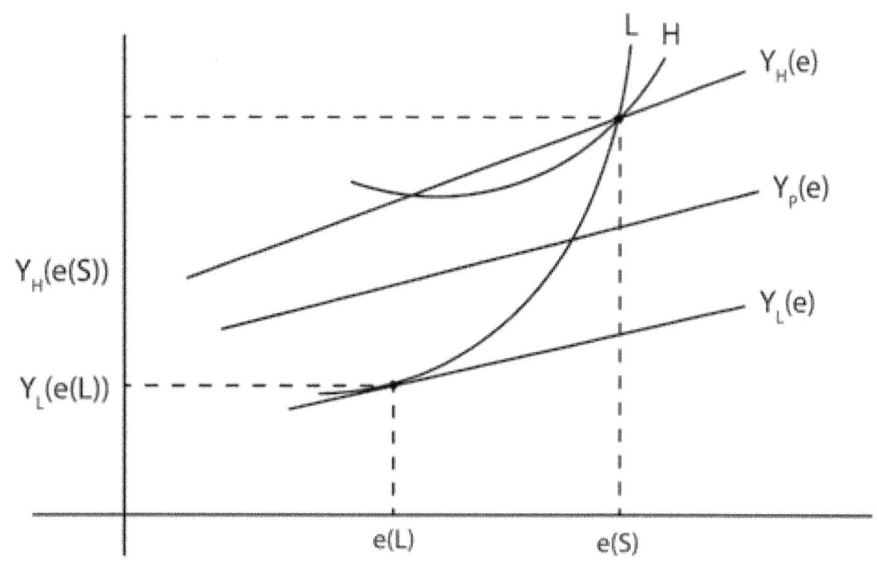

In this scenario, the wage function $Y_q(e)$ lies below the H- indifference curve through the point (e(S), Y_H(e)). In this case,

(1) There is no Pooling equilibrium can satisfy Requirement 5 because in such equilibrium, the high-ability employee cannot achieve the payoff of

$Y_H(e(S)) – C_H(e(S))$ as we shown that this is a testing condition for

Requirement 5 stated in (a).

(2) There is no Hybrid equilibrium satisfies Requirement 5 in which the low-ability employee does the randomizing because the (e, w(e)) point at which the Pooling equilibrium occurs must be on the low-ability indifference curve through the point (e(L), $Y_L(e(L))$ and so it must lie below the high-ability indifference curve passing through the point (e(S), $Y_H($ e(S)). Refer to Figure 7.30 for the hybrid equilibrium e(H_h), as shown in the figure, point A lies below the high-ability indifference curve passing through the point C (e(S), $Y_H(e(S))$, and so the corresponding pooling cannot satisfy Requirement 5 from (1).

(3) There is no Hybrid equilibrium where the high-ability employee does the randomizing satisfies Requirement 5 because the point (e, w(e)) at which pooling occurs lies below the wage function $Y_q(e) = q\, Y_H(e) + (1 - q)\, Y_H\, (e\ )$. This is because if e were the hybrid equilibrium when the high-ability does the randomizing, then the low-ability employee will choose e all the time, and the high-ability will randomize between e and e(H), so the employer's degree of belief u(H | e) that the employee is high-ability is less than q. So, pooling cannot happen follows from similar reasoning in (1).

7.7 Summary

We introduced the fourth and the last equilibrium concept, perfect Bayesian Nash equilibrium—or the study of dynamic games of incomplete information. This equilibrium concept can be considered as a refinement to the concept of Bayesian Nash equilibrium and subgame-perfect Nash equilibrium, which were introduced in the previouschapters.

Compared to subgame-perfect Nash equilibrium, the perfect Bayesian Nash equilibrium incorporates an additional element: the degree of belief assigned to tree nodes both on and off the equilibrium path. This added structure allows for a more precise analysis of strategic interactions with uncertainty.

We apply this refined equilibrium concept to a fundamental class of dynamic games of incomplete information known as Signaling games. In Example 7.4.12, we explore two practical applications:

1. Limit pricing and entry deterrence
2. The Spence Model. Example 7.4.13 explores the existence of the Separating, Pooling, and Hybrid equilibriums of this model

In Section 7.5, we introduced two additional requirements (Requirement 5&6), which allow for the study of more complex and sophisticated Signaling games. Finally, in Section 7.6, we examine the existence of perfect Bayesian Nash equilibria in the Spence Model.

7.8 Technical Key Takeaway

In dynamic games with incomplete information, Type Incentive Compatibility Constraints are essentially used for identifying separating and pooling equilibria. These constraints ensure that each type maximizes its own payoff by adhering to its prescribed strategy, rather than mimicking others. By first determining the maximum or minimum achievable profit for each type—referred to as type-specific constant constraints—we can then establish the feasible ranges for each type's prescribed strategies for the players. Like in the pricing limiting and entrance deterring Example 7.4.11, we first compute these constraints MHC1 and MLC1 for the two H and L types from the problem setup, then used them to establish the ranges for the strategies q_1 and q_2.

Chapter 8

Perfect Bayesian Nash Equilibrium and Cheap Talk games

8.0 Introduction

In Chapter 7, we examined a type of dynamic game with incomplete information known as Signaling Games. In this chapter, we explore another important class of dynamic game with incomplete information called Cheap-Talk Games.

A cheap-talk game is a two-player game involving a Sender and Receiver, where the messages transmitted from the sender to the receiver are costless and non-verifiable. This contrasts with signaling game, where message transmission may involve costly signals that influence strategic interactions.

We define Cheap-Talk Games in Section 8.1 and introduce the concept of Perfect Bayesian Equilibrium (PBE) in this setting in Section 8.2. A key result we establish in this chapter is that every CHeap-Talk Game has a trivial Perfect Bayesian Equilibrium, known as the Babbling Equilibrium (Definition 8.2.1), in which messages carry no meaningful information. In Section 8.5 , we present Crawford and Sobel's approach to systematically finding Perfect Bayesian Equilibria in Cheap-Talk games with specific payoff functions of the form $U_S(t, a) = -[a - (t + b)]^2$ and $U_R(t, a) = -(a - t)^2$. In Section 8.6, we present three MATLAB programs, myCrawfordSobel1.m, plotSenderReceiverCombine.m, and myCheapTalk4.m as computation tools to analyze various scenarios in Cheap-Talk games.

8.1 What is a Cheap-Talk game?

First, let us give a definition of what a simple Cheap-Talk game is.

Assume there are two players, Player 1, we also called it the Sender, and Player 2, we also called it the Receiver. Like the Signaling Games we described before, there are many feasible types for Player 1, denote these types for Player 1 by the set T,

$T = \{t_1, t_2, t_3, \ldots t_l\}$, where the number of types n could be finite, or infinite in which case Player 1's type space is a continuum like the unit interval $T = [0, 1]$.

Type information is considered as private information for Player 1 because only Player 1 himself (herself) knows which specific type he (she) currently belongs, Player 2 only knows about the probability distribution $p(t_i)$

of these types, but not which exact type of Player 1 belongs at any given time. In other words, we only have $p(t_1) + p(t_2) +. . . p(t_i) + . . . p(t_I) = 1$

Like any game, each player has a set of strategies, in the Cheap-Talk Game, the strategies Player 1 (Sender) is a set of messages $M = \{m_1, m_2, m_3, . . . m_J\}$ to be sent to Player 2 (Receiver). The strategy for Player 2 (Receiver) is a set of feasible actions $A = \{a_1, a_2, a_3, . . . a_K\}$.

The payoffs of both players are given by known functions $U_S(t, a)$ and $U_R(t, a)$.

Note here that payoff of Player 1(Sender), $U_S(t, a)$ does not depend on any of his (her) messages $m_1, m_2, m_3, . . . m_J$, it only depends on the type t that he (she) currently belongs to and depends on the action "a" which he(she) has no control of, because the action "a" is chosen by the other Player 2 (Receiver). Thus, Player 1 (Sender) can only prefer Player 2 to choose, or he is at the mercy of Player 2 to choose the action a so that Player 1's payoff $U_S(t, a)$ is optimal.

Because of the given restriction on the payoff functions $U_S(t, a)$ and $U_R(t, a)$ that they only depend on player 1's type t and player 2's action a, we can see that this is a key feature of Cheap-Talk games, that the message has no direct effect on either player's payoff. There is, however, a message can have an indirectly effect on players' payoff in the sense that because of the message's information content, it could change the Player 2's belief about Player 1's type, and Player 1's message can also change Player 2's action, and so indirectly effect both players' payoffs.

The sequential nature of the actions taken by the three players in a Cheap-Talk game is:

4. Nature picks up a random type of Player 1 (Sender) form T.
5. Player 1 (Sender) observes which type t_I he belongs to and chooses a message m_j from M.
6. Player 2 (Receiver) observes m_j but not t_I then chooses an action a_k from A
7. Players 1 and 2 then receive payoffs $U_S(t, a)$ and $U_R(t, a)$ respectively.

The general timing diagram of the Cheap-Talk game is shown in Figure 8.1 below:

Figure 8.1

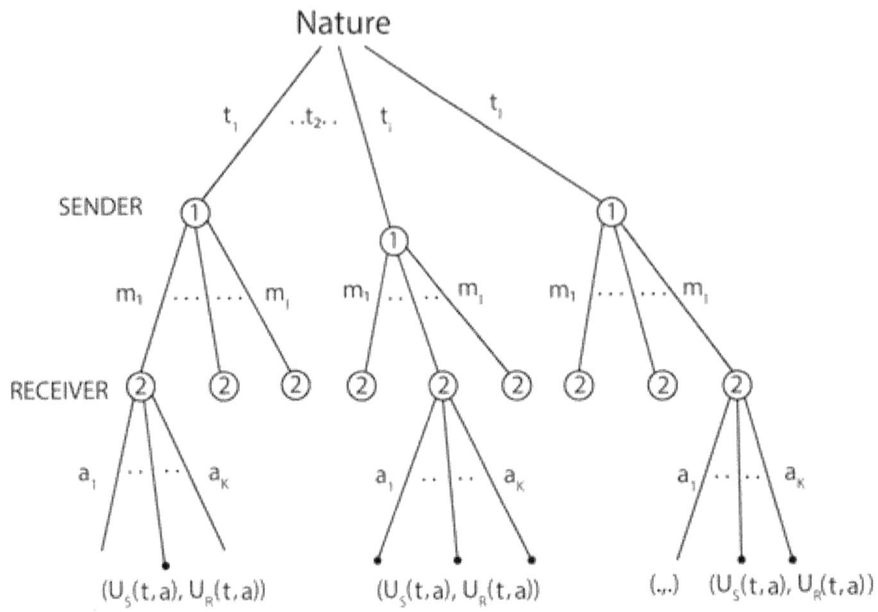

For simplicity, we assume that there are as many messages that Player 1 can send to Player 2 as there are different types for Player 1, and the messages can be thought of as a relabeling of the types; we assume M = T. From the payoff functions, we shall see that in any Cheap-Talk game,

(1) Receiver prefers different actions 'a' depending on the Sender's type 't'
(2) Different Sender type prefers different Receiver's actions.
(3) The receiver's preference of actions cannot be completely opposed to the Sender's preference of actions.

8.2 Perfect Bayesian Equilibrium in a Cheap-Talk game

Given a Cheap-Talk game, our immediate goal is to find the possible perfect Bayesian Equilibrium for the game. Why? Because only in equilibrium can the sender reveal his (hers) type of information to the Receiver and transmit useful information to the Receiver. This important observation will become clear when we discuss the Crawford and Sobel Procedure in the later sections. But for now, let us recall from before in the Signaling game and state clearly what it means to have a perfect Bayesian Nash Equilibrium in a Cheap-Talk game.

Example 8.2.1 Babbling Equilibrium always exists in any Cheap -talk games

A pair of strategies $(m^*(t_I), a^*(m_j))$ and a belief $u(t_i \mid m_j)$ satisfying the four Signaling Requirements as we described earlier constitute a perfect Bayesian Equilibrium. Because of the specific assumption on the payoff $U_R(t, a)$ that it does not depend on any of Player 1's message, we can demonstrate that in any Cheap-Talk game, a Pooling perfect Bayesian equilibrium exits for the Sender to play any pooling strategy m_I for any message in the message set M. This is because when the sender plays the same message m_I, for each type t_1, t_2, . . . t_I, the Receiver can play the single action a* in the receiver action set A such that a* will maximize the Receiver's expected payoff over all the type $t_1, t_2, \ldots t_I$. That is to say a* is the action a_k that maximizes the following expression :

$$\underset{a_k}{Max} \sum_{t_i} p(t_i) U_R(t_i,\ a_k)$$

The degree of belief u (t_i | m_j) is the same as the prior probability $p(t_i)$ for all messages on and off the equilibrium path by the Bayes Theorem as shown before, for instance, in Example 7.4.7. Thus, we see that in a Cheap-Talk game, a Polling perfect Bayesian equilibrium always exists. This is called the Babbling equilibrium in any Cheap-Talk game. Player 2's choice of action does not depend on any specific type of Player 1, and so Player 1 does not reveal any information to Player 2. So, this Babbling equilibrium does not provide any useful information transmission in the Cheap-Talk game. The question is, is there any other perfect Bayesian Equilibrium that is useful for information transmission? The answer is YES, in fact there is a lot, as we will see in the Crawford and Sobel procedure described later. But first, we will present a few examples below.

8.3 Examples of Cheap Talk Games

Example 8.3.2 A Simple Cheap-Talk game

We will refer to Figure 8.2 below for the game tree of this Example 8.3.2 and Example 8.5.4 later.

Figure 8.2

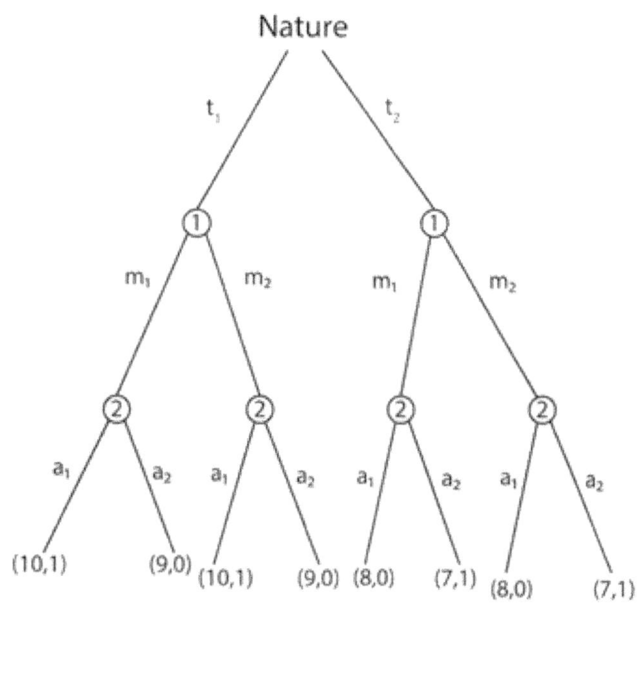

As stated before, in any Cheap Talk game, the payoffs are independent of the messages of the Sender (Player 1), we note that in Figure 8.2 above, the payoffs are identical for the messages m_1 and m_2 .

Assume that there are only two types t_1 and t_2, for Player 1 (Sender), the probability of t_1 is $p(t_1)$, two messages m_1 and m_2, and for Player 2 (Receiver) the action set $A = \{a_1, a_2\}$. Assume the payoffs of the sender $U_S(t, a)$ and receiver $U_R(t, a)$ are given as:

$U_S(t_1, a_1) = 10$ $\quad U_S(t_1, a_2) = 9$ $\quad U_S(t_2, a_1) = 8$ $\quad U_S(t_2, a_2) = 7$

$U_R(t_1, a_1) = 1$ $\quad U_R(t_1, a_2) = 0$ $\quad U_R(t_2, a_1) = 0$ $\quad U_R(t_2, a_2) = 1$

Note that the above payoffs are the same for both message m_1 and m_2. In this simple Cheap-Talk game, we want to know what the possible perfect Bayesian Equilibrium is. From the observation above, we know that the Bubbling Polling equilibrium always exists. But given the specific set of payoffs above, we see that a Separating equilibrium also exists because for type t_1, the Sender can choose message m_1, and for type t_2, the Sender can choose message m_2, the Receiver's belief can be chosen as $u(t_1 \mid m_1) = 1$, $u(t_1 \mid m_2) = 0$, and the Receiver's strategy is to choose $a^*(t_1) = a_1$ and $a^*(t_2) = a_2$.

We note in this simple example that when the Sender's type is t_1, the Receiver prefers the action a_1 and when the Sender's type is t_2. the Receiver prefers the action a_2 because $1 > 0$ from the payoffs of the receiver. As we mentioned earlier, in any Cheap-Talk game, different Sender's type prefers different Receiver's action, it is certainly true in this example. Suppose, on the contrary, both Sender types t_1 and t_2 choose the same Receiver's action a_1, then the Receiver would be led to believe that both Sender types is t_1 which is not possible. We also mentioned before that, in any Cheap-Talk game, the Receiver and the Sender choices of actions cannot be completely opposite. This is also true in the current example. Suppose on the contrary, the Sender's type t_1 prefers the action a_2 and the Sender's type t_2 prefers the action a_1, so that the Sender and the Receiver's choice of actions are completely opposite. This situation would lead the Receiver to believe that the type t_1 is t_2 and type t_2 is t_1 which is also impossible.

Before we give the formulation of the Crawford Sobel Model, let us first look at the preliminary example below. Recall that the payoff functions of the sender and receiver are :

$U_S(t, a) = -(a - (t + b))^2$ and $U_R(t, a) = -(t - a)^2$, where t is a continuous Type Space in the interval [0, 1] and it is uniformly distributed in the interval, the parameter 'b' is the bias of the Sender, 'a' is the response of the Receiver's action, "t" is the choice of the Sender's type. From the behavior of these payoff functions $U_S(t, a)$ and $U_R(t, a)$, the Sender would prefer the Receiver's action a to be close to $t + b$, and the Receiver prefers his action a to match t or as close as possible.

Example 8.3.3

In this example, we assume the bias $b = 0.15$ and see the effect of this value in determining the pooling equilibriums. First assume the Type Space [0, 1] is divided into two equal half intervals, call it the low Type Space I_L and the high Type Space I_H, where $I_L = [0, 0.5]$ and $I_H = (0.5, 1]$, assume I_L is pooled with the message m(L) and I_H is pooled with message m(H), The Receiver will interpret that each message m(L) and m(H) as the expected value of the type t in the interval I_L and I_H respectively. Thus,

$$m(L) = E(t \mid I_L) = \frac{1}{length(I_L)} \int_0^{0.5} t\,dt = 0.25$$, where length(I_L) = length (I_H) = 0.25, and

$$m(H) = E\,(t \mid I_H) = \frac{1}{length(I_H)} \int_{0.5}^{1} t\,dt = 0.75$$

Thus, the Receiver's actions to the message m(L) and m(H) are respectively $a_L = 0.25$ and $a_H = 0.75$

Now, for a specific bias k = 0.15, the Sender's payoff when the message is m(L) and m(H) is respectively:

$U_S(t, a_L) = -(a_L - (t + b))^2$

$= -(0.25 - (t + 0.15))^2$

$= -(t - 0.1)^2$

$U_S(t, a_H) = -(a_H - (t + b))^2$

$= -(0.75 - (t + 0.15))^2$

$= -(t - 0.6)^2$

Thus, the low Type Space I_L chooses to pool with m(L) if $U_S(t, a_L) \geq U_S(t, a_H)$, which is equivalent to $-(t - 0.1)^2 \geq -(t - 0.6)^2$, which is equivalent to $t \leq 0.35$. Therefore, when type $t \leq 0.35$, the Sender chooses to pool with m(L), and when $t > 0.35$, it chooses to pool with m(H).

Similarly, for the high Type Space I_H chooses to pool with m(H) if $U_S(t, a_H) \geq U_S(t, a_L)$, which is equivalent to $-(t - 0.6)^2 \geq -(t - 0.1)^2$, which is equivalent to $t \geq 0.35$.

Thus, we have the following pooling equilibrium:

For types $t \leq 0.35$, Sender pools with message m(L), and Receiver responses with $a_L = 0.25$

For types $t \geq 0.35$, Sender pools with message m(H), Receiver responses with $a_H = 0.75$

8.4 Basic assumptions in the Crawford Sobel Model

Earlier, we stated that in any Cheap-Talk game, the payoff functions $U_S(t, a)$ and $R_S(t, a)$ are not dependent on the messages of the Sender. In the formulation of the Crawford Sobel Model, the model assumes that the types t of the Sender and the actions of the Receiver both are continuous in the interval [0, 1]. That is T = [0, 1] and A = [0, 1]. The message-space and the type-space of the Sender are the same, M = T. Further, the payoff functions not only do not depend on the messages but further take the specific forms:

$U_S(t, a) = - [a - (t + b)]^2$ and $U_R(t, a) = - (a - t)^2$,

where the constant parameter b is called the bias of the Sender toward the Receiver, the Sender's type 't' is assumed to be uniformly distributed over the interval [0, 1]. From these specific function forms, it is easy to see that when the Sender's type is 't', the Receiver's optimal action for 'a' is when $a = t$, but the Sender's optimal action is when $a = t + b$. Thus, different types of the Senders have different preferences over the Receiver's actions, and the preference of choice of actions for both the Sender and the Receiver cannot be completely opposite. Recall that in Example 1, we had demonstrated this. Note that the constant bias b in the model measures the similarity of both the Sender and Receiver's choice of actions. When b is small, closer to 0, both players choose the same action as $a = t + b$ becomes $a = t$ when b approaches 0, which means both Sender's and Receiver's choice of actions are closely aligned.

Based on these assumptions above, we are going to show how to partition the type-space [0, 1] into two pooling equilibria in Example 8.4.2 and partition the type-space [0, 1] into three pooling equilibria in Example 8.4.3.

Example 8.4.4 Special case of the Crawford Sobel Model with two equilibria

In Section 8.2, we showed that in any Cheap-Talk game, there is always a single Pooling Bubbling equilibrium, where for every type of the Sender, the Sender only send a single message to the Receiver, and the Receiver responds with the action 'a' that maximize the expected value of the Receiver's payoff over all the types t of the Sender.

Now, we are going to show that there can be two partial pooling equilibria pending on the value of the bias b. We show the construction of these two equilibria for the Sender and Receiver when the type-space $T = [0, 1]$.

$t_0 = 0$ $\qquad\qquad$ t_1 $\qquad\qquad\qquad\qquad$ $t_2 = 1$

In other words, the type-space [0, 1] is divided into 2 intervals $[0, t_1]$ and $[t_1, 1]$. We want to seek a proper value of t_1 so that there is an equilibrium in the interval $[0, t_1]$ and there is another equilibrium in the interval $[t_1, 1]$. Suppose the Sender only sends one message m_1 for every t in the first interval to the Receiver and only one message m_2 to the Receiver in the second interval. Since the Sender's type is uniformly distributed in the interval $[0, t_1]$, so after the Receiver receiving the message m_1, the Receiver's optimal action 'a' will be given by the action 'a' that maximize the expected value of the Receiver's payoffs over all the Sender's type 't' in the interval $[0, t_1]$, which is given by the following expression:

$$\int_0^{t_1} p(t)U_R(t,a)\,dt$$

As the probability density function $p(t) = 1$ for the uniform distribution of t over the interval [0, 1], and $U_R(t, a) = - (t - a)^2$, so the about integral becomes:

$\int_0^{t_1} -(1)(t-a)^2\,dt$, which is equal to (1/3) t_1^3 - a t_1^2 + a^2 t_1. We want to maximize this integral with respect to the action 'a', so take the derivative of the above expression with respect to 'a', set it to 0, and solve for a, we find: a = $t_1/2$

Similarly, after the Receiver receives the message m_2, the Receiver's optimal action will be the action 'a' that maximizes the expected value of the Receivers' payoff as before; this expected value is given by the integral:

$$\int_{t_1}^{1} -(1)(t-a)^2\,dt,$$

which is equal to 1 - 3a + $3a^2 - t_1^3 + 3t_1^2a - 3t_1a^2$. Thus, to maximize this integral with respect to 'a' by taking the derivative with respect to 'a', set the derivative to 0 and solve for 'a', we get a = ½(t_1 + 1).

To recap what we have done so far. If the Sender's type t is between the interval [0, t_1], the Receiver's best action is a = $t_1/2$, and if the Sender's type t is between the interval [t_1, 1], then the Receiver's best action to respond is

a = ½(t_1 + 1).

Thus, in general, we have the following Lemma 8.1

Lemma 8.4.1

If the Sender's type t is uniformly distributed in the interval between t_a and t_b, then the best response for the Receiver to do is to take the action a given by a = ½(t_a + t_b)

(The proof of this lemma for the general interval between t_a and t_b is like that for the intervals [0, t_1] and [t_1, 1] before.

Now, we want to determine the Sender's best choice of the type t_1 as stated above. But first ,we need to use the following fact summarized in the following Lemma 8.2

Lemma 8.4.2

If the Sender's type t_1 prefers sending message m_2 over message m_1, then any type $t > t_1$ will prefer m_2. Similarly, if t_1 prefers to send message m_1 over m_2, then any type $t < t_1$ will also prefer m_1.

Figure 8.3

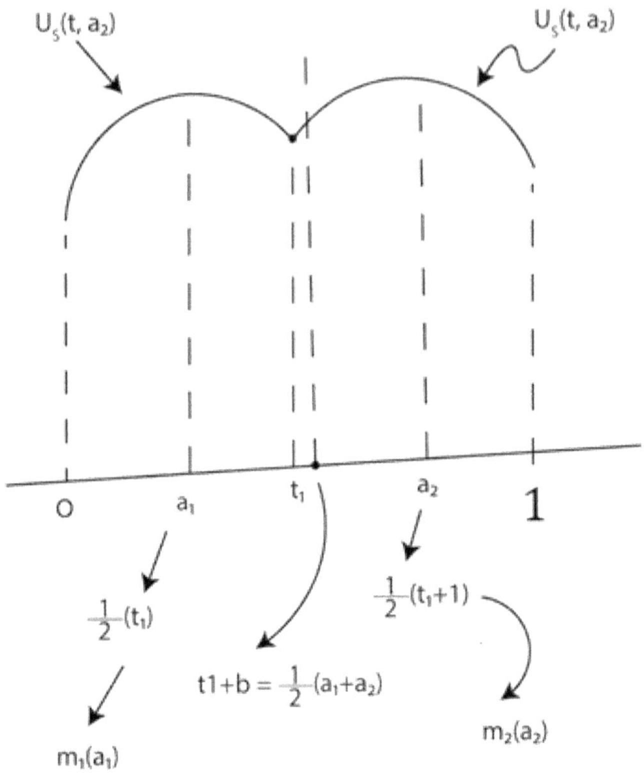

Since the payoff function of the Sender is given by $U_S(t, a) = -(t + b - a)^2$, so the Sender's choice of action is symmetric around his (hers) optimal action t + b. The Sender's type t prefers $a_1 = t_1/2$ to $a_2 = (1 + t_1)/2$ if the mid-point between the two actions a_1 and a_2 is larger than the type's optimal action t + b but prefers the action a_2 to a_1 if $t_1 + b$ is larger than the mid-point. Therefore, the optimal choice of type t for the Sender is when the Sender's optimal action,

$t_1 + b$, is exactly equal to the mid-point of actions a_1 and a_2. That is,

$t_1 + b = ½ \{½ t_1 + ½(1 + t_1)\}$

So, $t_1 = ½(1 - 4b)$

Therefore, for us to have 2 equilibria, t_1 must be greater than 0, which means,

$1 - 4b > 0$ or $b < ¼$

The observation stated above that $t_1 + b$ must be chosen to be the mid-point of actions a_1 and a_2 can also be proven algebraically as follows:

$t_0 = 0$ t $t_2 = 1$

$a_2(m_1)$ $a_2(m_2)$

$U_S(t, a_2(m_1)) = -(t + b - a_2(m_1))^2$

$U_S(t, a_2(m_2)) = - (t + b - a_2(m_2))^2$

If $t + b$ is chosen such that t prefers to send message m_1 over m_2, then,

$U_S(t, a_2(m_1)) > U_S(t, a_2(m_2))$.

Denote the extra gain D by choosing m_1 over m_2 by

$D = U_S(t, a_2(m_1)) - U_S(t, a_2(m_2))$, which is > 0. The derivative

$D = = - (t + b - a_2(m_1))^2 + (t + b - a_2(m_2))^2$ with respect to t is:

$2(a_2(m_1) - a_2(m_2))$, which is < 0. Thus, D is a monotonic decreasing function, and so, D is also > 0 for all $t_1 < t$.

Similarly, if $(t + b)$ is chosen such that t prefers to send message m_2 over m_1, this means that

$U_S(t, a_2(m_1)) > U_S(t, a_2(m_2))$, then this inequality will also be true for all $t_1 > t$.

Thus, to choose t such that it is indifferent in the preference of sending messages m_1 and m_2 we must have, for some $t = t_1$:

$U_S(t_1, a_2(m_1)) = U_S(t_1, a_2(m_2))$, so,

as $a_2(m_1) = - t_1 /2$ and $a_2(m_1) = - (t_1 + 1)/2$,

so, $t_1 + b - t_1 /2 = - (t_1 + b - (t_1 + 1)/2)$

or $t_1 + b = ½ \{t_1 /2 + (t_1 + 1)/2\}$, which means $t_1 + b$ should be chosen as the mid-point of the two actions a_1 and a_2.

Example 8.4.5 Special case of the Crawford Sobel Model with three equilibria

$t_0 = 0$ $\qquad$ t_1 $\qquad$ t_2 $\qquad$ $t_3 = 1$

In this case, the type-space [0, 1] is divided into three intervals, we need to determine the partition types t_1 and t_2 and the actions a_1, a_2, a_3 which are the optimal actions for the Receiver to respond for the Sender's type intervals $[0, t_1]$, $[t_1, t_2]$, and $[t_3, 1]$ respectively. Like what we did in Example 8.4.2, and because Lemma 8.4.1, we find:

$a_1 = ½(t_1)$, that $a_2 = ½(t_1 + t_2)$, $a_3 = ½(t_2 + t_3)$.

Refer to Figure 8.4 below, as we did in Example 8.4.2, the Sender must choose t_1 such that it is indifferent in the preference of actions between a_1 and a_2, and he (she) must choose t_2 such that it is indifferent in the preference of actions between a_2 and a_3.

Figure 8.4

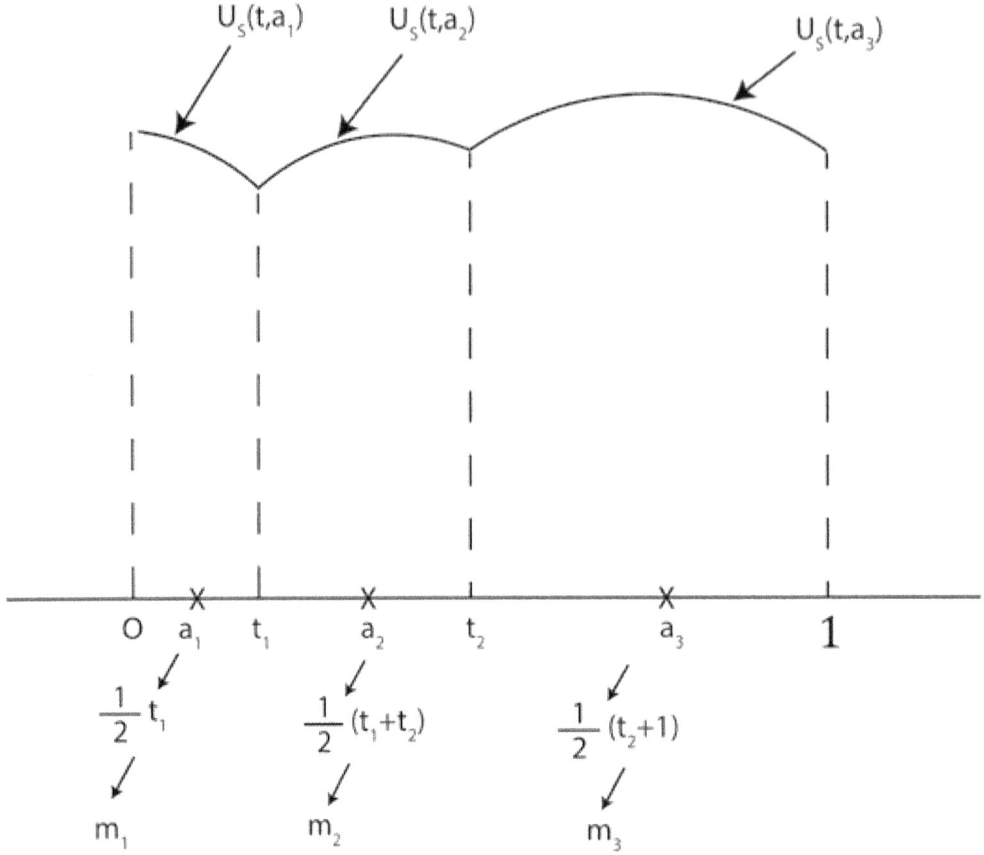

Therefore, we must have:

$t_1 + b = ½\{½(t_1) + ½(t_1 + t_2)\}$ (1)

$t_2 + b = ½\{½(t_1 + t_2) + ½(t_2 + 1)\}$ (2)

From (1), we have: $t_2 = 2t_1 + 4b$ (3)

From (2), we have: $2t_2 = t_1 - 4b + 1$ (4)

From (3) and (4), we get: $t_1 = 1/3(1 - 12b)$ (5)

$t_2 = 2/3(1 - 6b)$ (6)

Therefore, for a 3-equilibria to exist, t_1 must be greater than 0, which means $1 - 12b > 0$, or $b < 1/12$.

8.5 Formulation of the Crawford-Sobel model

Before we give the general formulation of the Crawford Sobel Model, we first establish the following Lemma 8.5.3, which tells us about the relationship between two consecutive intervals in the partitions of the type-space [0, 1].

$t_0 = 0$ t_1 t_2 t_k t_{k+1} t_{N-1} $t_N = 1$

Refer to the partition of the Sender's type-space [0,1] above, there are N-intervals, the first interval from $[t_0, t_1]$, the kth interval $[t_k, t_{k+1})$, and the Nth interval $[t_{N-1}, t_N)$.

Lemma 8.5.3

For any two consecutive intervals $[t_{k-1}, t_k]$ and $[t_k, t_{k+1}]$, if the first interval $[t_{k-1}, t_k]$ has length c, then the second interval $[t_k, t_{k+1}]$ has interval c + 4b.

Proof: As observed in the previous Lemma 8.4.1, the Receiver responded for the first interval $[t_{k-1}, t_k]$ with the action a_{k-1},

$a_{k-1} = ½ (t_{k-1} + t_k)$, and responded for the second interval $[t_k, t_{k+1}]$

$a_k = ½(t_k + t_{k+1})$.

From Lemma 8.4.2, the Sender chooses t_k such that:

$t_k + b = ½(a_{k-1} + a_k)$

Thus, $t_k + b = ½\{½(t_{k-1} + t_k) + ½(t_k + t_{k+1})\}$,

from which we get:

$2t_k + 4b = t_{k-1} + t_{k+1}$, which means $t_{k+1} - t_k = t_k - t_{k-1} + 4b$, or,

$t_{k+1} - t_k = c + 4b$.

The next lemma gives the maximum number of intervals N we can partition the Sender's type-space [0, 1] in terms of the bias 'b' in the Sender's payoff function $U_S(t, a) = -(t + b - a)^2$

Lemma 8.5.4

The maximum number of equilibria for the type-space [0, 1] is given by the largest integer N(b) less than the following real number:*

$r = ½[1 + \sqrt{1 + 2/b}]$

Proof: Let 'l' be the length of the first interval $[0, t_1]$, then by Lemma 9.3 above, the length of the second, third, fourth, … , the Nth interval are:

$l + 4b, l + 8b, l + 12b, \ldots, l + (N - 1)(4b)$.

Since the lengths of all these N intervals sum up to the length of the entire interval [0, 1], so,

$l + l + 4b + l + 8b + \ldots\ldots + l + (N - 1)(4b) = 1,$

$N*l + \{1 + 2 + \ldots. + (N - 1)\}\ (4b) = 1,$

$N*l + N(N - 1)(2b) = 1$

Thus, solving l from the above equation, the length of the first interval is:

$l = \{1 - N(N - 1)\ 2b\}/N.$ (1)

3333333Since for the equilibria to exist, the first interval length l must be larger than 0,

So, $1 - N(N - 1)\ 2b > 0$, which means,

$b < 1/\{2N\ (N - 1)\}.$ (2)

When N = 2, we have b < ¼ and when N = 3, b < ¹⁄₁₂. This is consistent with what we see from Example 2 and Example 3 before.

Also, since $1 - N(N - 1)\ 2b > 0$, we can solve for N from this Quadratic inequality, which gives a real number:

$r = ½[1 + \sqrt{1 + 2/b}]$. Denote N(b) as the largest integer that is less than r. Then N(b) is the maximum number of equilibria we can have for a given bias level b.

$N(b) = ½[1 + \sqrt{1 + 2/b}]$ (3)

, and the length of the first interval l is given from (1) by:

$l = \{1 - N(b)(N(b) - 1)\ 2b\}\ /N(b)$ (4)

Lemma 8.5.5

Refer to the partition points $t_0, t_1, ..., t_{N-1}$ of the type-space [0, 1] below:

$t_0 = 0$ t_1 t_2 t_k t_{k+1} t_{N-1} 1

The location of the ith partition point t_i can be expressed iteratively as the location of the first partition point t_1,

$t_i = t_1(i) + 2b\ (i)\ (i - 1)$ (5)

Proof:

We prove (5) by mathematical induction.

First, when I = 1, then $t_1 = t_1$ from (5), there is only one interval.

When I = 2, (5) gives $t_2 = 2t_1 + 4b$, or $t_2 - t_1 = t_1 + 4b$, which is true by Lemma 8.3. So, (5) is true for i = 1, and 2.

By induction hypothesis, we now assume that (5) is true for I = 1, 2, ... i - 1 and i. That is, we assume that $t_i = t_1(i) + 2b(i)(i - 1)$, $t_{i-1} = t_1(i - 1) + 2b(i - 1)(i - 2)$, and we want to show that:

$$t_{i+1} = t_1(i + 1) + 2b(i + 1)(i) \tag{6}$$

But from Lemma 8.3, we know:

$$t_{i+1} = t_i + (t_i - t_{i-1}) + 4b, \tag{7}$$

and $t_i - t_{i-1} = t_1 + 2b(i - 1)(2)$ from the induction assumption given above.

So, (7) becomes,

$t_{i+1} = t_i + t_1 + 2b(i - 1)(2) + 4b$

$= t_1(i) + 2b(i)(i - 1) + t_1 + 2b(i - 1)(2) + 4b$

$= t_1(i + 1) + 2b(i - 1)(i + 2) + 4b$

$= = t_1(i + 1) + 2b\, i(i + 1)$, which is what we want to show (6).

We are now familiar with the Crawford Sobel Model when the type-space is a continuous interval [0, 1] and the size of the bias 'b' determines the number of equilibria. The next Example 8.5.4 below shows how the size of 'b' can have an effect on the existence of equilibrium when the type-space is discrete and when the payoffs are continuous quadratic function $U_R(t, a)$ and $U_S(t, a)$ as was used in the Crawford Sobel Model.

Example 8.5.6: We repeat Example 8.3.2 when the type-space is {x, y}, where x, y are two real numbers and $x \leq y$, and when the payoff for the Sender is $U_1(t, a) = -(t + b - a)^2$ and for the Receiver is $U_2(t, a) = -(t - a)^2$ instead of discrete values. We want to show a truthful separating equilibrium exists when $2b \leq y - x$ and vice versa.

Proof :

We can refer to Figure 8.2 of Example 8.3.2, where the type t can take on two real numbers, x and y.

First, note that for the Receiver, the best strategy is $a_2 = t$, where t can be x, or t can be y, and in both cases, $U_2(a_2, a_2) = 0$, and note that 0 is the maximum value that player 2 can achieve. For the Sender, when t = x, and when a = x, $U_1(x, x) = - b^2$, and similarly, when t = y and a = y, $U_1(y, y) = - b^2$. Let the two messages of the Sender be a_1^1 and a_1^2 corresponds to the type x and y. We want to show that when the Sender chooses a_1^1 the Receiver chooses x, and when the Sender chooses a_1^2 the Receiver chooses y is a truthful separating equilibrium. This means that, when t = x, the Sender does not want to deviate from choosing the message a_1^1 to the message a_1^2, but this can only happen when the Sender's payoff for the message a_1^1, which is $U_1(x, x)$ is larger than the Sender's payoff for message a_1^2, which is $U_1(x, y)$. This means that, $U_1(x, x) \geq U_1(x, y)$, which means,

$-b^2 \geq - (x + b - y)^2$, which reduces to $y - x \geq 2b$

For the associations (a_1^1, x) and (a_1^2, y) to be a perfect Bayesian equilibrium, we also need to specify the Player 2's or the Receiver's beliefs. In our case,

$P(t = x \mid a_1 = a_1^1) = 1,$

$P(t = x \mid a_1 = a_1^2) = 0,$

$P(t = y \mid a_1 = a_1^2) = 1,$

$P(t = y \mid a_1 = a_1^1) = 0.$

8.6 A MATLAB Program for the Crawford Sobel model

Based on the analysis of Lemma 8.5.4 and Lemma 8.5.5 above, we are going to present a MATLAB Program (myCrawfordSobel1.m) with the source code and some sample output. An application of the Crawford Sobel Model to a discrete scenario is shown in another MATLAB program myCheapTalk4.m

8.6.A myCrawfordSobel1.m

```
myCrawfordSobel1.m:

fprintf('This program performs the following tasks: \n');
fprintf('(1) For a given number of partitions n, it computes the upper bound for the bias b \n');
fprintf('(2) For a given level of bias b, it computes the max partitions n \n');
fprintf('(3) The user input any k, where k < = n, the program computes the (k-1) partition points \n');
fprintf('(4) Show the plots of the payoffs of the Sender and Receiver for any number of equilibria \n');
prompt = "what is the number of partitions you want?";
n = input(prompt)
n;
b = 1/(2*n*(n-1));
fprintf('When the number of partitions n is: %d, the upper bound of the bias b is : %.4f\n', n, b);
prompt = "what is the level of bias you want?";
b = input(prompt)
```

```
b;
r = (1/2)*(1 + sqrt ( 1 + 2/b));
n = floor(r);
fprintf('For a given level of the bias b,%.4f, the maximun number of partition n is %d\n', b, n);
prompt = "Enter the number of intervals you want";
k = input(prompt);
if(k > n) fprintf('*******  STOP!!! The number of intervals you entered is too large !!!  *******\n');
else
    t1 = (1-2*b*k*(k-1))/k;
    fprintf('The first partion point is located at: % .4f \n', t1);
    ti = t1;
    tim1 = 0;
    ai = (tim1 + ti)/2;
    fprintf('The first action for receiver is : % .2f\n', ai);
    tim1 = ti;
    for i = 2:k
        ti = t1* i + 2 *b*i* (i-1);
        fprintf('the next partition point is : % .2f\n',ti);
        ai = (tim1 + ti)/2;
        fprintf('the next action for the receiver is : % .2f\n',ai);
        tim1=ti;
    end
end
```

The above program computes for a given number of intervals N for the partition of the continuous type space [0, 1], the corresponding upper bound of the bias 'b'. For example, when N = 2, 3, 4, 5, 6, 7… the program computes the respective upper bounds for the bias 'b' are 1/4, 1/12, 1/24, 1/40, 1/60, 1/84. The program also computes for a given bias level 'b' the corresponding maximum number of partitions N on the type -space [0,1]. A sample output of the program is shown below:

>> mycrawfordsobel1

This program performs the following tasks:

(1) For a given number of partitions n, it computes the upper bound for the bias b

(2) For a given level of bias b, it computes the max partitions n

(3) The user inputs any k, where k < = n, and the program computes the (k - 1) partition points

(4) Show the plots of the payoffs of the Sender and Receiver for any number of equilibria

What is the number of partitions you want? 7

n = 7

When the number of partitions n is: 7, the upper bound of the bias b is 0.0119

What is level of bias you want? 0.05

b = 0.0500

For a given level of the bias b ,0.0500, the maximum number of partition n is 3

Enter the number of intervals you want 3

The first partition point is located at: 0.1333

The first action for receiver is: 0.07

The next partition point is: 0.47

The next action for the receiver is: 0.30

The next partition point is: 1.00

The next action for the receiver is: 0.73

Based on the receiver's actions computed from the Crawford-Sobel Program above, we can plot the payoff functions of both the Sender and Receiver below.

We make these plots by a MATLAB Program (PlotSenderReceiverCombine.m) with the source code shown below:

8.6.B plotSenderReceiverCombine.m

```
PlotSenderReceiverCombine.m:

n = input('Enter the number of action points for the Receiver:');
for i = 1:n
   ar(i) = input('Enter the action point:');
end
ar
b = 0.05;
t = 0: 0.002:1;
a = ar(1);
ys1 = -(a-t-b).^2;
yr1 = -(a-t).^2;

plot(t, ys1,'*', "MarkerIndices", 1:9:length(ys1))
title('combine plots of Sender and Receiver')
xlabel('continuous type-spae between 0 and 1')
ylabel('payoff values of the Receiver')
```

```
ylim([-0.3 0.0])
hold on
plot(t, yr1)

a = ar(2);

ys2 = -(a-t-b).^2;
yr2 = -(a-t).^2;
plot(t,yr2)
plot(t, ys2,'*', "MarkerIndices", 1:9:length(ys2))
a = ar(3);
ys3 = -(a-t-b).^2;
yr3 = -(a-t).^2;
plot(t, yr3)
plot(t, ys3,'*', "MarkerIndices", 1:9:length(ys3))
hold off
```

A sample program output of plotSenderRecieverCombine.m is shown below: This is the plot of the Sender and Receivers' payoff functions when the Receiver has three action points listed below. In this case, the number of the partition intervals of [0, 1] n is 3, and the receiver's actions are a1 = 1/15, a2 = 3/10, and a3 = 11/15 from the output of myCrawfordSobel1.m

```
>> plotSenderRecieverCombine
Enter the number of action points for the Receiver:  3
Enter the action point:  1/15
Enter the action point:  3/10
Enter the action point:  11/15

ar = 0.0667    0.3000    0.7333
```

The curves with the '*' symbols are the Sender's payoff functions, while the curves without symbols are the Receiver's payoff functions corresponding to the Receiver's action a2 = 1/15, 3/10, and 11/15.

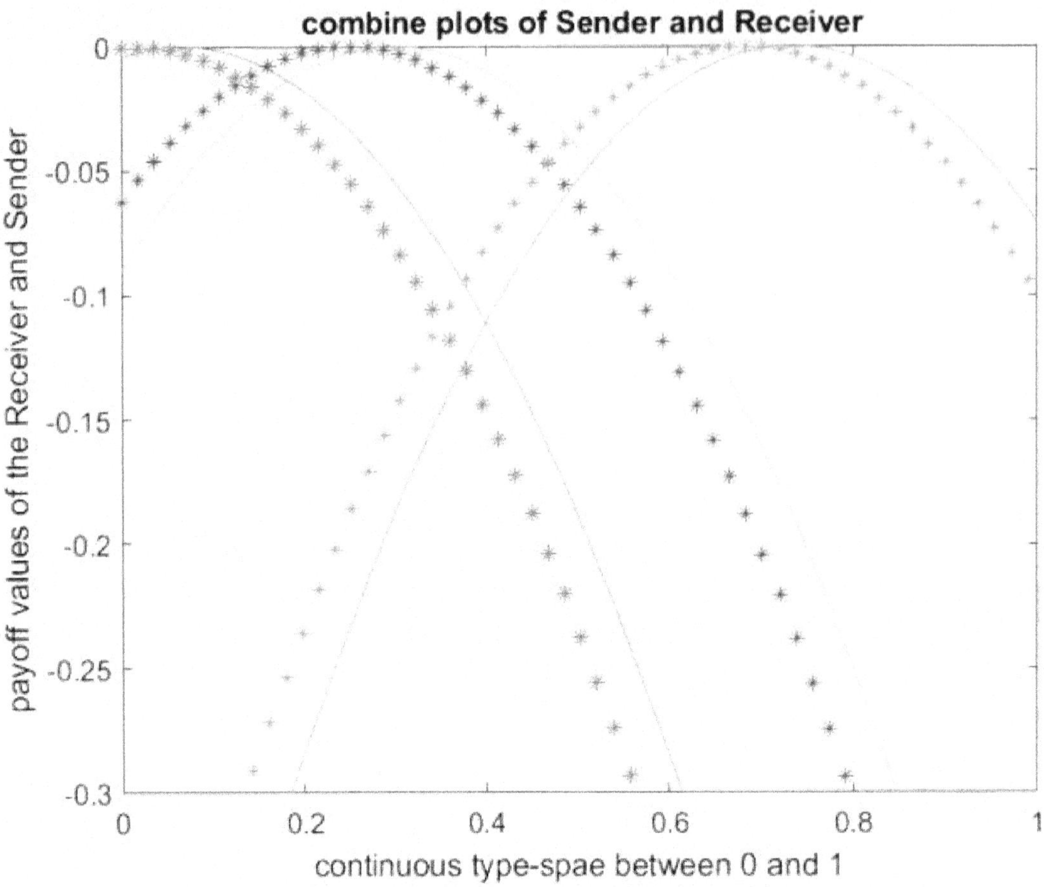

Now we run the program myCrawfordSobel1.m for the same input bias level b = 0.05 as before, but now we take the number of partition intervals of [0, 1] n = 2, the output of the program then gives the receiver's two actions to be a1 = 0.2, a2 = 0.7. The plotSenderRecieverCombine.m then gives us the following plot:

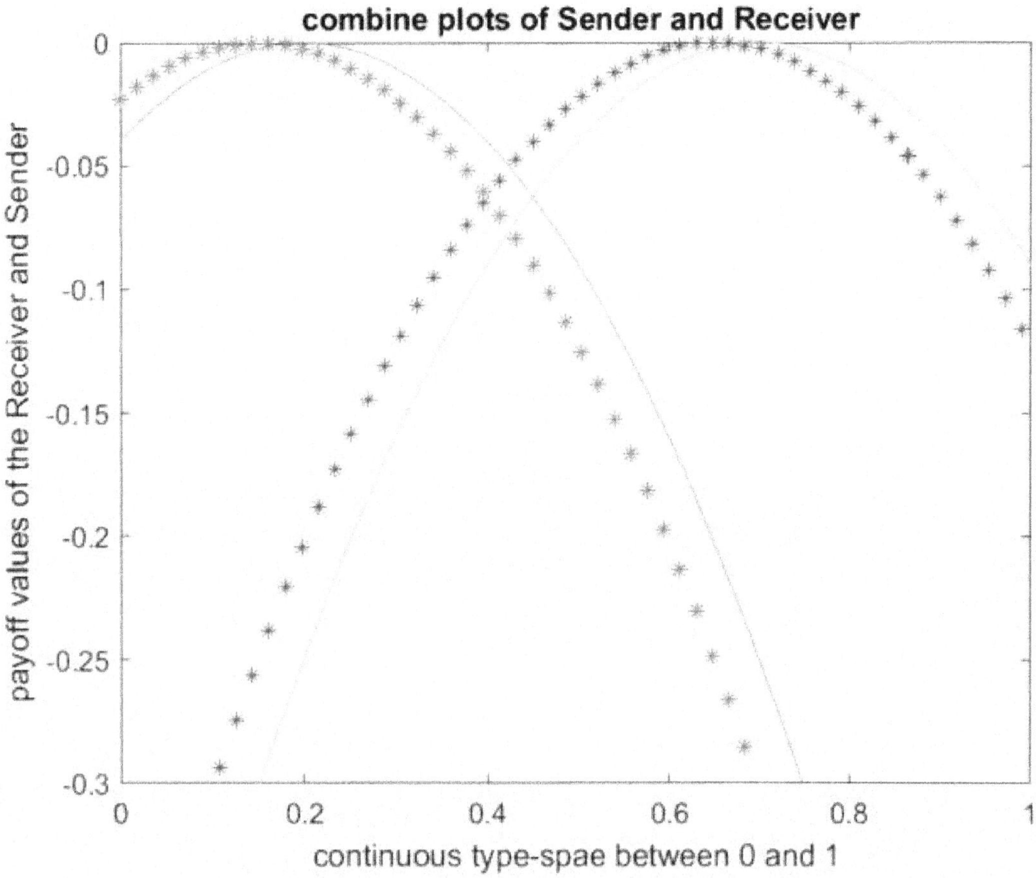

Finally, we run the program myCrawfordSobel1.m for the same input bias level b = 0.05 as before, but now we take the number of partition intervals of [0, 1] n = 1, the output of the program then gives the receiver's only action a1 = 0.5. The plotSenderRecieverCombine.m, in this case, then reduces to two simple curves over the entire type-interval [0, 1], which coincides with our analysis in Lemma 8.2 and Lemma 8.4.1 earlier.

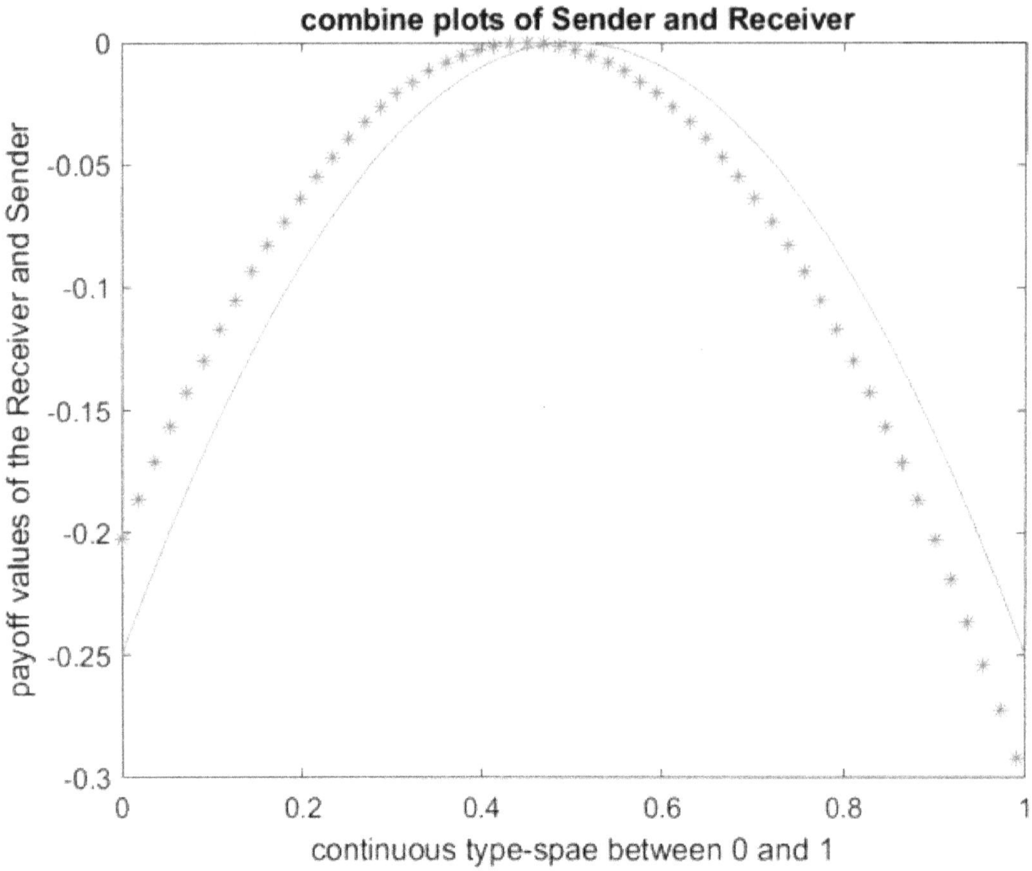

The interested reader can run the myCrawfordSobel1.m with the bias input b = 0.0119 (this is the maximum bias level for partitioning [0, 1] into 7 intervals) and verify that, if we partition [0, 1] only into 6 intervals, the receiver will response with the actions: 0.02, 0.10, 0.21, 0.38, and 0.60.

In our last example of the chapter, we are going to look at the Crawford Sobel model discussed before from a very general discrete scenario in which the type-space, the actions of the Sender and the Receiver are all represented by discrete Real or Integer values.

Example 8.6.5 In this example, we assume the type-space t is {1, 3, 5, 7}, and the actions a_2 that Player 2 (the Receiver) can take is {1, 2, 3, 4, 5, 6, 7, 8}, and the bias b = 1.2. Recall that there is no direct action that Player 1 (the Sender) can choose to maximize his payoff. The Sender can only prefer which action the Receiver chooses so that he (the Sender) can maximize his payoff. Suppose the payoff the Sender and Receiver are $U_R(a, t) = 5 - (t - a)^2$ and $U_S(a, t) = 5 - (t - a + b)^2$ We want to show that there exists a perfect Bayesian equilibrium with two equilibrium: When t = 1, Receiver selects a = 1, and when t = 3, 5, or 7, the Receiver selects a = 5.

Proof:

Since we need to compute the payoff functions for both the Sender and Receiver for each of the type t = {1, 3, 5, 7} and the actions of the receiver a_2 = {1, 2, 3, 4, 5, 6, 7, 8} for the evaluation of the perfect Bayesian equilibria in this example, we write the following MATLAB Program myCheapTalk4.m for this purpose and at the same time it can be used as a tool to solve any Cheap-Talk problem in discreate cases.

Inputs to this myCheapTalk4.m are the discrete type values theta, discrete values a1 of the Sender, the discrete values a2 of the actions of Receiver, the bias b, and the payoff function constant pc. The program outputs the payoff matrices, v1arraypayoff, v2arraypayoff for both players. Source codes of myCheapTalk4.m is listed below:

8.6.C myCheapTalk4.m

```
myCheapTalk4.m:

fprintf('This program computes two possible equilibriums from the user inputs:     \n');
    l1 = input('Enter the length of theata:');
for i = 1: l1
    theata(i) = input('Enter your theata type values:');
end
theata
l2 = input('Enter the length of a1:');
for i = 1:l2
    a1(i) = input('Enter your a1 values:');
end
a1
l3 = input('Enter the length of a2:');
for i = 1:l3
    a2(i) = input('Enter your a1 values:');
end

a2

prompt = "Input the bias level:       ";
bias = input(prompt)

bias;

prompt = "Input the constant for the payoff functions:";
pc = input(prompt)
pc

length(theata)


prob_theata = 1/length(theata);

for k = 1:length(theata)
```

```
for i = 1:length(a1)

    v1arraypayoff(k,1) = theata(k);
    v2arraypayoff(k,1) = theata(k);

    for j = 1:length(a2)
        v1 = payoff1(a2(j), theata(k), bias, pc);
        v2 = payoff2(a2(j), theata(k), pc);


        v1arraypayoff(k,j+1) = v1;

        v2arraypayoff(k,j+1) = v2;

        end

    end
end

fprintf('now printing out the payoff for both player 1 and player 2:  \n');

    for j = 1:length(a2)

        sum=0;
        for k= 1: length(theata)
        sum = sum + prob_theata * payoff2( a2(j), theata(k), pc );
        end

        expected_value2(j)= sum;

end


v1arraypayoff
v2arraypayoff


fprintf('For each type, optimal actions that player 1 and player 2 prefer are as follows: \n');

    for k=1: length(theata)
        [value1, idx1]= max ( v1arraypayoff( k, 2:length(a2) + 1));
        [value2, idx2] = max ( v2arraypayoff (k, 2:length(a2) + 1));
        fprintf('when the type is %d, optimal action for player 1 is %d, optimal action for player 2 is
        %d\n', theata(k), idx1, idx2);

    end

expected_value2

[val,idx] = max(expected_value2);

fprintf('Player 2 has the maximum expected value %.2f when he chooses strategy %d\n' ,val, idx);
```

```
fprintf('This is for the bubbling equilibrium');
fprintf('****** \n');
fprintf('Next, we want to find the second equilibrium as follows: \n');

    prob_theata =1/(length(theata) - 1);

    for j = 1:length(a2)

        sum = 0;
        for k = 2:length(theata)
        sum = sum + prob_theata * payoff2(a2(j), theata(k), pc);
    end

        expected_value2(j) = sum;

end

expected_value2
[val,idx] = max(expected_value2);
fprintf('Player 2 has the maximum expected value %.2f when he chooses strategy %d\n' ,val, idx);

function v1 = payoff1(a2, theata, bias,pc)
v1 = pc - (theata + bias - a2)^2;
end

function v2 = payoff2(a2, theata ,pc)
v2 = pc - (theata - a2)^2;

end
```

In our current example, we input theta = {1, 3, 5, 7}, a1 = {1, 3, 5, 7}, a2 = {1, 2, 3, 4, 5, 6, 7, 8}, b = 1.4, pc = 10.

The program output is shown below:

>> mycheaptalk4

This program computes two possible equilibriums from the user inputs:

Enter the length of theta: 4

Enter your theta type values: 1

Enter your theta type values: 3

Enter your theta type values: 5

Enter your theta type values: 7

theata = 1 3 5 7

Enter the length of a1: 4

Enter your a1 values: 1

Enter your a1 values: 3

Enter your a1 values: 5

Enter your a1 values: 7

a1 = 1 3 5 7

Enter the length of a2: 8

Enter your a1 values: 1

Enter your a1 values: 2

Enter your a1 values: 3

Enter your a1 values: 4

Enter your a1 values: 5

Enter your a1 values: 6

Enter your a1 values: 7

Enter your a1 values: 8

a2 = 1 2 3 4 5 6 7 8

Input the bias level: 1.4

bias = 1.4000

Input the constant for the payoff functions: 10

pc = 10

pc = 10

ans = 4

Now, we print out the payoff for both player 1 and player 2:

v1arraypayoff =

1.0000	8.0400	9.8400	9.6400	7.4400	3.2400	-2.9600	-11.1600	-21.3600
3.0000	-1.5600	4.2400	8.0400	9.8400	9.6400	7.4400	3.2400	-2.9600
5.0000	-19.1600	-9.3600	-1.5600	4.2400	8.0400	9.8400	9.6400	7.4400

7.0000 -44.7600 -30.9600 -19.1600 -9.3600 -1.5600 4.2400 8.0400 9.8400

v2arraypayoff =

1 10 9 6 1 -6 -15 -26 -39

3 6 9 10 9 6 1 -6 -15

5 -6 1 6 9 10 9 6 1

7 -26 -15 -6 1 6 9 10 9

For each type, optimal actions that player 1 and player 2 prefer are as follows:

when the type is 1, optimal action for player 1 is 2, optimal action for player 2 is 1

when the type is 3, optimal action for player 1 is 4, optimal action for player 2 is 3

when the type is 5, optimal action for player 1 is 6, optimal action for player 2 is 5

when the type is 7, optimal action for player 1 is 8, optimal action for player 2 is 7

expected_value2 =

-4 1 4 5 4 1 -4 -11

Player 2 has the maximum expected value 5.00 when he chooses strategy 4

This is for the bubbling equilibrium ******

Next, we want to find the second equilibrium as follows:

expected_value2 =

-8.6667 -1.6667 3.3333 6.3333 7.3333 6.3333 3.3333 -1.6667

Player 2 has the maximum expected value 7.33 when he chooses strategy 5

Some interpretation of the above program output is in order. This program saves us some efforts to compute the Sender and the Receiver's payoff values needed for the analysis of perfect Bayesian equilibrium. These two matrices are shown in the array v1arraypayoff and the v2arraypayoff:

V1arraypayoff:

1.0000 8.0400 9.8400 9.6400 7.4400 3.2400 -2.9600 -11.1600 -21.3600

3.0000 -1.5600 4.2400 8.0400 9.8400 9.6400 7.4400 3.2400 -2.9600

5.0000 -19.1600 -9.3600 -1.5600 4.2400 8.0400 9.8400 9.6400 7.4400

7.0000 -44.7600 -30.9600 -19.1600 -9.3600 -1.5600 4.2400 8.0400 9.8400

V2arraypayoff:

1	10	9	6	1	-6	-15	-26	-39
3	6	9	10	9	6	1	-6	-15
5	-6	1	6	9	10	9	6	1
7	-26	-15	-6	1	6	9	10	9

In both payoff matrices above, the 1st column is the Sender's type, and the 2nd to the 9th column are the actions of the Receiver, we call the Receiver's action names 1,2,3.... 8 by their actual values, which is also assumed to be 1, 2, 3...8 for simplicity.

The program outputs the bubbling equilibrium when the Receiver chooses the action a2 = 4. This is the strategy that the Receiver will choose assuming that he does not know any information about the type theta other than its uniform probability distribution, that is P(theta) = ¼ and a2 must be the strategy that the Receive will choose to maximize his expected payoff. The code in the program computes the expected value of each of the Receiver's actions and selects the action that gives the maximum expected value. When a2 = 4, the Receiver has the highest expected value.

The program selects a second equilibria by first examining the Receiver's payoff matrix V2arraypayoff:

When theta = 1, the first row of V2arraypayoff shows when a2 = 1 gives the Receiver the largest payoff of 10.

When theta = {3, 5, 7}, the program finds from the payoff matrix V2arraypayoff that when the Receiver selects action a2 = 5 will give him the largest expected value 7.33. So, in summary, when theta = 1, the best action for the Receiver is a2 = 1; when theta = {3, 5, 7} the best action for the Receiver is a2 = 5. To have an equilibrium, we must verify that this is also the Sender's optimal strategy. Yes, this is the case by examining the Sender's payoff matrix V1arraypayoff.

When theata = 1, the payoff of the Sender is 8.04 when a2 = 1, and the payoff of the Sender is 3.24 when a2 = 5, so the Sender will prefer the choice of a2 = 1 rather than a2 = 5.

When theta = {3, 5, 7}, the payoff of the Sender is {9.64, 8.04, -1.56} respectively when a2 = 5, which is larger than {-1.56, -19.16, -44.76} respectively when a2 = 1, so the Sender will prefer the choice of a2 = 5 rather than a2 = 1.

Finally, we note that the size of the bias 'b' can also affect the existence of the equilibriums as we saw in the original continuous case of the Crawford Sobel Model. The pair of strategies stated above in this example is no longer an equilibrium when the bias 'b' is selected b = 2.4 rather than b = 1.4. To see this, we simply run myCrawfordSobel4.m by holding all inputs the same except changing the bias to b = 2.4. The Sender's payoff matrix becomes :

v1arraypayoff =

1 -0.7600 3.0400 4.8400 4.6400 2.4400 -1.7600 -7.9600 -16.1600

3 -14.3600 -6.5600 -0.7600 3.0400 4.8400 4.6400 2.4400 -1.7600

5 -35.9600 -24.1600 -14.3600 -6.5600 -0.7600 3.0400 4.8400 4.6400

7 -65.5600 -49.7600 -35.9600 -24.1600 -14.3600 -6.5600 -0.7600 3.0400

In this case, the Sender would prefer the Receiver to choose a2 = 5 rather than a2 = 1 as 2.44 is larger than -0.76. So, the original pair of strategies cannot be in equilibrium. This brings us to an interesting question: Is there an upper bound b* for the bias b so that when b > b*, an existing equilibrium will no longer be an equilibrium? Before coming up with an analytical solution, the interested reader can run the MATLAB PROGRAM myCheapTalk4.m with various input choices of values of b to confirm a reasonable numerical solution.

8.7 Summary

After we explained the concept of Perfect Bayesian Equilibrium in Cheap-Talk games, we presented a powerful tool, the Crawford Sobel model, to systematically find these Perfect Bayesian Equilibria in specific Cheap-Talk games where the payoff functions for the sender and receiver are of the form $U_S(t, a) = -[a - (t + b)]^2$ and $U_R(t, a) = -(a - t)^2$, where t is the type-space of the sender, and a is the action response of the receiver. The main content of the Crawford Sobel model is summarized in the five lemmas: Lemma 8.4.1, Lemma 8.4.2, Lemma 8.5.3, Lemma 8.5.4, and Lemma 8.5.5. In the last section, Section 8.6, three MATLAB programs, myCrawfordSobel1.m, plotSenderReceiverCombine.m, and myCheapTalk4.m, were presented. The myCrawfordSobel1.m was written based on the information from the five lemmas above, this program can be used as a computation tool to analyze Cheap-Talk games that describe various scenarios in general; the plotSenderReceiverCombine.m presents the results obtained from myCrawfordSobel1.m in an informative graphical way; the myCheapTalk4.m is a discrete version of the original Crawford Sobel model, it is written to find up to 2 equilibria for discrete input data from the sender and the receiver, the program outputs the discrete payoff matrices for both players and the shows the perfect Bayesian Equilibrium if any.

Key technical takeaway of the Crawford-Sobel Model discussed in this chapter:

- In cheap talk games with conflicting interests, equilibrium communication is typically partially informative and coarse meaning that the sender's messages do not fully reveal the true state, especially when preferences are misaligned. The Crawford-Sobel model shows that strategic communication under conflict results in equilibria where information is selectively revealed, often in a "coarse" manner, depending on the degree of bias. Specifically, there is a trade-off between informativeness and incentives in the sense that the model captures how strategic incentives prevent fully truthful communication, leading to equilibrium messages that balance the sender's desire to influence the receiver against the cost of misrepresentation.

- Partial pooling equilibria: The sender partitions the state space into intervals and sends the same message for all states within each interval, resulting in coarse or blurry communication rather than perfect revelation.

- Dependence on preference alignment: The degree of informativeness depends on the bias parameter b, which measures the divergence in preferences between sender and receiver. When preferences are perfectly aligned (b = 0), full revelation (truth-telling) is an equilibrium. As b increases, equilibria become coarser, with more pooling.

Chapter 9

Game Theory and Feedback Control System

9.0 Introduction

In this Chapter, we demonstrate how classical control systems in Engineering can be formulated as a zero-sum two-person game, as we discussed in previous chapters. This formulation allows us to translate the problem of finding an optimal control strategy into the **problem of determining optimal strategies** for two players in a **properly constructed game-theoretic framework**.

Naturally, we ask: Who are the two players in this game? And what are their payoff values? Before addressing these questions, and without delving into the technical details of control system theory—which is a vast topic of its own—we provide a top-level block diagram representation. This will illustrate how the system's goal can be expressed as each player selecting an appropriate strategy in a two-person game.

In the present formulation, we introduce the H^{∞} method, along with the concept of H∞ (H infinitive) norm of a matrix, and that of a transformation from l-dimensional space R^{l} to a q-dimensional space R^{q}. The H^{∞} method in control theory was first introduced in:

[1] (G. Zames,1981),

[2] Bruce A. France (1987)

[3] T. Basar, P. Bernhard (2008).

[4] MATLAB Robust Control Toolbox.

By the end of this chapter, readers will see how game-theoretic principles provide a natural framework for optimal control problems, bridging the fields of engineering and strategic decision-making. In Example 9.7.3, we use numerical data to show how the MATLAB build-in function "hinfsyn" ([14]) is used to perform the H-infinitive optimal control synthesis of a control system in general.

9.1 A quick look at Feedback Control System

A typical State- Space Design of a Control System can be represented by the following equations:

$$\dot{X} = AX + Bu \qquad (1)$$

$$y = CX + Du \tag{2}$$

where X is the state vector, a (n X 1) matrix, $\dot{X}$ is its time derivative.

Y = output signal (k X 1) column matrix.

u = control signal (m X 1) column matrix.

A is a n X n real- or complex- valued matrix.

B = n X m real- or complex- valued matrix.

C= k x n real- or complex- valued matrix.

D = kX m real-or complex -valued matrix .

The control signal u is chosen as

$$u = -KX \tag{3}$$

where K is a m x n state feedback gain matrix yet to be determined from the system specification.

A block diagram of the control system described by the equations above (1) , (2), and (3) is shown below:

Figure 9.1

State - Space Representation

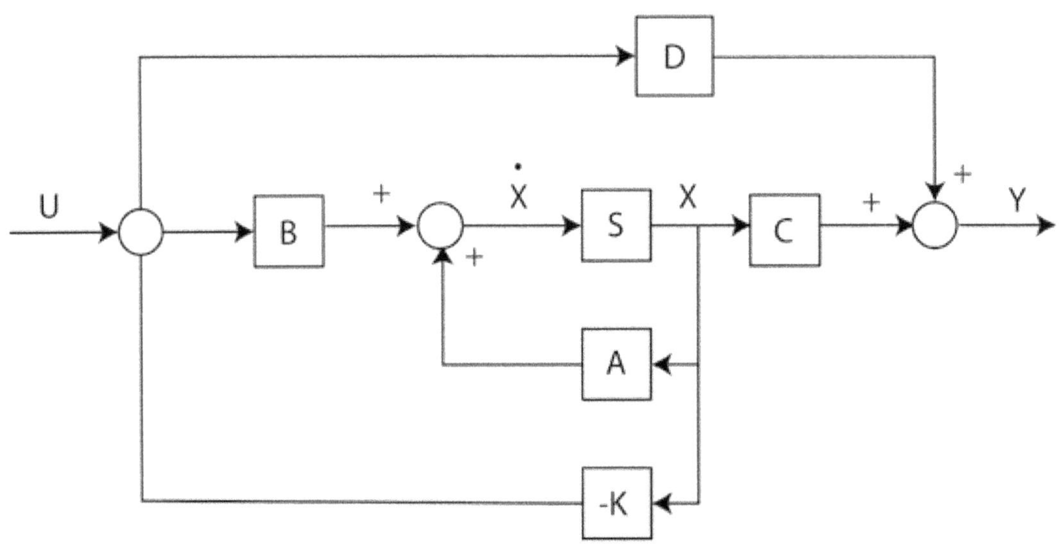

A Control System can also be represented by transfer functions as shown in the diagram below:

Figure 9.2

Transfer - Function Representation

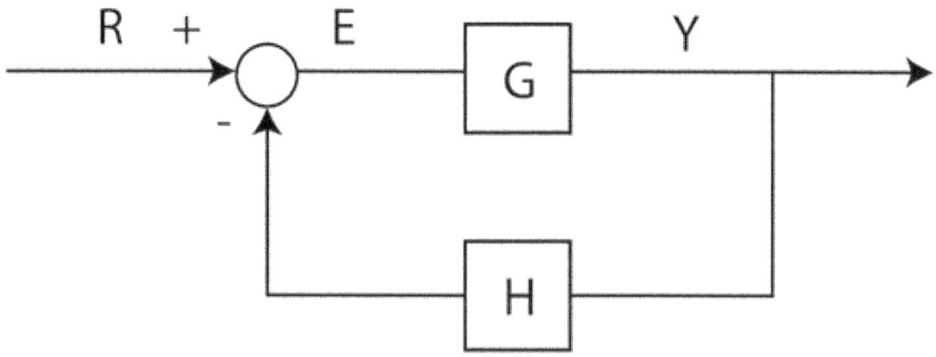

Where the blocks G, H are matrices of transfer functions, these transfer functions are referred to as the feed forward and backward transfer functions for the Plant and its Feedback Loop, respectively. The variable R is referred to as the system reference input, E is the system error, and Y is the system output.

From the diagram above, the following equations hold:

$$E = R - H\,Y \tag{4}$$

$$Y = G\,E \tag{5}$$

Where R is a (nX1) column vector, Y is a (kX1) column vector, G is a (kxn) matrix, and H is a (nxk) matrix, both are of real or complex values.

Now we can derive the transfer function between the system error and reference input as follows:

Substitute Y from (5) into (4), we get: $E = R - HGE$, so $E(I + HG) = R$, so

$$E/R = 1/(I + HG) \tag{6}$$

To derive the transfer function between the output and reference input as follows:

Multiply (4) both sides by the matrix G, we get,

$$G\,E = G\,R - G\,H\,Y$$

Substitute the term GE by Y from equation (5), we have:

$$Y = GR - GH\,Y$$

So, $$Y(I + GH) = GR$$

So, $$Y/R = G/(1 + GH) \tag{7}$$

Expressions (6) and (7) are very important in Control Theory, they allow us to compute system error and output given the system input.

A fundamental problem in classical control theory is that for a given plant G. We want to find a feedback controller K (H = K in the above diagram) with input U(reference input R in the diagram above), U = KY such that the so-called sensitivity transfer function T given by:

$$T = I/(I + GK) \tag{8}$$

will be insensitive against model errors in G, and it will also be desirable for the complementary sensitivity function T_c defined by:

$$T_c = GK/(I + GK) \tag{9}$$

to be insensitive to measurement error.

Figure 9.3

Control System with Disturbance & Noise

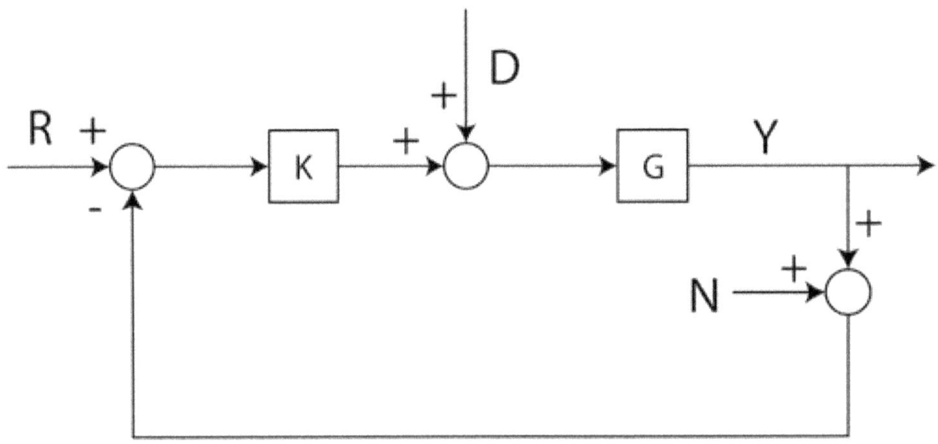

To find an operator T with these properties described and to be able to quantify the sensitivity level of T, we need to create a meaningful numerical quantity attached to T, call it the infinite norm of T, denoted by $<T>_{\infty}$ or by $<< T >>$. The exact definition of $<T>_{\infty}$ is given below.

9.2 What is the H-infinitive norm of a transformation T associated with a transfer matrix M?

Suppose we are given a (mxn) transfer matrix M, then we can view M as a transformation from n-dimensional Euclidean space to a m-dimensional Euclidean space, we denoted this transformation by T_M

$T_M : R^n \rightarrow R^m$ and $z = T_M(d)$

That is T_M takes a (nx1) column vector $d(t) \in R^n$ into a (mx1) vector $z(t) \in R^m$ by this transfer matrix M. That is z = M* d is the usual matrix multiplication between the matrix M and the column vector d resulting in a column vector z.

We use the well-known L^2 norm defined below for the vectors d and z in the time domain as follows:

$$\| z \| = [\int_{-\infty}^{\infty} | z(t) |^2 \, dt \,]^{1/2}$$

and $\| d \| = [\int_{-\infty}^{\infty} | d(t) |^2 \, dt \,]^{1/2}$

Then the supremum of all the ratios $\underset{d}{Sup} \frac{\| z \|}{\| d \|}$ is called the H^∞- norm of the transformation T associated with the transfer matrix M.

We denote the H^∞- norm of the transformation T associated with the transfer matrix M as:

$$<< T_M >> = \underset{d}{Sup} \frac{\| z \|}{\| d \|} \tag{10}$$

Now, we define the H^∞- norm of the transfer matrix M as follows:

Since M is a (mxn) matrix, so $M^t M$ is a square (nxn) square positive definite matrix, we can then compute the largest eigenvalue of the matrix $M^t M$, denoted this eigenvalue by $\sigma(M)$, define << M >> to be $\sigma(M)$. Assume the matrix components of M are rational functions in s with all their poles in the left-half complex plane, and therefore, they are differentiable at every point P in the right-hand-plane and differentiable at every point within a neighborhood of P. We define the H^∞ norm of the transfer matrix M as

$\| M \|_\infty = \underset{\omega}{Sup} << M(j\omega >>$ where ω could range all over R

To recap, we have thus defined two norms, one for the transfer matrix M(s), denoted by $\| M \|_\infty$ and the other for the transformation T induced by the matrix M, denoted by $<< T_M >>$.

A very useful fact for our formulation of a control system in terms of a two-person zero sum game is that the two norms defined above turn out to be equal. That is to say, $\ll T_M \gg = \| M \|_\infty$ Note that, intuitively, $\| M \|_\infty$ is computable in terms of getting the eigenvalue of some matrix, whereas $\ll T_M \gg$ is expressed in terms of the ratio of two L^2 norms, then taking the supremum of an infinitely many ratios are something not easily manageable. We shall, however, use the concept of $\ll T_M \gg$ to formulate our two-person game in the next section. We are going to prove this important and fundamental fact summarized in Lemma 9.1 below. But first, we need to state and understand the need for Parseval's Identity in the proof.

Lemma 9.0: Parseval's Identity :

Let d(t) and z(t) be the disturbance and output functions, and let T_M be the linear transformation induced by the (m x n) transfer matrix M(s) as described before,

$$z = T_M d$$

Let $F_{d(t)}(\omega)$ and $F_{z(t)}(\omega)$ be the Fourier Transform of the function d(t) and z(t) in the time domain t to the frequency domain ω. Then the following is true:

$$\int | z(t) |^2 \, dt = \int | F_{z(t)}(\omega) |^2 \, d\omega = \int | M(j\omega) F_{d(t)}(j\omega) |^2 \, d\omega$$

Note that the integrand $| M(j\omega) F_{d(t)}(j\omega) |^2$ can be written as:

$$| M(j\omega) F_{d(t)}(j\omega) |^2 = F'_{d(t)}(-j\omega) M'(j\omega) M(j\omega) F_{d(t)}(j\omega)$$

$$\leq \sigma(M(j\omega))^2 F'_{d(t)}(-j\omega) F_{d(t)}(j\omega)$$

because $M'(j\omega) M(j\omega) \leq \sigma(M(j\omega))^2$ by definition of $\sigma(M(j\omega))$

$$\leq \sigma(M(j\omega))^2 | F_{d(t)}(j\omega) |^2$$

$$\leq \| M \|_\infty^2 | F_{d(t)}(j\omega) |^2 \tag{11}$$

Lemma 9.1: $\ll T_M \gg = \| M \|_\infty$

We shall demonstrate the above equality in two steps.

Step 1: To show $\ll T_M \gg \leq \| M \|_\infty$

$\| z \|^2 = \int | z(t) |^2 \, dt = \int | M(j\omega) F_{d(t)}(j\omega) |^2 \, d\omega$ by Parseval's Identity.

By (11), $\| z \|^2 \leq \| M \|_\infty^2 \int | F_{d(t)}(j\omega) |^2 \, d\omega$

$$= \| M \|_\infty^2 \int | d(t) |^2 \, dt$$

because $F_{d(t)}$ is the Fourier transform of d(t)

$= || M ||_{\infty}^{2} || d ||^2$

So, $\frac{|| z ||}{|| d ||} \leq || M ||_{\infty}$ (12)

Since (12) holds true for any d, so it holds true for the supremum over d,

so $<< T_M >> \leq || M ||_{\infty}$

Step 2: To show $<< T_M >> \geq || M ||_{\infty}$

The proof, goes by first assuming that the maximum eigenvalue $|| M ||_{\infty}$ is greater than some arbitrary real number γ, and then we prove that $<< T_M >>$ is also greater than or equal to γ. As $|| M ||_{\infty}$ is always greater than or equal to $<< T_M >>$ from step 1, as we choose γ approaching $|| M ||_{\infty}$, we would have shown that $<< T_M >> \geq || M ||_{\infty}$.

So, let us start by assuming $|| M ||_{\infty} > \gamma$. This means that there is some frequency Ω_0 such that σ M(jΩ_0) > γ. Since the function σ $M(j\omega)$ is a continuous function in ω, so there must exist a small interval of length η surrounding the frequency $\Omega_{0,}$ and $-\Omega_{0,}$ such that σ M(jΩ_0) $\geq \gamma$ for all ω in the frequency interval A = $[\Omega_0 - \eta, \Omega_0 + \eta]$ and B = $[-\Omega_0 - \eta, = -\Omega_0 + \eta]$. Now select a disturbance $d^*(t)$ such that $F_{d^*(t)}(j\omega)$ is 0 when ω is outside these two intervals A and B and that inside the two intervals A and B, $M'(-j\omega)\, M(j\omega)$ gives the largest eigenvalues. Then, the output z that corresponds to the disturbance d* under the transformation T_M can be computed as follows:

$$||z|| = \int_{-\infty}^{\infty} | z(t) |^2 \, dt = \int_{-\infty}^{\infty} | F_{z(t)}(j\omega) |^2 \, d\omega = \int_A | F_{z(t)}(j\omega) |^2 \, d\omega + \int_B | F_{z(t)}(j\omega) |^2 \, d\omega$$

$$\geq \int_A \gamma^2 \, | F_{d(t)}(j\omega) |^2 \, d\omega + \int_B \gamma^2 \, | F_{d(t)}(j\omega) |^2 \, d\omega$$

$$= \gamma^2 \{ \int_A | F_{d(t)}(j\omega) |^2 \, d\omega + \int_B | F_{d(t)}(j\omega) |^2 \, d\omega \}$$

$$\geq \gamma^2 \int_{-\infty}^{\infty} | F_{d(t)}(j\omega) |^2 \, d\omega$$

$$= \gamma^2 \int_{-\infty}^{\infty} | d(t) |^2 \, dt$$

$$= \gamma^2 \, || d ||^2$$

So $\frac{|| z ||}{|| d ||} \geq \gamma$. Therefore, $<< T_M >> \geq \gamma$.

9.3 *H-infinitive norm of a controller*

To recap, we have defined the H^∞- norm of a transformation T_M from R^n to R^m associated with the transfer (mxn) matrix M. We now apply this infinitive norm concept to the control system shown in the following block diagram:

Figure 9.4

Control System with Disturbance and Control Input

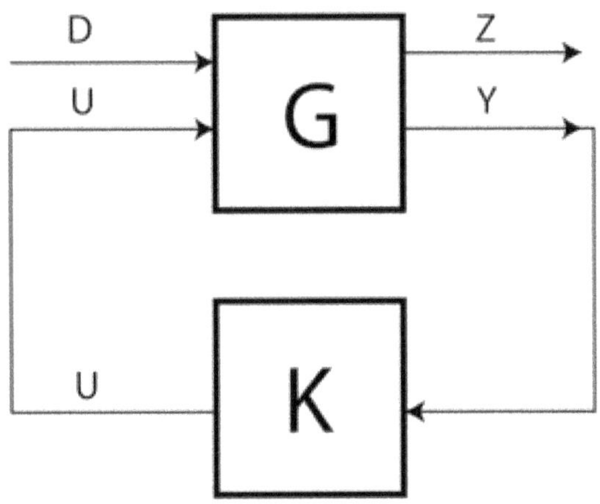

In the diagram above, d is the uncontrolled input disturbance, u is the controlled input, z is the controlled output, and y is the measured output. Note that for convenience, we have used the small letters for the variables d, u, z, y for D, U, Z, Y from the diagram, while the blocks G and K will always remain capital letters.

Further, we assume the input and output relationship satisfies the following linear relationships:

$$z = G_{11}(d) + G_{12}(u) \qquad 10(a)$$

$$y = G_{21}(d) + G_{22}(u) \qquad 10(b)$$

$$u = K(y) \qquad 10(c)$$

where G_{11}, G_{12}, G_{21}, G_{22} are appropriate linear bonded operators, and K(y) is a suitable function of the output y, called the controller. We claim that the system of equations (10) above can be viewed as a linear operator T that maps any uncontrolled input disturbance d into the controlled output z, that is to say,

T: {set of disturbances d} ------> {set of controlled output z} (*)

To show that (*) is true, we manipulate the equations (10) as follows.

From 10(b) and 10(c), we find the transfer function between y and d because

$y = G_{21}(d) + G_{22}(u)$

$\quad = G_{21}(d) + G_{22}(K(y))$, which gives

$y(I - KG_{22}) = G_{21}(d)$, or

$$y = (I - K\,G_{22})^{-1}\,G_{21}(d) \qquad 11(a)$$

From 10 (a), 10(c), we find the transfer function between z and d as follows:

$z = G_{11}(d) + G_{12}(K(y))$

$\quad = G_{11}(d) + G_{12}(K(y))$, and by 11(a),

$$= G_{11}(d) + G_{12}(I - K\,G_{22})^{-1}\,KG_{21}(d) \qquad 11(b)$$

Therefore, 11(b) shows that the transfer function T defined by

$$T = G_{11} + G_{12}(I - K\,G_{22})^{-1}\,KG_{21} \qquad 11(c)$$

is the required transfer function for the claim (*) we made above.

The following diagram shows, for a given controller K, the transfer function T between the disturbance d and the controlled output z.

Since the transformation T depends on the controller K, we denote T_K to be the transformation T given in 11(c).

Figure 9.5

Transfer Function Between Disturbance and Control Output

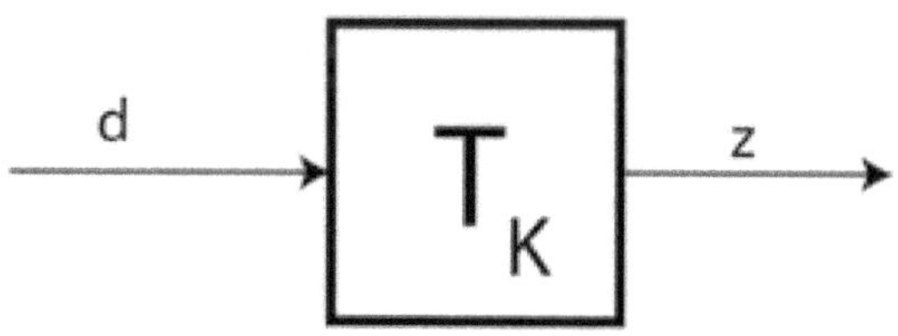

We defined the H∞ norm of the transfer function T_K denoted by $<< T_K >>$ as:

$$<< T_K >> = \underset{d}{Sup}\frac{\| T_K(d) \|}{\| d \|} \qquad 11(d)$$

, where the norms for d and z are the usual L^2 norms defined as follows:

$$\| d \| = [\int_{-\infty}^{\infty} | d(t) |^2 \, dt]^{1/2}$$

$$\| z \| = [\int_{-\infty}^{\infty} | z(t) |^2 \, dt]^{1/2}$$

We will show in the next sections how the classical control problem can be viewed as solving a two-person zero sum game problem using this concept of H_∞ norm of the transformation T_K. But first, we recall the two-person zero sum game and

9.4 Revisit of the two-person zero-sum game

Example 9.4.1 (Computation of L and U values for a game matrix)

Let us look at the following zero-sum game example and recall some basic terminologies we discussed earlier in Chapter 2. Denote the following matrix by C:

Payoff values

12	7	7	22	5	5
12	17	12	19	27	12
9	14	10	11	10	9
7	15	11	12	7	7
12	17	12	22	27	

The row player has four strategies, R1, R2, R3, R4, and the column player has 5 strategies C1, C2, C3, C4, C5. The payoff matrix is shown above.

Let C_{Ij} be the entries in the payoff matrix above when the Row Player chooses R_I and the Column Player chooses C_j, m is the number of rows, n the number of columns.

Let the Lower Value (denoted by L) and the Upper Value (denoted by U) of a two-person game be defined as:

$$L = \underset{1 \le i \le m}{Max} \; \underset{1 \le j \le n}{Min} \; c_{ij}$$

$$U = \underset{1 \le j \le n}{Min} \; \underset{1 \le j \le m}{Max} \; c_{ij}$$

To compute L for the game matrix in the example above, we first compute the minimum of each row, which are {5, 12, 9, 7}, then take the maximum of all the minimums, which is 12.

To compute U, we first compute the maximum of each column, which are respectively 12, 17, 12, 22, 27, then take the minimum of all the maximums, which is 12.

In general, L and U need not be equal, but it is always true that $L \leq U$. When $L = U$, this common value is called the value of the game. In our example, $L = U = 12$.

It is easy to see why $L \leq U$ always. We can demonstrate this as follows. Suppose the value L is the entry $C_{jk,}$ the value of U is the entry C_{mn}. Then clearly C_{jk}, which is L, is $\leq C_{jn}$ because C_{jk} was chosen as the minimum on row j, and now $C_{jn} \leq C_{mn}$ because C_{mn} was chosen as the maximum on column n, and so $C_{jk} \leq C_{mn}$ by transitivity. So, $L \leq U$.

Denote $F(R_I, C_j)$ be cost function F with the Row-Player playing R_i and the Column-Player playing C_j. For the game matrix C above, F(1, 3) = 7, F(2, 4) = 19 … etc.

Example 9.4.2 (Computation of L and U values for a game with a payoff function)

Instead of using the row and column numbers as the players' choice of strategies above, we can use the elements $r \in R$ and elements $c \in C$ for the various strategies from the strategy set R and C, respectively. The cost function F is then denoted by F(r, c). Then the L and U values of the game associated with the cost function F(r,c) are given by the following expression:

$$L = \underset{r\in R}{Max}\ \underset{c\in C}{Min}\ F(r,c)$$

$$U = \underset{c\in C}{Min}\ \underset{r\in R}{Max}\ F(r,c).$$

As before, $L \leq U$ in general.

We consider two cost functions $F(u, w) = u^2 - w$ and $F(u, w) = (u - w)^2$ in this example, the first one gives $L = U = 2$, and the second one gives $L = 0$, $U = 1$ for u, w in the interval [0, 2]

In what follows, we present two MATLAB programs, MyCombineMinMaxMatrix.m and MyMinMaxSaddleFunction1.m, to compute the L and U values of a two-person matrix game and a matrix game associated with the cost function F(c, r), respectively.

9.5 Computation of the L and U values of a game

9.5.1 A simple MATLAB program to compute L and U value of a game matrix

The user inputs the number of rows and columns of the matrix, then the user types in the matrix entries on the screen. The program outputs the upper value U of the matrix (Min Max) and the lower value L of the matrix (Max Min). The program also outputs the row and column numbers at the location of the U and L values. If both L and U are equal, the common value is called the value of the game. The location of this common value is called the saddle point of the matrix.

Below is the m-code of MyCombineMinMaxMatrix.m :

```
nrows = input('Enter number of rows: \n');
ncolumns = input('Enter number of columns: \n');
mymatrix = input('Enter a matrix: \n');
mymatrix
for l = 1: nrows
   for m = 1: ncolumns
    yw(m) = mymatrix(l, m);
end
    [minval, index] = min(yw);
   minyu (l) = minval;
end
length(minyu)
minyu
[maxval, index] = max(minyu)
fprintf('The Max of Min values is : % .4f located at row% 4d \n', maxval, index);
fprintf('**************************************************** \n');
for m = 1: ncolumns
   for n = 1: nrows
    yu (n) = mymatrix (n,m);
end
    [maxval, index] = max (yu);
   maxyu (m) = maxval;
end
   maxyu
   [minval, index] = min(maxyu)
fprintf('The Min of Max values is: % .4f located at column % 4d \n', minval, index);
fprintf('**************************************************** \n')
```

Below is the output of MyCombineMinMaxMatrix.m with the above example input matrix :

```
>> mycombineMinMaxMatrix
Enter number of rows :
4
Enter number of columns:
5
Enter a matrix :
[12 7 7 12 5;12 17 12 19 27;9 14 10 11 10;7 15 11 12 7]
mymatrix =
   12    7    7   12    5
   12   17   12   19   27
    9   14   10   11   10
    7   15   11   12    7
ans =
5
minyu =
[5, 12, 9, 7, 0]
maxval =
12
index =
2
The Max of Min values is : 12.0000 located at row 2
****************************************************
maxyu =
[12, 17, 12, 19, 27]
minval =
12
index =
1
The Min of Max values is :   12.0000  located at column   1
****************************************************
```

9.5.2 A simple MATLAB program to compute L and U of a game with cost function F(r,c)

This program computes the two MaxMin and MinMax values of a cost function f(u, w) entered by the user. If both values are equal, then (u, w) is a saddle point, and the common value is the value of the game defined by the function f(u, w).

Below is the m-code of MyMinMaxSaddleFunction1.m

```
syms f(u,w)

% Various functions f(u,w) was used as inputs to check out for saddle point

%f(u,w) = (u-w)^2          % not saddle
%f(u,w) = 4*u*w            % saddle
%f(u,w) = w^2- u^2         % saddle
%f(u,w) = u^2 -w^2         % saddle
                           % When u is constant w concave down and when w
                           % is constant u concave up
%f(u,w) = (u+w)^2          % saddle
%f(u,w) = w^2 - u          % not saddle
%f(u,w) = - w^2 + u        %saddle: when u is constant and w concave down

f(u,w)= u^2 - w            % saddle: when u is constant and w concave down and when
                           % w is constant, u concave up

delta = 0.5
u = 0: delta: 2;
w = 0: delta : 2;

maxpt = 2/delta

for k = 2: maxpt + 1
 arrayu (k) = delta * (k-1);
end
 arrayu

for k = 2: maxpt + 1
 arrayw(k) = delta * (k-1);
end
 arrayw
```

```
for l = 1: maxpt + 1
    u = arrayu (l);
    for m = 1: maxpt + 1
        w = arrayw (m);
        yw(m) = f(u, w);
    end
yw;
[minval, index] = min(yw);
minyu(l) = minval;
end
fprintf('Below is the array MINYU: \n');
minyu
[maxval, index] = max(minyu);
fprintf('The Max of Min values is: % .4f located at row % 4d \n', maxval, index);


fprintf('************************************************** \n');
for m = 1: maxpt + 1
    w = arrayw(m);
    for l = 1: maxpt + 1
        u = arrayu(l);
        yu(l) = f(u, w);
    end
    yu;
[maxval, index] = max(yu);
maxyu(m) = maxval;
end
fprintf('BELOW IS THE ARRAY MAXYU \n');
maxyu
[minval, index] = min(maxyu)
fprintf('The Min of Max values is: % .4f located at column % 4d \n', minval, index);
```

We are checking out below for the outputs of two input cost functions. One will give us a saddle point and the other does not by comparing the U-value and the L-value of the input cost functions.

The output of the program for f(u, w) = u^2 – w is shown below:

```
myminmaxsaddlefunction1
f(u, w) =
u^2 - w
delta =
   0.5000
maxpt =
   4
arrayu =
   0    0.5000    1.0000    1.5000    2.0000
arrayw =
   0    0.5000    1.0000    1.5000    2.0000
Below is the array MINYU :
minyu =
[-2, -7/4, -1, 1/4, 2]
The Max of Min values is : 2.0000  located at row   5
*************************************************

BELOW IS THE ARRAY MAXYU
maxyu =
[4, 7/2, 3, 5/2, 2]
minval =
2
index =
5
The Min of Max values is : 2.0000  located at column    5
```

The output of the program for f(u, w) = (u – w)2 is shown below:

```
f(u, w) =

(u - w)^2

delta =

0.5000

maxpt =

4

+

arrayu =

   0    0.5000    1.0000    1.5000    2.0000
```

```
   0    0.5000    1.0000    1.5000    2.0000

Below is the array MINYU  :

minyu =

[0, 0, 0, 0, 0]

The Max of Min values is :    0.0000  located at row   1

*************************************************

BELOW IS THE ARRAY MAXYU

maxyu =

[4, 9/4, 1, 9/4, 4]

minval =

1

index =

3

The Min of Max values is :    1.0000  located at column   3
```

9.6 *Formulation of a control system by a two-person zero- sum game*

After we learned the meaning of the Min Max(U-value) and the Max Min (L-value) of the cost function F(u, w) for a two-person game, we are now in a position to formulate the control system problem defined by the transformation :

$$T_K = G_{11} + G_{12}\,(I - KG_{22})^{-1}\,KG_{21} \qquad 11(c)$$

Our objective is to find a controller K such that the largest rate of growth over all the disturbances d measured by the ratio of $\| T_K(d) \|$ to $\| d \|$ is bounded by some finite value, say γ. And we also want the smallest such γ.

$$<< T_K >>= \sup_{d} \frac{\| T_K(d) \|}{\| d \|} \qquad 11(d)$$

That is to say, $\gamma = \inf_{K} << T_K >>$. We can restate this control problem as follows:

Given any γ, can we find K such that $<< T_K >> \leq \gamma$

That is to say,

$\| T_K(d) \|^2 \leq \gamma^2 \| d \|^2$ for all disturbances d (12)

Now we can view this optimization problem as a two-player zero-sum game, one player being the controller K that does the minimizing, and the other player being the disturbance that does the maximizing.

The above inequality (12) is equivalent to an objective function F parameterized by γ with $F\gamma(K, d) \leq 0$ where $F\gamma(K, d)$ is defined by:

$$F\gamma(K, d) = \| T_K(d) \|^2 - \gamma^2\| d \|^2 \text{ for all K and d}$$

So, the control problem is now equivalent to find (1) a smallest $\gamma \geq 0$ such that $F\gamma(K, d)$ is bounded above by zero, that is to say, $F\gamma(K, d) \leq 0$, and to find (2) the corresponding controller K to achieve the U-value for $F\gamma(K, d)$.

The U-value and the L-value for the cost function $F\gamma(K, d)$ is defined below:

$$U = \underset{K}{Min}\, \underset{d}{Max}\, F_\gamma(d, K)$$

$$L = \underset{d}{Max}\, \underset{K}{Min}\, F_\gamma(d, K)$$

We have thus demonstrated that solving a classical control problem as described before can be treated as a two-person zero sum game where one of the players is the disturbance and the other player is the controller and where the controller-player chooses the control strategies to minimize the U value, and the disturbance - player chooses the worst case disturbance to maximize the L value. The objective of the original control problem is solved when both the controller player and the disturbance player are choosing the right strategies so that both the L and U values of the game are equal.

Example 9.6.3

This example demonstrates the use of the built-in function "hinfsyn" from the Robust Control Toolbox in MATLAB. Input to the function is the plant P of the underlying control system, the number of inputs and outputs. The plant P is described by the state-space model with input parameters nx (the number of states), nu (the number of inputs), ny (the number of outputs), and the content of the 4 user input matrices A, B, C ,D described in the previous sections. Outputs of hinfsyn(P, ny, nu) are the performance level gamma described in Section 9.6, the controller K, and the closed loop CL. The source listing of myHinfiniteTesting.m is shown below with user inputs nx = 9, nu = 4, and ny = 5, and matrices A, B, C, D

Source Code of myHinfinitiveTest.m

```
% Define system dimensions
nx = 9;  % number of states
nu = 4;  % number of inputs
ny = 5;  % number of outputs

% Define specific system matrices
```

```
A = [0.5, 0.1, 0, 0, 0, 0, 0, 0, 0;
     0, 0.6, 0.2, 0, 0, 0, 0, 0, 0;
     0, 0, 0.7, 0.3, 0, 0, 0, 0, 0;
     0, 0, 0, 0.8, 0.4, 0, 0, 0, 0;
     0, 0, 0, 0, 0.9, 0.5, 0, 0, 0;
     0, 0, 0, 0, 0, 1.0, 0.6, 0, 0;
     0, 0, 0, 0, 0, 0, 1.1, 0.7, 0;
     0, 0, 0, 0, 0, 0, 0, 1.2, 0.8;
     0, 0, 0, 0, 0, 0, 0, 0, 1.3];

B = randn(nx, nu)*0.1;  % Small input influence
C = randn(ny, nx);      % Output matrix
D = zeros(ny, nu);      % Assume direct transmission is zero

% Create the plant system
plant = ss(A, B, C, D)

% For simplicity, define weights as identity matrices
% In practice, these should reflect your control objectives
Wp = ss(eye(ny));  % Weight on outputs
Wu = ss(eye(nu));  % Weight on inputs

% Create weighted plant for H-infinity synthesis
% using 'augw' to include weights
% Note: 'augw' is part of Robust Control Toolbox
P = augw(plant, Wp, Wu)

% Number of measurements and controls
n_meas = ny;  % number of measured outputs
n_ctrl = nu;  % number of control inputs

% Perform H-infinity synthesis
[K, CL, gamma, info] = hinfsyn(P, n_meas, n_ctrl);

% Display the results
disp(['H-infinity controller gamma:', num2str(gamma)]);

% Optional: view the controller
disp('Controller K:');
disp(K);
```

```
% You can analyze the closed-loop system
closed_loop = feedback(plant, K)
step(closed_loop)
title('Closed-Loop Step Response')
```

Output of myHinfinitiveTest.m

>> myHinfinitiveTest

plant =

A =

	x1	x2	x3	x4	x5	x6	x7	x8	x9
x1	0.5	0.1	0	0	0	0	0	0	0
x2	0	0.6	0.2	0	0	0	0	0	0
x3	0	0	0.7	0.3	0	0	0	0	0
x4	0	0	0	0.8	0.4	0	0	0	0
x5	0	0	0	0	0.9	0.5	0	0	0
x6	0	0	0	0	0	1	0.6	0	0
x7	0	0	0	0	0	0	1.1	0.7	0
x8	0	0	0	0	0	0	0	1.2	0.8
x9	0	0	0	0	0	0	0	0	1.3

B =

	u1	u2	u3	u4
x1	-0.112	-0.233	-0.1362	-0.006787
x2	0.2526	-0.1449	0.0455	-0.01952
x3	0.1655	0.03335	-0.08487	-0.02176
x4	0.03075	0.03914	-0.03349	-0.03031
x5	-0.1257	0.04517	0.05528	0.002305
x6	-0.08655	-0.01303	0.1039	0.005129
x7	-0.01765	0.01837	-0.1118	0.08261
x8	0.07914	-0.04762	0.1261	0.1527
x9	-0.1332	0.0862	0.06601	0.04669

C =

x1 x2 x3 x4 x5 x6 x7 x8 x9

y1	-0.2097	0.3071	-0.1623	-0.4838	1.53	-0.2296	-0.2612	-0.7411	-0.457
y2	0.6252	0.1352	-0.1461	-0.712	-0.249	-1.506	0.4434	-0.5078	1.242
y3	0.1832	0.5152	-0.532	-1.174	-1.064	-0.4446	0.3919	-0.3206	-1.067
y4	-1.03	0.2614	1.682	-0.1922	1.603	-0.1559	-1.251	0.01247	0.9337
y5	0.9492	-0.9415	-0.8757	-0.2741	1.235	0.2761	-0.948	-3.029	0.3503

D =

	u1	u2	u3	u4
y1	0	0	0	0
y2	0	0	0	0
y3	0	0	0	0
y4	0	0	0	0
y5	0	0	0	0

Continuous-time state-space model.

Input groups:

Name	Channels
U1	1,2,3,4,5
U2	6,7,8,9

Output groups:

Name	Channels
Y1	1,2,3,4,5,6,7,8,9
Y2	10,11,12,13,14

Continous-time state-space model.

Model Properties

H-infinity controller gamma: 400.3632

9.7 Formulation of a discrete time control system as a two-person zero-sum game

While we are not going to show the details of how both players are choosing the right strategies to solve the underlying control problem as this is the main subject in Control Theory and it is beyond the scope of this book on introduction to game theory, however, we state without proof a very fundamental theorem in Control Theory to show what each player can do to solve the control problem. The interested reader can refer to [4] for further understanding of the subject.

Basic background assumptions for Theorem 9.1

Suppose we have a discrete-time control system described by the following equations:

$$x_{k+1} = A_k x_k + B_k u_k + D_k d_k \qquad 13(a)$$

$$y_k = C_k x_k + E_k d_k \qquad 13(b)$$

$$z_k = H_k x_k + G_k u_k \qquad 13(c)$$

for k = 1,2 3…., (K+1)

Let define the following notations:

$$F\gamma = |x_{K+1}|^2 + \sum_{k=1..K} \{|x_k|^2 + |u_k|^2 - \gamma^2 \sum_{k=1..K} |d_k|^2\} \qquad 13(d)$$

$u_k = K_k(y_{[1,k]})$, where $y_{[1,k]} = \{y_1,\ldots, y_k\}$

$Q_k = H_{k'} H_{k'}$

$H_{k'} G_k = 0,$

$G_{k'} G_k = I,$

Where the norms of the state vector $x_{k'}$, $|x_K|$, k = 1... K, is defined according to the matrices $Q_{k'}$ for k = 1...K, and the norm $|x_{K+1}|$ of the final state vector x_{K+1} depends on a preassigned matrix Q_f. In what follows, we will state the main results for two simpler situations when the control is Open-loop (OL) and Closed-loop perfect state (CLPS). By OL, we mean the control only depends on the initial state, and by CLPS, we mean the control depends on the current state as well as the entire past values of the states.

Lemma 9.1 (A)

The objective function

$$F\gamma(u, d) = |x_{K+1}|^2 + \sum_{k=1..K} \{|x_k|^2 + |u_k|^2 - \gamma^2 \sum_{k=1..K} |d_k|^2\} \qquad 13(d)$$

is strictly concave in d for every open-loop policy u if and only if the following K matrices are positive, ie.

$$O_k \cong \gamma^2 I - D_{k'} S_{k+1} D_k > 0 \quad k = 1, 2, 3\ldots K \qquad (14)$$

Where the matrices $S_{2,} S_{3'}, \ldots, S_{K+1}$ are generated by the unknown Riccati equation:

$$S_k = Q_k + A_{k'} S_{k+1} A_k + A_{k'} S_{k+1} D_k[\gamma^2 I - D_{k'} S_{k+1} D_k]^{-1} D_{k'} S_{k+1} A_k \qquad (15)$$

$$S_{K+} = Q_f$$

In other words, we start with the last $S_{K+1} = Q_{f'}$ then apply equation (15) to recursively compute the matrices $S_{K'} S_{K...}, \ldots$ down to S_2. We can show that in this process, the matrices $O_k \cong \gamma^2 I - D_{k'} S_{k+1} D_k$ are invertible.

Lemma 9.1 (B)

We will construct the sequence of matrices M_k defined recursively by

$M_k = Q_k + A_{k'} M_{k+1} V_k^{-1} A_{k'}$ k = 1, 2, 3… (K + 1) (16)

, where we start with $M_{K+1} \cong Q_f$ and

$V_k \cong I + (B_k B_{k'} - \gamma^{-2} D_k D_{k'}) M_{k+1}$ (17)

, we recursively compute by (16) and (17) M_k and V_{k-1}, all the way down to M_2 and V_1. We can show that the sequence of matrices V_k constructed by (17) is invertible.

After we have defined the sequence of matrices $O_{k,}$ $S_{k,}$ $V_{k'}$ M_k by the equations (14), (15), (16), and (17), we are now in a position to state the main result :

Theorem 9.1

In the case of Open-loop Control, the game between the controller and the disturbance has a unique saddle point solution given by the following strategies:

$u_k^* = - B_{k'} M_{K+1} V_k^{-1} A_k x_k^*$ 18(a)

$d_k^* = \gamma^{-2} D_{k'} M_{K+1} V_k^{-1} A_k x_k^*$ 18(b)

where the sequence of state vectors $\{x^*_{k+1}\}$, k = 1 2 … K , is the corresponding state trajectory generated by

$x^*_{k+1} = V_k^{-1} A_k x_k^*$, and the initial $x_1^* \cong x_1$ (19)

The value of the game $F\gamma(u^*, d^*) = x_{1'} M_1 x_1$ (20)

If, however, one of the matrix O_k defined in (14) has a negative eigenvalue, then the upper value U of the game is unbounded.

A similar result for the closed-loop Control (CLPS) is stated as follows:

(1) There is unique feedback saddle-point solution if the sequence of matrices C_k is defined by:

$C_k \cong \gamma^2 I - D_{k'} M_{k+1} D_k$ k = 1, 2, 3… K (21)

are all positive.

(2) The sequence of matrices V_k defined by (17) are all invertible

(3) The unique feedback saddle-point strategies for the Controller and Disturbance are :

$u_k^* = - B_{k'} M_{K+1} V_k^{-1} A_k x_k$

$d_k^* = \gamma^{-2} D_{k'} M_{K+1} V_k^{-1} A_k x_k$

where the sequence of state vectors $\{x^*_{k+1}\}$, k = 1, 2 … K, is the corresponding state trajectory generated by

$x^*_{k+1} = V_k^{-1} A_k x_k^*$, and the initial $x_1^* \cong x_1$

The value of the game $F\gamma$ (u*, d*) = $x_1' M_1 x_1$

(4) If the sequence of matrices C_k defined in (21) has a negative eigenvalue for some k, then the game does not have a saddle point under CLPS and its upper value U is unbounded.

9.8 Summary

The concept of the H-infinitive norm of a general linear transformation T, denoted by << T >>, was introduced and defined in Section 9.2. For a given control system with a controller K, we show in Section 9.3 that we can always find a transformation, say T_K, that maps the set of disturbances {d} into the set of controlled outputs {z} (Equation 11(c)). By applying the concept of the H-infinitive norm to the transformation T_K, we show in Section 9.7 that the original control problem is equivalent to a two-person zero-sum game, one player is the controller K, and the other player is the disturbance d, the cost function for the game is given by $F\gamma(K, d) = || T_K(d) ||^2 - \gamma^2|| d ||^2$, it is bound above by 0, for all K and d, where γ is some positive value. The objective of the game is to find the smallest γ and to find a controller K that achieves the U-value of $F\gamma$(K, d). Recalling that the U and V values of the cost function $F\gamma$(K, d) are defined in Section 9.4 by

$$U = \underset{K}{Min}\,\underset{d}{Max}\,F_\gamma(d, K)$$

$$L = \underset{d}{Max}\,\underset{K}{Min}\,F_\gamma(d, K)$$

We state without proof the main result of solving the discrete-time control problem in Theorem 9.1 to enhance our understanding of the current approach using a two-player zero-sum game.

Take-away Technical:

Our basic objective is to design a control rule that minimizes the worst-case effect (maximum) of disturbances on the system modeled as a zero-sum game between the controller (player 1) and the disturbance (player 2). We do this by finding a controller K such that the H-infinitive norm $<< T_K >>$ of the closed-loop transfer function T_K as defined Section 9.6 is less than gamma for the smallest possible gamma γ, that is: $<< T_K >> \; < \gamma$

Chapter 10

Game Theory and Computer Science

10.0 Introduction

In this chapter, we introduce an alternative way to represent a game *G* using a function rather than its matrix form or tree structure. This function, denoted as $v(G)$, assigns a value from the continuum of real numbers *R* to each subset of players. Formally,

$$v(G)\colon 2^N \longrightarrow R$$

where N = {1, 2, …. , n} represents the set of players, and 2^N is the set of all possible subsets (coalitions) of *N*. Since there are 2^n such subsets, each coalition is assigned a specific numerical value by the characteristic function $v(G)$.

A key question in coalitional game theory is determining how to assign values to these coalitions in a way that accurately reflects the game's structure and the payoffs received by individual players. The function $v(G)$ plays a crucial role in capturing the dynamics of cooperation and resource distribution within the game.

This framework has significant applications in computer science, where various computational problems can be modeled as *Coalitional Games.* Many algorithms, particularly those related to optimization, distributed computeing, and artificial intelligence, leverage coalitional game theory for strategic decision-making and resource allocation.

To illustrate these concepts, this chapter provides a clear exposition of computer science algorithms that align with coalition games. We compliment this discussion with MATLAB programs that numerically demonstrate these algorithms and their relationship to corresponding coalition game structures. Through these examples, we bridge the theoritical foundations of game theory with practical computational applications.

10.1 The characteristic function of a coalitional game

As previously introduced, the characteristic function of a game is defined as:

$$v(G)\colon 2^N \longrightarrow R \qquad N = \{1, 2\ldots, n\}$$

Since there are 2^n possible coalitions, a coalitional game assigns 2^n values, one for each coalition. The empty coalition is always assigned a value of 0. The grand coalition, which includes all the players, receives either the largest (or smallest) values among all coalitions depending on the nature and objectives of the game. Generally, a coalition with more players is assigned a higher (or lower) value than a coalition within few players, though this depends on the strategic structure of the game.

The logic behind assigning values to coalitions follows a fundamental optimization principle. Each player has a set of available actions, and the players who do not join a given coalition are assumed to act in opposition to it.Consequently, the value assigned to a coalition reflects the worst-case outcome based on the interactions between coalition members and non-members. Specifically, the value of a coalition is determined as follows:

- It is computed as the **maximum** over the set of joint actions of all coalition members.
- It is simultaneously the **minimum** over the set of joint actions of all non-members.

Thus, coalition valuation can be framed as a maximin or a minimax optimization problem or a combination of both, depending on the game's structure and how coalition members and non-members interact.

We denote the value of a coalition *C* as *v(C),* and in the following sections, we will provide concrete examples to illustrate how this optimization process works in practice.

We will give examples below to illustrate this optimization process.

Example 10.1.1

Consider a two-players simultaneous game with payoff values shown below.

Figure 10.1

	X	Y	Z
A	(3,3)	(2,2)	(1,4)
B	(4,2)	(3,8)	(2,5)

Player 1 has actions A and B. Player 2 has actions X, Y ,and Z.

There are two players, so there are 4 coalitions, and each will be assigned to a specific value according to the maximin or minimax optimization process. The empty coalition will be assigned to 0, v(Ø) = 0. Let's show how v(1) is computed.

Player 1 has two actions A and B. So, v(1) is the maximum of the payoff values resulting from taking the action A and action B. If player 1 chooses action A, since player 2 is outside the coalition, the payoff of player 1 when playing A will be the minimum of {3, 2, 1} which is 1. Similarly, if player 1 chooses action B, the payoff of player 1 when playing B will be the minimum of { 4, 3, 2} which is 2. Thus v(1) = max {1, 2} = 2.

To compute v(2): V(2) is the maximum of payoff values resulting from taking the action X, Y ,and Z. Since player 1 is outside the coalition, the payoff of player 1. The payoff of player 2 when playing X will be the minimum of {3, 2} = 2. Similarly, the payoff of player 2 when playing Y will be the minimum of {2, 8} = 2; the payoff of player 2 when playing Z will be min{4, 5} = 4. Thus v(2) = max {2, 2, 4} = 4

Finally, to compute v(1, 2): In this case, both players 1 and 2 can each choose the best action as they are now in the same coalition so that v(1, 2) is chosen as the maximum of the sum of the payoff value of player 1 and that of player 2, this maximum is achieved by player 1 playing B and player 2 playing Y, which is 3 + 8 = 11, that is to say v(1, 2) =11.

Example 10.1.2

Now consider a two-player game with player 1 plays first

Figure 10.2

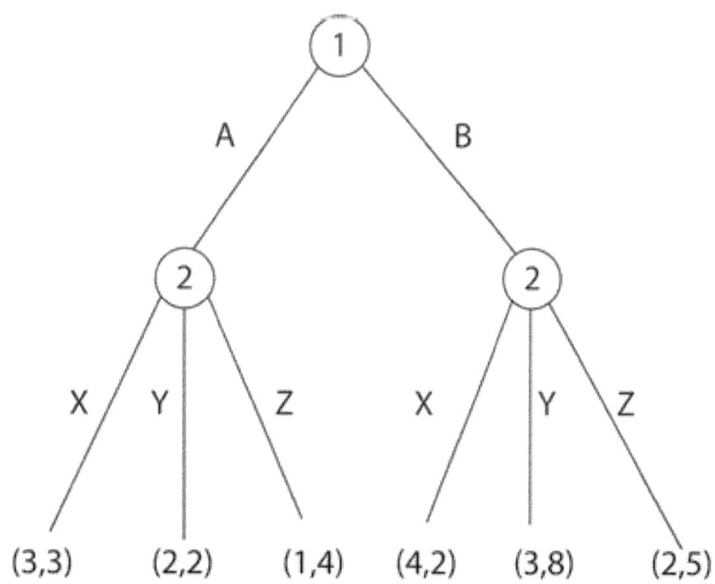

v(1) is computed similarly as in Example 10.1.1 v(1) = Max {1, 2}. where 1 and 2 are computed as 1 = Min {3, 2, 1} and 2 = Min{4, 3, 2}, so v(1) = 2, which we obtained from a Maximin process. On the other hand, v(2) is computed by a Minimax process. Since the flow of the game is that player 1 must play first by playing A or B and player 1 is not in the coalition c(2), so v(2) = min {4, 8}, where 4 = Max {3, 2, 4}, and 8 = Max{2, 8, 5}, thus v(2) = 4. To compute v(1, 2), both players could select the best actions such that the sum of payoffs from both players is a maximum. Thus v(1, 2) = 3 + 8 = 11 for player 1 to play B and player 2 to play Y.

Example 10.1.3

Now consider a three players game such that player 1 must play first with three possible actions A, B, or C. Player 1 plays A will end the game, plays B must be followed by player 2 to end the game by playing W or X, and player 1 plays C must be followed by player 3 to play Y or Z to end the game.

Figure 10.3

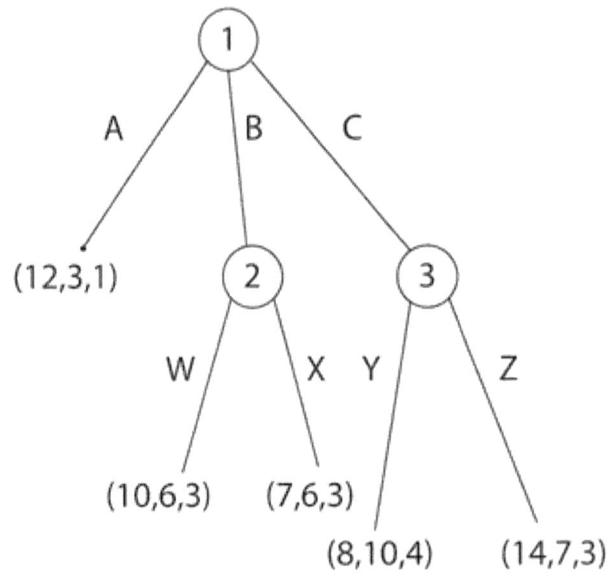

v(1) is the maximum of the payoffs from playing A or B or C. which is respectively 12, 7, 8. Thus v(1) = max {12, 7, 8} = 12.

To compute v(2): The maximum payoff of player 2 playing either W or X is 6. Since player 1 must play before player 2, player 1 playing A gives player 2 the payoff of 3, playing B gives player 2 the payoff of 6, and playing C gives player 2 the payoff of 7 Thus, v(2) = min {3, 6, 7} = 3. Similarly, v (3) = min {1, 3, 4} = 1.

To compute v(1, 2): v(1, 2) is the maximum of sum of payoffs for player 1 playing actions A, B, C, W, or X.

Thus v(1, 2) = maximum {12 + 3, 10 + 6, min(8 + 10, 14 + 7)}, we take the minimum on the sum of the payoffs for player 1 and player 2 is because player 3 is not in the coalition of C(1, 2). Thus v(1, 2) = maximum (15, 16, 18) = 18.

Similarly, as player 2 is not in the coalition C(1, 3),

v(1, 3) = maximum {12 + 1, min(10 + 3, 7 + 3), 14 + 3} = maximum {13, 10, 17} = 17

To compute the grand coalition C(1, 2, 3): We look for optimal combination of actions for players 1, 2, and 3 that will give the maximal total payoff. In this example, Player 1 plays C and player 3 plays Z gives the maximum total payoff of 14 + 7 + 3 = 24.

10.2 Application of algorithms in Computer Science as coalitional games

The computer algorithms we are going to use for the demonstration of coalition games in practical applications are very common in the field of computer science, as such, we are not going to discuss these algorithms in detail here, as explanations of these algorithms can be found in many of the computer algorithms books ([2], [16]), and the subject matter is usually covered in a one semester course for the students in Computer Science Major. What we are presenting here are MATLAB programs for each of these algorithms, a general description of the data structures used in the programs, and demonstration of the applications of coalition games from the output of these algorithms.

10.2.A Minimum Spanning Tree Algorithm

Figure 10.4

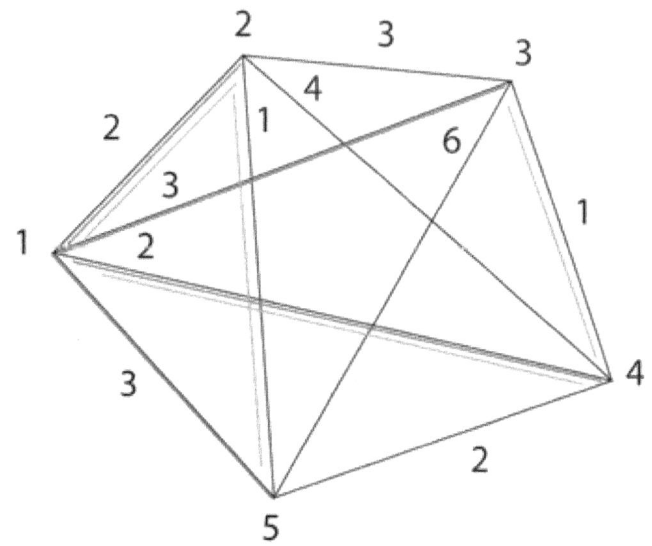

Example 10.2.1

We give an application of the Minimal Spanning Tree Algorithm in Figure 10.4 below. In the figure, node 1 or location 1 is the power source location that directly distributes power to other nodes or locations 2, 3, 4, and 5, with the corresponding cost represented by the cost units 2, 3, 2,and 3 respectively. Amongst these locations 2, 3, 4, and 5, there are, however, costs associated with sharing costs between any two locations, these costs are shown in the figure. If each location chose to receive power directly from the source location, the total cost would be 2 + 3 + 2 + 3 = 10 for all the locations as a whole. The question we are asking is if there is a sharing strategy amongst the locations so that the total cost is minimal and less than 10. Let us label the four players in this game as player 2, player 3, player 4, and player 5. In the language of coalition games discussed before, we want to know what the coalitions C(2), C(3), C(4), and C(5), C(2, 3),... C(4, 5) and C(2, 3, 4, 5) are and their corresponding values v(2), v(3), ... v(2, 3, 4, 5). Recall that the v-values of a coalition C is the minimum cost for the coalition C that can be directly or indirectly connected to the power source of node 1.

For a single player, there is no advantage to working individually, it is clear that v(2) = 2 , v(3)=3, v(4)=2, v(5)=3. For the two or more players team, there are advantages in forming coalitions as opposite to everyone receiving power from the source individually. The values are listed below.

v((2, 3) = 5, v(3, 4) = 3 ,v(4, 5) = 4, v(2, 4) = 4, v(2, 5) = 4, v(3, 5) = 6.

v(2, 3, 4) = 5, v(2, 3, 5) = 6, v(3, 4, 5) = 5, v(2, 4, 5) = 5

and finally, V(2, 3, 4, 5) = 6.

Note that the grand coalition C(2, 3, 4, 5) is the solution to the problem we are looking at. V(2, 3, 4, 5) is the minimal value (power cost) for the sum of all the locations 2, 3, 4 and 5. Note that each of the locations {2, 3, 4, 5} is connected directly or indirectly to a single source or a node {1} in the network, as such, this minimal value v(2, 3, 4, 5) can be computed by the Minimum Spanning Tree Algorithm.

In what follows, we run the MATLAB program myMinSpanTreeBV.m with the output shown below:

The output list of the program shows the graph matrix of the graph shown in Figure 9.4. It is shown here for reference, and it is the only input to the program.

The output list also shows the predecessor array Pdsr with the content:

Inf 1 4 1 2 0 0 0 0 0

This Pdsr array of 10 entries has the following interpretation:

Node 1 in the graph is the root node that has no predecessor, or it has predecessor Inf.

Node 2's predecessor is node 1.

Node 3's predecessor is node 4.

Node 4's predecessor is node 1.

Node 5's predecessor is node 2.

The rest of the 5 zero entries means the program is designed for a network a maximum of 10 notes, and the present network only consists of 5 nodes.

Knowing the predecessor of each node in the graph, we can construct the Minimum Tree Network with each edge shown in green color in Figure 10.4

Note that the edges shown in red color are the direct edges from the source node 1 to the nodes or locations 2, 3, 4, and 5.

The output list also includes the total cost of each selected edge (the green edges) shown in Figure 10.4 above. The value is 6.

As mentioned before, the total cost of the Minimum Spanning Tree network coincides with the value the grand coalition v(2, 3, 4, 5) which is also 6 as computed before.

:>> myMinSpanTreeBV

Output of the program are shown below:

graph =

0	2	3	2	3
2	0	3	4	1
3	3	0	1	6
2	4	1	0	2
3	1	6	2	0

Pdsr =

Inf 1 4 1 2 0 0 0 0 0

weight = 6

The weight of the Minimal Network is 6

Before we take a top-level look at how the MATLB program myMinSpanTreeBV.m works, we run the program on another scenario with the graph shown in Figure 10.5 below:

Figure 10.5

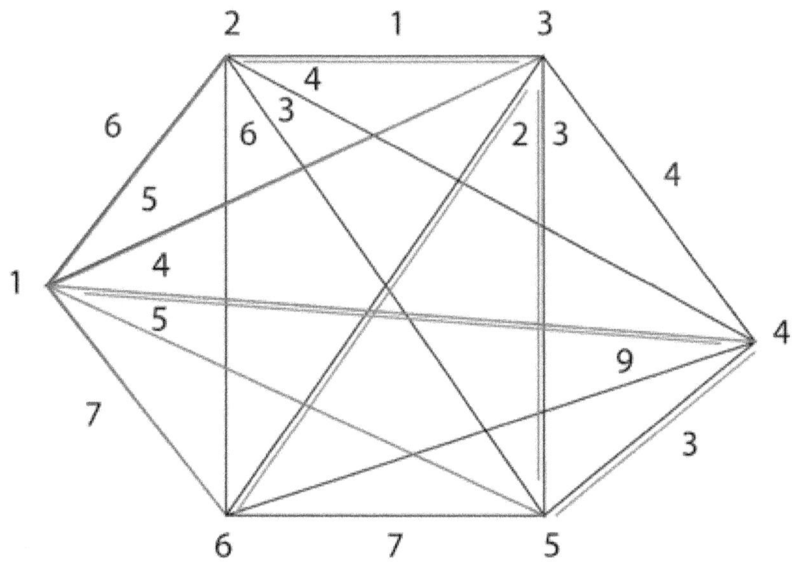

As before, the red edges are the directly connected edges from the source node 1 or the root node to other location nodes. The green edges form a Minimum Spanning Tree from the content of the predecessor array Pdsr computed by the Minimum Spanning Tree Algorithm with total cost 13. As noted, v(2, 3, 4, 5, 6) = 13. Output of myMinSpanTreeBV.m is shown below:

```
> myMinSpanTreeBV
graph =

        0   6   5   4   5   7

        6   0   1   4   3   6

        5   1   0   4   3   2

        4   4   4   0   3   9

        5   3   3   3   0   7

        7   6   2   9   7   0

Pdsr =

        Inf   3   5   1   4   3   0   0   0   0

weight =

        13

The weight of the Minimal Network is 13
```

Top-level description of myMinSpanTreeBV.m

1. The input to the program is the graph-matrix graph that has the direct distance between any two nodes in the graph. The output of the program is the predecessor array Pdsr that gives the composition nodes of the Minimum Spanning Tree and the total weight W of all the edges from the Minimum Spanning Tree.

2. The basic data structures used in the implementation of the Minium Spanning Tree Algorithm is the Priority Queues myPriority and myQueue represented by two arrays myQueue which consist of the nodes of the graph G and myPriority which consists of the priority of the corresponding nodes. We need the predecessor array Pdsr to keep track of the predecessors during the execution of the algorithm. Its final content defines the Minimum Spanning Tree. We need the array Visited to keep track of the nodes from the graph that have been looked at and each node visited will be added as a new node to the Minimum Span Tree during its tree creation.

3. The user picks any node from the graph as the root node to the Minimum Spanning Tree. This is the content of the variable headnode in the program.

4. Initially, we place headnode in myQueue and the program begins its execution by popping off the headnode from myQueue. Every time a node is popped off from myQueue, we look at all the other nodes that are connected

to the node that is being popped by the function findFollowers , and we push each of theCose connected nodes back to myQueue provided that is has not been visited. Each time new nodes are added to myQueue, it is always maintained in priority order, and the predecessor array and visited array are updated. All these actions mentioned before and carried by the functions initQueue, departQue, joinQue, sortPriority, adjustPriority, updatePdsr. Once the Minimum Spanning Tree or the final content of the Pdsr array is computed. The function ComputeMinWeight computes the total cost of the Minimum Spanning Tree.

5. The main program of the algorithm is outlined as follows:

```
initQueue;
while(myQueue is not empty) do
      departQue;
      findFollowers;
      sortPriority;
   for(each note that are sorted) do
      joinQueue;
      updatePdsr;
      adjustPriority;
   end % for-loop;
   sortPriority;
end % while-loop;
ComputeMinWeight;
```

6 To run myMinSpanTreeBV.m, the user can input any graph matrix by editing the content of the variable graph in line 1 of the source code. The user can also observe how the contents of myQueue, myPriority changes during program execution by un-commenting the output instructions of myQueue and myPriority within the while-loop.

The source code of myMinSpanTreeBV.m is shown below:

```
%graph = [0 2 0 0 0 7 3 0 0; 2 0 4 0 0 0 6 0 0; 0 4 0 2 0 0 0 2 0; 0 0 2 0 1 0 0 8 0; 0 0 0 1 0 6 0 0 2; 7 0 0 0 6 0 0 0 5; 3 6 0 0 0 0 0 3 1; 0 0 2 8 0 0 3 0 4; 0 0 0 0 2 5 1 4 0]
%graph = [0 4 5 0 0 0 7; 4 0 8 0 0 0 0; 5 8 0 2 0 0 1; 0 0 2 0 0 0 6; 0 0 0 0 0 2 8; 0 0 0 0 2 0 0; 7 0 1 6 8 0 0]

%graph = [0 10 8 9; 10 0 2 4; 8 2 0 3; 9 4 3 0]

%graph = [0 100 80 90; 100 0 30 40; 80 30 0 30; 90 40 30 0]
%graph = [0 6 5 4 5 7; 6 0 1 4 3 6; 5 1 0 4 3 2; 4 4 4 0 3 9; 5 3 3 3 0 7; 7 6 2 9 7 0]
%graph = [0 2 3 2 3; 2 0 3 4 1; 3 3 0 1 6; 2 4 1 0 2; 3 1 6 2 0]

% ************** Beginning of Main Program: **********************
```

```
numberOfnodes = size(graph,1); % number of nodes in the graph
ql = 10; % Queue length
myQueue = [0 0 0 0 0 0 0 0 0 0];
myPriority = [inf inf inf inf inf inf inf inf inf inf];
Pdsr = [0 0 0 0 0 0 0 0 0 0];
Visited = [0 0 0 0 0 0 0 0 0 0];

headnode = 1;
head = 1;

fptr = 1;
rptr = 1;

%fprintf('Calling initQueue: \n');

[myQueue, myPriority, fptr, rptr] = initQueue(myQueue, myPriority, head, fptr, rptr);

myQueue;
myPriority;

Pdsr(head) = inf;
Visited(head) = 1;

%fprintf('this is the content of myQueue before calling the while-loop: \n');

whileloopNum = 1;

while(fptr <= rptr)

%fprintf('********************** go into the while-loop ********************** %4d\n', whileloopNum);

myQueue;
myPriority;

%fprintf('************ calling departQue: **************************** \n')

[myQueue, myPriority, head, fptr, rptr] = departQue(myQueue, myPriority, fptr, rptr );

% fprintf('the head node %4d is now departed: \n', head);

Visited(head) = 1;
Visited;

%fprintf('Now calling findFollower: \n');

[head, fNodes, fPriority, k] = findFollowers(head, numberOfnodes, graph);

[Snode, Spriority] = sortPriority(fNodes, fPriority, 1, k);

nde = Snode;
pty = Spriority;
```

```
%fprintf('before calling joinQue: \n');

for lq = 1:k
        head;
        nx= nde(lq);
        px= pty(lq);

%fprintf('now looking at the node nx from the followers %4d\n', nx);

    if Visited(nx) == 0 && whileloopNum == 1

%fprintf('*********** calling joinQue ************: \n');

    [myQueue, myPriority, fptr1, rptr1] = joinQue(myQueue, myPriority, ql, fptr, rptr, nx, px);
    [Pdsr, Visited] = updatePdsr( Pdsr, Visited, head, nx);

    rptr= rptr1;


else
    if Visited(nx) == 1
    %fprintf('nx will not go back to the queue because it was popped from the queue before %4d\n', nx);
    Pdsr;
    end

    if Visited(nx) == 0

    %fprintf('************** now check if a node is already in the queue before joining the queue:
    ************** \n')

[yes, location, myQueue, myPriority, fptr,ql] = adjustPriority(myQueue, myPriority, fptr, ql, nx,px);


    if location~=0 && yes==1
          Pdsr(nx) = head;
          flag = 1;
    %fprintf('after priority adjustment and update Pdsr: \n')
    Pdsr;
          myQueue;
          myPriority;
    end

    if location==0
          Pdsr(nx) = head;
          [myQueue, myPriority, fptr, rptr] = joinQue(myQueue, myPriority, ql, fptr, rptr, nx, px);
    %fprintf('nx is not in myQueue and thus joined myQueue: %4d\n', nx)


          end
    end % if-end

end % if-else-end
```

```
end % end of the for-loop
%fprintf('********* end of each call to the for-loop ********************** \n');
myQueue;
myPriority;
Pdsr;
fptr;
rptr;

if flag == 1
[myQueue, myPriority] = sortPriority(myQueue, myPriority, fptr, ql);
end

myQueue;
myPriority;

whileloopNum = whileloopNum + 1;

end % of the while-loop

Pdsr

weight = ComputeMinWeight(graph, Pdsr, numberOfnodes, headnode)
fprintf('The weight of the Minimal Network is %4d\n:', weight);

%*********************************************************************%

% Procedures and functions defined below:

%*********************************************************************%

function[P, V] = updatePdsr(P, V, node1, node2)
    if V(node2) == 0
      P(node2) = node1;
    end

end

%*********************************************************************%
function[i, Qn, Qp, k] = findFollowers(i, numberOfnodes, W)

k = 1;
for j = 1: numberOfnodes
    if W(i,j) ~= 0
    Qn (k) = j;
    Qp (k) = W(i,j);
    k=k +1;

    end % end of if-statement

end % end of for-loop
k = k-1;

end % end of the function findFollowers
```

```
%*****************************************************************

function[Snode, SQ] = sortPriority(node, Q, fptr, rptr)

      fptr = 1;
      j = 1;

      while j <= rptr
        min = Q(fptr);
        sindex = fptr;

        for i = fptr + 1: rptr
          if min > Q(i)
            min = Q(i);
              sindex = i;
            end
          end % end of for loop
        SQ(j) = min;
        Q(sindex) = inf;
        Snode(j) = node(sindex);

        j = j+1;

        end % while loop

end % end of procedure SortPriority

%************************************************************** %

function[Qn1, Qp1, fptr1, rptr1] = joinQue(Qn, Qp, ql, fptr, rptr, nde, pty)
        x = fptr;
        fptr1 = fptr;
        rptr1 = rptr + 1;
        while Qp(x) < pty
        x = x + 1;
        end
        for i = fptr1: (x-1)
          Qp1(i) = Qp(i);
          Qn1(i) = Qn(i);
        end

        Qp1(x) = pty;
        Qn1(x) = nde;

        for i = x+1: ql
        Qp1 (i) = Qp (i-1);
        Qn1 (i) = Qn(i-1);
       end

end
```

```
%*****************************************************************%

function[Qn, Qp, head, fptr, rptr] = departQue(Qn, Qp, fptr, rptr)
        head = Qn(fptr);
        Qn(fptr) = 0;
        Qp(fptr) = 0;
        fptr = fptr + 1;

end

% End of Main Program

% Procedures and functions defined below:

%*****************************************************************%

function minWeight = ComputeMinWeight(W, Pdsr, nNodes, headnode)
        sum = 0;
        m = nNodes;
        for i = m: -1: 1
          if i ~= headnode
          sum = sum + W(i, Pdsr(i));
          end
        end
        minWeight = sum;
end

%*****************************************************************%

function [yes, location, Qn, Qp, fptr, rptr] = adjustPriority( Qn, Qp, fptr, rptr, node , pty)

    location = 0;
    for i = fptr: rptr
       if(Qn (i) == node)
       location = i;
       end
    end
    if location ~= 0 && (pty < Qp(location))
       yes = 1;
       Qp(location) = pty;
    else
       yes = 0;
       %fprintf('the priority of the node on the queue did not change: \n')
    end
end


%*****************************************************************

function [Qn, Qp, fptr, rptr] = initQueue(Qn, Qp, head, fptr, rptr)
         Qn = [0 0 0 0 0 0 0 0 0 0];
```

```
Qp = [inf inf inf inf inf inf inf inf inf inf];
Qn(1) = head;
Qp(1) = 0;
fptr = 1;
rptr = 1;
end
```

10.2.B The Maximum Flow Algorithm

Figure 10.6

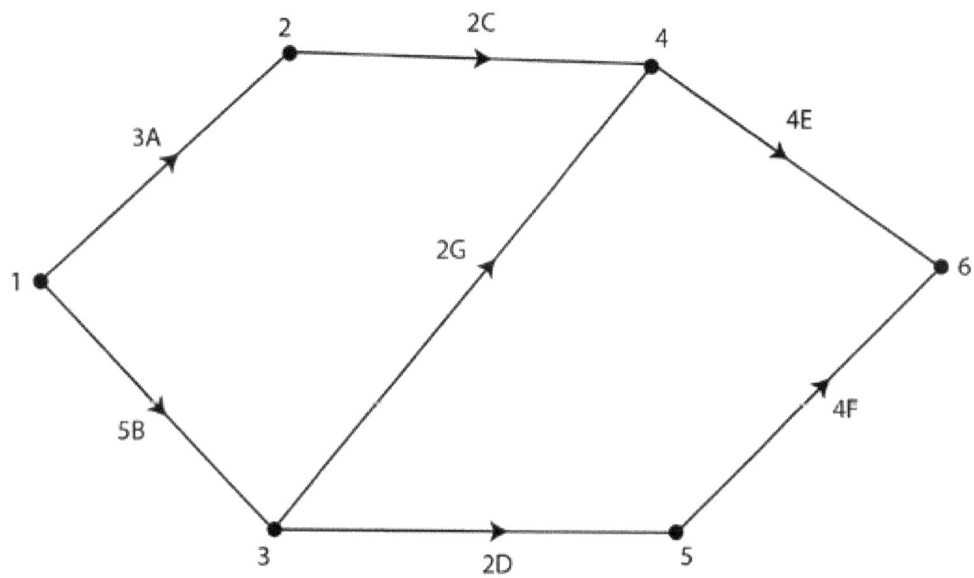

Example 10.2.2

We show an application of the Maximal Flow Algorithm as an example of coalition games. Figure 10.6 above is an example of a network flow diagram, where node 1 and node 6 are the source and sink respectively, and other connecting nodes 2, 3, 4, 5 nodes in between. The numbers in between any two nodes are the corresponding flow capacities. The objective of the Maximum Flow Algorithm is to find a set of flow values one for each connecting edge so that the sum total of the flows coming out of the sink is maximum subjected to only one basic requirement: for each one of the in-between nodes in the network flow diagram, the sum of flow values coming into that node is the same as the sum of flow values going out of that node. It is then proven that, with the above-mentioned requirement, the sum of all the flow values going out of the source node is the sum of all the flow values coming into the sink node. This common value is called the value of the flow network. To formulate this flow network problem as a coalition game, we assign each connecting edge of two nodes in the network as a player. Thus, we have 7 players for this network labelled as players A, B, C, D, E, F, and G with the corresponding flow capacity 3, 5, 2, 2, 4, 4, 2, and our objective is to find a set of flow values one value for each player so that for the grand coalition C (A, B, C, D, E, F, G), it's corresponding value v(A, B, C, D, E, F, G) is the value of flow network.

We are going to examine all possible coalitions C of the players above. For all single player and double players coalition combination, the corresponding v(x), v(x, x) are all zeros as they are not connecting the source and the sink.

For the triple players coalitions v(A, C, E) = 2, v(B, G, E) = 2, v(B, D, F) = 2, and for the grand coalition v(A, B, C, D, E, F, G) = 6.

Next, we are going to demonstrate the optimal value 6 of the grand coalition C(A, B, C, D, E, F, G) above is the same value 6 as the output of the Maximal Flow Algorithm.

The output of MymaximalFlowAlgorithm.m is shown below:

```
>> MymaximalFlowAlgorithm
capacityMatrix =
    0   3   5   0   0   0
    0   0   0   2   0   0
    0   0   0   2   2   0
    0   0   0   0   0   4
    0   0   0   0   0   4
    0   0   0   0   0   0
source =
        1
sink =
        6
numVertices =
        6
the path we are taking in the bfs is :
path =
    1   2   4   6
So far we have 1 path
flowMatrix =
    0   2   0   0   0   0
   -2   0   0   2   0   0
    0   0   0   0   0   0
    0  -2   0   0   0   2
```

```
    0    0    0    0    0    0
    0    0    0   -2    0    0
```

show the flow value after each successful path from source to sink

pathFlow =

```
        2
```

the path we are taking in the bfs is :

path =

```
    1    3    4    6
```

So far we have 2 path

flowMatrix =

```
    0    2    2    0    0    0
   -2    0    0    2    0    0
   -2    0    0    2    0    0
    0   -2   -2    0    0    4
    0    0    0    0    0    0
    0    0    0   -4    0    0
```

show the flow value after each successful path from source to sink

pathFlow =

4

the path we are taking in the bfs is :

path =

```
    1    3    5    6
```

So far we have 3 path

flowMatrix =

```
    0    2    4    0    0    0
   -2    0    0    2    0    0
   -4    0    0    2    2    0
    0   -2   -2    0    0    4
    0    0   -2    0    0    2
    0    0    0   -4   -2    0
```

show the flow value after each successful path from source to sink pathFlow =

6

maxFlow =

6

An explanation of the above output listing is in order.

The capacityMatrix is the matrix representation of the players' capacities shown in Figure 9.6. This matrix together with the source and sink notes are the only inputs needed for running the program. The program outputs the paths 1, 2 ,4 ,6 (ACE), 1, 3, 4, 6 (BGE) and 1, 3, 5, 6 (BDF) sequentially, and each time a path is outputted, the corresponding flow values are computed and updated in the flowMatrix. The pathFlow gives the value of the network based on all the existing paths from the source node to sink node so far, and maxFlow gives the optimal value of the network based on all possible paths that can exist in the network. By path, we mean that all the nodes in it, each one of them must satisfy the basic requirement as mentioned before. In our example maxFlow = 6, which coincides with the value v(A, B, C, D, E, F, G) with the grand coalition.

Example 10.2.3

As a second example to demonstrate the connection between coalition games and the Maximal Flow Algorithm, consider Figure 9.7 below.

Figure 10.7

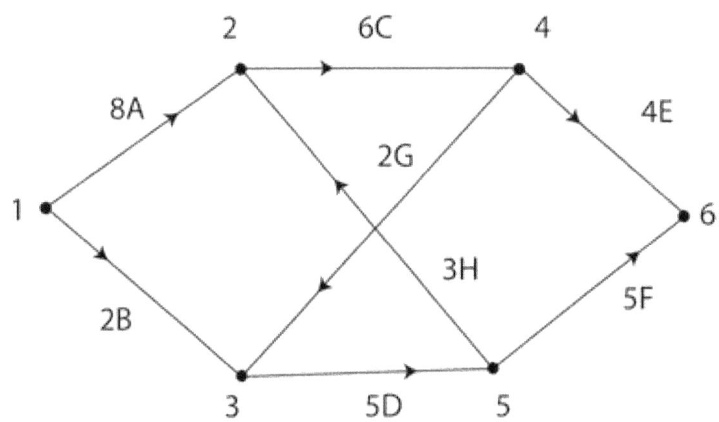

As before, all single player and double players coalitions have 0 values. The meaningful three players coalitions are C(A, C, E), C(B, D, F)

Output listing of the algorithm:

Running MymaximalFlowAlgorithm.m gives the following output listing:

```
>>MymaximalFlowAlgorithm
capacityMatrix =
    0   8   2   0   0   0
    0   0   0   6   0   0
    0   0   0   0   5   0
    0   0   2   0   0   4
    0   3   0   0   0   5
    0   0   0   0   0   0
source =
        1
sink =
        6
numVertices =
        6
the path we are taking in the bfs is :
path =
    1   2   4   6
So far we have 1 path
flowMatrix =
    0   4   0   0   0   0
   -4   0   0   4   0   0
    0   0   0   0   0   0
    0  -4   0   0   0   4
    0   0   0   0   0   0
    0   0   0  -4   0   0
show the flow value after each successful path from source to sink
pathFlow =
        4
the path we are taking in the bfs is :
path =
    1   3   5   6
```

So far we have 2 path

flowMatrix =

```
 0   4   2   0   0   0
-4   0   0   4   0   0
-2   0   0   0   2   0
 0  -4   0   0   0   4
 0   0  -2   0   0   2
 0   0   0  -4  -2   0
```

show the flow value after each successful path from source to sink

pathFlow =

```
     6
```

the path we are taking in the bfs is :

path =

```
 1   2   4   3   5   6
```

So far we have 3 path

flowMatrix =

```
 0   6   2   0   0   0
-6   0   0   6   0   0
-2   0   0  -2   4   0
 0  -6   2   0   0   4
 0   0  -4   0   0   4
 0   0   0  -4  -4   0
```

show the flow value after each successful path from source to sink

pathFlow =

```
     8
```

maxFlow =

```
     8
```

We see that the program outputs three paths in sequential order. The first path is going through the nodes 1, 2, 4, 6 with coalition C(A, C, E) and v(A, C, E) = 4

The second path is going through the nodes 1, 3, 5, 6 with coalition C(B, D, F) and v(B, D, F) = 6. The third path is going through the nodes 1, 2, 4, 3, 5, 6 with coalition C(A, C, G, D, F) and v(A, C, G, D, F) = 8. As for the grand coalition C(A, B, C, D, E, F, G), v(A, B, C, D, E, F, G) = 8

We see that each of the three paths the program outputted above, the direction flow of each segment in the path follows the direction flow from the source node to the sink node. We call these paths the forward paths. However, if a path consists of a segment whose flow direction is from the direction of sink to source node, then the path is called a backward path. In our present example, if the program was to output the following paths in the sequential order by a small modification of the source code:

Path 1: 1, 3, 5, 2, 4, 6 with C(B, D, H, C, E) and v(B, D, H, C, E) = 2

Path 2: 1, 2, 4, 3, 5, 6 with C(A, C, G, D, F) and v(A, C, G, D, F) = 4

Path 3: 1, 2, 4, 6 with C(A, C, E) and v(A, C, E) = 6

Path 4: 1, 2, 5, 6 with C(A, H, F) and C(A, H, F) = 8

Then Path 4 above is a backward path because the segment node 2 to node 5 is against the general direction of the flow, and the way we handle a backward segment is to decrease the amount of the flow value, which is 2, in the backward segment and increase the same amount to the flow value, which is also 2, in the next forward-segment in Path 4 which results in the total flow value of 4 for the player F. The amazing thing is, no matter what the order of these paths were generated, the final flow value of the network is always the same. The grand coalition C(A, B, C, D, E, F, G, H) has value v(A, B, C, D, E, F, G, H) = 8.

Finally, we can give a top-level description of MymaximalFlowAlgorithm.m

Top-level description of the algorithm:

The program MymaximalFlowAlgorithm.m has only the capacityMatrix and the source and sink nodes as inputs. The program computes and outputs the maximal flow for the whole flow network and the flow value of each edge in the flow network.

There are two basic functions (1) findPath, and (2) updateFlow.

(1) findPath finds a path consists of distinct intermediate nodes from the source to the sink node. The edge between each pair of consecutive nodes is required to have positive capacity slack. By capacity slack of an edge, we mean the difference between the capacity of the edge given by the capacityMatrix and the current flow value for the edge. The data structure used to store a path is an ordered Set path that stores each distinct nodes from the path in an orderly and iterative manner, path = [current, path], where initially path = [].

(2) updateFlow first computes the smallest capacity slack of all the capacity slacks along each of the edge segments on the path from the source node to the sink node. updateFlow then updates the flow values along each of the edge segments on the path using the smallest capacity slack that is computed.

(3) Top Level of the Maximal Flow Algorithm is shown below:

Initialize flowMatrix, MaxFlow, slackCapacityMatrix;

while (true) do

findPath;

if findPath indicates no more paths can be found

then break out of the while-loop;

end % if

updateFlow;

end (% while-loop)

Compute Max Flow of the network.

The source code of the MymaximalFlowAlgorithm.m is shown below:

```
%capacityMatrix = [0 3 0 3 0 0 0; 0 0 4 0 0 0 0; 0 0 0 0 0 0 5; 0 0 0 0 4 0 0; 0 4 0 0 0 2 0; 0 0 0 0 0 0 2; 0 0 0 0 0 0 0]
%source = 1
%sink = 7
    %capacityMatrix =

    capacityMatrix = [0 3 5 0 0 0; 0 0 0 2 0 0; 0 0 0 2 2 0; 0 0 0 0 0 4; 0 0 0 0 0 4; 0 0 0 0 0 0]
    source = 1
    sink = 6

    %capacityMatrix = [0 8 2 0 0 0; 0 0 0 6 0 0; 0 0 0 0 5 0; 0 0 2 0 0 4; 0 3 0 0 0 5; 0 0 0 0 0 0]
    %source = 1
    %sink = 6

    numVertices = size(capacityMatrix, 1)

    %Initialize the flow matrix
    flowMatrix = zeros(numVertices, numVertices);

    maxFlow = 0; %Initialize the maximum flow
    %Compute the residual capacity matrix
    slackCapacityMatrix = capacityMatrix - flowMatrix;

    %Main loop of Ford-Fulkerson algorithm with Edmonds-Karp implementation
    whileloopmain = 0;
    while true

    [isPath, path] = findPath(slackCapacityMatrix, source, sink, numVertices);

      if ~isPath;
        break; %No more augmenting path
      end
      whileloopmain = whileloopmain+1;
```

```
  fprintf('So far we have %4d path', whileloopmain);
  flowMatrix = updateFlow(flowMatrix, path, slackCapacityMatrix)
  slackCapacityMatrix = capacityMatrix - flowMatrix; %Update the residual capacity matrix

  fprintf('show the flow value after each successful path from source to sink \n')
  pathFlow = 0;
  for i = 1:numVertices
  pathFlow = pathFlow + flowMatrix(source, i);
end
  pathFlow
end

%Check for residual capacity along augmenting path
%Calculate the maximum flow
for i = 1:numVertices
  maxFlow = maxFlow + flowMatrix(source, i);
end
  maxFlow

%*******************************************************************************************
%Functions and procedures are defined below:
%*******************************************************************************************

function [isPath, path] = findPath(slackCapacityMatrix, source, sink, numVertices)
  prev = zeros(1, numVertices);
  visited = false(1, numVertices);
  q = zeros(1, numVertices);
  front = 1;
  rear = 1;
  visited(source) = true;
  q(rear) = source;
  rear = rear+1;
  isPath = false;
  whileloopbfs = 1;
  while front < rear
    whileloopbfs ;
    current = q(front);
    front = front+1;
    if current == sink
      isPath = true;
      break;
    end
    for next = 1:numVertices
      if ~visited(next) && slackCapacityMatrix(current, next) > 0
        q(rear) = next;
        rear = rear+1;
        prev(next) = current;
        visited(next) = true;
      end % end of if- loop
    end % end of for- loop
    whileloopbfs = whileloopbfs + 1;
  end % end of while- loop
```

```
    if isPath
      %Reconstruct the path
      path = [];
      current = sink;
      while current ~= source
        path = [current path];
        current = prev(current);
      end
      fprintf('the path we are taking in the bfs is :\n')
      path = [source path];
      path
    else
      path = [];
    end
  end

  %Augment the flow along the path
  function flowMatrix = updateFlow(flowMatrix, path, slackCapacityMatrix)
    delta = 999 ;
    for i = 1:numel(path)-1
      u = path(i);
      v = path(i+1);
      delta = min(delta, slackCapacityMatrix(u, v));
    end
    for i = 1:numel(path)-1
      u = path(i);
      v = path(i+1);
      flowMatrix(u, v) = flowMatrix(u, v) + delta;
      flowMatrix(v, u) = -flowMatrix(u, v);
      slackCapacityMatrix(u, v) = slackCapacityMatrix(u, v) - delta;
      slackCapacityMatrix(v, u) = slackCapacityMatrix(v, u) + delta;
    end
end
```

10.2.C The Assignment Algorithm

Example 10.2.4

We are going to show an application of the Assignment Algorithm in Computer Science that is commonly used in Industrial Engineering, this application can be formulated in the language of Coalition games. Consider the Assignment Matrix shown in Figure 10.8 below:

Figure 10.8

		Tasks		
		A	B	C
	1	5	7	9
Players	2	14	10	12
	3	15	13	16

Figure 10.8 is the content of a Cost Matrix that describes a situation where that are 3 players 1, 2 ,3 and their respectively tasks A, B and C. The cost associated with each player 1 to perform A is 5, player 2 to perform B is 10, and player 3 to perform C is 16. The rest of the entries in the matrix have similar meanings, for example, the cost for player 2 to perform C is 12 if players 2 and 3 are in coalition, the cost for player 3 to perform B is 13 if players 3 and 2 are in coalition. Any player can perform other players' own tasks if they are in coalition. The question is for each player how to choose the right task to perform, if not to perform his (her) own task, such that the total cost of the coalition is minimal. The objective of this coalition game is then to find the minimum total cost of the grand coalition C(A, B, C). In general, the coalition game can involve n-players and n tasks.

From Figure 10.8, we see that cost v(1) = 5 is, v(2) = 10, v(3) = 16 for the coalitions C(1), C(2), C(3) respectively.

For C(1, 2), v(1, 2) = min{5 + 10, 7 + 14} = 15 which occurs when player 1 chooses to perform A and player 2 chooses to perform B.

For C(1, 3), v(1, 3) = min{5 + 16, 15 + 9} = 21 which occurs when player 1 chooses A, and player 3 chooses C.

For C(2, 3), v(2, 3) = min{10 + 16, 13 + 12} = 25 which occurs when players 3 choose B and player 2 chooses to perform C.

Finally, to compute v(1, 2, 3) for the grand coalition C(1, 2, 3):

Since in our example, each player (3 in all) can choose to perform his own or the other players' tasks (3 in all), so there are 3x2x1 = 3! = 6 that many possibilities, so

v(1, 2, 3) is the minimum of the following 6 choices :

1A + 2B + 3C = 5 + 10 + 16 = 31

1A + 2C+ 3B = 5 + 12 + 13 = 30

1B + 2A + 3C = 7 + 14 + 16 = 37

1B + 2C + 3A = 7 + 12 + 15 = 34

1C + 2A + 3B = 9 + 14 + 13 = 36

1C + 2B + 3A = 9 + 10 + 15 = 34

The minimum {31, 30, 37, 34, 36, 34} = 30, which occurs when player 1 performs A, player 2 performs C, and player 3 performs B. Thus v(1, 2, 3) = 30

In general, we can see that for the grand coalition C that has n players $P_1, \ldots, P_n$, since there are n! that many permutations in n that many tasks, let σ be any of the permutations, $\sigma(i) = j$. Let k_{ij} be the cost for Player i to perform Task j = σ (i).

Then the value minimum value v(C) for the grand coalition C(1, 2,…., n) can be expressed as:

$v(C) = \underset{\sigma}{Min} \sum_{i \in C} k_{i\sigma(i)}$, where the minimum is taken over all the permutations .

Thus, we see that for any coalition C, the computation of v(C) using the expression shown above is a combinatorial optimization problem A celebrating observation is that this optimal value v(C) can also be computed from the output of the classical Assignment Algorithm, and any application of the Assignment Algorithm to a particular scenario can be formulated as a coalition game.

We are going to give a top-level description of the Assignment Algorithm below.

The only input to the algorithm is the cost matrix as shown in Figure 10.8. The output is the assignment matrix, which specifies for each player in grand coalition C what is the optimal task for the player to perform. The output also includes the optimal value v(C).

Running the MATLAB program MyAssignmentAlgorithmBV1.m for the above input shown in Figure 10.8 gives the output listing below:

```
>> myAssignmentAlgorithmBV1
originalC =
   5   7   9
  14  10  12
  15  13  16

show the size of the cost matrix m :

m =
   3
```

show the reduced cost matrix after row-min reduction:

```
C =

    0   2   4

    4   0   2

    2   0   3
```

show the reduced cost matrix after col-min reduction :

```
C =

    0   2   2

    4   0   0

    2   0   1

rzeros =

    1   2   1

czeros =

    1   2   1
```

Cover all the zeros of the modified cost matrix below:

```
coveredRows =

3×1 logical array

  0

  1

  0

coveredCols =

3×1 logical array

  1

  1

  0
```

check if we already have an assignment before further cost reduction:

```
Assign =

     1     0     0
     0     0     1
     0     1     0

yes =

     1

yes =

     1
```

yes, it is an assignment, go ahead to compute cost product of original matrix and the assignment matrix :

```
C =

     5     0     0
     0     0    12
     0    13     0

cost =

    30
```

As shown, original C is the input cost matrix as in Figure 10.8, Assign is the binary matrix computed by the algorithm, which in the case, shows that player 1 should perform A, player 2 should perform C, and player 3 should perform B. The total optimal cost for the grand coalition C(1, 2, 3) is 30.

We now give a second example with 4 players to further demonstrate the connection between this classical Assignment Algorithm and Coalition Games. The input to the algorithm is shown in Figure 10.9 below.

Figure 10.9

Tasks

Players				
1	1	4	6	3
2	9	7	10	9
3	4	5	11	7
4	8	7	8	5

As before, we can first list out the 4 one-player coalitions, the 6 two-players coalitions, the 4 three-players coalitions, and the single grand coalition. Then we compute each of the optimal value's v(i), v(i, j), v(i, j, k) and v(1, 2, 3, 4) for each coalition. It turns out that v(1, 2, 3, 4) = 21 which is the optimal value for the grand coalition. We can verify this result by running the MATLAB program myAssignmentAlgorithmBV1.m

Output listing of myAssignmentAlgorithmBV1.m is shown below:

```
>> myAssignmentAlgorithmBV1

originalC =

  1   4   6   3

  9   7  10   9

  4   5  11   7

  8   7   8   5

show the size of the cost matrix m :

m =

  4

show the reduced cost matrix after row-min reduction:

C =

  0   3   5   2

  2   0   3   2

  0   1   7   3

  3   2   3   0
```

show the reduced cost matrix after col-min reduction :

```
C =

     0     3     2     2
     2     0     0     2
     0     1     4     3
     3     2     0     0

rzeros =

     1     2     1     2

czeros =

     2     1     2     1
```

Cover all the zeros of the modified cost matrix below:

```
coveredRows =

  4×1 logical array

   0
   1
   0
   1

coveredCols =

  4×1 logical array

   1
   0
   0
   0
```

check if we already have an assignment before further cost reduction:

```
Assign =

     1     0     0     0
     0     0     1     0
     1     0     0     0
     0     0     0     1
```

```
yes =

     0

further modify the already modified cost matrix C below:

smallest =

     1

display the modified cost matrix C below:

C =

     0     2     1     1
     3     0     0     2
     0     0     3     2
     4     2     0     0

display the assignment to the modified cost matrix C below:

Assign =

     1     0     0     0
     0     0     1     0
     0     1     0     0
     0     0     0     1

yes =

     1

yes =

     1

product of original matrix and the assignment matrix :

C =

     1     0     0     0
     0     0    10     0
     0     5     0     0
     0     0     0     5
```

cost =

21

cost =

21

In the list above, the second Assign matrix passes the test because yes =1 so the optimal assignment is for player 1 to perform A, player 2 to perform C, player 3 to perform B, and player 4 to perform D with the total minimal cost of 21.

Note that the first Assign matrix does not pass the test because yes = 0, and thus the algorithm needs further processing to get to the second Assign matrix shown in the listing. The arrays coveredCols and coveredRows in the listing shows which row(s) and which column(s) are needed to cover the 0's in the reduced cost matrix with the minimal number of lines.

Top-level description of the Assignment Algorithm

We are finally able to give a top-level description of the Assignment Algorithm and the implementation MATLAB program MyAssignmentAlgorithmBV1.m

1. The algorithm starts with the cost matrix C, it then performs a min-row reduction, followed by a min-col reduction to compute a reduced cost matrix.
2. The algorithm then attempts to find an assignment from the reduced cost matrix by the function MyAssignmentFunction. The function myAssignmentCheck is used to check if an assignment as produced by MyAssignmentFunction is a valid assignment or not. If it is, the output is yes = 1, if it is not, the output is yes = 0. If yes = 1, the algorithm computes the total cost by the function computeCost and program ends.
3. If MyAssignFunction produces an assignment Assign that does not pass the myAssignmentCheck function, yes = 0, then the algorithm will further reduce the already reduced cost matrix C. This further cost reduction process is described in step 4 below.
4. First, we use the function MyCoveredZeros to identify which row(s) and column(s) can be selected to cover all the zeros in the reduced cost matrix C, and that the total number of such row(s) and column(s) used must be minimum. Then find the smallest value of all the entries that are not covered by the selected row(s) and column(s) in the reduced cost matrix C and call this value 'smallest'. Finally, add the value 'smallest' to every entry that is in the intersection(s) of the selected row(s) and column(s) and subtract from every entry that is not covered by the selected row(s) and column(s) by the value 'smallest'.
5. After step 4 above, we obtain a further reduced cost matrix C, we then apply the function myAssignmentFunction to C to come up with a new assignment Assign. This time, if we apply the function myAssignmentCheck to Assign, then the result must be yes = 1. The assignment Assign is the final solution that assign each player to perform the optimal task, and the function ComputeCost computes the optimal cost for the assignment and the program end.

Source codes for myAssignmentAlgorithmBV1.m

```
%originalC = [1 4 6 3; 9 7 10 9; 4 5 11 7; 8 7 8 5]

%C = [1 4 6 3; 9 7 10 9; 4 5 11 7; 8 7 8 5];

originalC = [5 7 9; 14 10 12; 15 13 16]
C = originalC;

%originalC = [3 8 2 10 3; 8 7 2 9 7; 6 4 2 7 5; 8 4 2 3 5; 9 10 6 9 10]
%C = originalC;

%originalC = [3 9 2 3 7; 6 1 5 6 6; 9 4 7 10 3; 2 5 4 2 1; 9 6 2 4 6]
%C = originalC;

%originalC = [1 2 3; 3 6 9;4 8 12]
%C = originalC;

fprintf('show the size of the cost matrix m: \n')
m = size(C, 1)
fprintf('show the reduced cost matrix after row-min reduction: \n')
%reduce cost matrix by subtracting each row by its row minimum
C = C - min(C, [], 2)
fprintf('show the reduced cost matrix after col-min reduction :\n')
%reduce cost matrix by subtracting each column by its column minimun
C = C - min(C, [], 1)

%Cover all zeros with minimum number of lines to the reduce cost matrix

[rzeros, czeros] = countZeros (C);
rzeros
czeros
fprintf('Cover all the zeros of the modified cost matrix below: \n')
[coveredRows, coveredCols] = MyCoverZeros(C, rzeros, czeros);

coveredRows
coveredCols

fprintf('check if we already have an assignment before further cost reduction: \n')
coveredRows1 = coveredRows;
coveredCols1 = coveredCols;
Assign = MyAssignmentFunction(C, m)

yes = myAssignmentCheck(Assign);
yes
if yes==1
    fprintf('yes, it is an assignment, go ahead to compute cost')
    cost = ComputeCost(originalC, Assign,m);
else
    fprintf('further modify the already modified cost matrix C below: \n')
    %Compute the smallest un-covered number. This smallest un-covered number
    %is subtracted from every un-covered number and this smallest un-covered
    %number is added to every number at the intersection of two lines
```

```
    smallest = min(C(~coveredRows1, ~coveredCols1), [], "all");
    smallest
    C(coveredRows,:)= C(coveredRows,:) + smallest;
    C(:, ~coveredCols) = C(:, ~coveredCols) - smallest;
    fprintf('display the modified cost matrix C below: \n')
    C
    fprintf('display the assignment to the modified cost matrix C below: \n')
    Assign = MyAssignmentFunction(C, m)
    yes = myAssignmentCheck(Assign)
    cost = ComputeCost(originalC, Assign, m)

end % end of if-else-end

%**********  The procedures are defined below:  *****************
%*************************************************************************

function Assign = MyAssignmentFunction(B, m)
    %Initialize variables
    dim = size(B, 1);

    Assign = zeros (dim, dim);

    for j = 1: dim
      for i = 1:dim
        if B(i, j) == 0

            if onlyrowzero(B, m, i, j) == 1
                B(i, j) = 999;
            else
                Assign(i, j) = 1;
                B(i, j) = 999;
            end % end of if-else

        end % end of the outer-if
      end % end of for i-loop

      j;
      B;
      Assign;

    end % end of for j-loop

end
% ******************************************************************
function [yes] = myAssignmentCheck(Assignment)
    m = size(Assignment);
    yes = 0;
    [rzeros, czeros] = countZeros(Assignment);
    %a = sum(rzeros, "all");
    %b = sum(czeros, "all");
    %if a==m & b==m
    %yes = 1;
    %end
```

```
if rzeros == czeros
    yes=1
    end % if

end

%*****************************************************************
function [rzeros, czeros] = countZeros(C)
m = size(C, 1);

for i = 1:m
    rcount = 0;
    for j = 1:m

         if C(i, j)==0
         rcount = rcount +1;

         end

    end % end of j-loop
    rzeros (i) = rcount;
end % end of i-loop

for j = 1:m
    ccount = 0;
    for i = 1:m

         if C(i, j)==0
         ccount = ccount +1;

         end
    end % end of i-loop
    czeros(j) = ccount;
end % end of j-loop

end

%*****************************************************************

function next = nextrowzero(B, m, row, column)

    next = 0;
    for i=(row +1):m
         if B(i, column) == 0
            next = i;

         end
    end
end
%*****************************************************************
function yesrow = onlyrowzero(B, m, row, column)
    yesrow = 0;
    for j = (column + 1):m
         if B(row , j) == 0
              yesrow = 1;
         end
    end
end
%*****************************************************************
function [cost] = ComputeCost(C, Assign, n)

for i = 1 : n
```

```
    for j = 1: n
         C(i, j) = C(i, j) * Assign(i, j);
    end
end
fprintf('product of original matrix and the assignment matrix :\n')
C
cost = sum(C, "all")
end
%******************************************************************
function[coveredRows, coveredCols] = MyCoverZeros(B, rzeros, czeros)
    % Initialize variables
    dim = size(B, 1);
    coveredRows = false(dim, 1);
    coveredCols = false(dim, 1);
    reminders = true( dim, dim);

    for i = 1:dim
         for j = 1:dim
              if B(i, j) == 0 && ~coveredCols(j) && ~coveredRows(i)

                   if rzeros(i) == 1 & czeros(j) == 1
                        coveredCols(j)=true;
                   end

                   if rzeros(i) > 1 & czeros(j) > 1
                   coveredRows(i) = true;
                   if all (reminders(:, j))

                        coveredCols(j) = true;
                   end % end of nested if
                end % end of if

                if rzeros(i) > 1 & czeros(j) == 1
                       coveredRows(i) = true;
                       reminders(i,:) = false;
                end

                if czeros(j) > 1 & rzeros(i) == 1
                       coveredCols(j) = true;
                end
                end % end of the outer-if
           end % end of for j-loop
      end % end of for i-loop
end
%******************************************************************
```

10.2.D The pairwise shortest path Algorithm

Like the last three computer algorithms where the concepts of Coalition Games can be applied as an alternative way to solving the underlying problems, we now look at another formulation of Coalition Games to a scenario related to the celebrated Floyd-Warshall Algorithm which is used to compute the shortest path between any pair of nodes in a digraph.

Example 10.2.5

Consider the following digraph Figure 10.10 and its adjacency matrix shown in Figure 10.11 below:

Figure 10.10

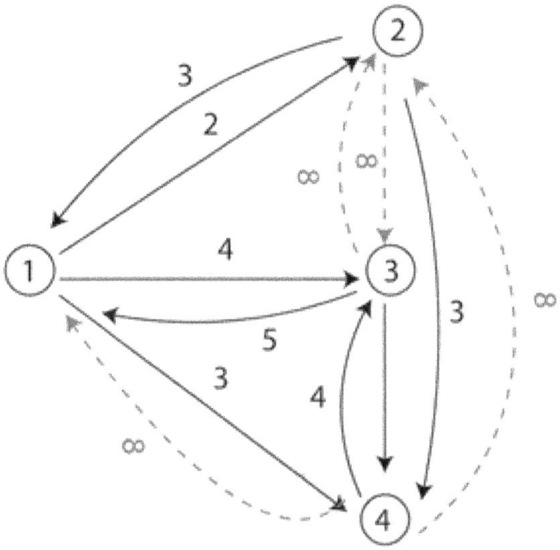

Figure 10.11

	1	2	3	4
1	0	2	4	3
2	3	0	∞	3
3	5	∞	0	3
4	∞	∞	4	0

There are four players labelled 1, 2, 3, 4. The direct distances between any two nodes in the digraph are shown. The distance marked in red symbol ∞ means that two nodes are not directly connected as shown, but they could be connected by other directly connected edges. Think of a scenario where the digraph above represents a network of interconnected operating control centers and the numbers between any two centers represents the timing cost of the communication between them. The edge marked ∞ the minimum timing cost yet to determine. In this application, we treat each node in the network as a player of a coalition game. The coalitions to be formed amongst the players. The objective of the game is to form a minimal coalition amongst the players 1, 2, 3, 4 so that the timing costs' ∞ s of the indirect connected nodes are determined, and each one is the minimum value possible. It turns out that the problem we just described is very well fitted for the FloydWarshall Algorithm to solve. As such we first give a top-level description of the algorithm below followed by running the algorithm MyFloydWarshallAlgorithm. m using the inputs of Figure 10.10 and followed by the interpretation of the outputs listing.

Top-level description of the FloyWarshall Algorithm

The only input to the algorithm is the adjacency-matrix D that gives the direct distances between two adjacency nodes, and where the entries marked with the symbol ∞ in the (i, j) location of D are replaced with a large numerical value 99. This numerical value 99 is used here to signify that before running the algorithm the distance between node i and node j are large, and as the algorithm begins its execution, the distance between node i and node j are getting to improve to a smaller value and this smaller value will be replacing the value 99 in the (i, j) location of the distance-matrix dist. The output of the algorithm gives the minimal values one each at the corresponding (i, j) locations where 99 were placed before, this value represents the indirect minimal distance between node i and node j that can be connected by the direct pair of nodes as shown in the adjacency-matrix.

Basic idea of the Floyd-Warshall Algorithm

The basic idea of the Floyd Warshall Algorithm is that for any pair of nodes (i, j), if node i is directly connected to node j, the distance is shown in the adjacent matrix; if node i is not directly connected to node j, we then ask ourselves, can they be indirectly connected in between by a subset of nodes taken from the total set of nodes labelled 1, 2, 3….n ?. The algorithm starts with looking at node 1 as the intermediate node between any two nodes i and j are not directly connected, check from the adjacency matrix if node i is directly connected to node 1 and if node 1 is directly connected to node j; if both conditions are met, then we can update the distance d(i, j) from 99 to d(i, j) = d(i, 1) + d(1, j). The algorithm next will look at two nodes {1, 2} as intermediate nodes between node i and node j. Check now if node i is directly connected to node 2 and if node 2 is directly connected to node j. If yes to both checks, then update the distance between node i and node j as d(i, j) = d(i, 2) + d(2, j). Note d(i, 2) and d(2, j) are the distances between node i and node 2, and node 2 to node j with only the node {1} in between, but d(i, j) is the distance between node i and node j with nodes {1, 2} are as the connecting nodes. The algorithm can now be described iteratively as follows. Assume now, between any pair of nodes (i, j), node i and node j can be directly connected using the nodes from the set of nodes labelled {1, 2….0.k}. We then look at the node labelled (k + 1), a node that is not in the set {1, 2….k}. To compute d(i, j), the indirect distance between node i and node j, we see that, d(i, j) = d(i, k+1) + d(k+1, j). Note that the two terms d(i, k+1) and d(k+1, j) on the right hand side are the respective indirect distances from node i to node (k+1), and from node (k+1) to node j whose in between connecting nodes are from the set {1, 2,….k}, and the distance d(i, j) on the left hand side is the distance between node i and node j whose in between nodes are from the set {1, 2,…k+1}. Note that based on this iterative process, the value of d(i, j) using the set of in-between nodes {1, 2….k+1} in the computation does not have to be smaller than the value of d(i, j) using the set of in-between nodes {1, 2…k}. To keep the smallest distance d(i, j) in the final output, the algorithm therefore always keeps the smaller value of d(i, j) by comparison in any two consecutive iterations. And the algorithm repeats itself until k = n then stops, where n is the total number of nodes.

Output listing of the algorithm:

Running the MyFloydWarshallAlgorithm.m , the output listing is as follows:

```
>> MyFloydWarshallAlgorithm

D =

     0     2     4     3
     3     0    99     3
     5    99     0     3
    99    99     4     0

For node 1

Using the set  of nodes labelled from 1 to 1 to connect any pair (i, j) of nodes with smallest

dist =

     0     2     4     3
     3     0     7     3
     5     7     0     3
    99    99     4     0

For node 2

Using the set  of nodes labelled from 1 to 2 to connect any pair (i, j) of nodes with smallest

dist =

     0     2     4     3
     3     0     7     3
     5     7     0     3
    99    99     4     0

For node 3

Using the set  of nodes labelled from 1 to 3 to connect any pair (i, j) of nodes with smallest

dist =

     0     2     4     3
     3     0     7     3
     5     7     0     3
     9    11     4     0
```

For node 4

Using the set of nodes labelled from 1 to 4 to connect any pair (i, j) of nodes with smallest

dist =

0	2	4	3
3	0	7	3
5	7	0	3
9	11	4	0

Let's examine the output listing. The input D matrix repeats what the given input matrix of direct distances between node i and node j other than those 4 locations at (2, 3), (3, 2), (4, 1) and (4, 2) that are marked with the 99. Using only node 1 as the interconnecting node, D is updated from 99 to 7 at the location (2, 3), and from 99 to 7 at the location (3, 2). However, the value of D at the location of (4, 1) and (4, 2) remains as 99 unchanged as we can see in the second matrix. When nodes {1, 2} are used as the interconnecting nodes between i and j, the values of D at locations (4, 1) and (4, 2) still remain 99 as we can see in the second D matrix. When {1, 2, 3} are used as the interconnection nodes, values of D at locations (4, 1) and (4, 2) are improved to 9 and 11 respectively as we can see in the third D matrix. When the algorithm performs its last iteration by using {1, 2, 3, 4} as the interconnection nodes, the value entries in D at the locations (4, 1) and (4, 2) remain unchanged as we can see from the fourth D matrix. In summary, given the current input data, forming the minimal coalitions C(1), C(1, 3) are helpful in achieving our objectives, forming other or larger coalitions such as C(1, 2, 3), C(1, 2, 3, 4) does not improve.

Example 10.2.6

As a second example, let us consider

Figure 10.12

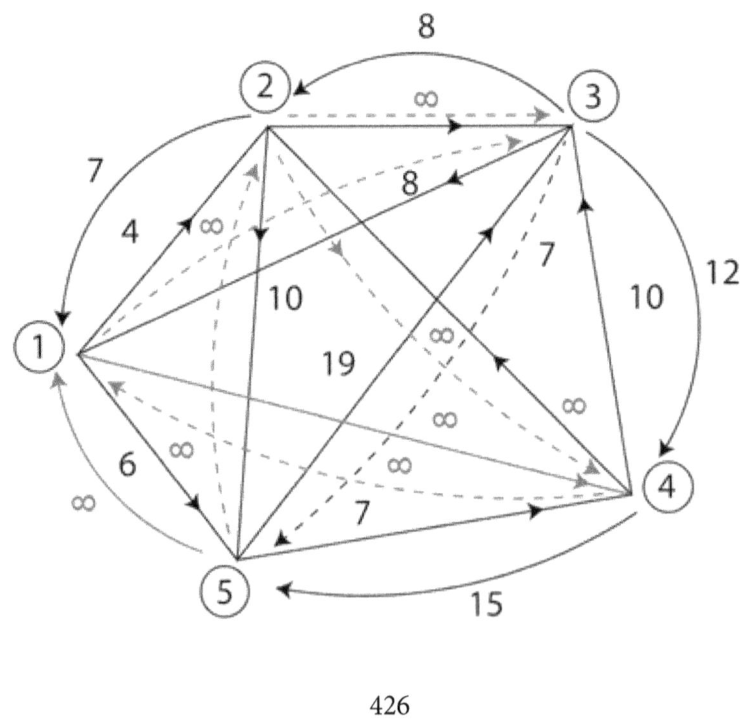

The corresponding adjacent matrix is shown in Figure 10.13 below.

Figure 10.13

	①	②	③	④	⑤
①	0	4	∞	∞	6
②	7	0	∞	∞	10
③	8	8	0	12	7
④	∞	∞	10	0	15
⑤	∞	∞	19	7	0

Running MyFloydWarshalAlgorithm.m with the above input adjacency matrix, the output listing is shown below:

Output listing of Example 10.2.6

```
>> MyFloydWarshallAlgorithm

D =

   0    4   99   99    6

   7    0   99   99   10

   8    8    0   12    7

  99   99   10    0   15

  99   99   19    7    0

For node 1

Using the set of nodes labelled from 1 to 1 to connect any pair (i, j) of nodes with smallest

dist =

   0    4   99   99    6

   7    0   99   99   10

   8    8    0   12    7

  99   99   10    0   15

  99   99   19    7    0
```

For node 2

Using the set of nodes labelled from 1 to 2 to connect any pair (i, j) of nodes with smallest

dist =

0	4	99	99	6
7	0	99	99	10
8	8	0	12	7
99	99	10	0	15
99	99	19	7	0

For node 3

Using the set of nodes labelled from 1 to 3 to connect any pair (i,j) of nodes with smallest

dist =

0	4	99	99	6
7	0	99	99	10
8	8	0	12	7
18	18	10	0	15
27	27	19	7	0

For node 4

Using the set of nodes labelled from 1 to 4 to connect any pair (i, j) of nodes with smallest

dist =

0	4	99	99	6
7	0	99	99	10
8	8	0	12	7
18	18	10	0	15
25	25	17	7	0

For node 5

Using the set of nodes labelled from 1 to 5 to connect any pair (i, j) of nodes with smallest dist =

0	4	23	13	6
7	0	27	17	10
8	8	0	12	7
18	18	10	0	15
25	25	17	7	0

Let's interpret the above output listing. The first matrix is the input matrix. Using node {1} or nodes {1, 2} as the interconnecting node or nodes, the algorithm does not update any of the eight entries marked 99 as the second matrix and the third matrix show. Using nodes {1, 2, 3}, the entries (4, 1), (4, 2), (5, 1), (5, 2) have updated to 18, 18, 27, 27 respectively, but the entries located at (1, 3), (1, 4), (2, 3), (2, 4) remain unchanged as shown in the fourth matrix. Using nodes {1, 2, 3, 4} as the interconnecting nodes, the situation is the same as before when using {1, 2, 3} as the interconnecting nodes. When the algorithm performs it final iteration, using nodes {1, 2, 3, 4, 5} as the interconnecting nodes, the values the entries at the locations (1, 3), (1, 4), (2, 3), (2, 4) are now updated to 23, 13, 27, 17 respectively. In summary, for the present input data, the minimal coalitions are C(1, 2, 3) and C(1, 2, 3, 4, 5), they are the key coalitions used to achieve our objectives. The other coalitions C(1), C(1, 2), C(1, 2, 3, 4) are redundant.

10.2.E The Dijkstra's Shortest Path Algorithm

Example 10.2.7

As we have seen before the Minimum Spanning Tree Algorithm (MST) provides useful information to certain scenarios described by coalition games, in particular, the algorithm computes the value to the grand coalition of the game. The Dijkstra's Shortest Path Algorithm, on the other hand, can also be helpful in determining the value(s) of the coalition (s) described in the game.

Consider a similar application scenario described in Figure 10.5 before, but this time, the original cost associated with distributing the power directly from the power source location 1 to each of the other locations 2, 3, 4, 5, 6 (lines shown in red in the figure) are much higher (Figure 10.14). The Dijkstra's Shortest Path Algorithm can help in identifying which of the direct line from 1 can be rerouted by other intermediate connection(s) between locations(s).

Figure 10.14

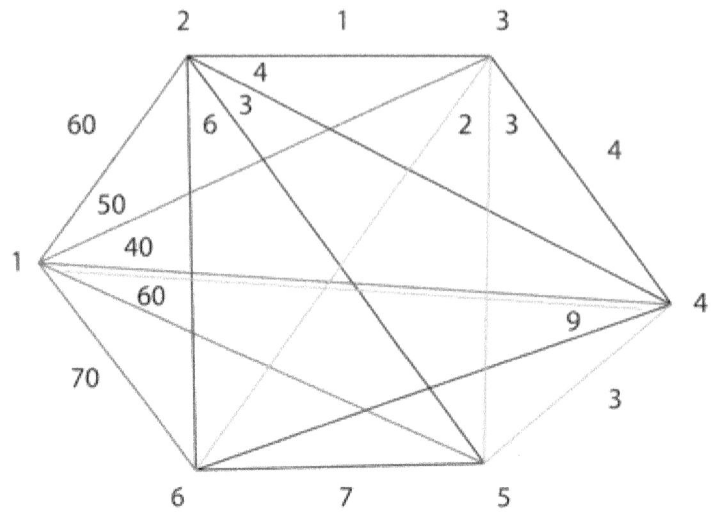

Running the Diikstra's Algorithm coded in MATLAB

Running the Dijkstra's Algorithm coded in MATLAB, myDijkstraAlgorithmBV.m, gives the output listing below:

Output listing of myDijkstraAlgorithmBV.m & myMinSpanTreeBV.m

```
>> myDijkstraAlgorithmBV

graph =

  0  60  50  40  60  70

 60   0   1   8   3   6

 50   1   0   4   3   2

 40   8   4   0   3   9

 60   3   3   3   0   7

 70   6   2   9   7   0

Show the final Pdsr, Visited, Distances arrays after all iteration(s):

***************************************************************

Pdsr =

 100   3   4   1   4   3   0   0   0

Visited =

   1   1   1   1   1   1   0   0   0
```

Distances =

0 45 44 40 43 46 0 0 0

Interpretation of myDijkstraAlgorithmBV.m outputs above are in order.

The graph matrix just repeats the adjacency graph shown in Figure 10.14

The entries in the Pdsr (predecessor array) array have the following meaning:

The first entry 1 has value 100 indicates node 1 has no predecessor, it is the root node. The subsequent entries 34143 means node 3 is the predecessor of node 2, node 4 is the predecessor of node 3….,and node 3 is the processor of node 6.

The entries in the Visited array have the following meaning:

Entry 1 means the node given by the array-index has been visited, and 0 has not. In our case, the Visited array shows all 6 nodes have been visited.

The Distances array gives the numerical distance from node 1 to the node given by the array index. In our case, 0 45 44 40 43 46 means node 1 is the root node, it has 0 distance from itself, node 2 has 45 from node 1, node 3 has 44 form node 1, and so on.

Note that, the Pdsr array and the Distances array together tells us not only the shortest distance from each node to the root node, but also gives the backward traversal of the shortest path from any node back to the root node. For example, the shortest path from the source node to node 4 is 40, which is a direct connection; the next shortest path from the source node to node 5 is 43 which goes through from source node to node 4, to node 5; the longest distance from source node to node 6 is 46 which goes through from source node to node 4, to node 3, then to node 6. Using the above information, we can form coalitions C(2), C(3), C(4), C(5), C(6), C(4, 3), C(4, 5), C(4, 3, 6), C(4, 3, 2) using the following values v(2) = 60, v(3) = 50, v(4) = 40, v(5) = 60, v(6) = 70, v(4, 3) = 44, v(4, 5) = 43, v(4, 3, 6) = 46, v(4, 3, 2) = 45, and v(2, 3, 4, 5, 6) = 49 for the grand coalition C(2, 3, 4, 5, 6).

Now running the myMinSpanTreeBV.m, we would have the same result for the grand coalition C(2, 3, 4, 5, 6) as the output listing shows:

```
>> myMinSpanTreeBV

graph =

 0 60 50 40 60 70
60  0  1  4  3  6
50  1  0  4  3  2
40  4  4  0  3  9
60  3  3  3  0  7
70  6  2  9  7  0
```

Pdsr =

Inf 3 5 1 4 3 0 0 0 0

weight =

49

The weight of the Minimal Network is 49

Example 10.2.8 (Incorporate MinSpanTree & Dijkstra's algorithm in coalition games)

We close this section with another example to show incorporating both the Minimal Network Spanning Tree and the Dijkstra's Shortest path Algorithms in determining the values of coalitions in Coalition Games.

Figure 10.15

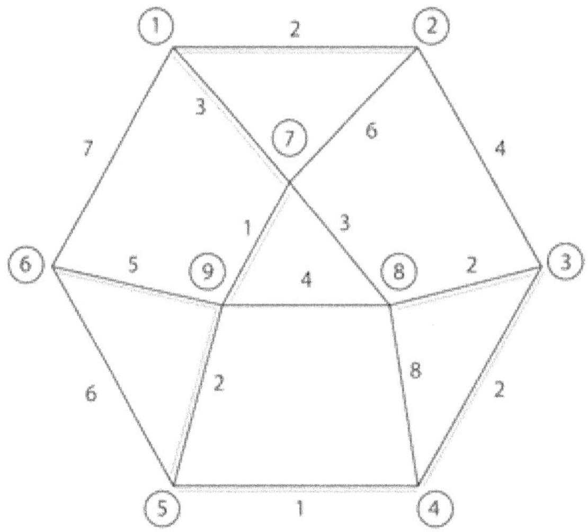

Running myMinSpanTreeBV.m shows that the edges shown in green form the Minimal Spanning Tree with the tree's root node at node 1. The content of the Pdsr array of the myMInSpanTreeBV.m also shows that the leave nodes are nodes 2, 6, and 8. Tracing back the path that consists of the most number of edges from leave node 8 to the root node 1, 8 < -3 < -4 < -5 < -9 < -7 < -3 < -1 gives the value v(3, 4, 5, 7, 9) = 11 for the coalition C(3, 4, 5, 7, 8, 9). Thus, for the grand C(2, 3, 4, 5, 6, 7, 8, 9), v(2, 3, 4, 5, 6, 7, 8, 9) = 11 + 2 + 5 = 18, where 2 and 5 are the respectively shortest distance from node 2, node 6 to the coalition C(3, 4, 5, 7, 9). Running myDijkstraAlgorithmBV.m gives:

v(1) = 0, v(2) = 2, v(6) = 7, v(7) = 3 are the non-zero single coalitions C(1), C(2) ,C(6), C(7) respectively. v(2, 3) = 6, v(4, 5, 7, 9) = 7, v(5, 7, 9) = 6, v(7, 8) = 6, v(7, 9) = 4 are the values for the coalitions C(2, 3), C(4, 5, 7, 9), C(5, 7, 9), C(7, 8), and C(7, 9) respectively. Note that for node 8 in the present graph, the Distances array in the Dijkstra's Algorithm outputs gives the shortest distance from node 1 to node 8 is 6, but the Pdsr array in the output of the Minimum Spanning Tree Algorithm gives the distance of 11. The output of both algorithms are shown below

```
>> myMinSpanTreeBV

graph =

     0     2     0     0     0     7     3     0     0
     2     0     4     0     0     0     6     0     0
     0     4     0     2     0     0     0     2     0
     0     0     2     0     1     0     0     8     0
     0     0     0     1     0     6     0     0     2
     7     0     0     0     6     0     0     0     5
     3     6     0     0     0     0     0     3     1
     0     0     2     8     0     0     3     0     4
     0     0     0     0     2     5     1     4     0

Pdsr =

   Inf     1     4     5     9     9     1     3     7     0

weight =

    18

The weight of the Minimal Network is   18

>> myDijkstraAlgorithmBV

graph =

     0     2     0     0     0     7     3     0     0
     2     0     4     0     0     0     6     0     0
     0     4     0     2     0     0     0     2     0
     0     0     2     0     1     0     0     8     0
     0     0     0     1     0     6     0     0     2
     7     0     0     0     6     0     0     0     5
     3     6     0     0     0     0     0     3     1
     0     0     2     8     0     0     3     0     4
     0     0     0     0     2     5     1     4     0

Show the final Pdsr, Visited, Distances arrays after all iteration:

*************************************************************
```

Pdsr =

100 1 2 5 9 1 1 7 7

Visited =

1 1 1 1 1 1 1 1 1

Distances =

0 2 6 7 6 7 3 6 4

10.3 Summary

This chapter introduces the concept of **coalition games**, a framework for analyzing how groups of players (coalitions) form, the actions they take, and the payoffs they receive. Instead of focusing on individual players, coalition games examine the strategic interactions between different subsets of players within a given set N.

A coalition game is formally defined by its characteristic function, which assigns a numerical value to each possible coalition, representing the collective payoff or strategic advantage of that group.

In Section 10.2, we explore practical applications of coalition games through five fundamental algorithms in computer science:

1. Minimum Spanning Tree Algorithm
2. Maximum Flow Algorithm
3. Assignment Algorithm
4. Pairwise Shortest Path Algorithm
5. Dijkstra's Shortest Path Algorithm

By modeling these algorithms as coalition games, we illustrate how cooperative decision-making principles apply to optimization problems, network structures, and resource allocation. Through these examples, we establish a clear connection between theoretical game mechanics and real-world computational strategies.

Technical Takeaway: **The value v(C) of a coalition C in a game with payoff values can be expressed as a maximin or minimax optimization problem. Specifically, it is calculated as the maximum over the set of joint actions of the members within coalition C, while simultaneously being the minimum over all joint actions of the members outside the coalition C. (Refer to examples 10.1.1, 10.1.2, and 10.1.3). The classic algorithms from computer science discussed in this chapter are valuable tools for computing coalitions and modeling various real-world situations as coalition games.**

Chapter 11

Computing tools for Coalition Games

11.0 Introduction

Recall from the last chapter, we formulate a coalition game as a function (the characteristic function) that assigns each coalition C from a set of n players {1, 2…n} a specific value v(C). This set of values v(C), however, only gives the payoff or the cost for the entire coalition C and it does not reveal any information about the cost or reward of each individual player inside the coalition C. In this chapter, we discuss two approaches, the imputation set approach and the Shapley Value approach, both of these methods are used to divide the cost or reward v(C) of the coalition C to each of its players in the coalition. Along with the concept of imputation set, we introduce a very important subset of the imputation set called the Core of the coalition game. Since Core is a set, we also introduce the concept of a very special element in the Core called the Nucleolus of the Core. We state without proof of this, but exemplify this, if the Core is not empty, the Nucleolus is always there (i.e. it always exists!). There are several MATLAB programs written to demonstrate the above concepts. These are (1) myShapleyvalueBV.m, (2) myComputeNucleolus112BV.m (3) myComputeNucleolusBV4444.m (4) myInsurancelExample.m (5) myMobilesExample.m (6) myManufacturingExample.m

11.1 Algebraic and Geometric descriptions of the Imputation Set and the Core of Coalition Games

Before we explain the concepts stated before, we need to establish some notations and definitions.

Definition 1.1 (Imputation Set)

Given a coalition game of n players {1, 2...n} with its characteristic function defined by the coalitions C and the corresponding values v(C). The imputation set, denoted by I(v), is a set of all n-tuples of values $(x_1, x_2, \ldots x_n)$ in R^n, where x_i is the value allocated to the i-th player, such that the following two conditions are satisfied:

(1) Efficiency Condition: All the values allocated to each player sum up to the value of the grand coalition given by the characteristic function.

$$\sum_{i\in N} x_i = v(N)$$

(2) Individual Rationality Condition: We assume that no two players benefit from quitting the grand coalition by playing alone. This condition is expressed in the inequality shown below:

$x_i \geq v(i) \quad \forall \ i \in N$

This is because x_i is the value allocated to player i when he(she) is with the other players in a grand coalition C(1, 2... i….n), this value is greater or equal to the value v(i) when player i is playing alone by assumption.

Note that the imputation set I(v) defined by the Efficiency Condition and the Individual Rationality Condition above may be empty. We want to impose an additional condition (3) below to guarantee that I (v) will not be empty.

(3) $v(N) \geq \sum_{i \in N} v(i)$

Example 1.1

Let us look intuitively at the imputation set I(v) when n = 3. Assume we have a coalition game whose characteristic function is given as:

v(1) = 1, v(2) = 2, v(3) = 3, v(1, 2, 3) = 10 for the coalitions C(1), C(2), C(3), and C(1, 2, 3).

Then the imputation set I(v) is a subset of points (x_1, x_2, x_3) in R^3 defined by 4 equality and inequalities,

$x_1 + x_2 + x_3 = 10$ (1)

$x_1 \geq 1$ (2)

$x_2 \geq 2$ (3)

$x_3 \geq 3$ (4)

The totality of these points (x_1, x_2, x_3) will be a subset of the simplex triangle ABC shown in Figure 11. 1 because of the equality (1) above. Intuitively, we can use the following procedure described below to create a convex hull to contain all the points defined by the set of equations and inequalities (1) through (4) above. The corner points of the convex hull, one corner point for each player, is computed as follows:

Figure 11.1

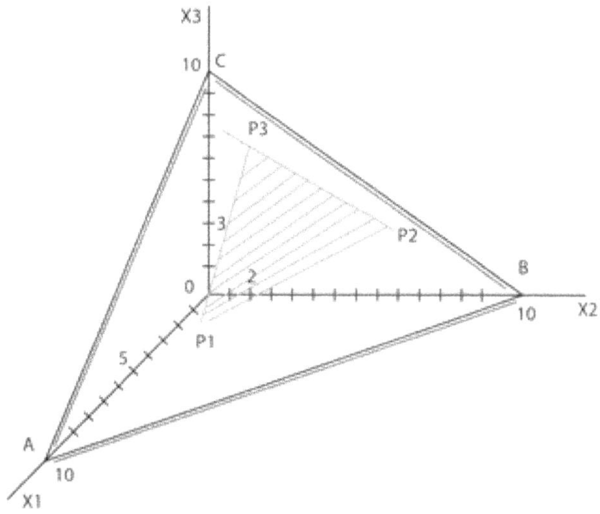

The corner point for Player 1 has coordinate P1 = (a, b, c) where b and c are the values of Player 2 and Player 3, which is v(2) = 2, v(3) = 3 respectively, and a is the value obtained from equation (1) which is v(1, 2, 3) - v(2) – v(3) = 10 - 2 - 3 = 5. Thus, the corner point for P1 is (5, 2, 3).

The corner point for Player 2 has coordinate P2 = (a, b, c) where a and c are the values of Player 1 and Player 3, which is v(1) = 1, v(3) = 3 respectively, and b is the value obtained from equation (1) which is v(1, 2, 3) - v(1) – v(3) = 10 - 1 - 3 = 6. Thus, the corner point for P2 is (1, 6, 3).

Similarly, the corner point for Player 3 has coordinate P3 = (a, b, c) where a and b are the values of Player 1 and Player 2, which is v(1) = 1, v(2) = 2 respectively, and c is the value obtained from equation (1) which is v(1, 2, 3) - v(1) – v(2) = 10 - 1 - 2 = 7. Thus, the corner point for P3 is (1, 2, 7).

Now the three corner points P1, P2, P3 form a triangle simplex as shown in Figure 11.2 shown. The points (x_1 , x_2 ,x_3) in the triangle simplex can be thought of as a geometric representation of the algebraic equation and inequalities (1) to (4), and it is the imputation set for the three players coalition game. We now show that this imputation set is not empty, and every point in the set satisfies the conditions (1), (2), (3) in Definition 11.1 as follows. As any point in this set is given by m1(5, 2, 3) + m2(1, 6, 3) + m3(1, 2, 7), which can be simplified as (1 + 4m1, 2 + 4m2, 3 + 4m3), for any m1, m2,m3 in [0, 1] such that m1 + m2 + m3 = 1. So, the three conditions (1), (2) and (3) are easily verified as V(1) = 1, v(2) = 2, V(3) = 3, v(1, 2, 3) = 10

In general, if there are n players, and v(1), v(2) …., v(n), v(1, 2… n) are the values corresponds to the single coalitions C(1), C(2) …. C(n)., C(1, 2… n). Then the implementation set for the coalition game is the convex hull of the n corner points P1, P2… Pn, where the corner point Pi for the i-th player is given by Pi = (x_1, x_2, x_i,…x_n), where for all j ≠ i, x_j = v(j) and the coordinate for the ith player is the value given by:

$$v(i) = v(N) - \sum_{j \neq i} v(j)$$

Definition 11.1.2 (The Core of a coalition game)

In addition to the conditions (1), (2), and (3) that defines the imputation set stated in Definition 11.1 above, if the following condition, known as "stability with respect to sub-coalitionv" also satisfies, then the collection values $(x_1, x_2, \ldots x_n)$ in R^n is defined as the Core of the game denoted by C(v).

(4) For every sub-coalition C in the game, $\sum_{i \in C} x_i \geq v(C)$

In other words, the Core C(v) is always a subset of the Imputation set I(v).

Example 11.1.2

If the game in Example 11.1 above, has in addition three coalitions C(1, 2) and C(2, 3), C(1, 3) with v(1, 2) = 4, v(2, 3) = 5, v(1, 3) = 4, then C(v) is a collection (x_1, x_2, x_3) satisfying the following:

$$x_1 + x_2 + x_3 = 10 \quad (1)$$

$$x_1 \geq 1 \quad (2)$$

$$x_2 \geq 2 \quad (3)$$

$$x_3 \geq 3 \quad (4)$$

$$x_1 + x_2 \geq 4 \quad (5)$$

$$x_2 + x_3 \geq 5 \quad (6)$$

$$x_1 + x_3 \geq 4 \quad (7)$$

In this example, it is easily checked that the points P1(5, 2, 3) and P2(1, 6, 3) from Example 11.1 satisfies all the equation and inequalities (1) to (6) above, thus P1 and P2 are in C(v), and C(v) is not empty. However, the general question of whether C(v) is empty or not for a game of n players is rather involved, for the intention of this book, we shall only state and without proof of the following Theorem due to Bondareva and Shapley.

11.2 When is the Core of a coalition game non-empty?

We are going to give a necessary and sufficient condition for a coalition game to have a non-empty Core. However, before doing so, we need to introduce the concept of a balanced map on a coalition game.

Definition 11.2.3 (A balanced map on a coalition game)

A balanced map on a coalition game is a function u that assigns each non-zero coalition of N = {1, 2… n} a value from R, u: $2^N \rightarrow R$ such that the following system of n linear equations are satisfied:

$$\sum_{C} u(C) 1_C = 1_N \qquad \text{(D4)}$$

where the summation is over all coalition C from the game, and where 1_C and 1_N are the two nX1 column-vectors.

1_N is a nx1 column-vector consisting of all n 1 entries.

1_C is a nx1 column-vector consisting of 0's and 1's, it is a 1 in the ith position only if player i is in the coalition C, and it is a 0 in the ith position if player i is not in the coalition C. The column vector 1_C defined in this way is referred to as the characteristic vector of coalition C.

Example 11.2.3

Using the coalition game shown in Example 11.1.2, we give an example of a balanced map u as follows. First, we represent the given coalitions C by their characteristic vectors,

$$1_{C(1)} = \begin{bmatrix} 1 \\ 0 \\ 0 \end{bmatrix}, \quad 1_{C(2)} = \begin{bmatrix} 0 \\ 1 \\ 0 \end{bmatrix}, \quad 1_{C(3)} = \begin{bmatrix} 0 \\ 0 \\ 1 \end{bmatrix}$$

$$1_{C(1,2)} = \begin{bmatrix} 1 \\ 1 \\ 0 \end{bmatrix}, \quad 1_{C(1,3)} = \begin{bmatrix} 1 \\ 0 \\ 1 \end{bmatrix}, \quad 1_{C(2,3)} = \begin{bmatrix} 0 \\ 1 \\ 1 \end{bmatrix}$$

$$1_N = \begin{bmatrix} 1 \\ 1 \\ 1 \end{bmatrix}$$

Let u(C(1)) = 0, u(C(2)) = 0, u(C(3)) = 0, u(C(1, 2, 3)) = 0, u(C(1,2)) = 1/2, u(C(1, 3)) = 1/2, u(C(2, 3)) = 1/2. Then u is a balanced map because

$$\tfrac{1}{2}\, 1_{C(1,2)} + \tfrac{1}{2}\, 1_{C(1,3)} + \tfrac{1}{2}\, 1_{C(2,3)} = \tfrac{1}{2} \begin{bmatrix} 1 \\ 1 \\ 0 \end{bmatrix} + \tfrac{1}{2} \begin{bmatrix} 1 \\ 0 \\ 1 \end{bmatrix} + \tfrac{1}{2} \begin{bmatrix} 0 \\ 1 \\ 1 \end{bmatrix} = \begin{bmatrix} 1 \\ 1 \\ 1 \end{bmatrix} = 1_N$$

So, condition D4 is satisfied.

Example 11.2.4

We can see from the definition of a balanced map D4 that they are not unique. In fact, if we assign u the value 1 to every singleton coalition, and assign u the value 0 to any other coalitions, then u will also be a balanced map because, trivially,

$$1 \begin{bmatrix} 1 \\ 0 \\ 0 \end{bmatrix} + 1 \begin{bmatrix} 0 \\ 1 \\ 0 \end{bmatrix} + 1 \begin{bmatrix} 0 \\ 0 \\ 1 \end{bmatrix} = \begin{bmatrix} 1 \\ 1 \\ 1 \end{bmatrix}$$

Now we are in position to state the following equivalent condition for a game to have a non-empty Core.

Theorem 11.2.1 (Bondareva and Shapley Theorem)

The Core C(v) is non-empty if and only if

$$\sum_{C} u(C)v(C) \leq v(N) \qquad \text{(D5)}$$

, where the summation is over all non-empty coalitions C.

Example 11.2.5

Referring to Example 11.2, we know that the points P1 and P2 are in the core, and so the core is not empty. We now verify that (D5) above is satisfied.

Using the balanced map u shown in Example 11.3, we have on the LHS of (D5)

½ v(1, 2) + ½ v(1, 3) + ½ v(2, 3)

= ½(4) + ½(4) + ½(5) = 6.5 ≤ 10, so D5 is verified.

If we were to use the trivial balanced map from Example 11.4, then condition D5 is also satisfied because,

1v(1) + 1v(2) + 1v(3)

= 1 + 2 + 3 = 6 ≤ 10.

11.3 What are the Shapley Values?

Recall that the characteristic function in a coalition game assigns each coalition C to a payoff value v(C), the formation of these coalitions can be thought of as starting from the singleton coalitions, by joining an additional player to form the 2-players coalitions; by joining an additional player to the 2-players coalitions to form 3-players coalitions, and so to come up with the final n-players grand coalition. Thus, for each player i, not already in an existing coalition, he(she) has the choice of joining which of the existing coalition that is most advantageous to him (her). Shapley in 1953 formulated the idea of 'value added' to the existing coalition C when the new player joins the coalition, and the player can use this 'value added to C' as his (hers) bargaining power for the acceptance of joining C.

We are going to describe and formulate the concept of Shapley value as follows:

Assume there are n players {1, 2… n}, denote a permutation of these n players by p to mean a coalition C is formed in sequential order by joining the coalition one player at a time. For example, when n = 3, the permutation p1 = (1 2 3) means player 1 started the coalition as a singleton, then player 2 joins, then player 3 joins to form the grand coalition. The permutation p2 = (3 1 2) means player 3 started the singleton coalition, then player 1 joins in, then player 2 joins in last to form the grand coalition. We also define and denote what is meant by the predecessor (s) P of a player in a permutation as follows. In p1, the predecessor of player 1 has none, the predecessors of player 2 is player 1, and predecessors of player 3 are players 1 and 2. We denote $P(1, p1) = \varnothing$, $P(2, p1) = \{1\}$, $P(2, p2) = \{1, 3\}$ and $P(1, p2) = \{3\}$. Thus, if p is any permutation in a game with n players, and i any player, then the notation P(i, p) is the set of players preceding player i in forming the coalition p. Now we can give the definition of what is meant by value added to a coalition by joining a new player.

Definition 11.3.4 (value added to a coalition and Shapley values)

Let n be the number of players, the added value $\Delta(i, p)$ of player i to a coalition C defined by the permutation p is defined as follows:

$$\Delta(i, p) = v(P(i, p) \cup \{i\}) - v(P(i, p)) \qquad \text{(D6)}$$

In other words, the added value $\Delta(i, p)$ is the difference between the value of the coalition after player i joined the coalition and the value of the coalition before player i joined the coalition.

For a given p, the vector of added value for each player is defined as:

$$\Delta(p) = (\Delta(1, p), \Delta(2, p), \ldots, \Delta(n, p)) \qquad \text{(D7)}$$

The Shapley value SV of a coalition game with characteristic function v is defined as the average of the vectors of added value for each player over all the n! permutations:

$$SV(v) = \frac{1}{n!}\sum_{p} \Delta(p) \qquad \text{(D9)}$$

11.3.1 Computing the Shapley Values

(myShapleyvalueBV.m)

Example 11.3.6

We are going to use the coalition game shown in Example 11.2 to illustrate the computation of the Shapley values using the above definitions D6 to D9 above.

First, let us list all the permutations:

p1 = (1 2 3)

p2 = (2 3 1)

p3 = (3 1 2)

p4 = (3 2 1)

p5 = (2 1 3)

p6 = (1 3 2)

Let's compute the SV value for player 1 by finding out player 1's value added to each of the coalition C from the permutations listed above.

In p1: Player 1 has no predecessor, P(1, p1) = ∅, from D6,

$\Delta(1, p1) = v(P(1, p1) \cup \{1\}) - v(P(1, p1)) = v(\{1\}) - 0 = 1$

In p2: Player 1 has two predecessors, P(1 ,p2) = {2, 3}, from D6,

$\Delta(1, p2) = v(P(1, p2) \cup \{1\}) - v(P(1, p2)) = v(\{1, 2, 3\}) - v(\{2,3\}) = 10 - 5 = 5$

In p3: Player 1 has one predecessor {3}, P(1 ,p3) = {3}, from D6,

$\Delta(1, p3) = v(P(1, p3) \cup \{1\}) - v(P(1, p3)) = v(\{1, 3\}) - v(\{3\}) = 4 - 3 = 1$

Similarly, for permutations p4, p5, and p6, we find Δ(1, p4) = 5, Δ(1, p5) = 2, and Δ(1, p6) = 1.

Thus, if we sum Δ(1, p) over p = p1 to p6, we find the total value added if player 1 joined each possible coalition is: 1 + 5 + 1 + 5 + 2 + 1 = 15. The Shapley value for player 1 is then SV(1) = 15/6 = 2.5.

As for Player 2, using D6 as before, Δ(2, p) = v(P(2 ,p) ∪ {2}) – v(P(2 ,p)) for any permutation p, we find Δ(2, p1) = 3, Δ(2, p2) = 2, Δ(2, p3) = 6, Δ(2, p4) = 2, Δ(2, p5) = 2, and Δ(2, p6) = 6. The Shapley value for player 2 is SV(2) = (3+ 2 + 6 + 2 + + 2 + 6)/6 = 21/6 = 3.5.

Similarly, for player 3, by D6, we find Δ(3, p1) = 6, Δ(3, p2) = 3, Δ(3, p3) = 3, Δ(3 ,p4) = 3, Δ(3, p5) = 6, and Δ(3, p6) = 3. The Shapley value for player 3 is SV(3) = (6 + 3 + 3 + 3 + 6 + 3)/6 = 24/6 = 4. These values were also computed in the Matlab Program myShapleyvalueBV.m, which will be explained in detail in the next section.

Note that in the present example, the point (2.5, 3.5, 4) with the Shapley values for each player as coordinates is a point in the Core of the game in Example 11.2, this is because if it satisfies the equations (1) to (7). However, it general, the point with Shapley values as coordinates do not always belong to the Core.

11.3.2 A Matlab program myShapleyvalue.m

This Matlab program implements the logic and computes the Shapley values based on the definition D9 described in the last section. The only inputs to the program are the list of permutations P, the list of coalitions C, and the values of the corresponding coalitions Cv. The output of the program first repeats what the inputs are, then shows, for each player, the value added of that player to each of the coalition from the permutations, and finally the program shows the Shapley value of each player.

The inputs to Example 11.5 are as follows:

```
P = [1 2 3; 2 3 1; 3 1 2; 3 2 1; 2 1 3; 1 3 2];
C = [1 0 0; 0 1 0; 0 0 1; 0 1 1; 1 0 1; 1 1 0; 1 1 1];
Cv = [1 2 3 5 4 4 10];
```

```
>> myShapleyvalueBV
```

Output listing is shown below:

```
P =

    1   2   3
    2   3   1
    3   1   2
    3   2   1
    2   1   3
    1   3   2

C =

    1   0   0
    0   1   0
    0   0   1
    0   1   1
    1   0   1
    1   1   0
    1   1   1

Cv =

    1   2   3   5   4   4   10

Outputs of the Shapleyvalue program:
Compute value added to a coalition for Player: 1
iplayer, permuation number, player location in the permutation , value added:
1, 1, 1, 1
```

1, 2, 3, 5

1, 3, 2, 1

1, 4, 3, 5

1, 5, 2, 2

1, 6, 1, 1

Compute value added to a coalition for Player: 2

iplayer, permuation number, player location in the permutation , value added:

2, 1, 2, 3

2, 2, 1, 2

2, 3, 3, 6

2, 4, 2, 2

2, 5, 1, 2

2, 6, 3, 6

Compute value added to a coalition for Player: 3

iplayer, permuation number, player location in the permutation , value added:

3, 1, 3, 6

3, 2, 2, 3

3, 3, 1, 3

3, 4, 1, 3

3, 5, 3, 6

3, 6, 2, 3

sum =

    15   21   24

svalue =

    2.5000   3.5000   4.0000

******* End of output listing **********

11.3.3 Important properties of Shapley Values

There are two important properties of the Shapley Values we will show and exemplify in this section.

Properties 11.3.1

The sum of all the Shapley Values of each player in {1, 2…, n} equals to the value of the characteristic function assigned to the grand coalition. In symbol,

$$\sum_{i=1..n} SV(i) = v(\{1,2...,n\}) \tag{P1}$$

We exemplify this property by Example 11.5 above. We computed SV(1) = 2.5, SV(2) = 3.5, and SV(3) = 4. Thus, SV(1) + SV(2) + SV(3) = 10, which is exactly equals to 10, the v-value of the grand coalition C(1, 2, 3).

Recall from definition 11.4, if the characteristic function of a coalition game of n players is defined by the v-values on each of the coalitions C in the game, the Shapley value of the game is a 1xn vector SV(v) defined by equation D9:

$$\text{SV(v)} = \frac{1}{n!}\sum_{p} \Delta(p) \tag{D9}$$

The ith component of SV (v) contains the Shapley value of the ith player.

Properties 11.3.2

If v1 and v2 are two different v-values each corresponds to a different characteristic function in a coalition game, then

$$\text{SV(v1 + v2) = SV(v1) + SV(v2)} \tag{P2}$$

11.3.4 Verification of properties of Shapley value

Instead of proofing the above property (P2) directly from definition (D9), we are going to exemplify (P2) by choosing two arbitrary different v1 and v2 and perform the following steps.

Step 1: We first compute SV(v1) by running myShapleyBV.m with inputs for

v1, then compute SV(v2) by running myShapleyBV.m with inputs for v2.

Compute the sum of the two vectors SV(v1) + SV(v2)

Step 2: Run myShapleyvalueBV.m with inputs for v1 + v2 to compute the vector

SV(v1+v2)

Step 3: Show that the two vectors are equal: SV(v1 + v2) = SV(v1) + SV(v2)

We proceed with these steps as follows.

Step 1: Use the following input for v1:

```
P  = [1 2 3; 1 3 2; 2 1 3; 2 3 1; 3 1 2; 3 2 1]
C  = [1 0 0; 0 1 0; 0 0 1; 1 1 0; 1 0 1; 0 1 1; 1 1 1]
Cv = [0 0 0 4 7 15 20]
```

Running myShapleyvalueBV.m gives the following output listing:

```
>> myShapleyvalueBV

P =

     1     2     3
     1     3     2
     2     1     3
     2     3     1
     3     1     2
     3     2     1

C =

     1     0     0
     0     1     0
     0     0     1
     1     1     0
     1     0     1
     0     1     1
     1     1     1

Cv =

     0     0     0     4     7    15    20
```

Inputs to the program:

```
P =

     1     2     3
     1     3     2
     2     1     3
     2     3     1
     3     1     2
     3     2     1

C =

     1     0     0
     0     1     0
     0     0     1
     1     1     0
```

```
1 0 1
0 1 1
1 1 1
```

```
Cv =

0 0 0 4 7 15 20
```

Outputs of the Shapleyvalue program:

```
sum =
  21  45  54
svalue =
  3.5000  7.5000  9.0000
```

******* End of output listing **********

Thus SV(v1) = (3.5, 7.5, 9.0). Note that we have suppressed the intermediate output listings for clarity of illustration to our objective.

Now run myShapleyvalueBV.m with the following input for v2:

```
P  = [1 2 3; 1 3 2; 2 1 3; 2 3 1; 3 1 2; 3 2 1]
C  = [1 0 0; 0 1 0; 0 0 1; 1 1 0; 1 0 1; 0 1 1; 1 1 1]
Cv = [1 0 -2 5 3 2 3]
```

```
>> myShapleyvalueBV

P =

1 2 3
1 3 2
2 1 3
2 3 1
3 1 2
3 2 1

C =

1 0 0
0 1 0
0 0 1
1 1 0
1 0 1
0 1 1
1 1 1
```

```
Cv =

   1   0   -2   5   3   2   3
```

Inputs to the program:

```
P =

   1   2   3
   1   3   2
   2   1   3
   2   3   1
   3   1   2
   3   2   1

C =

   1   0   0
   0   1   0
   0   0   1
   1   1   0
   1   0   1
   0   1   1
   1   1   1

Cv =

   1   0   -2   5   3   2   3
```

Outputs of the Shapleyvalue program:

```
sum =

   14   8   -4

svalue =

   2.3333   1.3333   -0.6667
```

******* End of output listing **********

Thus SV(v2) = (2.333, 1.333, -0.6667)

Thus,

SV(v1) + SV(v2)
= (3.5, 7.5, 9.0) + (2.333, 1.333, -0.6667)
= (5.833, 8.833, 8.333)

Step2: The inputs needed to run myShapleyvalueBV.m for v1 + v2 are:

P = [1 2 3; 1 3 2; 2 1 3; 2 3 1; 3 1 2; 3 2 1]
C = [1 0 0; 0 1 0; 0 0 1; 1 1 0; 1 0 1; 0 1 1; 1 1 1]
Cv= [1 0 -2 9 10 17 23]

Both the permutations P and the coalitions C are the same as that of v1's and v2's, but the coalition values Cv is obtained from taking the sum of the coalition values

Cv = [0 0 0 4 7 15 20] for v1 and the coalition values Cv = [1 0 -2 5 3 2 3] for v2.

```
>> myShapleyvalueBV

P =

     1     2     3
     1     3     2
     2     1     3
     2     3     1
     3     1     2
     3     2     1

C =

     1     0     0
     0     1     0
     0     0     1
     1     1     0
     1     0     1
     0     1     1
     1     1     1

Cv =

     1     0    -2     9    10    17    23
```

Inputs to the program:

```
P =

     1     2     3
     1     3     2
     2     1     3
     2     3     1
     3     1     2
     3     2     1
```

C =

```
 1  0  0
 0  1  0
 0  0  1
 1  1  0
 1  0  1
 0  1  1
 1  1  1
```

Cv =

```
 1  0  -2  9  10  17  23
```

Outputs of the Shapleyvalue program:

sum =

```
 35  53  50
```

svalue =

```
 5.8333  8.8333  8.3333
```

******* End of output listing **********

Thus, SV(v1 + v2) = (5.8333, 8.8333, 8.333)

Step 3. From Steps 1 and 2, we have verified that

SV(v1 + v2) = SV(v1) + SV(v2)

11.4 What is the Nucleolus of a game?

As we have mentioned in the introduction, the Nucleolus, denoted by N(v) , is a very special element in the Core of a coalition game because it always exists in the core if the core is not empty. Before we can give a precise definition of N(v), first we need to know what it meant by the Excess Vector of an imputation x and a coalition C, and the concept of lexicographical order of two vectors of equal length.

Definition 4.5 The excess of a coalition C and an imputation x, denoted by e(C, x).

We define e(C,x) as the difference between v(C) and $\sum_{i \in C} x_i$. That is,

$$e(C, x) = v(C) - \sum_{i \in C} x_i \qquad \text{(D10)}$$

Since there are 2^n - 1 that many non-zero coalitions C in a coalition game of n players, the vector e(x), formed by the 2^n - 1 components e(C, x), where C ranges from each of the coalition in the game, is called the excess vector the imputation x. We give an example of excess vectors for the coalition game shown in Example 11.2 before.

Example 11. 4. 7

The coalitions in Example 11.2 are C_1= {1}, C_2= {2}, C_3= {3}, C_4= {1, 2}, C_5= {2, 3}, C_6 = {1, 3}, C_7 = {1, 2, 3}, where $v(C_1) = v(1) = 1$, $v(C_2) = v(2) = 2$, $v(C_3) = =v(3) = 3$, $v(C_4) = v(1, 2) = 4$, $v(C_5) = v(2, 3) = 5$, $v(C_6) = v(1, 3) = 4$, $v(C_7) = v(1, 2, 3) = 10$, and $P_1 = (5, 2, 3)$, $P_2 = (1, 6, 3)$, $P_3 = (1, 2, 7)$ are three imputations. Now we show the computations of the excess vectors $e(P_1)$, $e(P_2)$ and $e(P_3)$ as follows:

To compute the vector $e(P_1)$, let us first compute its components:

$e(C_1, P_1)$, $e(C_2, P_1)$, $e(C_3, P_1)$, $e(C_4, P_1)$, $e(C_5, P_1)$, $e(C_6, P_1)$, $e(C_7, P_1)$.

Now $e(C_1, P_1) = v(C_1) - x_1 = 1 - 5 = -4$, $e(C_2, P_1) = v(C_2) - x_2 = 2 - 2 = 0$,

$e(C_3, P_1) = v(C_3) - x_3 = 3 - 3 = 0$, $e(C_4, P_1) = v(C_4) - (x_1 + x_2) = 4 - (5 + 2) = -3$,

$e(C_5, P_1) = v(C_5) - (x_2 + x_3) = 5 -(2 + 3) = 0$,

$e(C_6, P_1) = v(C_6) - (x_1 + x_3) = 4 - (5 + 3) = -4$, and $e(C_7, P_1) = v(C_7) - (x_1 + x_2 + x_3) = 10 - (5 + 2 + 3) = 0$. Thus, the excess vector $e(P_1)$ is $e(P_1) = (-4, 0 , 0, -3, 0, -4, 0)$

Similarly, for P_2 and P_3, we have $e(P_2) = (0,-4, 0, -3,-4, 0, 0)$, and

$e(P_3) = (0, 0, -4, 1, -4, -4, 0)$. To recap, each of the vector components are arranged in the order of the coalitions C_1, C_2, … C_6 above, Now, if we rearrange each of the vector components in the order of "less than or equal to $\leq x$" of the excess values, then

$e(P_1) = (0, 0, 0 , 0, -4, -4, -3)$,

$e(P_2) = (0, 0, 0, 0, -4, -4, -3)$,

$e(P_3) = (0, 0, 0 , -4, -4, -4, 1)$

We are going to define what is meant by a lexicographic order of the three vectors $e(P_1)$, $e(P_2)$, $e(P_3)$ below.

Definition 11.4.6 Lexicographical order of two vectors of equal length denoted by {x ≤ y (lex)}, where the length is the number of non-zero coalitions

Suppose $x = (x_1, x_2\dots ,x_m)$, $y = (y_1, y_2\dots,y_m)$ are two vectors of length m, where m is the total number of non-empty coalitions, and the x_i and y_j are the excess values. We say that $x \leq y$ (lex) if there exists an index k in between 1 and m such that $x_i = y_i$ for all $i < k$, and $x_k < y_k$. Based on this definition, we see that $e(P_1) \leq e(P_2)$ (lex) and $e(P_3) \leq e(P_1)$ (lex). Now we can define what is meant by the nucleolus N(v) of a coalition game with characteristic function v(C).

Definition 11.4.7 Nucleolus N(v) in Core(v)

Roughly speaking, we can think of the Nucleolus N(v) as the unique imputation whose excess vector e(N(v)) is lexicographically smaller or equal to the excess vector of any other imputation vector x. In symbol, the Nucleolus is a particular imputation x^* such that: for all x in I(v),

$$e(x^*) \leq e(x) \text{ (lex)} \qquad \text{(D11)}$$

Next , we will present an algorithm to find this special imputation $x^* = N(v)$ satisfy the condition (D11) above.

Algorithm 11.4. 1 Computation for the nucleolus N(v)

Before we present an algorithm based on linear programming approach, we are going to show that $x^* = N(v)$ is an imputation that can be obtained in the following way. Suppose $C_1, C_2, \ldots C_k$ are the non-empty coalitions. Recall that for x in the Core (v), $\sum_{i\in C} x_i \geq v(C)$ for $C = C_1, C_2, \ldots C_k$ this means that $v(C) - \sum_{i\in C} x_i \leq 0$ or $e(C, x) \leq 0$ for all $C = C_1, C_2, \ldots C_k$. We define the first, second , third…,k^{th} - largest excess of a coalition game below.

We start with the definition of the 1st -largest excess of a coalition game. It is an imputation vector x from the Core, together with a coalition excess upper bound t satisfying the following Optimization Problem:

$$\underset{t}{Min}\ \underset{x}{Max}\{e(C_1,x) \leq t, e(C_2,x) \leq t, \ldots \quad , e(C_k,x) \leq t\} \qquad \text{(D12)}$$

subject to x is an imputation vector in the Core.

(D12) above can be interpreted as follows:

For each vector x in the Core, we compute its k coalition excesses e(C,x), where $C = C_1, C_2, \ldots C_k$, we then select the largest of these excesses at $C = C_{max}$ where max is some index between 1 and k. We let $t_{max} = e(C_{max}, x)$, then $e(C_i, x) \leq t_{max}$ for all $i \neq max$. That is for each x, we can find a t_{max} that is the least upper bound of all the k coalitions $e(C, x) \leq t_{max}$. Finally, amongst all t_{max}, we find the smallest value, call it $t_{min(max)}$ by looking up the corresponding vector x_{min} in $e(C_{max}, x) \leq t_{min(max)}$. Conceptually, the vector x_{min} and the excess $t_{min(max)}$ is what we call the 1st-largest excess of a coalition game. To find numerically the excess $t_{min(max)}$, we translate the above min-max optimization process into a linear programming formulation as follows.

Let z_1 be the minimum of the 1st-largest excess, let $x = (x_1, x_2, \ldots, x_n)$, then the linear programming problem is expressed as :

$Z_1 = \min t_{max}$ subjected to the following three groups of inequalities and equalities (LP1, LP2, LP3).

$$e(C, x) \leq t_{max} \text{ for all non-empty coalitions } C = C_1, C_2, \ldots C_k \qquad \text{(LP1)}$$

$$x_i \geq v(i) \text{ for } i = 1 \ldots n \qquad \text{(LP2)}$$

$x_1 + x_2 + .. + x_n = v(1, 2,... n)$ (LP3)

We are going to describe the repetitive process of modifying the above inequalities and equalities (LP1, LP2, LP3) below to find the 2nd, 3rd and 4th … largest excess below.

After we solve the above LP program, the value of the minimum of the 1st-largest excess Z_1 and the corresponding vector $x = z_1$ are determined. We then use the value Z_1 and z_1 to find the minimum of the 2nd- largest excess Z_2 by removing some of the inequalities from (LP1) and including them to the group of equalities in (LP3) in the manner described next. For each of the k inequalities in(LP1), one for each coalition, compute $e(C_k, z_1)$ and check if $e(C_k, z_1)$ and Z_1 are equal. We then put all those coalitions C_k that can attain the value Z_1, i.e. $e(C_k, z_1) = Z_1$, into a collection set, call it Ω. In other words, either $C_k \in \Omega$ or $C_k \notin \Omega$. Now we can formulate the LP solution to compute the minimum of the 2nd-largest excess Z_2 below:

$Z_2 = \min t_{max}$ subjected to the following three groups of inequalities and equalities (LP1, LP2, LP3).

$e(C, x) \leq t_{max}$ for all non-empty coalitions $C_k \notin \Omega$. (LP1)

$x_i \geq v(i)$ for $i = 1…n$ (LP2)

$x_1 + x_2 + .. + x_n = v(1, 2,…n)$ (LP3)

$e(C, x) = t_{max}$ for all non-empty coalitions $C_k \in \Omega$

From the LP formulation for computing the 1st- largest and the 2nd- largest excess described above, we can see the recursive nature of formulating the computation of the 3rd- largest excess and in general, the kth -largest excess in a coalition game. We note that in computing the 1st -largest excess Z_1, we employ all k coalitions C which gives k inequalities in (LP1), and there is only a single equality in (LP3). In computing the 2nd largest excess Z_2, the number of inequalities in (LP1) is getting less and the number of equalities is getting more in (LP3). Iteratively, we use the result of the $(k - 1)^{th}$ - largest excess Z_{k-1} to eliminate some of the inequalities in (LP1) and add more equalities in (LP3) in formulating the linear programming to compute the kth - largest excess Z_k. Thus, the computation of the kth – largest excess Z_k other than the computation of Z_1 requires solving a sequence of LP problems. We now give an example to demonstrate this sequential process.

11.5 A MATLAB program myComputeNucleolus112BV.m

Example 11.5.8

We now present a MATLAB program, myComputeNucleolus112BV.m, to compute the Nucleolus of the coalition game described in Example 1.2. A description of the inputs to the program is in order. As the program is making use of the build-in function $x = \text{linprog}(f, A, b, A_{eq}, b_{eq})$ in MATLAB, a description of the input parameters f, A, b, A_{eq}, b_{eq} is shown below.

Refer to the inputs of Example 1.2, for the group of inequalities (LP1)

$e(C, x) \leq t_{max}$ is given by:

$1 - x_1 \leq t$ (1)

$2 - x_2 \leq t$ (2)

$3 - x_3 \leq t$ (3)

$4 - (x_1 + x_2) \leq t$ (4)

$4 - (x_1 + x_3) \leq t$ (5)

$5 - (x_2 + x_3) \leq t$ (6)

$10 - (x_1 + x_2 + x_3) \leq t$ (7)

, for the group of inequalities (LP2) $x_i \geq v_I$:

$x_1 \geq 1$ (8)

$x_2 \geq 2$ (9)

$x_3 \geq 3$ (10)

, for the group of equalities (LP3), it has only a single equality initially:

$x_1 + x_2 + x_3 = 10$ (11)

The inequalities (1) to (10) are translated into:

$-x_1 - t \leq -1$ (1)'

$-x_2 - t \leq -2$ (2)'

$-x_3 - t \leq -3$ (3)'

$-x_1 - x_2 - t \leq -4$ (4)'

$-x_1 - x_3 - t \leq -4$ (5)'

$-x_2 - x_3 - t \leq -5$ (6)'

$-x_1 - x_2 - x_3 - t \leq -10$ (7)'

$-x_1 \leq -1$ (8)'

$-x_2 \leq -2$ (9)'

$-x_3 \leq -3$ (10)'

Thus, the inequalities (1)' to (10)' are represented by the A matrix and the column vector b. Each row in A consists of the coefficients of the four variables x_1, x_2, x_3, t, and each entry in b is the constant at the right-hand side of each inequality.

```
A = [-1  0  0  -1;      % corresponds to coalition C1
      0 -1  0  -1;      % corresponds to coalition C2
      0  0 -1  -1;      % corresponds to coalition C3
     -1 -1  0  -1;      % corresponds to coalition C4
     -1  0 -1  -1;      % corresponds to coalition C5
      0 -1 -1  -1;      % corresponds to coalition C6
     -1 -1 -1  -1;      % corresponds to coalition C7
     -1  0  0   0;      % corresponds to (8)'
      0 -1  0   0;      % corresponds to (9)'
      0  0 -1   0]      % corresponds to (10)'
     b = [ -1 -2 -3 -4 -4 -5 -10 -1 -2 -3]'
```

The single equality in (11) gives the matrix A_{eq} and the column vector b_{eq}, it has the initial contents:

```
Aeq = [ 1 1 1 0]        % corresponds to the left- hand side of (11)
beq = 10                % corresponds to the right-hand side of (11)
```

Since we are minimizing t, we let f = [0 0 0 1] and call the built-in function x = linprog(f, A b, A_{eq}, b_{eq}), we obtain the 1st - largest excess t = 0 and the corresponding imputation vector x = (1, 6, 3).

Now, to find the 2nd - largest excess, we first evaluate which coalition can attain the 1st - largest excess t = 0, that is to check which coalition C satisfies e(C, x) = t = 0, we check that the coalitions C_1, C_3, C_5, C_7 are the candidates. Follow the description before, we are going to remove the inequalities from the A-matrix that correspond to these coalitions and include the corresponding equalities to the A_{eq} - matrix.

```
A = [0  -1   0  -1;        % corresponds to coalition C2
    -1  -1   0  -1;        % corresponds to coalition C4
     0  -1  -1  -1;        % corresponds to coalition C6
    -1   0   0   0;        % corresponds to (8)'
     0  -1   0   0;        % corresponds to (9)'
     0   0  -1   0]        % corresponds to (10)'

b = [-2  -4  -5  -1  -2  -3]'

Aeq = [-1  0  0  0;        % corresponds to coalition C1
        0  0 -1  0;        % corresponds to coalition C3
       -1  0 -1  0;        % corresponds to coalition C5
       -1 -1 -1  0;        % corresponds to coalition C7
        1  1  1  0]        % corresponds to left hand side of (11)

beq = [-1; -3; -4; -10; 10]     first 4 entries correspond to C1, C3, C5, C7
                                % last entry corresponds to right hand side of (11)
```

Now the 2nd-largest excess t can be found by calling the built-in function x = linprog (f, A, b, A_{eq}, b_{eq}), we find t = -3 and the corresponding imputation vector x = (1, 6, 3).

As before, to find the 3rd-largest excess t, we first evaluate which coalition can attain the 2nd-largest excess t = -3, that is to check which coalition C will give e(C, x) = t = -3, we find that there is only one coalition C_4 that does this. So, we remove the inequality that corresponds to coalition C_4 in matrix A and add it to matrix A_{eq}. Now since $e(C_4, x) = -3$, $v(C_4) - (x_1 + x_2) = -3$, so $-x_1 - x_2 = -7$, the corresponding entry -4 in b and add -7 to the corresponding entry in b_{eq}, we obtain:

```
A = [0  -1   0  -1;        % corresponds to coalition C2
```

```
     0  -1  -1  -1;         % corresponds to coalition C6
    -1   0   0   0;         % corresponds to (8)'
     0  -1   0   0;         % corresponds to (9)'
     0   0  -1   0]         % corresponds to (10)'
b = [-2  -5  -1  -2  -3]'
Aeq = [-1  -1   0   0;      % corresponds to coalition C4
       -1   0   0   0;      % corresponds to coalition C1
        0   0  -1   0;      % corresponds to coalition C3
       -1   0  -1   0;      % corresponds to coalition C5
       -1  -1  -1   0;      % corresponds to coalition C7
        1   1   1   0]      % corresponds to left hand side of (11)
beq = [-7; -1; -3; -4; -10; 10]   % entries correspond to C4,C1, C3,C5,C7
                                  % last entry corresponds to right hand side of (11)
```

Now we find the 3rd-largest excess t by calling x = linprog(f, A, b, A_{eq}, b_{eq}), we get t = -4, and the corresponding imputation vector x = (1, 6, 3).

Continue to find the 4th-largest excess t, again we first check which coalition C will give e(C, x) = -4, we find coalitions C_2 and C_6 will do. So, $2 - x_2 = -4$ and $5 - x_2 - x_3 = -4$, or $-x_2 = -6$ and $-x_2 - x_3 = -9$. Next, we remove the inequalities correspond to C_2 and C_6 in A and add them to A_{eq}, remove the corresponding entries in b and add the entries -4 and -9 in b_{eq}. We get:

```
A = [-1   0   0   0;        % corresponds to (8)'
      0  -1   0   0;        % corresponds to (9)'
      0   0  -1   0]        % corresponds to (10)'
b = [-1  -2  -3]'
Aeq = [0  -1   0   0;       % corresponds to coalition C2
```

```
  0  -1  -1   0;          % corresponds to coalition C6
 -1  -1   0   0;          % corresponds to coalition C4
 -1   0   0   0;          % corresponds to coalition C1
  0   0  -1   0;          % corresponds to coalition C3
 -1   0  -1   0;          % corresponds to coalition C5
 -1  -1  -1   0;          % corresponds to coalition C7
  1   1   1   0]          % corresponds to left hand side of (11)

beq = [ -6; -9; -7; -1; -3; -4; -10; 10]

                          % correspond to C2,C6, C4,C1,C3, C5,C7

                          % last entry corresponds to right hand side of (11)
```

Now call x = linprog(f, A, b, A_{eq}, b_{eq}) to find the 4th - largest excess; however, this time, the linprog function returns the message "Problem is unbounded" and so we end up with getting the imputation vector corresponds to the 3rd-largest excess, the nucleolus imputation vector is thus x = (1, 6, 3).

11.6 A MATLAB program myComputeNucleolusBV4444.m

Example 11.6.9

In this example, we consider a 4-players coalition game such that the value v(C) of each single player coalition C is zero, and the value v(C) of all non-singleton coalitions C is the square of the number of players in C. We are going to use the MATLAB program myComputeNucleolusBV4444.m to find the Nucleolus imputation vector in this game. Like the last example, Example 5.8, the inputs to myComputeNucleolusBV4444.m are the coalitions C_1, C_2,… C_{15} and the corresponding values of each coalition C_i are shown below:

```
C =
   1  0  0  0
   0  1  0  0
   0  0  1  0
   0  0  0  1
   1  1  0  0
   1  0  1  0
```

```
     1     0     0     1
     0     1     1     0
     0     1     0     1
     0     0     1     1
     1     1     1     0
     1     1     0     1
     1     0     1     1
     0     1     1     1
     1     1     1     1

Cv =

     0     0     0     0     4     4     4     4     4     4     9     9     9     9    16
```

The output of the program shows, step by step, the contents of A, b, A_{eq}, and b_{eq} at each iteration and the program stops util an unbounded solution is reached.

```
A =

    -1     0     0     0    -1
     0    -1     0     0    -1
     0     0    -1     0    -1
     0     0     0    -1    -1
    -1    -1     0     0    -1
    -1     0    -1     0    -1
    -1     0     0    -1    -1
     0    -1    -1     0    -1
     0    -1     0    -1    -1
     0     0    -1    -1    -1
    -1    -1     1     0    -1
    -1    -1     0    -1    -1
    -1     0    -1    -1    -1
     0    -1    -1    -1    -1
    -1    -1    -1    -1    -1
    -1     0     0     0     0
```

```
     0    -1     0     0     0
     0     0    -1     0     0
     0     0     0    -1     0

b =

     0     0     0     0    -4    -4    -4    -4    -4    -4    -9    -9    -9    -9   -16     0     0     0     0

Aeq =

     1     1     1     1     0

beq =

    16

A =

    -1     0     0     0    -1
     0    -1     0     0    -1
     0     0    -1     0    -1
     0     0     0    -1    -1
    -1    -1     0     0    -1
    -1     0    -1     0    -1
    -1     0     0    -1    -1
     0    -1    -1     0    -1
     0    -1     0    -1    -1
     0     0    -1    -1    -1
    -1    -1    -1     0    -1
    -1    -1     0    -1    -1
    -1     0    -1    -1    -1
```

```
    0  -1  -1  -1  -1

   -1   0   0   0   0

    0  -1   0   0   0

    0   0  -1   0   0

    0   0   0  -1   0

b =

    0   0   0   0  -4  -4  -4  -4  -4  -4  -9  -9  -9  -9   0   0   0   0

Aeq =

   -1  -1  -1  -1   0

    1   1   1   1   0

beq =

   -16

    16

f =

    0   0   0   0   1

Optimal solution found.

x =

    4.0000

    4.0000

    4.0000

    4.0000

   -3.0000
```

The value of the largest excess vector is -3.0000:

```
x =

    4.0000

    4.0000

    4.0000

    4.0000

   -3.0000

t =

    -3
```

Coalition Number with excess equal to t : 11

Coalition Number with excess equal to t : 12

Coalition Number with excess equal to t : 13

Coalition Number with excess equal to t : 14

Include coalitions 11, 12, 13, 14 :

```
A =

    -1   0   0   0  -1

     0  -1   0   0  -1

     0   0  -1   0  -1

     0   0   0  -1  -1

    -1  -1   0   0  -1

    -1   0  -1   0  -1

    -1   0   0  -1  -1
```

```
     0   -1   -1    0   -1

     0   -1    0   -1   -1

     0    0   -1   -1   -1

    -1    0    0    0    0

     0   -1    0    0    0

     0    0   -1    0    0

     0    0    0   -1    0

b =

     0     0     0     0    -4    -4    -4    -4    -4    -4     0     0     0     0

Aeq =

    -1   -1   -1    0    0

    -1   -1    0   -1    0

    -1    0   -1   -1    0

     0   -1   -1   -1    0

    -1   -1   -1   -1    0

     1    1    1    1    0

beq =

   -12

   -12

   -12

   -12

   -16

    16
```

```
f =

     0     0     0     0     1

Optimal solution found.

x =

     4

     4

     4

     4

    -4

t =

    -4

Coalition Number with excess equal to t :  1

Coalition Number with excess equal to t :  2

Coalition Number with excess equal to t :  3

Coalition Number with excess equal to t :  4

Coalition Number with excess equal to t :  5

Coalition Number with excess equal to t :  6

Coalition Number with excess equal to t :  7

Coalition Number with excess equal to t :  8

Coalition Number with excess equal to t :  9

Coalition Number with excess equal to t :  10

A =
```

```
    -1     0     0     0     0

     0    -1     0     0     0

     0     0    -1     0     0

     0     0     0    -1     0

b =

     0     0     0     0

Aeq =

    -1     0     0     0     0

     0    -1     0     0     0

     0     0    -1     0     0

     0     0     0    -1     0

    -1    -1     0     0     0

    -1     0    -1     0     0

    -1     0     0    -1     0

     0    -1    -1     0     0

     0    -1     0    -1     0

     0     0    -1    -1     0

    -1    -1    -1     0     0

    -1    -1     0    -1     0

    -1     0    -1    -1     0

     0    -1    -1    -1     0

    -1    -1    -1    -1     0

     1     1     1     1     0
```

beq =

-4 -4 -4 -4 -8 -8 -8 -8 -8 -8 -12 -12 -12 -12 -16 16

f =

0 0 0 0 1

The problem is unbounded.

x =

[]

11.7 Source codes of the MATLAB programs

(1) myShapleyvalueBV.m

```
% P = [1 2 3; 2 3 1; 3 1 2; 3 2 1; 2 1 3; 1 3 2];
% C = [1 0 0; 0 1 0; 0 0 1; 0 1 1; 1 0 1; 1 1 0; 1 1 1];
% Cv = [1 2 3 5 4 4 10];

% P = [1 2 3; 2 3 1; 3 1 2; 3 2 1; 2 1 3; 1 3 2];
% C = [1 0 0; 0 1 0; 0 0 1; 0 1 1; 1 0 1; 1 1 0; 1 1 1];
% Cv = [0 1 1 3 4 5 16];

%P = [1 2 3; 1 3 2; 2 1 3; 2 3 1; 3 1 2; 3 2 1]
%C = [1 0 0; 0 1 0; 0 0 1; 1 1 0; 1 0 1; 0 1 1; 1 1 1]
%Cv = [0 0 0 4 7 15 20]

%P = [1 2 3; 1 3 2; 2 1 3; 2 3 1; 3 1 2; 3 2 1]
%C = [1 0 0; 0 1 0; 0 0 1; 1 1 0; 1 0 1; 0 1 1; 1 1 1]
%Cv = [1 0 -2 5 3 2 3]

P = [1 2 3; 1 3 2; 2 1 3; 2 3 1; 3 1 2; 3 2 1]
C = [1 0 0; 0 1 0; 0 0 1; 1 1 0; 1 0 1; 0 1 1; 1 1 1]
Cv = [1 0 -2 9 10 17 23]

fprintf('Inputs to the program: \n');
    P
```

```
    C
    Cv


sum = [0 0 0];
svalue = [0 0 0];
fprintf('Outputs of the Shapleyvalue program: \n');

for iplayer = 1:3
%fprintf('Compute value added to a coalition for Player: %d \n',iplayer);
%fprintf('iplayer, permuation number, player location in the permutation, value added: \n');
for p= 1:6

     % we are going to locate player i in each of the permuataion by the
    % index pl
  for pl= 1:3
  if P(p, pl) == iplayer
     q = pl;
  end
end

  if q == 1
x = [0 0 0];
Y = [0 0 0];
m = P(p, q);
  Y(m) = 1;
  % fprintf('modify Y block \n');
  Y;
  for k = 1:7
t = compare(C(k,:), Y);
  if t == 1
       iysave = k;
  end
  % fprintf (' value of iysave %d\n', iysave);
  end
     mydelta = Cv(iysave);
    sum (iplayer) = sum (iplayer) + mydelta;

   % fprintf( ' %d, %d, %d, %d\n' , iplayer, p, q, mydelta);
  end % end of if q =1
```

```
if q == 2

   Y = [0 0 0];
   X = [0 0 0];
   m = P(p, q);
   mp = P(p, q-1);

   X(mp) = 1;
   Y(m) = 1;
   Y(mp) = 1;

   % fprintf('modify the X, Y blocks: \n');
   % X
   % Y

   for i = 1:7

   t = compare (C(i,:), Y);
   if t == 1
       iysave = i;
   end
   t = compare (C(i,:), X);
     if t == 1
       ixsave =i;
     end
   end

   mydelta = Cv(iysave) - Cv(ixsave);
   sum (iplayer) = sum (iplayer) + mydelta;

   % fprintf('%d, %d, %d, %d\n', iplayer, p, q, mydelta);

end % end of if q = 2

if q == 30
    X = [0 0 0];
      Y = [0 0 0];
      m = P(p, q);
      mp = P(p, q-1);
      mpp = P(p, q-2);
      X(mp) = 1;
```

```
    X(mpp) = 1;
    Y(m) = 1;
    Y(mp) = 1;
    Y(mpp) = 1;

    for i = 1:7
        tx = compare(C(i,:), X);
        ty = compare(C(i,:), Y);

    if tx == 1
        ixsave = i;
    end
    if ty == 1
        iysave = i;
    end
    mydelta = Cv(iysave) - Cv(ixsave);
    sum (iplayer) = sum(iplayer) + mydelta;

  % fprintf('%d, %d, %d, %d\n', iplayer, p, q, mydelta);
end % end of if q = 3
% fprintf('iplayer, permuation number, player location, mydelta value \n');
% fprintf('%d, %d, %d, %d\n', iplayer, p, q, mydelta);
end % end of the for-loop for each permuatation for player iplayer


sum(iplayer);
svalue(iplayer) = sum(iplayer) /6;


end % end of the for-loop for each player iplayer


  sum
  svalue
fprintf('*******  End of output listing  ********** \n');

%% ****************************************************************************

%               Functions for the Main Program are shown below:
```

```
%*****************************************************************************
function y = compare (X, Y)

  if (X(1) == Y(1))&(X(2) == Y(2))&(X(3) == Y(3))
    y = 1;
  else
    y=0;
end

end % of function compare
```

(2) myComputeNucleolus112BV.m

```
% Original linear program

C = [1 0 0; 0 1 0; 0 0 1; 1 1 0; 1 0 1; 0 1 1; 1 1 1]
Cv = [1 2 3 4 4 5 10]

% X1 >= 1
% x2 >= 2
% x3 >= 3
% X1 + X2 >= 4
% x2 + X3 >= 5
% x1 + X3 >= 4
% X1 + X2 + X3 = 10

% e(C, x) <= t
% e(C, x) = v(C) - summation x
% 1 - x1 <= t {-1   0    0    -1    -1}
% 2 - x2 <= t { 0  -1    0    -1    -2}
% 3 - x3 <= t { 0   0   -1    -1    -3}
% 4 - (x1 + x2) <= t {-1  -1    0    -1    -4}
% 4 - (x1 + x3) <= t {-1   0   -1    -1    -4}
% 5 - (x2 + x3) <= t { 0  -1   -1    -1    -5}
% 10 -(X1 + x2 + x3) <= t;  {-1  -1  -1   -1  -10}

% x1 + x2 + x3 = 10; {1    1    1    0    10}
% x1 >= 1; {-1    0    0    0    -1}
% x2 >= 2; { 0   -1    0    0    -2}
% x3 >= 3; { 0    0   -1    0    -3}

% Converting into the matlab LP format:

fprintf('Compute the largest excess vector for the game: \n');

A = [-1 0 0 -1; 0 -1 0 -1; 0 0 -1 -1; -1 -1 0 -1; -1 0 -1 -1; 0 -1 -1 -1; -1 -1 -1 -1; -1 0
      0 0; 0 -1 0 0; 0 0 -1 0]

b = [-1 -2 -3 -4 -4 -5 -10 -1 -2 -3]'

Aeq = [1 1 1 0]
beq = 10

f = [0 0 0 1]
```

```
x = linprog(f, A, b, Aeq, beq)
fprintf('The vlaue of t is % .4f: \n', x(4));
x
t = x(4);
sum = 0;
for i = 1:7
   sum = C(i, 1) * x(1) + C(i, 2) * x(2) + C(i, 3) * x(3);
    ev(i) = Cv(i) - sum;
     if (ev(i) == t)
       fprintf('Coalition Number with excess equal to t: %d\n', i);
     end
end
ev


fprintf('Compute the Second largest excess vector for the game: \n');

A = [0 -1 0 -1; -1 -1 0 -1; 0 -1 -1 -1; -1 0 0 0; 0 -1 0 0; 0 0 -1 0]

b = [-2 -4 -5 -1 -2 -3]'


Aeq = [-1 0 0 0; 0 0 -1 0; -1 0 -1 0; -1 -1 -1 0; 1 1 1 0]
beq = [-1; -3; -4; -10; 10]


P = [5 2 3; 1 6 3; 1 2 7]


f = [0 0 0 1]
x = linprog(f, A, b, Aeq, beq)
fprintf('The vlaue of t is %.4f: \n', x(4));

t = x(4);
sum = 0;
for i = 1:7
   sum = C(i, 1) * x(1) + C(i, 2) * x(2) + C(i, 3) * x(3);
    ev(i) = Cv(i) - sum;
     if (ev(i) == t)
       fprintf('Coalition Number with excess equal to t: %d\n', i);
     end
end
x
ev

fprintf('Compute the third largest excess vector for the game: \n');

A = [0 -1 0 -1; 0; -1 -1 -1; -1 0 0 0; 0 -1 0 0; 0 0 -1 0]

b = [-2 -5 -1 -2 -3]'


Aeq = [-1 -1 0 0; -1 0 0 0; 0 0 -1 0; -1 0 -1 0; -1 -1 -1 0; 1 1 1 0]
beq = [-7; -1; -3; -4; -10; 10]
f = [0 0 0 1]
x = linprog(f, A, b, Aeq, beq)
fprintf('The vlaue of t is %.4f: \n', x(4));

t = x(4);
sum = 0;
```

```
for i = 1:7
   sum = C(i, 1) * x(1) + C(i, 2) * x(2) + C(i, 3) * x(3);
    ev(i) = Cv(i) - sum;
     if (ev(i) == t)
       fprintf('Coalition Number with excess equal to t: %d\n', i);
     end
end
x
ev

fprintf('Compute the Fourth largest excess vector for the game: \n');

A = [-1 0 0 0; 0 -1 0 0; 0 0 -1 0]

b = [ -1  -2  -3]'


Aeq = [0 -1 0 0; 0 -1 -1 +0; -1 -1 0 0; -1 0 0 0; 0 0 -1 0; -1 0 -1 0; -1 -1 -1 0; 1 1 1 0]
beq = [-6;-9 ;-7; -1; -3; -4; -10; 10]
f = [0 0 0 1]
x = linprog(f, A, b, Aeq, beq)

if x ~= []
   fprintf('The vlaue of t is %.4f: \n', x(4));

   t = x(4);
   sum=0;
for i = 1:7
   sum = C(i, 1) * x(1) + C(i, 2) * x(2) + C(i, 3) * x(3);
    ev(i) = Cv(i) - sum;
     if (ev(i) == t)
       fprintf('Coalition Number with excess equal to t: %d\n', i);
     end
end
x
ev
else
   fprintf('Compare the excess vectors for the three points P1, P2, and P3:\n');
   for j = 1:3
     x = P(j,:)

     for i = 1:7
       sum = C(i,1) * x(1) + C(i, 2) * x(2) + C(i, 3) * x(3);
         ev(i) = Cv(i) - sum;

     end % i-loop
    ev
   end % j-loop
      fprintf('End of the Program \n');
end % if
```

(3) myComputeNeuclusBV4444.m

```
% Original linear program

C = [1    0    0    0;

     0    1    0    0;

     0    0    1    0;
```

```
    0    0    0    1;
    1    1    0    0;
    1    0    1    0;
    1    0    0    1;
    0    1    1    0;
    0    1    0    1;
    0    0    1    1;
    1    1    1    0;
    1    1    0    1;
    1    0    1    1;
    0    1    1    1;
    1    1    1    1]

Cv = [0 0 0 0 4 4 4 4 4 4 9 9 9 9 16]
x(1) = 4;
x(2) = 4;
x(3) = 4;
x(4) = 4;

for i = 1:15
    sum= C(i,1)*x(1) + C(i,2)* x(2)+ C(i,3)* x(3) + C(i,4)* x(4);
   ev(i)= Cv(i)-sum;

end
ev



A =[-1  0  0  0 -1;
     0 -1  0  0 -1;
     0  0 -1  0 -1;
     0  0  0 -1 -1;
    -1 -1  0  0 -1;
    -1  0 -1  0 -1;
    -1  0  0 -1 -1;
     0 -1 -1  0 -1;
     0 -1  0 -1 -1;
     0  0 -1 -1 -1;
    -1 -1 -1  0 -1;
    -1 -1  0 -1 -1;
    -1  0 -1 -1 -1;
```

```
     0 -1 -1 -1 -1;
    -1 -1 -1 -1 -1;
    -1  0  0  0  0;
     0 -1  0  0  0;
     0  0 -1  0  0;
     0  0  0 -1  0]

b = [0 0 0 0 -4 -4 -4 -4 -4 -4 -9 -9 -9 -9 -16 0 0 0 0]

Aeq = [1 1 1 1 0]
beq = 16
% *******************************************************************

A = [-1    0    0    0   -1;
      0   -1    0    0   -1;
      0    0   -1    0   -1;
      0    0    0   -1   -1;
     -1   -1    0    0   -1;
     -1    0   -1    0   -1;
     -1    0    0   -1   -1;
      0   -1   -1    0   -1;
      0   -1    0   -1   -1;
      0    0   -1   -1   -1;
     -1   -1   -1    0   -1;
     -1   -1    0   -1   -1;
     -1    0   -1   -1   -1;
      0   -1   -1   -1   -1;
     -1    0    0    0    0;
      0   -1    0    0    0;
      0    0   -1    0    0;
      0    0    0   -1    0]

b = [0 0 0 0 -4 -4 -4 -4 -4 -4 -9 -9 -9 -9 0 0 0 0]

Aeq = [-1 -1 -1 -1 0; 1 1 1 1 0]

beq = [-16; 16]

f = [0 0 0 0 1]
```

```
x = linprog(f, A, b, Aeq, beq)
fprintf('The vlaue of the largest excess vector is %.4f: \n', x(5));

x
t = x(5)
sum = 0;
for i = 1:15
   sum = C(i, 1) * x(1) + C(i, 2) * x(2) + C(i, 3) * x(3) + C(i, 4) * x(4);
      ev(i) = Cv(i) - sum;
       if (ev(i) == t)
        fprintf('Coalition Number with excess equal to t: %d\n', i);
       end
end

fprintf('Include coalitions 11, 12, 13, 14: \n');
A = [-1  0  0  0  -1;
      0 -1  0  0  -1;
      0  0 -1  0  -1;
      0  0  0 -1  -1;
     -1 -1  0  0  -1;
     -1  0 -1  0  -1;
     -1  0  0 -1  -1;
      0 -1 -1  0  -1;
      0 -1  0 -1  -1;
      0  0 -1 -1  -1;

     -1  0  0  0   0;
      0 -1  0  0   0;
      0  0 -1  0   0;
      0  0  0 -1   0]


b = [0 0 0 0 -4 -4 -4 -4 -4 -4 0 0 0 0]

Aeq = [-1  -1  -1   0   0;
       -1  -1   0  -1   0;
       -1   0  -1  -1   0;
        0  -1  -1  -1   0;
       -1  -1  -1  -1   0;
        1   1   1   1   0]
beq = [-12; -12; -12; -12; -16; 16]
```

```
f = [0 0 0 0 1]
x = linprog(f, A, b, Aeq, beq)

t = x(5)
  for i = 1:15
     sum = C(i, 1) * x(1) + C(i, 2) * x(2) + C(i, 3) * x(3) + C(i, 4) * x(4);
       ev(i) = Cv(i) - sum;
     if (ev(i) == t)
       fprintf(' Coalition Number with excess equal to t :  %d\n', i);
     end
  end

A = [

     -1    0    0    0    0;
      0   -1    0    0    0;
      0    0   -1    0    0;
      0    0    0   -1    0]

b = [0 0 0 0]

Aeq = [-1   0   0   0   0;
        0  -1   0   0   0;
        0   0  -1   0   0;
        0   0   0  -1   0;
       -1  -1   0   0   0;
       -1   0  -1   0   0;
       -1   0   0  -1   0;
        0  -1  -1   0   0;
        0  -1   0  -1   0;
        0   0  -1  -1   0;
       -1  -1  -1   0   0;
       -1  -1   0  -1   0;
       -1   0  -1  -1   0;
        0  -1  -1  -1   0;
       -1  -1  -1  -1   0;
        1   1   1   1   0]
beq = [-4; -4; -4; -4; -8; -8; -8; -8; -8; -8; -12; -12; -12; -12; -16; 16]'

f = [0 0 0 0 1]
x = linprog(f, A, b, Aeq, beq)
```

11.8 Industrial applications of the Nucleolus

The concept of Nucleolus has numerous applications in industry. In this section, we present three examples—Examples 11.8.10, 11.8.11, and 11.8.12—drawn from various sources- Papers ([14], [7], [22], [6]). The industry data used in these examples are sourced from these papers. Recall from Definition 11.4.7 that the Nucleolus is the imputation vector that minimizes the maximum dissatisfaction among all coalitions. Consequently, determining this minimum value is of practical importance. [14] examines the collaboration among three insurance companies investing in common funds. [7] involves four mobile companies sharing broadcast transmission costs for streaming a sporting event. [22] addresses cost-sharing issues among three joint network-centric manufacturing firms. Mario Guajardo and Kurt Jörnsten [6] pointed out that the calculation of the Nucleolus vectors in these studies was incorrect. Specifically, when multiple coalitions have the same maximum excess during an iteration of the Nucleolus computation algorithm (described in Algorithm 11.4.1), the authors failed to properly update the linear program before proceeding to the next iteration.

In contrast, our computations—using three MATLAB programs named `myInsuranceExample.m`, `myMobilesExample.m`, and `myManufacturingExample.m`—correctly implement the process. We applied the Duality Theorem (2.7.2) and the Slackness Theorem (2.7.3) within these programs to identify the coalition with the maximum excess at each step. This information was then used to update the linear program appropriately before continuing to subsequent iterations of Algorithm 11.4.1. Our results align with those presented in [6].

Below, we provide detailed descriptions of each example.

Example 11.8.10

This example examines the collaboration among three insurance companies {1, 2, 3} investing in common funds. The data taken from [14] are C(1) = {1}, C(2) = {2}, C(3) = { 3}, C(4) = {1, 2}, C(5) = {1, 3}, C(6) = {2, 3}, C(7) = {1, 2, 3}. The corresponding benefits are v(1) = 46125.0, v(2) = 17437.5, v(3) = 5812.5, v(4) = 69187.5, v(5) = 53812.5, v(6) = 30750.0, v(7) = 90000.0. The initial LP for this problem is shown below:

```
max e1 subject to:
x1 + x2 + 0 x3 - e1 >= 69187.5
x1 + 0 x2 + x3 - e1 >= 53812.5
0 x1 + x2 + x3 - e1 >= 30750.0

x1 + x2 + x3 = 90000
x1 - e1 >= 46125.0
x2 - e1 >= 17437.5
x3 -e1  >= 5812.5
, where we denote the excess by e1 above.
```

The program myInsuranceExample.m computes the 1st and 2nd largest excess to be t = 6562.5 and t= 7031.2 respectively , and the corresponding imputation vectors are x= (52687, 24000, 13313) and x = (52687, 24469,12844) respectively.

Source code of myInsuranceExample.m is shown below:

```
% this is the Insurance Example
% max e1
% x1 + x2 + 0 x3 - e1 >= 69187.5
% x1 + 0 x2 + x3  - e1 >= 53812.5
% 0 x1  + x2 + x3 - e1 >= 30750.0

% x1 + x2 + x3 = 90000
% x1 - e1 >= 46125.0
% x2 - e1 >= 17437.5
% x3 -e1 >= 5812.5

C = [1 0 0; 0 1 0; 0 0 1; 1 1 0; 1 0 1; 0 1 1; 1 1 1]
Cv = [46125.0 17437.5 5812.5 69187.5 53812.5 30750.0 90000.0 ]


A = [1  0  0 -1;
     0  1  0 -1;
     0  0  1 -1;
     1  1  0 -1;
     1  0  1 -1;
     0  1  1 -1;]

b = [46125.0 17437.5 5812.5 69187.5 53812.5 30750.0  ]'
Aeq = [1  1  1  0]
beq = 90000.0

lb =[0; 0; 0; 0]

f = [0 0 0 1]
x = linprog(-f, -A, -b, Aeq, beq,lb,[])
fprintf('The Maximum of the Primal Program is: \n')
t = x(4)

sum = 0;
for i = 1:7
  sum = C(i, 1) * x(1) + C(i, 2) * x(2) + C(i, 3) * x(3);
```

```
    ev(i) = Cv(i) - sum;
      if (ev(i) == -t)
         fprintf('Coalition Number with excess equal to t: %d\n', i);
      end
end
ev
fprintf('***********************************************************: \n')
fprintf('Compute the the dual program below: \n')

f_dual = [-46125.0; -17437.5; -5812.5; -69187.5; -53812.5; -30750.0; 90000.0]
A_dual = [1   0   0   1   1   0  -1;
          0   1   0   1   0   1  -1 ;
          0   0   1   0   1   1  -1;
         -1  -1  -1  -1  -1  -1   0]

b_dual = [0; 0; 0; -1]

lb_dual = [0; 0; 0; 0; 0; 0; -inf]

y = linprog(f_dual, A_dual, b_dual, [], [], lb_dual, [])

dual_min_value = y'*f_dual

fprintf('************************************************************ \n');

fprintf(' Compute the Second largest excess vector for the game: \n');

% Thus from the positive values of y1 and y6, we know that coaltions C1 and
% C6 we must have: x1 - t = 46125.0 and x2 + x3 - t = 30750.0, which gives
% the two equalities:

% 1 X1 + 0x2 + 0x3 -1t = 46125.0
% 0x1 + 1 X2 + 1x3 - 1t = 30750.0
% t =6562.5 46125 + t = 52687.5 30750 + t = 37312.5

A = [0  1  0 -1;
     0  0  1 -1;
     1  1  0 -1;
     1  0  1 -1]

b = [17437.5 5812.5 69187.5 53812.5]'
Aeq = [1 1 1 0; 1 0 0 0; 0 1 1 0]
```

```
beq = [90000.0; 52687.5; 37312.5]

lb = [0; 0; 0; 0]

f = [0 0 0 1]
x = linprog(-f, -A, -b, Aeq, beq, lb, [])
fprintf('The Maximum of the Primal Program is: \n')
t = x(4)

sum = 0;
for i = 1:7
     sum = C(i, 1) * x(1) + C(i, 2) * x(2) + C(i, 3) * x(3);
       ev(i) = Cv(i) - sum;
          if (ev(i) == -t)
             fprintf('Coalition Number with excess equal to t: %d\n', i);
          end
end
ev
fprintf('**********************************************************: \n')

fprintf(' Compute the above dual program the second time: \n')

f_dual = [-17437.5; -5812.5; -69187.5; -53812.5; 90000.0; 52687.5; 37312.5]
A_dual = [0  0  1  1 -1 -1  0;
           1  0  1  0 -1  0 -1;
           0  1  0  1 -1  0 -1;
          -1 -1 -1 -1  0  0  0]

b_dual = [0; 0; 0; -1]

lb_dual = [0; 0; 0; 0; -inf; -inf; -inf]

y = linprog(f_dual, A_dual, b_dual, [], [], lb_dual, [])

dual_min_value = y'*f_dual

fprintf('Compute the Third largest excess vector for the game: \n');

% Thus from the positive values of y1 and y6, we know that coaltions C2 and
% C3 with t = 7031.2  we must have: x2 - t = 17437.5 and x3 - t = 5812.5, or x2 =
% 24468.7 and x3 = 12844.2 which gives the two equalities:
```

```
% 0 X1 + 1 x2 + 0x3 -0t = 24468.7

% 0x1 + 0 X2 + 1x3 - 0t = 12843.7


A = [1  1  0 -1;
     1  0  1 -1]

b = [69187.5 53812.5]'
Aeq = [1 1 1 0; 1 0 0 0; 0 1 1 0; 0 1 0 0; 0 0 1 0]
beq = [90000.0; 52687.5; 37312.5; 24468.7; 12843.7]


lb = [0; 0; 0; 0]

f = [0 0 0 1]
x = linprog(-f, -A, -b, Aeq, beq, lb, [])

fprintf('***********************************************************: \n')
```

Output listing of myInsuranceExample.m is shown below:

```
>> myInsuranceExample

C =

    1   0   0
    0   1   0
    0   0   1
    1   1   0
    1   0   1
    0   1   1
    1   1   1

Cv =

    1.0e+04 *

    4.6125   1.7437   0.5813   6.9188   5.3812   3.0750   9.0000

A =

    1   0   0  -1
    0   1   0  -1
```

```
     0     0     1    -1
     1     1     0    -1
     1     0     1    -1
     0     1     1    -1
b =
     1.0e+04 *
     4.6125

     1.7437

     0.5813

     6.9188

     5.3812

     3.0750

Aeq =
     1     1     1     0

beq =
     90000

lb =
     0
     0
     0
     0
f =
     0     0     0     1

Optimal solution found.

x =
     1.0e+04 *
     5.2687
```

```
    2.4000
    1.3313
    0.6562
The Maximum of the Primal Program is:

t =

    6.5625e+03

Coalition Number with excess equal to t:  1

Coalition Number with excess equal to t:  2

Coalition Number with excess equal to t:  6

ev =

    1.0e+04 *

    -0.6562   -0.6562   -0.7500   -0.7500   -1.2188   -0.6562    0

****************************************************************:

Compute the the dual program below:

f_dual =

    1.0e+04 *

    -4.6125
    -1.7437
    -0.5813
    -6.9188
    -5.3812
    -3.0750
    9.0000

A_dual =

     1     0     0     1     1     0    -1
     0     1     0     1     0     1    -1
     0     0     1     0     1     1    -1
    -1    -1    -1    -1    -1    -1     0
```

```
b_dual =

     0
     0
     0
    -1

lb_dual =

     0
     0
     0
     0
     0
     0
  -Inf

Optimal solution found.

y =

    0.5000
         0
         0
         0
         0
    0.5000
    0.5000

dual_min_value =

    6.5625e+03
```

****

Compute the Second largest excess vector for the game:

```
A =

     0     1     0    -1
```

```
    0    0    1   -1
    1    1    0   -1
    1    0    1   -1

b =

    1.0e+04 *
    1.7437
    0.5813
    6.9188
    5.3812

Aeq =

    1    1    1    0
    1    0    0    0
    0    1    1    0

beq =

    1.0e+04 *
    9.0000
    5.2687
    3.7313
lb =
    0
    0
    0
    0
f =
    0    0    0    1

Optimal solution found.

x =

    1.0e+04 *
    5.2687
    2.4469
```

1.2844

0.7031

The Maximum of the Primal Program is:

t =

7.0312e+03

Coalition Number with excess equal to t: 2

Coalition Number with excess equal to t: 3

ev =

1.0e+04 *

-0.6562 -0.7031 -0.7031 -0.7969 -1.1719 -0.6562 0

***:

Compute the above dual program the second time:

f_dual =

1.0e+04 *

-1.7437

-0.5813

-6.9188

-5.3812

9.0000

5.2687

3.7313

A_dual =

```
 0   0   1   1  -1  -1   0
 1   0   1   0  -1   0  -1
 0   1   0   1  -1   0  -1
-1  -1  -1  -1   0   0   0
```

b_dual =

0

0

0

-1

lb_dual =

0

0

0

0

-Inf

-Inf

-Inf

Optimal solution found.

y =

0.5000

0.5000

0

0

0

0

0.5000

dual_min_value =

7.0312e+03

Compute the Third largest excess vector for the game:

A =

1	1	0	-1
1	0	1	-1

b =

1.0e+04 *

```
    6.9188
    5.3812

Aeq =

    1   1   1   0
    1   0   0   0
    0   1   1   0
    0   1   0   0
    0   0   1   0

beq =

    1.0e+04 *

    9.0000
    5.2687
    3.7313
    2.4469
    1.2844

lb =

    0
    0
    0
    0

f =

    0   0   0   1

No feasible solution found.

Linprog stopped because no point satisfies the constraints.

x = []
```

**:

Example 11.8.11

This example deals with four mobile companies sharing broadcast transmission costs for streaming a sporting event . The data taken from [7] are C(1) = {1}, C(2) = {2}, C(3) = {3}, C(4) = {4}, C(5) = {1, 2}, C(6) = {1, 3}, C(7) = {1, 4}, C(8) = {2, 3}, C(9) = {2, 4}, C(10) = {3, 4}, C(11) = {1, 2, 3}, C(12) = {1,2,4},

C(13) = {1, 3, 4}, C(14) = {2, 3, 4}, C(15) = {1, 2, 3, 4}. The corresponding cost are v(1) = 8, v(2) = 1, v(3) = 10, v(4) = 11, v(5) = 8, v(6) = 10, v(7) = 19, v(8) = 10, v(9) = 12, v(10) = 12, v(11) = 10, v(12) = 19,v(13) = 20, v(14) = 13, v(15) = 20

The initial LP for this problem is shown below:

```
max e1
x1 + 0 x2 + 0 x3 + 0 x4 + e1 <= 8
0 x1 + x2 + 0 x3 + 0 x4 + e1 <= 1
0 x1 + 0 x2 + x3 + 0 x4 + e1 <= 10
0x1 + 0x2 + 0x3 + 1 x4 + e1 <= 11

x1 + x2 <= 8
x1 + X3 <= 10
X1 + x4  <= 19
x2  + x3 <= 10
x2 + x4 <=12
x3  + x4 <= 12

X1 + X2 + X3 <= 10
X1 + x2 + x4 <= 19
x1 + x3 + x4 <=20
x2 + x3 + x4 <= 13

x1 +x2 +x3 +x4 =20
```

In `myMobilesExample.m`, as with the other two examples, we compute the values of the dual variable y of the dual LP, positive value of y indicates which coalition contributes to the optimal solution of original LP. In the present case, we find y = 0.5 corresponds to the coalition 5 and 10, others have 0 or negative values, the imputation vector x = (7.5, 0.5, 1.5, 10.5), and excess x(4) = 0.5 (See Output of `myMobilesExamples.m` below)

Source code of myMobilesExample.m is shown below:

```
% this is the Mobiles Example
% max e1
```

```
% x1 + 0 x2 + 0 x3 + 0 x4 + e1 <= 8
% 0 x1 + x2 + 0 x3 + 0 x4 + e1 <= 1
% 0 x1 + 0 x2 + x3 + 0 x4 + e1 <= 10
% 0x1 + 0x2 + 0x3 + 1 x4 + e1 <= 11

% x1 + x2 <= 8
% x1 + X3 <= 10
% X1 + x4 <= 19

% x2 + x3 <= 10
% x2 + x4 <=12
% x3 + x4 <= 12
% X1 + X2 + X3 <= 10
% X1 + x2 + x4 <= 19
% x1 + x3 + x4 <= 20
% x2 + x3 + x4 <= 13
% x1 + x2 + x3 + x4 =20

C = [1 0 0 0; 0 1 0 0; 0 0 1 0; 0 0 0 1; 1 1 0 0; 1 0 1 0; 1 0 0 1; 0 1 1 0; 0 1 0 1; 0 0 1 1; 1 1 1 0;
     1 1 0 1; 1 0 1 1; 0 1 1 1; 1 1 1 1]
Cv = [8; 1; 10; 11; 8; 10; 19; 10; 12; 12; 10; 19; 20; 13; 20]

A = [1   0   0   0   1;
     0   1   0   0   1;
     0   0   1   0   1;
     0   0   0   1   1;
     1   1   0   0   1;
     1   0   1   0   1;
     1   0   0   1   1;
     0   1   1   0   1;
     0   1   0   1   1;
     0   0   1   1   1;
     1   1   1   0   1;
     1   1   0   1   1;
     1   0   1   1   1;
     0   1   1   1   1;]

b = [8 1 10 11 8 10 19 10 12 12 10 19 20 13]'
Aeq = [1  1  1  1  0]
beq = 20
```

```
lb = [0; 0; 0; 0; 0]

f = [0 0 0 0 1]
x= linprog(-f, A, b, Aeq, beq, lb, [])

x

fprintf('The Maximum of the Primal Program is: \n')
t = x(5)

sum = 0;
for i = 1:15
    sum = C(i, 1) * x(1) + C(i, 2) * x(2) + C(i, 3) * x(3) + C(i, 4) * x(4);
     ev(i) = Cv(i) - sum;
      if (ev(i) == t)
         fprintf('Coalition Number with excess equal to t: %d\n', i);
      end
end
ev
fprintf('Compute the the dual program below: \n')


f_dual = [8; 1; 10; 11; 8; 10; 19; 10;12; 12; 10; 19; 20; 13; 20]
A_dual = [1 0 0 0 1 1 1 0 0 0 1 1 1 0 1;
          0 1 0 0 1 0 0 1 1 1 1 1 0 1 1;
          0 0 1 0 0 1 0 1 0 1 1 0 1 1 1;
          0 0 0 1 0 0 1 0 1 1 0 1 1 1 1;
          1 1 1 1 1 1 1 1 1 1 1 1 1 1 0]

b_dual = [0; 0; 0; 0; 1]

lb_dual = [0; 0; 0; 0; 0; 0; 0; 0; 0 ; 0; 0; 0; 0; 0; -inf]

y = linprog(f_dual, -A_dual, -b_dual, [], [], lb_dual, [])

dual_min_value = y'*f_dual

fprintf('************************************************************* \n');

fprintf(' Compute the Second largest excess vector for the game: \n');

% Thus from the positive values of y5 and y10, we know that coalitions C3 and
% C4 we must have: x1 + x2 + t = 8 and x3 + x4 + t = 12, which gives
```

```
% the two equalities:
% x1 + x2 = 8 - t and x3 + x4 = 12 - t
% Now the Max LP of the original LP becomes:


A = [1  0  0  0  1;
     0  1  0  0  1;
     0  0  1  0  1;
     0  0  0  1  1;
     1  0  1  0  1;
     1  0  0  1  1;
     0  1  1  0  1;
     0  1  0  1  1;
     1  1  1  0  1;
     1  1  0  1  1;
     1  0  1  1  1;
     0  1  1  1  1;]

b = [8 1 10 11 10 19 10 12 10 19 20 13]'
Aeq = [1 1 1 1 0; 1 1 0 0 0; 0 0 1 1 0]
beq = [20; 8; 12]

lb = [0; 0; 0; 0; 0]

f = [0 0 0 0 1]
x = linprog(-f, A, b, Aeq, beq, lb, [])

x;

fprintf('The Maximum of the Primal Program is: \n')
t = x(5)
```

Output Listing of myMobilesExample.m is shown below:

```
>> myMobilesExample

C =

    1   0   0   0

    0   1   0   0
```

0 0 1 0

0 0 0 1

1 1 0 0

1 0 1 0

1 0 0 1

0 1 1 0

0 1 0 1

0 0 1 1

1 1 1 0

1 1 0 1

1 0 1 1

0 1 1 1

1 1 1 1

Cv =

8

1

10

11

8

10

19

10

12

12

10

19

20

13

20

A =

1 0 0 0 1

```
     0     1     0     0     1
     0     0     1     0     1
     0     0     0     1     1
     1     1     0     0     1
     1     0     1     0     1
     1     0     0     1     1
     0     1     1     0     1
     0     1     0     1     1
     0     0     1     1     1
     1     1     1     0     1
     1     1     0     1     1
     1     0     1     1     1
     0     1     1     1     1

b =

     8
     1
    10
    11
     8
    10
    19
    10
    12
    12
    10
    19
    20
    13

Aeq =

     1     1     1     1     0
```

```
beq =

  20

lb =

     0
     0
     0
     0
     0

f =

     0     0     0     0     1

Optimal solution found.

x =

     7
     1
     2
    10
     0

x =

     7
     1
     2
    10
     0
The Maximum of the Primal Program is:

t =

     0

Coalition Number with excess equal to t:  2

Coalition Number with excess equal to t:  5
```

Coalition Number with excess equal to t: 10

Coalition Number with excess equal to t: 11

Coalition Number with excess equal to t: 14

Coalition Number with excess equal to t: 15

ev =

1 0 8 1 0 1 2 7 1 0 0 1 1 0 0

Compute the dual program below:

f_dual =

8

1

10

11

8

10

19

10

12

12

10

19

20

13

20

A_dual =

1 0 0 0 1 1 1 0 0 0 1 1 1 0 1

0 1 0 0 1 0 0 1 1 1 1 1 0 1 1

0 0 1 0 0 1 0 1 0 1 1 0 1 1 1

0 0 0 1 0 0 1 0 1 1 0 1 1 1 1

1 1 1 1 1 1 1 1 1 1 1 1 1 1 0

b_dual =

0

0

0

0

1

lb_dual =

0

0

0

0

0

0

0

0

0

0

0

0

0

0
-Inf

Optimal solution found.

y =

0

0

0

0

0.5000

0

0

0

0

0.5000

0

0

0

0

-0.5000

dual_min_value =

0

****

Compute the Second largest excess vector for the game:

A =

```
1 0 0 0 1
0 1 0 0 1
0 0 1 0 1
0 0 0 1 1
1 0 1 0 1
1 0 0 1 1
0 1 1 0 1
0 1 0 1 1
1 1 1 0 1
1 1 0 1 1
1 0 1 1 1
0 1 1 1 1
```

b =

8

1

10

11

```
    10

    19

    10

    12

    10

    19

    20

    13

Aeq =

    1    1    1    1    0

    1    1    0    0    0

    0    0    1    1    0

beq =

    20

    8

    12

lb =

    0

    0

    0

    0

    0

f =

    0    0    0    0    1

Optimal solution found.

x =

    7.5000

    0.5000

    1.5000
```

10.5000

0.5000

x =

7.5000

0.5000

1.5000

10.5000

0.5000

The Maximum of the Primal Program is:

t =

0.5000

Example 11.8.12

This example addresses cost-sharing issues among three joint network-centric manufacturing firms {1, 2, 3} in [22]. The data they used for the computation of nucleolus is C(1) = {1}, C(2) = {2}, C(3) = {3}, C(4) = {1, 2}, C{5} = {1, 3}, C(6) = {2, 3}, C(7) = {1, 2, 3}, v(1) = 375144, v(2) = 245280, v(3) = 211239, v(4) = 568232, v(5) = 551055, v(6)= 452411, v(7)= 772500

The initial LP is shown as:

```
max e1
x1 + 0 x2 + 0 x3 + e1 <= 375144.0
0 x1 + x2 + 0 x3 + e1 <= 245280.0
0 x1 + 0 x2 + x3 + e1 <= 211239.0
x1 + x2 + e1 <= 568232.0
x1 + X3 + e1 <= 551055.0
X2 + x3 + e1 <= 452411.0
x1 + x2 + x3 = 772500
```

Like the approach used in the last two examples, myManufacturingExample.m computes nucleolus vector x = (331700, 233050, 207750) and excess = 11610. (See Output Listing of myManufacturingExample.m)

Source code of myManufacturingExample.m is shown below:

```
% this is the Manufactruing Firm Example
% max e1
% x1 + 0 x2 + 0 x3 + e1 <= 375144.0
```

```
% 0 x1 + x2 + 0 x3 + e1 <= 245280.0
% 0 x1 + 0 x2 + x3 + e1 <= 211239.0

% x1 + x2 <= 568232.0
% x1 + X3 <= 551055.0
% X2 + x3 <= 452411.0

% x1 + x2  + x3 = 772500

C = [1 0 0; 0 1 0; 0 0 1; 1 1 0; 1 0 1; 0 1 1; 1 1 1]
Cv = [375144.0  245280.0  211239.0  568232.0  551055.0  452411.0  772500.0 ]

A = [1  0  0  1;
     0  1  0  1;
     0  0  1  1;
     1  1  0  1;
     1  0  1  1;
     0  1  1  1;]

b = [375144.0  245280.0  211239.0  568232.0  551055.0  452411.0]'
Aeq = [1  1  1  0]
beq = 772500.0

lb = [0; 0; 0; 0]

f = [0 0 0 -1]
x = linprog(f, A, b, Aeq, beq, lb, [])

x;

fprintf('The Maximum of the Primal Program is: \n')
t = x(4)

sum = 0;
for i = 1:7
     sum = C(i, 1) * x(1) + C(i, 2) * x(2) + C(i, 3) * x(3);
       ev(i) = Cv(i)-sum;
         if (ev(i)==t)
           fprintf('Coalition Number with excess equal to t: %d\n', i);
         end
end
ev
```

```
fprintf('Compute the the dual program below: \n')

f_dual = [375144; 245280; 211239; 568232; 551055; 452411; 772500]
A_dual = [1  0  0  1  1  0  1;
          0  1  0  1  0  1  1;
          0  0  1  0  1  1  1;
          1  1  1  1  1  1  0]

b_dual = [0; 0; 0; 1]

lb_dual = [0; 0; 0; 0; 0; 0; -inf]

y = linprog(f_dual, -A_dual, -b_dual, [], [], lb_dual, []);

dual_min_value = y'*f_dual

fprintf('************************************************************ \n');

fprintf('Compute the Second largest excess vector for the game: \n');

% Thus, from the positive values of y3 and y4, we know that coalitions C3 and
% C4 we must have: x3 + t = 211239.0 and x1 + x2 + t = 568232.0, which gives
% the two equalities:
%  x3 = 211239.0 -t and x1 + x2 = 568232.0 -t
% Now the Max LP of the original LP becomes:

A = [1  0  0  1;
     0  1  0  1;
     1  0  1  1;
     0  1  1  1;]

b = [375144.0  245280.0  551055.0  452411.0]'
Aeq = [1  1  1  0;
       0  0  1  0;
       1  1  0  0]
beq = [772500.0;  211239.0-t;  568232.0-t]

lb = [0; 0; 0; 0]

f = [0 0 0 -1]
x = linprog(f, A, b, Aeq, beq, lb, [])

x;
```

```
fprintf('The Maximum of the Primal Program is: \n')
t = x(4)
```

Output Listing of myManufacturingExample.m is shown below:

```
>> myManufacturingExample

C =

    1     0     0
    0     1     0
    0     0     1
    1     1     0
    1     0     1
    0     1     1
    1     1     1

Cv =

    375144      245280      211239      568232      551055      452411      772500

A =

    1     0     0     1
    0     1     0     1
    0     0     1     1
    1     1     0     1
    1     0     1     1
    0     1     1     1

b =

    375144
    245280
    211239
    568232
    551055
    452411

Aeq =

    1     1     1     0

beq =

    772500
```

```
lb =

     0
     0
     0
     0

f =

     0     0     0    -1

Optimal solution found.

x =

   1.0e+05 *

    3.2357
    2.4117
    2.0775
    0.0349

x =

   1.0e+05 *

    3.2357
    2.4117
    2.0775
    0.0349

The Maximum of the Primal Program is:

t =

   3.4855e+03

Coalition Number with excess equal to t:  3
Coalition Number with excess equal to t:  4
Coalition Number with excess equal to t:  6

ev =

   1.0e+04 *

    5.1570    0.4108    0.3486    0.3486    1.9727    0.3486         0    0.0007    0.0001         0         0    0.0001
    0.0001         0         0
```

```
Compute the the dual program below:

f_dual =

    375144
    245280
    211239
    568232
    551055
    452411
    772500

A_dual =

    1    0    0    1    1    0    1
    0    1    0    1    0    1    1
    0    0    1    0    1    1    1
    1    1    1    1    1    1    0

b_dual =

    0
    0
    0
    1

lb_dual =

    0
    0
    0
    0
    0
    0
    -Inf

Optimal solution found.

dual_min_value =

    3.4855e+03

************************************************************

Compute the Second largest excess vector for the game:
```

```
A =

    1     0     0     1
    0     1     0     1
    1     0     1     1
    0     1     1     1

b =

    375144
    245280
    551055
    452411

Aeq =
    1     1     1     0
    0     0     1     0
    1     1     0     0

beq =

    1.0e+05 *

    7.7250
    2.0775
    5.6475

lb =

    0
    0
    0
    0

f =

    0     0     0    -1

Optimal solution found.

x =

    1.0e+05 *
```

```
    3.3170

    2.3305

    2.0775

    0.1161

x =

    1.0e+05 *

    3.3170

    2.3305

    2.0775

    0.1161

The Maximum of the Primal Program is:

t =

    1.1606e+04
```

11.9 Summary

In this chapter, we presented two approaches, the imputation set and the Shapley Value approaches to reveal the payoff information of each individual player within the coalitions in coalition games. Three MATLAB programs, myShapleyvalueBV.m, myComputeCoreBV.m, and myComputeNucleolusBV.m were written to compute the Shapley values, the Core and the Nucleolus of a coalition game. These programs are very helpful computation tools and they were used to exemplify various concepts and definitions introduced in this chapter. Section 11.8 explores the industrial applications of the nucleolus in coalition games. To facilitate the computation of the nucleolus from given data, we developed three MATLAB programs: myInsuranceExample.m, myMobilesExample.m, and myManufacturingExample.m. Each program demonstrates the use of the Duality and Slackness Theorems to effectively address situations where multiple coalitions share the same maximum excess. Since the nucleolus is defined as the imputation vector that minimizes the worst dissatisfaction among all coalitions, a key technical insight from this chapter is that calculating the nucleolus involves solving a sequence of linear programs. These programs iteratively reduce the maximum coalition excess, requiring careful handling of ties and overlapping coalitions to ensure accurate results.

Chapter 12

Cooperative Games and Nash Equilibrium

12.0 Introduction

In this final chapter, we explore a distinct class of games known as cooperative games, where players can communicate and negotiate agreements before making their strategic choices. Unlike non-cooperative games, where players act independently, cooperative games allow for pre-game communication, enabling players to form binding agreements that lead to mutually beneficial outcomes.

This additional element of cooperation differentiates cooperative games from competitive or zero-sum games, where one player's gain is another's loss. However, despite this shift in dynamics, many of the key principles and results established in previous chapters still apply to cooperative game scenarios.

Real-World Applications of Cooperative Games

Cooperative game theory provides a powerful framework for modeling conflict resolution and negotiations in various real-world situations, including:

- **Labor negotiations** – Agreements between upper management and labor unions over wages and work conditions.
- **Trade agreements** – Economic treaties and cooperative strategies among nations.
- **Political disputes** – Diplomatic negotiations and coalition formations among countries.
- **Profit-sharing conflicts** – Disputes over revenue distribution within organizations or between individuals.

Applying Zero-Sum Game Theory to Cooperative Games

In the following sections, we present simple examples of cooperative games to illustrate these concepts. We then revisit key results from zero-sum games and explore how they can be adapted to cooperative game settings. In particular, we apply these insights to the Nash two-step bargaining procedure, a fundamental approach for resolving strategic negotiations between players in cooperative games.

By bridging competitive and cooperative game theories, this chapter provides a comprehensive understanding of how rational players can leverage cooperation to achieve optimal outcomes in strategic decision-making.

12.1 Simple examples of pre-play cooperative games

For the rest of the chapter, we assume that there are two players, player A and player B with the respective payoff values (a_{ij}, b_{Ij}) for both players shown in matrix p below, where i = 1, m, j = 1, n, and where m is the number of pure strategies $x_1, x_2, \ldots. x_m$ for player A, n is the number of pure strategies $y_1, y_2, \ldots. y_n$ for player B.

$$p = \begin{bmatrix} (a_{11},b_{11}) & (a_{1j},b_{1j}) & (a_{1n},b_{1n}) \\ \ldots\ldots\ldots\ldots\ldots\ldots\ldots\ldots\ldots\ldots \\ (a_{i1},b_{i1}) & (a_{ij},b_{ij}) & (a_{in},b_{in}) \\ \ldots\ldots\ldots\ldots\ldots\ldots\ldots\ldots\ldots\ldots \\ (a_{m1},b_{m1}) & (a_{mj},b_{mj}) & (a_{mn},b_{mn}) \end{bmatrix} \qquad \text{(D1)}$$

$$a = \begin{bmatrix} a_{11} & a_{1j,} & a_{1n} \\ \ldots\ldots\ldots\ldots\ldots\ldots \\ a_{i1} & a_{ij,} & a_{in} \\ \ldots\ldots\ldots\ldots\ldots\ldots \\ a_{m1} & a_{mj,} & a_{mn} \end{bmatrix}, \text{ with m pure strategies } x_1, x_2, \ldots. x_m \text{ for player A} \qquad \text{(D2)}$$

$$b = \begin{bmatrix} b_{11} & b_{1j,} & b_{1n} \\ \ldots\ldots\ldots\ldots\ldots\ldots \\ b_{i1} & b_{ij,} & b_{in} \\ \ldots\ldots\ldots\ldots\ldots\ldots \\ b_{m1} & b_{mj,} & b_{mn} \end{bmatrix}, \text{ with n pure strategies } y_1, y_2, \ldots. y_n \text{ for player B} \qquad \text{(D3)}$$

Example 1.1

This example shows that players could employ threat strategies to their advantage if pre-play communication is allowed. Suppose the payoff matrix is given as:

$$p = \begin{bmatrix} (0,-100) & (0,-200) \\ (10,100) & (100,20) \end{bmatrix}$$

If this was a simultaneous game with no pre-play communication, then obviously player A would have chosen strategy x_2 instead of x_1 because x_2 dominates x_1, and similarly player B would have chosen strategy y_1 over y_2. Thus, with no pre-play communication, both players would settle on the payoff (10, 100), which means player A can only get 10 in this case. But with 12.1pre-play communication, player A can threaten player B to play y_2 otherwise he (player A) would play x_1 which would result in a worse loss of -100 or -200 to player B. Thus, both players would settle on the payoff of (100, 20) in this case. This means that player A can increase his payoff value from 10 to 100.

Example 1.2

This next example shows that, if pre-play agreement and side-payment is allowed, then it would be much more beneficial to both players than each player choosing their own optimal strategies. Suppose the payoff matrix is given as:

$$p = \begin{bmatrix} (100,0) & (1,\ 2) \\ (1,0\ ) & (1,2) \end{bmatrix}$$

If this was a simultaneous game, then player A would choose strategy x_1 and player B would choose y_2, thus resulting in the payoff (1, 2) for both players. But if pre-play communication was allowed, both players could agree on playing the strategies (x_1, y_1) resulting in the payoff (100, 0), and player A agrees to share his 100 units of utility in some way to player B (assuming more than 2), and thus results in a win-win solution for both players.

Example 1.3

$$p = \begin{bmatrix} (100,10) & (1,\ 2) \\ (1,2) & (100,10) \end{bmatrix}$$

If the above is a simultaneous game, there is no optimal strategy for each player because there is no dominating strategy for each player. However, if they agree to play strategy (x_1, y_1) or (x_2, y_2) in advance, then the optimal payoff would always be (100, 10) resulting in a win-win solution for both players.

12.2 Recapitulation of basic results for non-cooperative games

In this section, we give examples to recapitulate some of the key concepts and results for competitive zero-sum or non-zero-sum games discussed previously. We show these concepts and results such as security level and Nash Equilibrium are useful tools to analyze cooperative games as well. We also demonstrate some differences between cooperative and competitive games regarding these results.

Example 2.4

This example shows the computation of saddle point and the corresponding security levels for both players. Suppose the payoff matrices for both players are given below.

$$p = \begin{bmatrix} (12,21) & (21,12) \\ (11,1) & (11,2) \end{bmatrix} \quad a = \begin{bmatrix} 12 & 21 \\ 11 & 11 \end{bmatrix} \quad b = \begin{bmatrix} 21 & 12 \\ 1 & 2 \end{bmatrix}$$

We demonstrate this computation by using the Matlab function call mycomputesaddlepoint (a) and mycomputesaddlepoint (b’). This function inputs a, the payoff matrix for player A (transpose b’ of the payoff matrix for player B), outputs logical flag yes = 0 or 1 to indicate if the matrix has a saddle point or not, the security level, and the security level strategy. The result shows that player A has a saddle point, security level 12, and security level strategy (1, 0); player B has a saddle point, security level 2, and security level strategy (0, 1).

(See the output of the MATLAB program SecurityLevelandEquilibriumPairs1.m)

Example 2.5

This example shows that the security level strategies of two players in a non-zero-sum game are not necessarily in Nash Equilibrium. This is different from a key result with zero-sum games which states that the security levels strategies are in Nash Equilibrium. We demonstrate this by the MATLAB function testofEPairs and the same payoff matrices as in the last Example 12.4. Inputs to the function testofEPairs are the two payoff matrices for player A and player B, a strategy vector for A and a strategy vector for B. It outputs a logical value yes = 0 or 1 to indicate whether the two given strategy vectors are in Equilibrium or not. From Example 12.4, we know that Xa = (1 0), and Yb = (0 1) are security strategies for player A and player B respectively, however, calling testofEPairs(a, b, Xa, Yb) gives yes = 0 which means security level strategies are not necessary in equilibrium. Note that if we change Yb = (0 1) to Yb1 = (1 0) and call testofEPairs(a, b, Xa, Yb1), we get yes = 1. This means that equilibrium pairs are not necessarily security strategy pairs.

(See the output of the MATLAB program SecurityLevelandEquilibriumPairs1.m)

Example 2.6

This example shows that in general, if (x, y) is an equilibrium strategy pair for the two players A and B with respective payoff matrix a and b, then (x, y) provides upper bounds ubsl(A) and ubsl(B) for the security levels sl(A) and sl(B) of respective players A and B. Precisely,

sl(A) $\leq$ ubsl(A) and sl(B) $\leq$ sbsl(B), where ubsl(A) = x a y^t and ubsl(B) = x b y^t

The MATLAB function computeUpperbounds takes payoff matrices a and b, and the equilibrium pairs Xa and Yb as inputs, then it computes the upper bounds ubsl1 and ubsl2 for the security level of the respective players A and B.

Using the same payoff matrices a, b and the equilibrium pair Xa = (1 0) and Yb1 = (1 0) as in the last Example 12.5, calling the function computeUpperbounds (a, b, Xa, Yb1) gives us ubsl(A) = 12 and ubSl(B) = 21

(See the output of the MATLAB program SecurityLevelandEquilibriumPairs1.m)

Example 2.7

This example shows the computation of security levels for both players by the simplex-method when the saddle point in a payoff matrix does not exist and that the function mycomputesaddlepoint described before is not applied. Consider the following payoff matrices for both players A and B respectively,

$$a = \begin{bmatrix} 5 & 7 \\ 8 & 4 \end{bmatrix} \quad b = \begin{bmatrix} 5 & 7 \\ 4 & 8 \end{bmatrix}$$

First, we check if the matrix for player A has a saddle point by calling mycomputesaddlepoint(a). The function gives yes = 0, which means that matrix a does not have a saddle point. So, we use the simplex method to compute the security level and security strategy instead. The function computeSLbySimplexMethod takes the payoff matrix of a player as input and outputs the security level and the security strategy. In addition, to illustrate the Duality Principle, the function also outputs the security level and security strategy of the column player.

Calling the function computeSLSimplexMethod(a) gives the security value of 6, and the security strategy (0.6667 0.3333). The Duality Principle also confirms that the security level for the column player is also 6 with security strategy (0.5 0.5) for player A. Calling mycomputesaddlepoint (b') gives player B the security level of 7 and security level strategy (0 1).

(See the output of the MATLAB program SecurityLevelandEquilibriumPairs1.m)

12.3 Source code and output listing of MATLAB program SecurityLevelsandEquilibriumPairs1.m

A. Output listing of Example 1.1 to 2.7

```
>> SecurityLevelsandEquilibriumPairs1

a =

   12   21

   11   11

b =

   21   12

    1    2

Check if the player 1 matrix has a saddle point:

yes =

    1

slevel =

   12

svector =

    1    0

Xa =

    1    0

Check if the player 2 matrix has a saddle point:

yes =

    1

svalue =

    2
```

```
svector =

     0     1

Yb =

     0     1
```

check if the pair Xa and Yb are in equilibrium :

the given pair is not an equilibrium:

```
yes =

     0
```

check if we keep the security strategy Xa fixed but change Yb to another strategy [1 0] would result in equilibrium:

```
first =

     1     0

second =

     1     0
```

the given pair Xa and Yb is an equilibrium:

```
yes =

     1
```

Compute the upper bounds of security levels by an equilibrium pair:

```
ubsl1 =

    12

ubsl2 =

    21

a =

     5     7
     8     4

b =

     5     7
```

```
    4   8
```

Check if the player 1 matrix has a saddle point:

```
yes =

    0

svalue =

   999

svector =

    0   0
```

results of the simplex method for a payoff matrix a :

```
b =

    5   7

    8   4

b =

   -5  -8

   -7  -4

fx =

    1   1

bx =

   -1  -1
```

Optimal solution found.

```
x =

    0.1111

    0.0556

vx =

    0.1667

nu1 =
```

6.0000

xoriginal =

0.6667

0.3333

nu1 =

6.0000

sl2 =

6.0000

Verify the result by the duality principle :

fy =

-1 -1

by =

1 1

Optimal solution found.

B. Source code of SecurityLevelsandEquilibriumPairs1.m

```
n=2; % number of columns
m=2; % number of rows

%Given the two matrices a and b, the pair Xa and Yb. Check if (Xa, Yb) is
%an equilibrium pairs?
a = [12 21; 11 11]
b = [21 12; 1 2]

fprintf('Check if player 1 matrix has a saddle point: \n')
[yes, slevel, svector] = mycomputesaddlepoint(a)

Xa = svector
fprintf('Check if player 2 matrix has a saddle point:\n')
[yes, svalue, svector] = mycomputesaddlepoint(b')
Yb = svector

first = Xa;
second = Yb;

fprintf('check if the pair Xa and Yb are in equilibrium :\n')
yes = testofEPairs(a, b, first, second, m ,n)
```

```
%************************************************************
fprintf(['check if we keep the security strategy Xa fixed but change' ...'Yb to other strategy [1  0]
would result in equilibrium:\n'])
Yb = [1 0];
first

second = Yb
yes = testofEPairs(a, b, first, second, m ,n)

fprintf('Compute upper bounds of security levels by an equilibrium pair: \n')

[ubsl1, ubsl2] = computeUpperbounds(a, b, Xa, Yb);

a = [5 7; 8 4]
b = [5 7; 4 8]

fprintf('Check if player 1 matrix has a saddle point:\n')
[yes,svalue,svector] = mycomputesaddlepoint(a)

[sl1, sl2 ] = computeSLbySimplexMethod(a);

fprintf('Check if player 2 matrix has a saddle point:\n')
[yes,svalue,svector] = mycomputesaddlepoint(b')

%**************************************************************
%**************************************************************

function [yes, svalue ,svector] = mycomputesaddlepoint(Gmat)
d = size(Gmat);
nrows = d(1);
yes = 0;
svector = zeros(1, nrows);
svalue = 999;
x = min(Gmat, [], 2);
[slx, index] = max (x);

y = max(Gmat, [], 1);
sly = min(y);

if slx==sly
   yes = 1;
   svalue = slx;

   svector(index) = 1;
end

end

%************************************************************
function yes = testofEPairs(a, b, first, second,m, n)
yes = 1;
for i = 1: m
   if a(i,:)* second'> first * a * second'
         yes=0;
```

```
    end %if
end
for j = 1: n
    if first * b(:,j) > first * b * second'
          yes = 0;
    end %if
end

if yes == 1
    fprintf('the given pair Xa and Yb is an equilibrium: \n')
else
    fprintf('the given pair is not an equilibrium: \n')
end % if else

end
%************************************************************
function [ubsl1, ubsl2] = computeUpperbounds(a, b, Xa, Yb)
%fprintf('Compute upper bounds of security levels: \n')

ubsl1 = Xa* a * Yb'
ubsl2 = Xa* b * Yb'
end
%****************************************************************
function [sl1, sl2 ] = computeSLbySimplexMethod (a)
fprintf('results of the simplex method for a payoff matrix a :\n')

b = a
fx = [1 1];
b = -b'
fx = [1 1]
bx = [-1 -1]
[x, vx] = linprog(fx, b, bx)
nu1 = 1/(fx *x)
xoriginal = nu1*x
nu1 = 1/vx
sl2 = nu1

fprintf('Verify the result by the duality principle :\n')
fy = [-1 -1] % because converting the given max to the matlab formatted min

by = [1 1]
y = linprog(fy, a, by)

-fy* y
nu2 = 1/(-fy*y)
yoriginal = nu2*y
sl1 = nu2
end
```

12.4 Concept of cooperative payoff set, and Negotiation set

The cooperative payoff set P of a cooperative game is a graphical representation of all possible payoff values that each player can possibly obtain when both players are taking all possible mixed strategies. Note that the finite number of payoff values (a_{ij}, b_{ij}) shown in the matrix p(Section 12.1) are only a small subset of the infinitely many possible payoff values when the mixed strategies are played out. For example, if player A and player B are playing a mixed strategy x_i and y_j respectively, then the payoff pairs of A and B would be (x_i a y_j^t, x_i b y_j^t), where x_i and y_j are infinitely many.

Definition 12.4.1: Cooperative Payoff Set P of a Cooperative Game

We formally define the cooperative payoff set P consisting of the infinitely many pairs (v_1, v_2) defined as follows:

$$P = \{(v_1, v_2)\}, \text{ where } (v_1, v_2) = \sum_{i,j} k_{ij}(a_{ij}, b_{ij}), \text{ and where} \qquad \text{(D4)}$$

i is one of player A's strategy ranging from 1 to m,

j is one of player B's strategy ranging from 1 to n,

k_{ij} are weighting factors such that $0 \leq k_{ij} \leq 1$

Example 4.8

Consider the following three cooperative games with payoff matrix p1, p2, p3 respectively:

$$p_1 = \begin{bmatrix} (2,2) & (3,\ 12) \\ (12,3) & (2,2) \end{bmatrix} \quad p_2 = \begin{bmatrix} (5,10) & (6,\ 4) \\ (4,6) & (10,5) \end{bmatrix} \quad p_3 = \begin{bmatrix} (5,10) & (7,3) & (12,7) \\ (4,6) & (10,12) & (7,7) \end{bmatrix}$$

The cooperative payoff sets for the respective payoff matrix p1, p2, p3 are within the blackened-edges polygons shown in Figure 12.1, Figure12.2, Figure 12.3 below.

Figure 12.1

Payoff Set of p_1

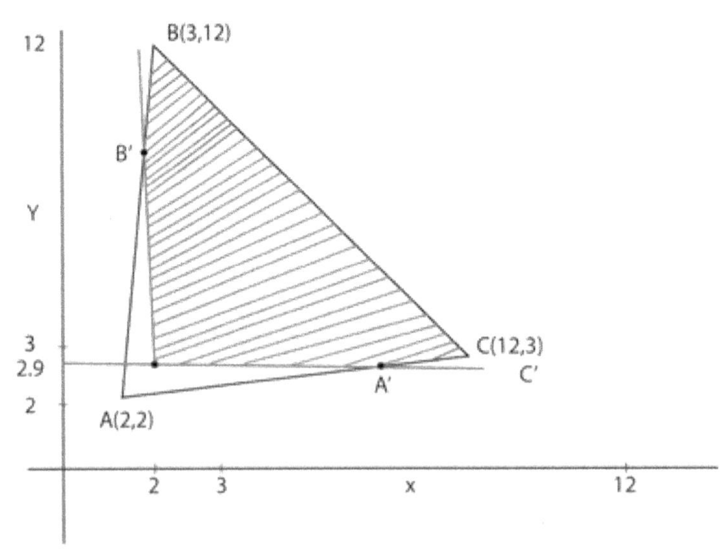

Figure 12.2

Payoff Set of p_2

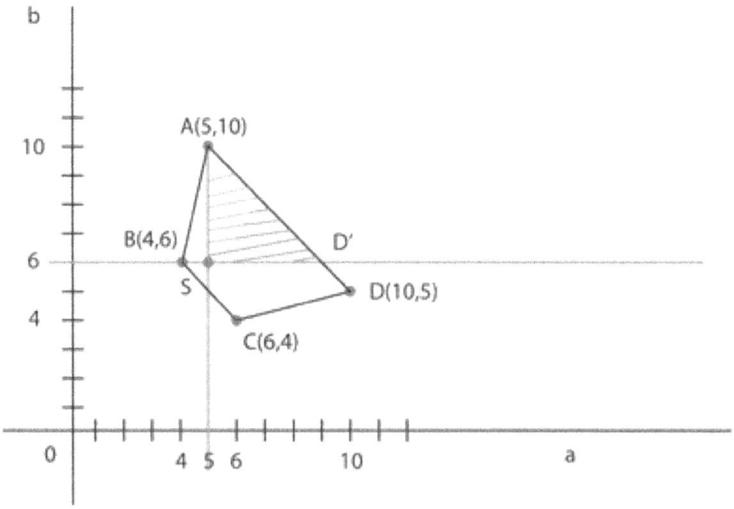

Figure 12.3

Payoff Set of p_3

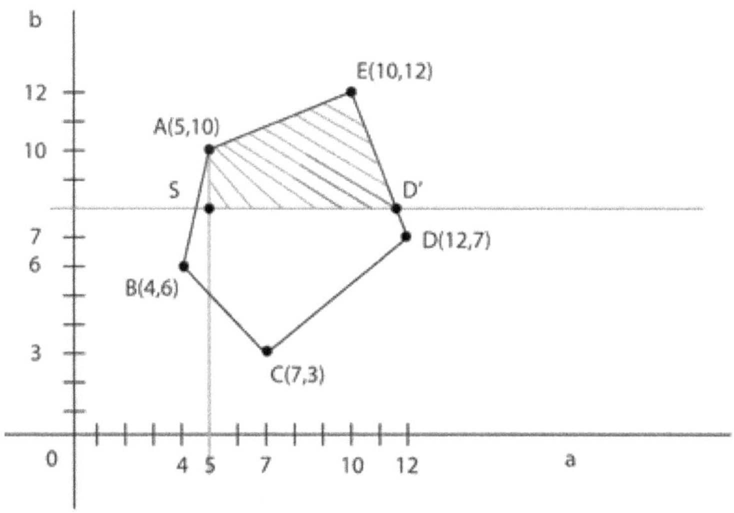

Note that the payoff sets P shown are the respective simplex polygons made of the corner points corresponding to each of the payoff value pairs of both players. Any point within and on this simplex polygon could be a solution to the game if both players agree. The question is how we select a reasonable point from this simplex polygon that can be accepted by both players. Next, we look at some possible acceptable points. Note that if two interior points $p_1(x_1, y_1)$ and $p_2(x_2, y_2)$ in simplex polygon P, where x_1, x_2 are the payoff values for player A, and y_1, y_2 are the payoff values for player B, if $x_1 \geq x_2$ and $y_1 \geq y_2$, then we say point A dominates point B. Obviously, both players would not accept any dominated interior points in P($p_2(x_2, y_2)$ in this case) because both players with their payoff value pair $p_2(x_2, y_2)$ can attain a better payoff value $p_1(x_1, y_1)$ through the cooperation of pre-play communication (See Definition 12.4.1) in Section 12.4). Thus, the players would agree only on the boundary points of the simplex polygon. But recall from the definition of the security level of each player, each player would prefer those payoff value pairs on the boundary for which his (hers) payoff value is greater than his(hers) security level. Thus, we are now restricting our Payoff Set to a smaller Negotiation Set for a possible agreeable solution between both players. This set establishes the basis for negotiation between the players and is defined as follows.

Definition 12.4.2: Negotiation Set N of a Cooperative Game

Let (sl1), sl2))be the security level payoff value pair, the Negotiation Set N of a Cooperative Game is a collection of payoff value pairs (x, y) on the boundary of the simplex polygon P such that x ≥ sl1 and y ≥ sl2.

N = {(x,y) | (x,y) is a boundary point of P such that x ≥ sl1 and y ≥ sl2} (D5)

We are going to show graphically the Negotiation Set N of each game shown in Example 12.8.

Example 4.9

[A] For the payoff set p1 shown in Figure 12.1, the payoff matrix of player A and player B are respectively

$$a = \begin{bmatrix} 2 & 3 \\ 12 & 2 \end{bmatrix} \quad b = \begin{bmatrix} 2 & 12 \\ 3 & 2 \end{bmatrix}$$

Calling the function mycomputesaddlepoint(a) and mycomputesaddlepoint(b) gives yes = 0 for both matrices a and b, indicating that neither a nor b has a saddle point, so we call computeSlbySimplexMethod(a) and computeSLbySimplexMethod(b) give us the security level for a sl1 = 2.9091 with security strategy (0.0909, 0.9091), and security level for b, sl2 = 2.9091 with security strategy (0.9091, 0.0909). (See the output listing of SecurityLevelandSimplexMethodFigure1.m). The payoff value pair for both players is the interior point S (2.9091, 2.9091) in the solution set P. The negotiation set N consists of all possible pairs (x, y) on the boundary BB'SA'C with x ≥ 2.9091 and y ≥ 2.9091, which then restricts only to the points in the segment BC with the largest possible x and y as shown in Figure 1. The equation of line AB is y=10x – 18, that of line BC is y= -x + 15, and that of line AC is $y = \frac{1}{10}x + \frac{9}{5}$. Note that the coordinates of A', B', C', S are A'(10.091, 2.9091), B'(2.9091, 11.091), C'(12.0909, 2.9091), S(2.9091, 2.9091) To recap, the negotiation set N is the entire segment BC.

Example 4.10

[B] For the payoff set p2 shown in Figure 12.2, the payoff matrix of player A and player B are respectively

$$a = \begin{bmatrix} 5 & 6 \\ 4 & 10 \end{bmatrix} \quad b = \begin{bmatrix} 10 & 4 \\ 6 & 5 \end{bmatrix}$$

We first make the function calls mycomputesaddlepoint(a) and mycomputesaddlepoint(b) gives the security levels sl1=5 for player A and security level sl2=6 for player B. (See output listing of SecurityLevelsandSimplexMethodFigure2.m).

The security level for each player is represented by point S (5, 6) shown in Figure 2. The negotiation set N is the segment AD', where D' has coordinate (9, 6), and the equation of line AD is y = -x + 15.

Example 4.11

[C] For the payoff set p3 shown in Figure 12.3, the payoff matrix of player A and player B are respectively

$$a = \begin{bmatrix} 5 & 7 & 12 \\ 4 & 10 & 7 \end{bmatrix} \quad b = \begin{bmatrix} 10 & 3 & 7 \\ 6 & 12 & 7 \end{bmatrix}$$

The function called mycomputesaddlepoint(a) gives yes = 1 and the security level for player A is 5, sl1 = 5 , with security strategy (1, 0). The call mycomputesaddlepoint(b) gives yes = 0, so player B does not have a saddle point, so we try to call computeSLbySimplexMethod (b). However, we cannot make the call directly because b is not a square matrix, and the current setup for the simplex method is based on a square matrix. In any game with an n by m payoff matrix, there are optimal strategies for both players with at most the minimum of m and n non-zero components. As b is a 2 by 3 matrix, we are going to break down b into 3 matrices b_1, b_2, b_3 :

$$b_1 = \begin{bmatrix} 10 & 3 \\ 6 & 12 \end{bmatrix} \quad b_2 = \begin{bmatrix} 3 & 7 \\ 12 & 7 \end{bmatrix} \quad b_3 = \begin{bmatrix} 10 & 7 \\ 6 & 7 \end{bmatrix},$$

, we then compute the security level of each of the above 2 X 2 matrix and select the one with the maximum value. From the output listing of computeSLbySimplexMethod.m the result is as follows:

b_1 has a security level of 7.8462, with security strategy (0.4615, 0.5385) for player A and (0.6923, 0.3037) for player B.

b_2 has security level 7, with security strategy (0.5556, 0.4444) for player A and (0, 1) for player B.

b_3 has security level 7, with security strategy (0.25, 0.75) for player A and (0,1) for player B.

As b_1 has the largest security level 7.8462 amongst that of b_1, b_2, b_3, the security level of player B is 7.8462, sl2 = 7.8462 with security strategy (0.6923, 0.3037, 0). The last entry is assigned 0 because b_1 only selected the first two columns from b. To summarize the above results in Figure 3, point S has the coordinates of security levels for both players, it has coordinates S (5, 7.8462), the segment ED' on the shaded area SAD' is the negotiation set N, where D' has coordinate D'(9, 6), and the equation of line ED is $y = -\frac{5}{2}x + 37$.

12.5 Solution of the cooperative game and Nash bargaining function F

In the last section, we derived the Negotiation Set N from the payoff value matrix p of the game. However, N still consists of infinitely many possible payoff value pairs (x y) that both players are undecided on which one to accept and agree on. Nash[17] came up with a method that allows the players to decide which payoff value pair (x, y) on the Negotiation Set N is fair and better for both players if they mutually agree. This method we are going to describe makes use of a function $F(N,(x_0, y_0))$ that depends on the Negotiation Set N and a point (x_0, y_0) inside the Cooperative Payoff Set P, the output of this function F is a payoff value pair (x_1, y_1) in N which represents a fair solution to both players. In notation

$$(x_1, y_1) = F(P,(x_0, y_0)) \quad \text{(D6)}$$

The point (x_0, y_0) henceforth we refer to as the status quo point, and the function F we refer to as Nash Bargaining Function. happy to employ this function F to determine their final payoff value pair (x_1, y_1) because it satisfies the following four properties which seems fair to them. Henceforth we refer to these four properties as the Axioms of Nash and are stated below.

The Axioms of Nash

Axiom 1: Feasibility

The payoff pair (x_1, y_1) must be an undominated point in the Negotiation Set N and that $x_1 \geq x_0$ and $y_1 \geq y_0$

Axiom 2: Linearity

If T is a linear transformation in the x-y plane such that

$T(x, y) = (c_1x + d_1, c_2x + d_2)$, where c_1, d_1, c_2, d_2 are constants, then

$F[T(P), T(x_0, y_0)] = T(x_1, y_1)$

Axiom 3: Symmetry

If (x_0, y_0) is any point in P implies (y_0, x_0) is also in P, and if $x_0 = y_0$ then $x_1 = y_1$

Axiom 4: Independence irrelevant payoffs

Suppose the payoff set P_1 is extended to a larger payoff set P_2, and if (x_0, y_0) is any point in P_1 and if the solution for P_2, $F[P_2, (x_0, y_0)] = (x_1, y_1)$ is a point in the smaller subset P_1, then (x_1, y_1) is also a solution for P_1, that is to say, $F[P_1, (x_0, y_0)] = (x_1, y_1)$ as well.

Next, we will show such a function F does exist and we will show it does satisfy the four axioms above in our next Section 12.6 [Verification of the Nash Axioms]. But first, we will define and construct F and demonstrate how it is applied to the solutions of the three games shown in Figures 1, 2 and 3 before.

Definition and the construction of $F[P, (x_0, y_0)]$:

As we recall, the objective of the function F is to find a point (x_1, y_1) on the Negotiation Set N such that it would satisfy both players. As the status quo (x_0, y_0) is already a point that both players agree on, it is reasonable to select (x_1, y_1) such that the product $(x_1 - x_0)(y_1 - y_0)$, call it A, is maximized. Thus, we apply differentiation below to find (x_1, y_1). As (x_1, y_1) is a point on a N, which is a line segment described by the equation $y = m x + b$, so it boils down to maximize the expression A = $(x_1 - x_0)(m x_1 + b - y_0)$. Differentiate A with respect to x_1 and set the derivative to 0, we find the solution of (x_1, y_1) is:

$$x_1 = \frac{1}{2m}(y_0 + mx_0 - b) \quad \text{(D8)}$$

$$y_1 = \frac{1}{2}(y_0 + mx_0 + b)$$

Example 5.12

Refer to Figure 12.1 for Example 12.9, we choose the point (x_0, y_0) as the status quo point S(2.9091, 2.9091), and the Negotiation Set N has the equation y = -x + 15, thus, from (D8) with m = - 1 and b = 15 gives the coordinates of the solution point S' (x_1, y_1) on N for both players with payoff values

$x_1 = 7.5$ and $y_1 = 7.5$.

Example 5.13

Refer to Figure 12.2 for Example 12.10, choose the point (x_0, y_0) as the status quo point S (5, 6), and the Negotiation Set N has the equation y = -x + 15, from (D8) above, with m = -1 and b = 15, we find $x_1 = 7$ and $y_1 = 8$ for the coordinates of the solution point S' (x_1, y_1) on N for both players.

Example 5.14

Refer to Figure 12.3 for Example 12.11, with (x_0, y_0) as the status quo point S (5,7.8462), and the Negotiation Set ED' has the equation $y = -\frac{5}{2}x + 37$ then from (D8) above, with m = $-\frac{5}{2}$ and b = 37, we find $x_1 = 8.34$ and $y_1 = 16.17$ or the coordinates of the solution point S' (x_1, y_1) on ED' for both players.

We have just seen from the previous three games how to use the Nash Bargaining Function defined in (D8) to come up with a reasonable proposed solution for both players. We now demonstrate that this Bargaining Function does satisfy the four axioms stated before.

12.6 Verification of the Nash Axioms

Axiom 1: Feasibility

This is obvious from the construction of $(x_1, y_1) = F(P,(x_0, y_0))$

$x_1 = \frac{1}{2m}(y_0 + mx_0 - b)$ and $y_1 = \frac{1}{2}(y_0 + mx_0 + b)$ where (x_1, y_1) is on the boundary N of P and thus dominated by no other payoff pairs, for otherwise, the pair would lie outside the boundary, and also each coordinate is at least acceptable as the status quo point (x_0, y_0).

Axiom 2: Linearity

Suppose the linear transform $T(x, y) = (c_1x + d_1, c_2y + d_2)$, where c_1, d_1, c_2, d_2 are constants, we want to prove

F[T(P), T(x_0, y_0)] = T(x_1, y_1),where (D9)

$$x_1 = \frac{1}{2m}(y_0 + mx_0 - b)$$

$$y_1 = \frac{1}{2}(y_0 + mx_0 + b)$$

The right-hand side of (D9) is

$$T(x_1, y_1) = \left[\frac{c_1}{2m}(y_0 + mx_0 - b) + d_1 \, , \frac{c_2}{2}(y_0 + mx_0 + b) + d_2\right] \quad \text{(D10)}$$

To show $T(x_1, y_1)$ is the same as the left-hand side of (D9), we make use of the fact that if a line L whose equation is y = mx + b that has slope m and y-intercept b, then under the linear transformation T, the line L' would have new slope and intercept

$$m' = m\frac{c_2}{c_1} \quad \text{(D11)}$$

$$b' = \frac{c_1(c_2 b + d_2) - mc_2 d_1}{c_1}$$

Since T(x_0, y_0) = (c_1x_0 + d_1, c_2y_0 + d_2),

F[T(P), T(x_0, y_0)] = F[T(P), (c_1x_0 + d_1, c_2y_0 + d_2)] = (x', y'),

where x' = $\frac{1}{2m'}[(c_2y_0 + d_2) + m'(c_1x_0 + d_1) - b']$ and

y' = $\frac{1}{2}[(c_2y_0 + d_2) + m'(c_1x_0 + d_1) + b']$ by (D9)

Thus, linearity is proven if x' is the x-coordinate of T(x_1, y_1) and y' is the y-coordinate of T (x_1, y_1) in (D10) above. We show both x-coordinates of (x', y') and T(x_1, y_1) matches below and similarly for the y-coordinates.

x' = $\frac{1}{2m'}[(c_2y_0 + d_2) + m'(c_1x_0 + d_1) - b']$, so, by (D11),

$$= \frac{c_1}{2mc_2}[c_2y_0 + mx_0c_2 + 2mc_2d_1 - c_2b]$$

$$= \frac{c_1}{2m}[y_0 + mx_0 + 2md_1 - b]$$

$$= \frac{c_1}{2m}[y_0 + mx_0 - b] + d_1$$

Similarly, y' = $\frac{c_2}{2}[y_0 + mx_0 + b] + d_2$

The above algebraic manipulations demonstrate that the operations T and F are commutative:

Figure 12.4

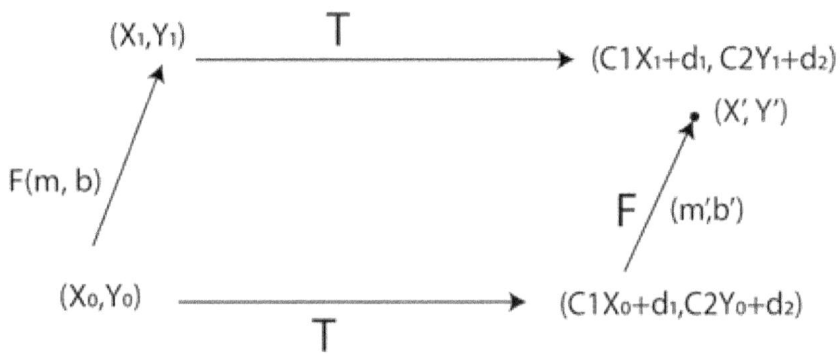

Axiom 3: Symmetry

If (x_0, y_0) is any point in P implies (y_0, x_0) is also in P, and if $x_0 = y_0$ then $x_1 = y_1$

To show symmetry is preserved, let the cooperative payoff set P be symmetric to the line with slope 1 shown in Figure 12.5 below:

Figure 12.5

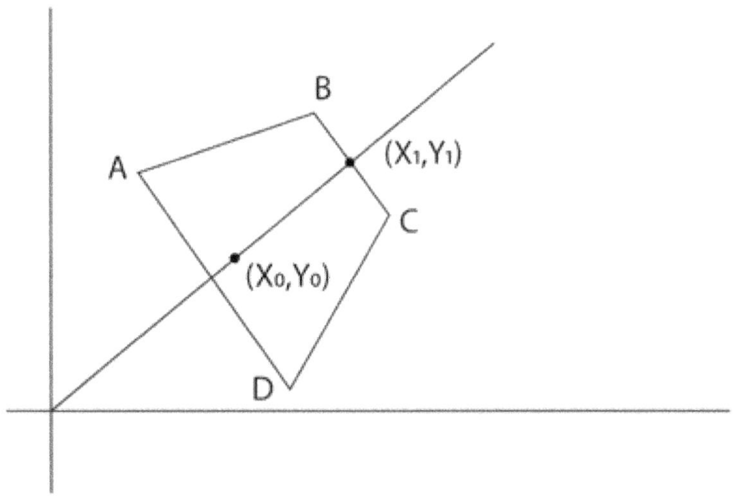

Assume $x_0 = y_0$, (x_0, y_0) is on the line of symmetry, we want to show (x_1, y_1) is on the line of symmetry as well. (D9) implies,

$$x_1 = \frac{1}{2m}(x_0 + mx_0 - b) \;=\; \frac{b}{2} \quad \text{m = -1 as the axis of symmetry for P has slope 1}$$

$$y_1 = \frac{1}{2}(x_0 + mx_0 + b) \;=\; \frac{b}{2}$$

Thus, $x_1 = y_1$ as well.

Axiom 4: Independence irrelevant payoffs

Figure 12.6

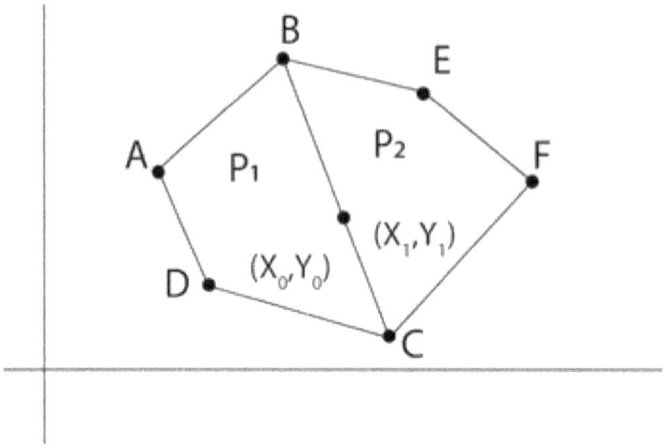

Suppose the payoff set P_1 is the polygon ABCD, it is now extended to a larger payoff set P_2 with two additional payoff pairs E and F for both players. If (x_0, y_0) is any point in P_1 and assume that the solution for P_1 is $(x_1, y_1) = F[P_1, (x_0, y_0)]$ is in P_1, then the solution for P_2, $F[P_2, (x_0, y_0)]$ is either the same point (x_1, y_1) before or a new point in the larger payoff set P_2. When $F[P_2, (x_0, y_0)]$ is the same as (x_1, y_1), it means that the two additional payoff pairs E and F are irrelevant to the solution of the game with payoff set P_1 or P_2.

12.7 The threat strategies of the cooperative game

Thus far, we have chosen the status quo point (x_0, y_0) in P as a pair of security levels for both players in the evaluation of the Nash Bargaining Function $F(P,(x_0, y_0))$ with solution point (x_1, y_1). In the event that the players are not happy with the payoff outcomes (x_1, y_1) such as they want to differentiate their payoff values, Nash [99] proposed yet another way to determine the status quo point (x_0, y_0) by means of a threat strategy pairs (X_0, Y_0), where X_0 is player A's strategy giving threat to player B, and Y_0 is player B's strategy giving threat to player A. This pair of mutually threat strategies to each other seems reasonable and is accepted by both players as a starting point in determining the status quo point (x_0, y_0). In what follows, we are going to show how this threat strategy pair (X_0, Y_0) is determined.

12.7.1 Formulation of the threat strategy pair strategy pair (X_0, Y_0)

As before (refer to section 12.1), let P be the cooperative payoff set of a cooperative game, and let a, b be the payoff set of Players A and B below.

$$a = \begin{bmatrix} a_{11} & a_{1j,} & a_{1n} \\ \multicolumn{3}{c}{\dots\dots\dots\dots} \\ a_{i1} & a_{ij,} & a_{in} \\ \multicolumn{3}{c}{\dots\dots\dots\dots} \\ a_{m1} & a_{mj,} & a_{mn} \end{bmatrix}$$, with m pure strategies $x_1, x_2, \dots. x_m$ for player A (D2)

$$b = \begin{bmatrix} b_{11} & b_{1j,} & b_{1n} \\ \multicolumn{3}{c}{\cdots\cdots\cdots\cdots} \\ b_{i1} & b_{ij,} & b_{in} \\ \multicolumn{3}{c}{\cdots\cdots\cdots\cdots} \\ b_{m1} & b_{mj,} & b_{mn} \end{bmatrix}$$, with n pure strategies $y_1, y_2, \ldots y_n$ for player B (D3)

Let S_A and S_B be the set of all strategies for player A and player B respectively. If X and Y are typical strategies for A and B respectively, then the matrix multiplications XaY^t and XbY^t are a pair of expected payoffs for players A and B respectively, and so (XaY^t, XbY^t) is a typical point in P. Now using any of these typical points (XaY^t, XbY^t) as our status quo point (x_0, y_0), we can compute the solution (x_1, y_1) for P in the Nash Bargaining Function :

$$(x_1, y_1) = F[P, (XaY^t, XbY^t)]$$

Thus, for each X in S_A and each Y in S_B, the pair (x_1, y_1) can be thought of as payoffs for a new game for players A and B with strategy pair (X, Y). We refer to this new game as *the threat game T associated with the original cooperative game with payoff matrix P*. We call it a threat game because player A is going to choose a proper strategy X_0 and player B a proper strategy Y_0 that will be most secure for himself (herself) without knowing what the other player's strategy which serves as a mutual threat to each other. The existence of this strategy pair (X_0, Y_0) is described below. Note that both x_1, y_1 are functions of the strategy pair X, Y, that is $x_1 = x_1(X, Y)$ and $y_1 = y_1(X, Y)$. Note that T is a m by n payoff matrix where each entry is indexed by the strategy pairs (X_i, Y_j).

$$T = \begin{bmatrix} (x_1(X_1,Y_1), y_1(X_1,Y_1)) & & \\ \cdots\cdots\cdots\cdots & & \\ & (x_1(X_i,Y_j), y_1(X_i,Y_j)) & \\ \cdots\cdots\cdots\cdots & & \\ & & (x_1(X_m,Y_n), y_1(X_m,Y_n)) \end{bmatrix} \quad \text{(D12)}$$

Now we can compute the security levels $sl(a_1)$ and $sl(b_1)$ for both players in T, where the matrices a_1 and b_1 correspond to the first and second components of payoffs in the entries of T.

$$sl(a_1) = \underset{X \in S_A}{Max}\ \underset{Y \in S_B}{Min}\ x_1(X,Y) = \underset{Y \in S_B}{Min}\ x_1(X_0,Y) \quad \text{(D13)}$$

$sl(b_1) = \underset{Y \in S_B}{Max}\ \underset{X \in S_A}{Min}\ y_1(X,Y) = \underset{X \in S_A}{Min}\ y_1(X,Y_0)$ where X_0 and Y_0 are the security level strategies for the game T with the payoff matrix shown in (D12) above. So, we can directly determine X_0 and Y_0 from (D13) above. But by the fundamental theorem of Nash, we know there always exists at least one equilibrium strategy pair (X_0', Y_0') for the game T so we can use (X_0', Y_0') to replace (X_0, Y_0) if we wish because the expected payoff values of the pair (X_0', Y_0') gives even higher payoffs than the pair (X_0, Y_0) does. In other words, we show that $sl(a_1) \leq X_0' a_1 Y_0'^t$ and $sl(b_1) \leq X_0' b_1 Y_0'^t$ .We are going to show

$sl(a) \leq X_0' a Y_0'^t$ and $sl(b) \leq X_0' b Y_0'^t$ using the fact that X_0' and Y_0' are in Nash Equilibrium.

$$sl(a_1) = \underset{X \in S_A}{Max}\ \underset{Y \in S_B}{Min}\ Xa_1Y' \leq \underset{X \in S_A}{Max}\ Xa_1Y_0^{'t} \leq X_0^{'}a_1Y_0^{'t} \qquad \text{(D14)}$$

$$sl(b_1) = \underset{Y \in S_B}{Max}\ \underset{X \in SA}{Min}\ Yb_1{}'X' \leq \underset{Y \in S_B}{Max}\ Yb_1{}'X_0^{'t} \leq \underset{Y \in S_B}{Max}\ X_0^{'}b_1Y' = X_0^{'}b_1Y_0^{'t} \qquad \text{(D15)}$$

Thus, instead of using the security levels of the players as status quo point (x_0, y_0) in the evaluation of $F[P, (x_0, y_0)]$, we can use point $(X_0^{'}aY_0^{'t}, X_0^{'}bY_0^{'t})$ as another status quo (x_0, y_0) to come up with another solution to the cooperative game.

In the next example, we are going to show how we can compute the threat strategies X_0, Y_0 from the thread game T described above. We then compute the new status quo point based on the threat strategies.

Example 7.15

Refer to the same game shown in Figure 12.2 for Example 12.10. For easy and later references, we redraw Figure 12.2 and rename it as Figure 12.7 below.

Figure 12.7

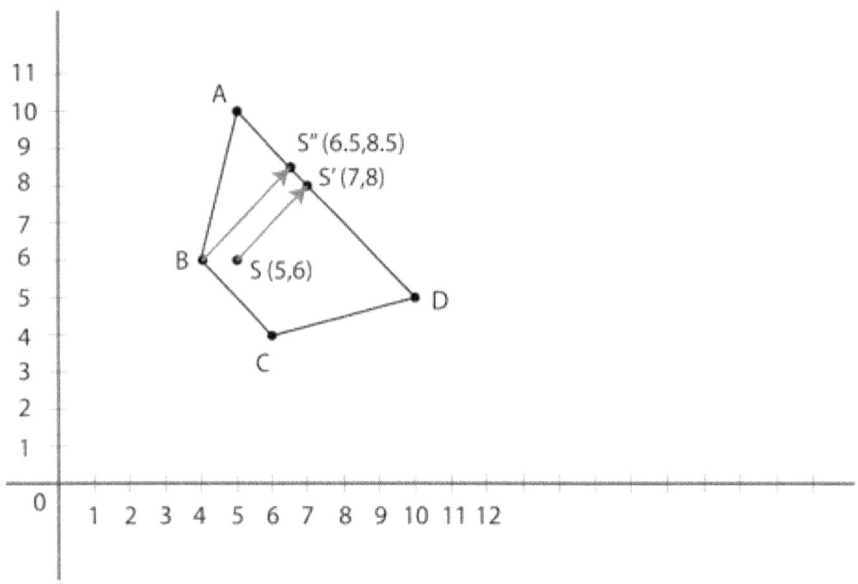

To compute the thread game T, we compute the outputs of the Nash Bargaining Function of $(x_1, y_1) = F(P, (x_0, y_0))$, where the various status quo points (x_0, y_0) are now taken as the corner points A(5, 10), B((4, 6), C((6,4), D(10,5), and P is the bargaining set N described by the equation y = -x + 15. Using the formula (D8) developed before, and with m = -1, b = 15,

$$x_1 = \frac{1}{2m}(y_0 + mx_0 - b)$$

$$y_1 = \frac{1}{2}(y_0 + mx_0 + b)$$

we find:

for A, with $(x_0, y_0) = (5, 10)$, $(x_1, y_1) = (5, 10)$

for B, with $(x_0, y_0) = (4, 6)$, $(x_1, y_1) = (6.5, 8.5)$

for C, with $(x_0, y_0) = (6,4)$, $(x_1, y_1) = (8.5, 6.5)$

for D, with $(x_0, y_0) = (10,5)$, $(x_1, y_1) = (10, 5)$

Thus, the threat game T associated with the game $p_2 = \begin{bmatrix} (5,10) & (6,\ 4) \\ (4,6) & (10,5) \end{bmatrix}$ is $T = \begin{bmatrix} (5,10) & (8.5,\ 6.5) \\ (6.5,\ 8.5) & (10,5) \end{bmatrix}$ with strategies X_1, X_2 for player 1, strategies Y_1, Y_2 for player 2. Now the respective payoff matrix for player 1 and player 2 is $a_1 = \begin{bmatrix} 5 & 8.5 \\ 6.5 & 10 \end{bmatrix}$ and $b_1 = \begin{bmatrix} 10 & 6.5 \\ 8.5 & 5 \end{bmatrix}$, and the respective security level strategies for player 1 and player 2 are $X_0 = (0, 1)$, $Y_0 = (1, 0)$ respectively. As stated above, X_0 and Y_0 are mutual threat strategies for both players in the game T, and when we compute the expected payoff value for both players in the original payoff matrices $a = \begin{bmatrix} 5 & 6 \\ 4 & 10 \end{bmatrix}$ and $b = \begin{bmatrix} 10 & 4 \\ 6 & 5 \end{bmatrix}$ we find the new status quo point is $(X_0aY_0^t, X_0bY_0^t) = (4, 6)$. Now we apply the Nash Bargaining Function

$(x_1, y_1) = F[P, (x_0, y_0)]$ with $(x_0, y_0) = (4, 6)$ and the negotiation set N given by the line AD with equation $y = -x + 15$, we find from (D8),

$(x_1, y_1) = (6.5, 8.5)$, which is the point S"(6.5, 8.5) shown above the point S'(7,8) as shown in Figure 7. It is interesting to note that both S" and S' are reasonable solutions from the solution set N both solutions are accepted by both players.

12.8 Nash's two-step bargaining procedure

Before we leave Nash's fundamental approach to the solution of cooperative games, let us recap what the approach does. It is based on the function $(x_1, y_1) = F[P, (x_0, y_0)]$ (Section 12.5) which was shown to be reasonably accepted by both players in Section 12.6. Since the procedure depends also on the choice of the status quo point (x_0, y_0), so other than choosing the security levels for both players as a reasonable candidate for (x_0, y_0), Nash also proposed a pair of threat strategies (Section 12.7) for both players from which a new status quo point (x_0, y_0) can be determined. Thus, Nash's two-step bargaining procedure involves first making the choice of the status quo point and second, computing the solution to the game (x_1, y_1) by the equation (D8)

$$x_1 = \frac{1}{2m}(y_0 + mx_0 - b) \text{ and } y_1 = \frac{1}{2}(y_0 + mx_0 + b)$$

12.9 Approaches other than the Nash Procedure

(a) Raiffa's approach
(b) Braithwaite's approach

Both other approaches make use of the non-cooperative game "d", whose payoffs are the differences between the payoffs of the two players from the original game, and make use of the concept of the constant payoff difference contour lines. For illustration, we look at the game shown in Figure 12.7 with the game matrix p_2, $p_2 = \begin{bmatrix} (5,10) & (6,\ 4) \\ (4,6) & (10,5) \end{bmatrix}$, then game d has the corresponding entries $d = \begin{bmatrix} -5 & 2 \\ -2 & 5 \end{bmatrix}$, where each entry is the difference between the payoff of player 1 and the payoff of player 2. If P is any point inside the payoff set ABCD with payoff (u_1, u_2), the difference $dP = u_1 - u_2$ is called the relative advantage to player 1. Any 45^0-degree line passing through P will consist of all other points P' will also have the same relative advantage to player 1, i.e., dP' = dP.

This is because if P' has the payoff pair (u_1', u_2'), then $dP' = u_1' - u_2' = u_1 - u_2$. The second equality holds because $u_2' - u_2 = u_1' - u_1$ as the slope of the line PP' is 1. Thus every 45^0 - degree line is a contour line of certain relative advantage v of any point on the line. These contour lines form a partition of all the points inside the payoff set P. In both the Raiffa and Braithwaite approaches, the solution of the cooperative game is a point S chosen from the negotiation set N with payoff (u_1, u_2) and where $S(u_1, u_2)$ lies on a contour line passing through a properly chosen point T in P such that both S and T have the same relative advantage, i.e., $dS = dT = u_1 - u_2$.

Now let's look at the associated non-cooperative game d closely. Every entry in d is associated with a strategy X from player 1 and a strategy Y from player 2, and it is the difference d(X, Y) = a(X, Y) – b(X, Y). In Figure 12.8 below, suppose F(sl1, sl2) is the point with the security levels of players 1 and 2 as coordinates in the payoff set P, CD is the negotiation set N, suppose the 45° degree line FS is meeting BC at S, then S is a reasonable candidate for the solution of the cooperative game if

(a) player 1 has a strategy X_0 such that no matter what strategy player 2 chooses, player 1's payoff is on or below the contour line passing S

(b) player 2 has a strategy Y_0 such that no matter what strategy player 1 chooses, player 2's payoff is on or above the contour line passing S

Figure 12.8

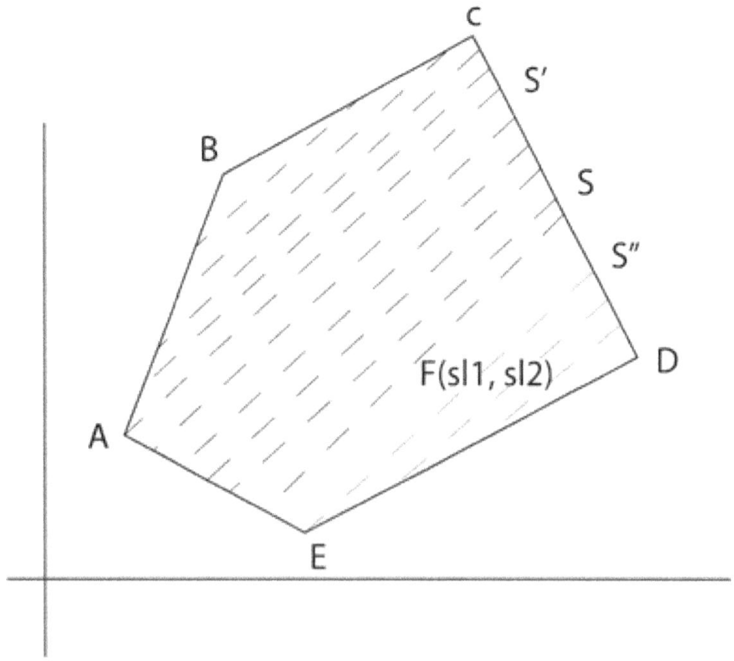

Under the circumstances stated above, neither player would want to deviate from point $S(u_1, u_2)$ because then he (she) would receive a lower payoff value than u_1 for player 1 and u_2 for player 2. The strategy pair (X_0, Y_0) can be thought of as a mutual-threat strategy between player 1 and player 2. The existence of this pair of strategies ensures that no player would deviate from S. Does (X_0, Y_0) always exist? The answer is Yes. This is because the difference payoff matrix d is a zero-sum game, and from the Fundamental Theorem of Game Theory discussed in previous chapters, there always exist strategies X_0 and Y_0 and a value v such that $d(X_0, y) \geq v$ for all y and $d(x, Y_0) \leq v$ for all x, the value v is called the value of the zero-sum game. We apply this strategy pair (X_0, Y_0) back to the original cooperative games in the discussions below.

Next, we give an example using the same payoff set shown in Figure 12.7

Example 9.16 (Raiffa's approach)

The Raiffa's approach starts with transforming the entries of the utility matrix of the original game p_2 to a new utility matrix p_2' whose entries are in the ranges between 0 and 1. This can be achieved by a linear mapping T which maps the best payoff M to 1 and the worst payoff m to 0, then any payoff value v in between m and M would be mapped into a new value v', T(v) = v'.

$v' = \dfrac{v - m}{M - m}$ and the inverse transformation is given by $v = m + v'(M - m)$

In p_2, M = 10, m = 4 happens to be the same for both players, the above transform gives

$$p_2' = \begin{bmatrix} (1/6,\ 1) & (1/3,\ 0) \\ (0,\ 1/3) & (1,\ 1/6) \end{bmatrix}$$

With this new scale, the payoff set P is shown in Figure 12.9 below.

Figure 12.9

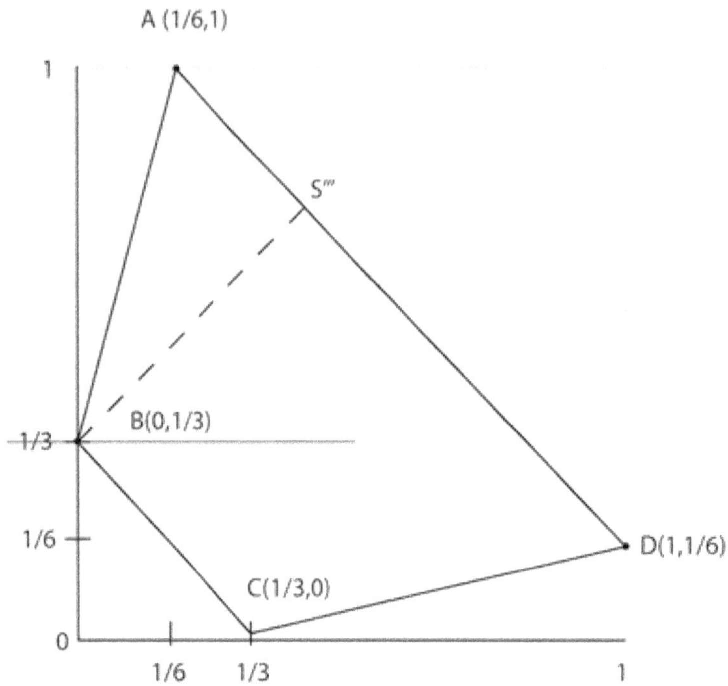

Keep in mind that, in p_2, the pure strategies corresponding to the rows are X_1, X_2 and the columns are Y_1, Y_2. If player 1 chooses X_2, then regardless of which strategy player 2 chooses player 1's payoffs are at B and D which are on or below the 45° degree line, if player 2 chooses Y_1, then regardless of which strategy player 1 chooses, player 2's payoffs are at A and D which are on or above the 45° degree line. Thus, the solution to the game would be on the 45° degree line intersecting with the negotiation set which is the point S''' on the line AD (Figure 12.9).

To find the location of S''', since it is in between A(1/6,1) and D(1, 1/6), it must be p(1/6, 1) + (1 - p)(1, 1/6) for some p between 0 and 1, which is (1 - 5/6p, 5/6p + 1/6)

Since both B and S''' are on the contour line, both have the same advantage, so

-1/3 = 1 - 5/6p – {5/6p + 1/6}, which gives p = 7/10, which gives the location of the solution (5/12, 9/12). Using the inverse transformation, we find the solution to the game is S''' (6.5, 8.5), which completely matches the solution by using the Nash Bargaining Function (Figure 12.7).

Example 9.17 (Braithwaite's approach)

In this example, we consider a new game p_4 this payoff matrix shown below.

$$p_4 = \begin{bmatrix} (1,3) & (10,\ 4) \\ (5,12) & (3,\ 0) \end{bmatrix} \qquad a = \begin{bmatrix} 1 & 10 \\ 5 & 3 \end{bmatrix} \qquad b = \begin{bmatrix} 3 & 4 \\ 12 & 0 \end{bmatrix}$$

Braithwaite's approach also starts with transforming the entries of the utility matrix of the original game p_4 to a new utility matrix p_4' but the transformation is much more involved. We outline this process below. For each player, we compute the change of payoff value from his (hers) maximin strategy to the min-max strategy assuming the adversary is holding to his(hers) maximin strategy. Denote dX and dY to be these incremental changes of payoff values for player 1 and player 2 respectively. We then compute the ratio dX/dY and round off to an integer ratio of say m : n, for some integers m and n. The entries in p_4' are then normalized by m for player 1, by n for player 2, to come up with the new payoff matrix

$$p_4' = \begin{bmatrix} (1/m, 3/n) & (10/m,\ 4/n) \\ (5/m,\ 12/n) & (3/m, 0) \end{bmatrix}$$

To describe the computation of dX and dY, we make use of the following strategies:

X_1 ~ Player 1's maximin strategy that gives his(hers) security level sl1

Y_2 ~ Player 2's minimax strategy that ensures Player 1's payoff is at most sl1

Y_1 ~ Player 2's maximin strategy that gives his (hers) security level sl2

X_2 ~ Player 1's minimax strategy that ensures player 2's payoff is at most sl2

In terms of these strategies defined above, then

$dX = X_2 a Y_1 - sl1$ and $dY = X_1 b Y_2 - sl2$, where a, b are the payoff matrices of p_4 for player 1 and player 2 respectively.

We now compute dX, dY for the respective payoff matrices a and b by calling the function computeSlbySimplexMethod(a) and computeSLbySimplexMethod(b). The first function call gives for player 1 or player A:

$X_1 = (0.1818, 0.8182)$

sl1 = 4.2727

$Y_2 = (0.6364, 0.3636)$

The second function call gives for player 2 or player a:

$X_2 = (0.9231, 0.0769)$

sl2 = 3.6923

$Y_1 = (0.3077, 0.6923)$

(see output listing of SecurityLevelsandSimplexMethodBV.m)

Thus,

$$
\begin{aligned}
dX &= X_2 a\, Y_1 - sl1 \\
&= (0.9231,\ 0.0762.689) \begin{bmatrix} 1 & 10 \\ 5 & 3 \end{bmatrix} (0.3077,\ 0.6923)^t - 4.2727 \\
&= 2.68
\end{aligned}
$$

$$
\begin{aligned}
dY &= X_1 b Y_2 - sl2 \\
&= (0.1818,\ 0.8182) \begin{bmatrix} 3 & 4 \\ 12 & 0 \end{bmatrix} (0.6364,\ 0.3636)^t - 3.6923 \\
&= 3.17
\end{aligned}
$$

The ratio dX to dY is approximately 4 to 5, and the new payoff matrix $p_4^{'}$ becomes

$$
p_4^{'} = \begin{bmatrix} (1/4,\ 3/5) & (5/2,\ 4/5) \\ (5/4,\ 12/5) & (3/4,\ 0) \end{bmatrix}
$$

Figure 12.10

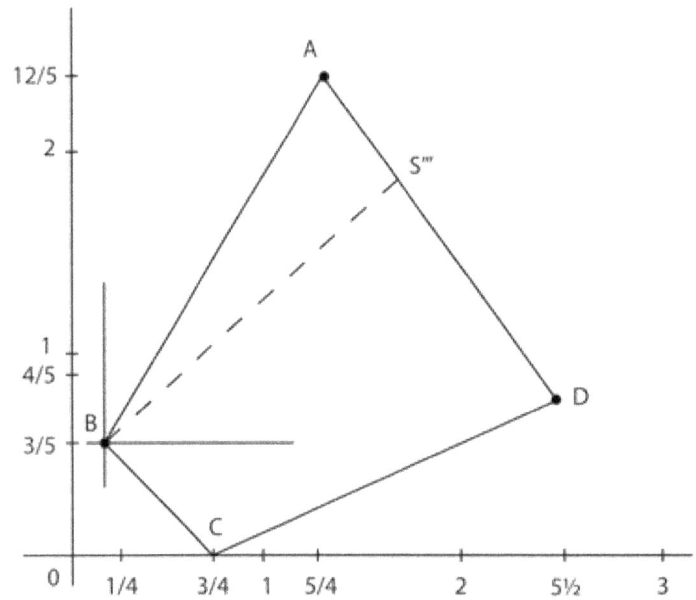

Coordinates of the respective corner points are A(5/4, 12/5), B(¼, 3/5), C(3/4, 0), D(5/2, 4/5). Now recall that X_1, X_2 are the pure strategies of player 1 and Y_1, Y_2 for player 2. If player 1 chooses X_1, then no matter what strategies player 2 chooses, the payoff values for player 2 at points B and D are on or below the 45° line; and if player 2 chooses Y_1, then no matter what strategies player 1 chooses, the payoff values for player 1 at points B and A are on or above the 45° line. Thus, the solution to the game $p_4^{'}$ is the point S''' lying at the intersection of the line AD and the 45° line. As in the last example, to find the coordinates of S''', since it is in between A(5/4, 12/5) and D(5/2, 4/5), it must be p(5/4, 12/5) + (1 - p)(5/2, 4/5) for some p between 0 and1, which is (5/2 - 5/4p, 4/5 + 8/5p). Since both B and S''' are on the contour line, both points have the same advantage, so, ¼ - 3/5 = 5/2 — 5/4p - {4/5 + 8/5p}, which gives p = 41/57, which gives the coordinates of S'''(1.6, 1.95). Using the ratio of 4 to 5 found before, the solution to the original game p_4 is S'''(6.4, 9.75).

12.10 Summary

In this chapter, we explored cooperative games and key concepts such as the Cooperative Payoff Set, the Negotiation Set, and Nash's Bargaining Function. These concepts provide a foundation for understanding how players negotiate and distribute payoffs in a cooperative setting.

While the Nash Procedure remains a fundamental method for solving cooperative games, we also introduced two alternative approaches: Raiffa's Approach and Braithwaite's Approach, discussed in **Section 12.9**. These methods offer different perspectives on negotiation, emphasizing fairness, rational decision-making, and iterative compromise.

By examining multiple strategies, this chapter highlights the flexibility of cooperative game theory in resolving conflicts and achieving mutually beneficial agreements across various real-world applications.

The key technical insights in this chapter are the concepts of the Negotiation Set and Nash's Bargaining Function. The Negotiation Set defines the feasible boundary within which players can negotiate, while Nash's Bargaining Function determines a unique, optimal solution that maximizes their joint utility while respecting individual disagreement points. This solution serves as a reference point for both players. Other negotiation approaches discussed include those of Raiffa and Braithwaite, which are iterative methods based on feedback between players, rather than directly computing an optimal solution like Nash's Bargaining Function.

Concluding Thoughts

In this book, we have explored a variety of concepts, definitions, and models within the realm of game theory. We hope that students and readers find the subject both stimulating and fascinating, inspiring them to apply these concepts to tackle their daily challenges. The guiding motto is: "Think Game Theory!"

However, it is essential to remember that selecting the appropriate model—or the most relevant one you have learned—is crucial for accurately representing the specific situation you are analyzing. A general guideline for choosing the right model is to carefully assess the scenario at hand, identifying the key players, their possible strategies, and the potential payoffs or outcomes as discussed in Chapters 1 through 4. Additionally, consider the dynamic nature of interactions between the players, including whether decisions are made simultaneously or sequentially (Chapters 5 through 8) and the nature of those interactions as either competitive or cooperative (Chapters 11 and 12).

Once the appropriate model is selected, it is important to consider: what insights can the Nash Equilibrium strategies—one of the key concepts we've focused on throughout the book—provide us? I believe these insights can lead to satisfying solutions for the real-life problems you may encounter.

As outlined earlier, the applications of game theory span a broad spectrum of fields, including artificial intelligence (AI). AI has garnered significantly more attention since the completion of this book, driven by advances in machine learning, natural language processing, autonomous systems, and more. This growing intersection between AI and game theory is fostering a vibrant area of research, especially as AI systems become more strategic, interactive, and autonomous. AI will undoubtedly impact on our daily lives in the very near future.

While this book is not solely focused on AI, I feel it is important to highlight some immediate areas within AI that rely heavily on game-theoretic principles. These include:

- **Multi-Agent Systems (MAS)**
- **Reinforcement Learning (RL) in Multi-Agent Environments**
- **Mechanism Design and Auction Theory**
- **Adversarial AI and Security**
- **Negotiation and Autonomous Decision-Making**

Currently, many AI companies are concentrating on the following topics:

- **Large Language Models (LLMs) and Chatbots:** Examples include ChatGPT, GPT-4, and Bard, focusing on natural language understanding, generation, and conversational AI.

- **Foundation Models:** Developing large-scale models trained on vast datasets for applications like translation, summarization, and coding.
- **Deep Search and Retrieval Systems:** Improving search relevance through embeddings, knowledge graphs, and contextual understanding.
- **Multimodal AI:** Combining text, images, audio, and video—examples include DALL·E (image generation from text) and CLIP (image-text alignment).
- **AI for Scientific Discovery:** Automating research, drug discovery, and materials science, exemplified by DeepMind's AlphaFold for protein folding.
- **Reinforcement and Self-supervised Learning:** Training models with minimal supervision in complex environments.
- **AI Safety, Ethics, and Alignment:** Ensuring AI systems act in accordance with human values and prevent misuse.
- **AI in Robotics and Autonomous Systems:** Self-driving cars, drones, and industrial automation.

Recall the two game-theoretic tools discussed in this book: the Revelation Principle (Chapter 5) and Folk's Theorem (Chapter 6). These concepts are actively influencing AI, particularly in multi-agent systems, mechanism design, and repeated interaction models. Many other game-theoretic ideas—such as Nash equilibrium, mechanism design, and Bayesian games—are foundational in developing AI systems that involve strategic interactions, cooperation, and decision-making under uncertainty.

Many AI systems—like autonomous vehicles and collaborative robots—involve multiple agents that need to coordinate effectively. Game theory provides powerful models and optimization tools for enhancing these interactions.

While recent AI breakthroughs often focus on models, data, and algorithms, game theory offers a fundamental framework for understanding strategic behavior, incentives, and interactions in multi-agent environments. Its application is crucial for ensuring AI systems are safe, cooperative, and efficient, especially as AI becomes more autonomous and integrated into society.

To conclude, I wish you the best of luck in applying game theory to solve your everyday challenges—and, even more ambitiously, in integrating these principles into current and future AI systems!

Bibliography

A. Books

[1] W. Adams: Fundamentals of Mathematics for Business, Social, and Life Sciences, Prentice Hall 1979

[2] S. Baase, A. Gelder: Computer Algorithms, Addison-Wesley 2000

[3] T. Basar, G. Olsder: Dynamic Noncooperative Game Theory, SIAM 1999

[4] T. Basar, P. Bernhard: H_∞ - Optimal Control and Related Minimax Design Problems, Birkhauser Publisher, Boston., 2008

[5] D. Bauso: Game Theory with Engineering Applications, SIAM 2016

[6] A.T. Berztiss: Data Structures, theory and practice, Academic Press, New York 1975

[7] T. Elbert: Estimation and Control of Systems, VanNostrand Reinhold Inc., 1984

[8] B. Francis: A course in H_∞ Control Theory. Lecture Notes on Control and Information Sciences, Springer-Verlag.

[9] R. Gibbons: Game Theory for Applied Economists, Princeton University Press., Princeton, New Jersey.,1992

[10] R. Gardner (2003): Games for Business and Economics, John Wiley & Sons, New York.

[11] F. Hillier, G. Lieberman: Operations Research, Holden-Day Inc. 1974

[12] D. Luce, H. Raiffa: Games and Decisions, John Wiley & Sons, New York 1985

[13] MathWorks: Robust Control Toolbox User's Guide

[14] _______________: Control System Toolbox User's Guide

[15] J. McKinsey: Introduction to the Theory of Games, McGraw-Hill, New York, 1952

[16] G. Owen: Game Theory, 2nd ed. Academic Press, Inc., New York 1982.

[17] R. Prather: Discrete Mathematical Structure for Computer Science ,Houghton Mifflin Company, Boston, 1976

[18] A. Rapoport, Two-person Game Theory, The Essential Ideas, University of Michigan Press, Ann Arbor, 1960

[19] A. Rapoport: N-person Game Theory, Concepts and Applications., University of Michigan Press, Ann Arbor., 1970

[20] W. Spaniel: Game Theory 101, the complete textbook, You Tube Series.

[21] P. Thie, G. Keough: An Introduction to Linear Programming and Game Theory, John Wiley& Sons, Inc., Hoboken, New Jersey., 2008

[22] S. Tadelis: Game Theory, An Introduction, Princeton University Press, 2013

[23] H.A. Taha: Operations Research, An Introduction, 4th ed., 1987, Macmillan Publishing

B. Papers

[1] d'Aspremont, C., Galszewicz, J. J., and Thisse, J.F., 1979: "On Hotelling's Stability in competition", Econometrica, 47(5), 1145-1150

[2] J. S. Bain (1949): "A note on Pricing in Monopoly and Oligopoly", American Economic Review 39(2): 448-464

[3] P. Bernhard: "A Lecture on the game theoretic approach to H∞ optimal control, "INRIA- Sophia Antipolis, France, August 21, 1999

[4] K. Chatterjee, R. Chaturvedi: "Integrating Raiffa and Nash approaches to bargaining using interim agreements", Science Direct., Games and Economic Behavior, Vol. 146, July 2024, pp/ 105-120

[5] I.K. Cho, D.K. Kreps (1987): “Signaling Games and Stable Equilibria, “Quarterly Journal of Economics 102: 179-222

[6] J. Friedman(1971): A Noncooperative Equilibrium for Super games. Review of Economics Studies, 38(1), 1-12.

[7] C. Hasan, E. Altman, J. Gore(2011): A coalition game approach io the association problem of mobiles in broadcast transmission. In Modeling and Optimization in Mobile, Ad Hoc and Wireless Networks (WiOpt), 2011 International Symposium on, pp.236-240. IEEE.

[8] ____________, “Common Mistakes in Computing the Nucleolus”, Discussion Series of the Department of Business and Management Science, ,NHH Norwegian School of Economics, March 2014

[9] Hotelling, H. (1929). Stability in competition. Economic Journal, 39(1953), 41-57.

[10] Khan, Koubaa, Benjdira, Boulila: “A game theory approach for smarty traffic management”, Science Direct., Computer and Electrical Engineering , Vol. 110, September 2023, 108825

[11] E. Kohlberg, J. F. Mertens (1986): “On the Strategic Stability of Equilibria”, Econometrica. 54: 1003-10038

[12] D. Kreps, R. Wilson (1982): “Sequential Equilibrium”, Econometrica 50: 863-894

[13] “Spatial Competition Models”, June 2020, Ester Iglesias Maso, Universitat de Barcelona

[14] J. Lemaire (1991). Cooperative game theory and its insurance applications. Astin Bulletin 21(1), 17-40

[15] P. Milgram, J. Roberts (1982): “Limit Pricing and Entry under Incomplete Information: An Equilibrium Analysis”, Econometrica 50(2), 443-459

[16] R. Myerson (1979): “Incentive Compatibility and Bargaining Problem.”, Econometrica 47: 61-73

[17] J. Nash: “Two-Person Cooperative Games”, Econometrica, Vol.21, No.1, pp.128-140, 1953.

[18] _______: “The Bargaining Problem”, Econometrica, Vol.18, pp.155-162, 1950.

[19] _______: “Non-Cooperative Games”, Annals of Mathematics(2), Vol. 54, pp.268-295,1951.

[20] ____________: “Equilibrium points in n-person games”, Proc. Natl. Acad. Sci. USA,36:48-49,1950

[21] Nobel Seminar, December 8, 1994: “The Work of John Nash in Game Theory”, Journal of Economic Theory 69, 143-185(1996) Article No. 0042

[22] S. Oh and J. Shin (2012): "A semantic e-kanban system for network-centric manufacturing : technology, scale-free convergence, value and cost-sharing considerations." Journal of Production Research 50(19), 5292-5316.

[23] J. Rogers: "The Game Theory Glitch in A Beautiful Mind", Essay April 27, 2018

[24] Schmeidler, D. (1969). The nucleolus of a characteristic function game. SIAM Journal on Applied Mathematics 17(6), 1163-1170

[25] Schelling T.C. (1960): The Strategy of Conflict, Harward University Press, Cambridge MA.

[26] C. de Souza, U. Shaked, M. Fu: "Robust H∞ Tracking : A game theory Approach", International Journal of Robust and Nonlinear Control, Vol 5., 223-238 (1995)

[27] M. Spence (1973): "Job Market Signaling", The Quarterly Journal of Economics, Vol. 87, No.3, pp.355-374

[28] "A Nash Game Approach to Mixed H2 / H_∞ Control for Input-Affine Nonlinear Systems", Science Direct., IFAC - Papers On Line 49-18(2016), pp.1024-1029

[29] G. Zames: "Feedback and Optimal Sensitivity: Model Reference transformation, multiplicative semi-norms and approximate inverses. IEEE Trans. Automat. AC-26: 301-320, 1981

C. Online

[1] M. Machado: "Hotelling Model"

[2] B. Olken: "The Median Voter Theorem"

[3] Wikipedi: "Coase Theorem"

[4] ____________: "H-infinity methods in Control Theory"

[5] ____________: "Traffic Lights and Nash Equilibrium"

[6] ___________: "Median Voting Rule"

[7] ____________: "Median Voter Theorem"

[8] ___________: "Game Theory"

www.ingramcontent.com/pod-product-compliance
Lightning Source LLC
LaVergne TN
LVHW070844210926
848482LV00016B/865
9798892284240